Landolt-Börnstein / New Series

Springer

Berlin
Heidelberg
New York
Barcelona
Budapest
Hong Kong
London
Milan
Paris
Tokyo

Landolt-Börnstein

Numerical Data and Functional Relationships
in Science and Technology

New Series
Editor in Chief: W. Martienssen

Units and Fundamental Constants in Physics and Chemistry

Nuclear and Particle Physics (Group I)

Atomic and Molecular Physics (Group II)

Solid State Physics (Group III)

Macroscopic Properties of Matter (Group IV)

Geophysics (Group V)

Astronomy and Astrophysics (Group VI)

Biophysics (Group VII)

Landolt-Börnstein
Numerical Data and Functional Relationships in Science and Technology
New Series / Editor in Chief: W. Martienssen

Group III: Solid State Physics
Volume 24

Physics of Solid Surfaces

Subvolume C
Interaction of Charged Particles and Atoms with Surfaces

Editor:
G. Chiarotti

Contributors:
P. Alkemade, V. Celli, G. Chiarotti, M. Rocca, E. Zanazzi

Springer

ISSN 0942-7988 (Solid State Physics)

ISBN 3-540-56071-8 Springer-Verlag Berlin Heidelberg New York
ISBN 0-387-56071-8 Springer-Verlag New York Berlin Heidelberg

Library of Congress Cataloging in Publication Data
Zahlenwerte und Funktionen aus Naturwissenschaften und Technik, Neue Serie
Editor in Chief: W. Martienssen
Vol. III/24C: Edited by G. Chiarotti
At head of title: Landolt-Börnstein. Added t.p. : Numerical data and functional relationships in science and technology.
Tables chiefly in English.
Intended to supersede the Physikalisch-chemische Tabellen by H. Landolt and R. Börnstein of which the 6th ed. began publication in 1950 under title: Zahlenwerte und Funktionen aus Physik, Chemie, Astronomie, Geophysik und Technik.
Vols. published after v. 1 of group I have imprint: Berlin, New York, Springer-Verlag
Includes bibliographies.
1. Physics--Tables. 2. Chemistry--Tables. 3. Engineering--Tables.
I. Börnstein, R. (Richard), 1852-1913. II. Landolt, H. (Hans), 1831-1910.
Physikalisch-chemische Tabellen. IV. Title: Numerical data and functional relationships in science and technology.
QC61.23 502'.12 62-53136

Product Liability: The data and other information in this handbook have been carefully extracted and evaluated by experts from the original literature. Furthermore they have been checked for correctness by the authors and the editoral staff before printing. Nevertheless the publisher can give no guarantee for the correctness of the data and information provided. In any individual case of application the respective user must check the correctness by consulting other relevant sources of information.

Production: PRODUserv Springer Produktions-Gesellschaft, Berlin
Printing: Druckhaus Langenscheidt KG, Berlin
Binding: Lüderitz & Bauer, Berlin

SPIN 10086066 63/3020 - 5 4 3 2 1 0 - Printed on acid-free paper

Editor
of Volumes III/24 a ... d

G. Chiarotti
Dipartimento di Fisica, Università degli Studi di Roma "Tor Vergata",
I-00133 Roma, Italy

Contributors
to Volumes III/24 a ... d

P.F.A. Alkemade
Delft Institute for Micro-Electronics and Submicron-Technology, Delft University of Technology,
2600 GA, Delft, The Netherlands
Elastic and inelastic scattering of ions

A.M. Bradshaw
Fritz-Haber-Institut der Max-Planck-Gesellschaft,
D-14195 Berlin, Germany
Photoemission and inverse photoemission

C. Calandra
Dipartimento di Fisica, Università degli Studi di Modena,
I-41100 Modena, Italy
Electronic structure of surfaces: semiconductors

V. Celli
Department of Physics, University of Virginia,
Charlottesville, Virginia 22901, USA
Interaction of atoms with surfaces

P. Chiaradia
Dipartimento di Fisica, Università degli Studi di Roma "Tor Vergata",
I-00133 Roma, Italy
Optical properties of surfaces

G. Chiarotti
Dipartimento di Fisica, Università degli Studi di Roma "Tor Vergata",
I-00133 Roma, Italy
General introduction

R. Colella
Physics Department, Purdue University,
West Lafayette, Indiana 47907, USA
X-ray diffraction of surface structures

A. Fasolino
International School for Advanced Studies (ISAS),
I-34014 Trieste, Italy
Dipartimento di Fisica, Università degli Studi di Modena
I-41100 Modena, Italy (present address)
Surface reconstruction and relaxation

U. Gradmann
Physikalisches Institut der Technischen Universität Clausthal,
D-38678 Clausthal-Zellerfeld, Germany
Magnetic properties of single crystal surfaces

R. Hamers
Department of Chemistry, University of Wisconsin,
Madison, Wisconsin 53706, USA
Scanning tunneling microscopy

R. Hemmen
Fritz-Haber-Institut der Max-Planck-Gesellschaft,
D-14195 Berlin, Germany
Photoemission and inverse photoemission

M. Henzler
Institut für Festkörperphysik der Universität Hannover,
D-30167 Hannover, Germany
Structural defects at surfaces

K. Jacobi
Fritz-Haber-Institut der Max-Planck-Gesellschaft,
D-14195 Berlin, Germany
Electronic structure of surfaces: metals

G.L. Kellogg
Sandia National Laboratories, Organization 1114,
Albuquerque, New Mexico 87185, USA
Field emission, field ionization and field desorption

F. Manghi
Dipartimento di Fisica, Università degli Studi di Modena,
I-41100 Modena, Italy
Electronic structure of surfaces: semiconductors

A.A. Maradudin
Department of Physics and Institute for Surface and Interface Science, University of California,
Irvine, California 92717, USA
Surface polaritons

J.F. Nicholas
Unit 35, Fountain Court , 100 Station Street,
Burwood 3125, Victoria, Australia
(formerly of the now no longer existent CSIRO Division of Tribophysics, Melbourne, Australia)
The structure of ideal surfaces

W. Ranke
Fritz-Haber-Institut der Max-Planck-Gesellschaft,
D-14195 Berlin, Germany
Structural defects at surfaces

D. Ricken
Fritz-Haber-Institut der Max-Planck-Gesellschaft,
D-14195 Berlin, Germany
Photoemission and inverse photoemission

M. Rocca
Centro di Fisica delle Superfici e delle Basse Temperature del C.N.R. and INFM,
Dipartimento di Fisica, Università di Genova,
I-16146 Genova, Italy
Inelastic scattering of electrons

Th. Schedel-Niedrig
Fritz-Haber-Institut der Max-Planck-Gesellschaft,
D-14195 Berlin, Germany
Photoemission and inverse photoemission

A. Selloni
International School for Advanced Studies (ISAS),
I-34014 Trieste, Italy
Department of Physical Chemistry, University of Geneva,
CH-1211 Geneva, Switzerland (present address)
Surface reconstruction and relaxation

A. Shkrebtii
Institute of Semiconductors, Academy of Sciences of Ukraine,
Kiev 252028, Ukraine
Dipartimento di Fisica, Università degli Studi di Roma "Tor Vergata"
I-00133 Roma, Italy (present address)
Surface reconstruction and relaxation

S.Y. Tong
Department of Physics and Laboratory for Surface Studies, University of Wisconsin-Milwaukee,
Milwaukee, Wisconsin 53201, USA
Surface phonons

R.F. Wallis
Department of Physics and Institute for Surface and Interface Science, University of California, Irvine, California 92717, USA
Surface phonons

E. Zanazzi
Dipartimento di Chimica, Università degli Studi di Firenze, I-50121 Firenze, Italy
Elastic scattering and diffraction of electrons and positrons

Preface

Surface physics has grown in the last two decades into a mature field of research with data of the highest accuracy and reproducibility. A large amount of results and a great number of reliable surface models represent the "scenario" that calls for the present Landolt-Börnstein volume.

Volume III/24 is concerned with the "physics of solid surfaces" (clean surfaces). It is divided into four subvolumes, a: structure, b: electronic and vibrational properties, c: interaction of charged particles and atoms with surfaces, d: interaction of radiation with surfaces and electron tunneling.

The list of the chapters is given on the inside of the cover. The scope and content of the various chapters are briefly outlined in the general introduction.

Each chapter consists of an introduction aimed at giving a brief presentation of the subject and a data section of results in the form of tables, figures, and pictures. This "format" has been chosen to increase the "readability" of the volume and to facilitate the access to scientists of different fields. The extreme latitude of the subjects covered by surface physics was also taken into account. A general index in which all data are ordered alphabetically according to the material and to the surface is reported at the end of the last subvolume.

The volume is limited to the so called "clean" surfaces i.e. to surfaces atomically clean and well characterized, leaving the more complex field of the contaminated surfaces and overlayers to a later occasion.

Acknowledgements: It is a pleasure to acknowledge the contribution of Prof. A.M. Bradshaw of the Fritz-Haber Institute of the Max-Planck Society of Berlin to the early development of this work when the volume acquired the present format.

Thanks are also due to Dr. C. Goletti of the University of Rome who compiled the "General index" and to the editorial staff of the Landolt-Börnstein in Darmstadt, especially to Dr. H. Seemüller and Mrs. H. Hämmer for accurate work and several comments and suggestions.

Some of the authors (A.A. Maradudin and R. Colella) acknowledge the support of various research grants (NSFDMR-8918184 and NSFDMR-871503A1, DEFG02-89ER45380) during the compilation of the respective chapters.

Rome, February 1995 **The Editor**

Contents

Physics of solid surfaces

Subvolume c: Interaction of charged particles and atoms with surfaces

1 General introduction

1.1 Motivations for a Landolt-Börnstein volume on surface physics

Surface physics has experienced in the last two decades an explosive expansion caused by several concomitant events, namely:

(i) The technical advancements that allowed the attainment of atomically clean surfaces and the control of chemical contamination. This was primarily due to the development of ultra high vacuum (10^{-8} Pa) techniques and the commercialization of UHV apparatus.
(ii) The development and/or substantial improvement of surface sensitive techniques like for example: angularly resolved ultraviolet photoemission and photoemission spectroscopy with a continuous source of radiation (synchrotron radiation); the various techniques based on electron excitation like inverse photoemission or electron energy loss spectroscopy; the use of atomic and ionic beams for studying diffraction and scattering from surfaces; the extension of LEED to intensity vs. voltage analysis; the optical techniques etc.
(iii) The development of microscopy techniques that allowed direct observation of surfaces at the atomic scale, like scanning tunneling microscopy or field-ionization microscopy, and improved dramatically our knowledge of the local structure and properties of surfaces.
(iv) The availability of realistic theoretical calculations and models for the electronic and vibrational properties of surfaces and their relevance for the structure, reconstruction and relaxation of surfaces.
(v) The stimulus to further research given by the great variety of surface phenomena: the various reconstruction structures of the surface proper and of the chemisorbed and physisorbed layers, their transformation upon annealing and contamination; the great theoretical challenge of surface instabilities, the defective properties of the surface etc.
(vi) The great importance of surface physics for technological applications like control of the work-function of metals and semiconductors, interface physics, field effect devices, molecular beam epitaxy, chemisorption and catalysis, corrosion, surface hardening etc.

Because of the improvements of the experimental techniques, surface physics, when applied to atomically clean or atomically controlled surfaces, has reached a remarkable degree of accuracy and reproducibility. Models of reconstructed surfaces, once simply guessed on the ground of theoretical considerations, are now tested with great accuracy by various techniques and in many cases have attained a high reliability. Notable examples are the Si(111) 7×7 surface reconstructed according to the DAS (dimer-adatom-stacking fault) model [84T], the Si (111) 2×1 π-bonded chain model [81P], the Au(110)1×2 missing-row model [79M2, 83B2] etc.

This is the "scenario" that called for the compilation of the data presented in this volume "Physics of Solid Surfaces", entirely devoted to *clean surfaces* i.e. to crystalline surfaces that are atomically clean and well characterized. This choice was aimed at presenting a homogeneous set of data and results with the highest consistency and reliability. Besides its intrinsic interest, the knowledge of the clean surface is the basis for the understanding of the more complex fields of chemisorption and catalysis.

A few exceptions to this rule have been allowed when the process of contamination or the growth of an overlayer plays an important role for understanding the properties of the clean surface or when the length-scale of the phenomenon makes the influence of the first atomic layers less crucial.

1.2 Outline of the volume

The volume III/24, "Physics of Solid Surfaces" is divided into four subvolumes a, b, c, and d. The first two are concerned with the properties of clean surfaces (structural, electronic, vibrational, magnetic) irrespective of the method of investigation, while the last two deal with results that are proper to a specific technique or for which the technique is relevant (diffraction of electrons and positrons, inelastic scattering of electrons, elastic and inelastic scattering of atoms and ions, optical properties of surfaces, photoemission and inverse photoemission, diffraction of X-rays by surfaces, field emission and field ionization microscopies, scanning tunneling microscopy). The unavoidable superpositions have been kept to a minimum and are in any way indicated in the single articles.

In addition to the present general introduction, the volume consists of 17 "chapters" and a general index. The chapters are grouped according to scientific homogeneity into 9 larger areas or fields (example: 2 The structure of surfaces. 2.1 The structure of ideal surfaces. 2.2 Surface reconstruction and relaxation. 2.3 Structural defects at surfaces). Each chapter is written by one or more authors of undiscussed authority in that specific field and consists of

(1) an introduction,
(2) a compilation of data in the form of tables and figures,
(3) a bibliography.

The introduction reviews in a succinct form the problems of the specific field, presents the relevant formulas, a hint to their derivation, a definition of the quantities reported in the data section. The scope of having such an introduction is to offer a volume that is more "readable" than usual data-handbooks so to be used also by scientists working in a different area or approaching for the first time surface physics.

In the data section the results are ordered according to elements or compounds and, for each substance, according to specific surfaces, listed in the order of increasing sum of Miller indices, or in the order of increasing complexity of the surface structure.

In order to facilitate the comparison of the various results obtained with different techniques and reported in different chapters and to make quicker the search for a specific property of a given surface, a general index has been included at the end of the last subvolume. In it materials are listed in alphabetic order and according to the various surfaces.

A table of bulk lattice parameters of crystals quoted in the present volume is added at the end of this general introduction, so to make them available for quick reference in each subvolume.

A major problem in the compilation of a data-handbook is the space to be given to theoretical results. In the present volume the attitude has been taken to include theoretical results to the extent they bear direct relation to data reported in the volume. This decision is supported by the importance of theoretical results in the interpretation of many experimental data, especially in certain areas. Common examples are models for reconstruction of surfaces and band structure calculations.

In the following the single chapters of the volume are briefly introduced:

2 The structure of surfaces

This part of the volume deals with the atomic structure of crystalline surfaces: in the ideal case (chapter 2.1), for reconstructed or relaxed surfaces (2.2) and for defective surfaces (2.3).

2.1 The structure of ideal surfaces, by J.F. Nicholas.

This article presents the relevant crystallographic formulas, symmetries and notations for an ideal half-crystal i.e. a structure formed by dividing a crystal by a plane and removing all atoms on one side, leaving all other atoms in their original positions An ideal surface is then the surface of an ideal half-crystal.

The second part of the article is entirely devoted to an Atlas of models of surface structures (approximately one hundred and seventy), i.e. to a pictorial representation of atoms in the surface and first sublayers. The relative positions of atoms are given in the tables of atomic positions in terms of lattice parameters.

The lattice parameters of all crystals quoted in the volume are reported in a table at the end of the general introduction.

2.2 Surface reconstruction and relaxation, by A. Fasolino, A. Shkrebtii, and A Selloni.

Surface reconstruction is one of the most challenging problems of surface physics. Understanding the great variety of surface structures, both commensurate and incommensurate to the bulk, and their transformation upon annealing and contamination is a formidable task. At present a general theory does not seem to be available, though "ab initio" calculations done by the Car-Parrinello molecular-dynamical method seem to reproduce the observed structures for Si(111) 2×1 [90A].

Evidence of surface reconstruction came through LEEDS around 1960 for Si(111) 7×7 and Si(111) 2×1 [59S, 63L1, 63L2]. The first direct observation in real space of the same surfaces became possible in the 80's by the use of scanning tunneling microscopy [83B1, 2, 87S].

In metals, the energy gain associated to reconstruction is only a few tens of meV so that even small amounts of impurities or overlayers alter the reconstruction. The tendency to reestablish the bulk charge density at the surface by removal of atoms seems to cause the reconstruction of some of the noble metals. Relieving of surface stress may also be important.

In semiconductors reconstruction energies are much larger so that reconstruction seems to be universally present. It is generally assumed that the driving force is the tendency to minimize the number of unsaturated dangling bonds with ensuing re-hybridization.

The article discusses the various models of reconstruction and relaxation, the criteria, the notations, the experimental techniques etc. The results are presented in form of tables for the various crystallographic faces of metals, semiconductors and insulators. For each element or compound and for each surface the table reports: the symmetry of the surface as given by LEED; the technique used in that specific investigation; the method for surface preparation; the model of the reconstruction; a comment on the results; references; figures or diagrams if necessary.

Special structural tables give numerical parameters characteristic of the various reconstructed or relaxed surfaces, like the change of interlayer distances, the vertical and lateral displacements of single atoms, the coordinates of atoms or adatoms, the buckling parameters, the tilt angles etc..

2.3 Structural defects at surfaces, by M. Henzler and W. Ranke.

The local structure of a solid surface may differ from the ideal, reconstructed or relaxed one because of the presence of randomly distributed defects like steps, terraces, vacancies, kinks, domain boundaries etc. If such defects are periodically arranged they contribute to the reconstruction pattern (e.g. adatoms and vacancies in the 7×7 reconstruction of the (1 1 1) face of Si) and are discussed primarily in chapter 2.2. The case of surfaces that are intentionally tilted from the low Miller index planes so to give rise to regular step arrays are, however, presented in this chapter.

The presence of steps was first recognized in cleaved surfaces of silicon through the splitting of integer order spots in LEED patterns along certain crystallographic directions [70H].

After the definition of surface defects and their relation to the ideally reconstructed or relaxed surface, the chapter discusses briefly the various techniques that have been used for detecting surface defects, with the aim of giving evidence to the respective merits and difficulties. The introduction of STM has allowed the observation of local structures in real space with atomic resolution, determining a great advance of our knowledge of surfaces. Diffraction techniques, however, using electrons, atoms, ions or X-rays have maintained their interest since very often they allow the determination of the properties of the defects with great accuracy, though averaged over large portions of the sample.

A section of the chapter is entirely devoted to structural transformations occurring at the surface i.e. roughening transitions at stepped surfaces and phase transitions of low index metal surfaces.

The data section presents results on the stability of the various surfaces mainly in the form of tables in which are given, for each face, the angular displacement from a low Miller index surface, the technique used for the observation and the surface preparation, comments on the type of local structure, step heights etc..

When the surface is known in a more quantitative way, the article reports models for the stepped surface, the restructuring of the macroscopic face into low index facets, models for the adsorption of atoms at steps etc.

3 Electronic structure of surfaces

The concept of electronic surface states has determined the birth of Surface Physics through the work of I. Tamm in 1932 and of W. Shockley in 1939 [32T, 39S]. They found that the termination of the periodic potential at the surface introduces additional states that, in the 1D case, have energies in the gaps of the bulk solid. Such states correspond to Bloch states with imaginary or complex wave-vectors and are therefore localized at the surface. For 3D potentials the $\boldsymbol{k}$-vector has two real components (corresponding to the motion of the electron along the surface) so that the states are not necessarily localized in the gaps.

The subject of electronic surface states is covered in this volume by two chapters dealing respectively with metals and semiconductors.

3.1 Electronic structure of surfaces: metals, by K. Jacobi.

In the introductory part various problems characteristic of metallic surfaces are discussed.

The jellium model [70L] i.e. a model solid in which a degenerate electron gas is neutralized by a continuous distribution of positive charge, is treated in great detail. Despite its simplicity it has contributed to the understanding of many surface properties of metals. The article reports theoretical results associated to the jellium model including the charge density distribution near the surface, the effective exchange and correlation potentials, the position of the image plane etc.

Considerable attention is given to the problem of surface states (both Tamm-like and Shockley-like) and of image states [83J, 79M1], i.e. electron states that arise through the long range nature of the image potential and are confined into the potential well between the surface and the image barrier. The comparison between image and Shockley states is also discussed. Other subjects of great interest for metal surfaces like surface Fermi-surfaces, surface plasmons, surface core-level shifts etc. are considered in detail.

The data-section presents results on work function, surface core-level shifts and surface band structure for the various metals ordered according to the periodic table.

A large fraction of the reported results, especially those on band structure determination, are obtained by means of photoemission spectroscopy. Some of them are then also discussed in chapter 8.2 "Photoemission and inverse photoemission" by A:M. Bradshaw et al.. Superposition has been kept to a minimum, though in some cases a few figures have been duplicated in order to facilitate the comparison between experimental results and theory as well as to increase the "readability" of the volume. In such cases the figure number is followed by an asterisk (∗) to remind the reader that additional information can be found in the other chapter.

3.2 Electronic structure of surfaces: semiconductors, by C. Calandra and F. Manghi.

The electronic surface structure of semiconductors and insulators presents some distinct properties with respect to metals and is discussed in a separate chapter. As already mentioned the surface of semiconductors is very often reconstructed or rather strongly relaxed. Surface states lay often in the forbidden gap (or gaps) and can be considered as the most direct evidence of Tamm or Shockley states. The existence of states in the gap causes in many cases the "pinning" of the Fermi level at the surface and a surface bending of the bulk bands (i.e. a compensating space-charge region with a characteristic thickness called Debye length) [65M, 67F].

The large energy gained upon reconstruction determines considerable displacement of atoms from equilibrium positions and a transfer of charge among the surface atoms. Moreover the change of symmetry often splits the surface bands, thus allowing the direct observation of transitions between the surface states. Such properties are discussed separately in chapters 6.2 and 8.1.

The results reported in chapter 3.2 refer principally to photoemission and inverse photoemission; the same remarks made at the end of the presentation of chapter 3.1 hold as well in this case.

A considerable space is given to the various methods of surface band structure calculations and to comparison with models and experimental results.

4 Vibrational and excitational properties of surfaces

Electronic surface states and surface phonons bear some similarities: both are states localized at the surface having a $\boldsymbol{k}$-vector with an imaginary (or complex) component in the direction perpendicular to the surface and energies into the gaps of the bulk structure. They are, however, substantially different in many respects, like the nature and the energy of the excitation, the type of interaction, the statistical behavior etc..

Two chapters deal with the vibrational properties of surfaces in this part of the volume: surface phonons (4.1) and surface polaritons (4.2).

4.1 Surface phonons, by R.F. Wallis and S.Y. Tong.

The article initially discusses the case of a bounded elastic continuum, a model valid for wavelengths much larger than the atomic spacing (Rayleigh waves). Follows then a microscopic approach with the definition of the 2D Brillouin zone and the solution of the dynamical equations for the atomic displacements in the case of the harmonic approximation.

The specific problems of the various materials (ionic crystals, semiconductors, metals) are discussed in detail as well as the experimental techniques (inelastic He scattering, electron energy loss spectroscopy) commonly used for studying surface phonons. In this respect, the reader can profitably consult chapters 6.2 and 7 that deal specifically with those methods.

In the data section the materials are arranged alphabetically and for each of them the specific frequencies of the surface phonons, the methods used for their determination and the dispersion curves are given.

Most of the results on surface-phonon dispersion are obtained with the technique of atom (mainly He) scattering. There is then a limited superposition with the data section of Chapter 7 on elastic and inelastic scattering of atoms by surfaces, by V. Celli. In some cases, especially when comments accompanying the spectra are extensive or relevant, figures have been repeated in the two chapters. In such cases an asterisk (∗) has been added to the figure number to remind the reader that additional information can be found in the other chapter.

4.2 Surface polaritons, by A.A. Maradudin.

Surface polaritons are electromagnetic waves that propagate along the interface of two semi-infinite dielectric media (one of which can be as well the vacuum) and whose amplitudes decay exponentially into each medium. They have then a true surface character even if the decay length, being of the order of the wavelength, is much larger than the atomic spacing. For this reason the properties of surface polaritons do not depend much on the atomic structure of the surface (or interface) plane. As a consequence the rule of this volume, to report only results of atomically clean surfaces, has been somewhat relaxed.

In ionic crystals (or in other solids with more than one atom per unit cell) there is a coupling between the optical phonons and the electromagnetic field, that gives rise to the so called surface phonon-polariton. There is then a strict connection between chapters 4.1 and 4.2.

The article presents a description of the microscopic nature of the excitation, i.e. the cases of surface phonon-polaritons, plasmon-polaritons and magnon-polaritons. The theory is developed in great detail for variously bounded interfaces (isotropic-isotropic, anisotropic-isotropic media, magnetic media). The experimental methods used for obtaining the dispersion curves are also briefly discussed.

In the data section the experimental results are listed according to the various systems (isotropic-isotropic media, uniaxial/biaxial crystals-isotropic medium, magnetic media etc.) and in each system the materials are ordered alphabetically.

5 Magnetic properties of single crystal surfaces, by U. Gradmann.

Magnetic moments of atoms at surfaces may differ from those in the bulk. Also magnetic ordering near the surface is modified by reduced magnetic co-ordination.

Surface magnetism is at present a very active field of physics. Techniques sensitive to the magnetic structure of surfaces, like for example spin-polarized photoemission or spin-polarized LEED have been recently developed [76E, 80K, 81F].

Data on thin films and interfaces have been reviewed by the same author in a previous volume of this series (see LB III/19g). Here data are reported only for well characterized clean surfaces of bulk single crystals or thick single crystal films. Theoretical results are reported provided they are directly connected with the experimental data.

The data are presented in the form of tables and figures that emphasize the surface behavior and/or the comparison with the bulk.

The following subjects are considered: surface magnetization at a given temperature; magnetic hyperfine fields near surfaces; temperature dependence of surface magnetic ordering; critical behavior of surface magnetization and the formation of "live layers" (ferromagnetic surface sheets); magnetic surface anisotropies.

Chapters from 6.1 to 9.2 deal with the various techniques commonly used in surface physics. They are divided into four partitions:

6 Interaction of charged particles with surfaces,
7 Interaction of atoms with surfaces,
8 Interaction of electromagnetic radiation with surfaces,
9 Electron tunneling at surfaces.

6 Interaction of charged particles with surfaces

Because of the strong Coulomb interaction, charged particles are a very effective tool for the study of surfaces. The particles used are electrons and positrons elastically (chapter 6.1) and inelastically (6.2) scattered and ions (6.3).

6.1 Elastic scattering and diffraction of electrons and positrons, by E. Zanazzi.

Electron diffraction by crystals was discovered in 1927 by C.J. Davisson and L.H. Germer and was intimately connected with the development of quantum mechanics. It was only after the war, however, that low energy electron diffraction, LEED, became a major tool for investigating the structure of surfaces.

The great relevance of LEED as a surface tool stems from the small escape depth of electrons (typically 5 ··· 10 Å for electrons of 100 eV) that makes it extremely surface sensitive. The appearance of fractional order spots in LEED patterns marked, on the other hand, the discovery of surface reconstruction [59S, 63L1, 63L2].

LEED patterns give directly the translational symmetry of the surface layer and are routinely used to investigate surface reconstruction. In principle the study of the energy dependence of the spot intensities (I/V curves) should allow the determination of atomic positions through a Fourier transform. However, multiple scattering caused by the high value of the electronic cross-section requires the development of dynamical theories.

The article consists of an introduction where the various techniques relevant for electron and positron diffraction, namely LEED, VLEED, SPLEED, RHEED, LEPD, are discussed in great detail and a data-section where the results for the various elements and compounds are presented for each surface. When available, the displacements of surface atoms from their ideal position are given, as well as quantitative models for some reconstructed surfaces. The elements and compounds are listed alphabetically.

Surface defects like steps, dislocations, disorder, roughening etc. give rise to broadening or splitting of the LEED spots. This subject is discussed in chapter 2.3.

For the results that make use of spin-polarized electrons, discussed in section 6.1.1.4.3 of this article, the reader can profitably consult Chapter 5 on the magnetic properties of surfaces.

6.2 Inelastic scattering of electrons, by M. Rocca.

Electrons passing through a solid loose energy by interacting with other electrons and with the atoms of the crystal. Because of the short escape depth, electrons of relatively low energy (< 1 keV) reflected by a crystal interact with the excitational spectrum of its surface.

The interaction is conventionally schematized into two main processes: dipole and impact scattering. At low energies (< 20 eV) the first process predominates.

Dipole scattering is caused by the interaction of the electromagnetic field generated by the moving primary particle with the elementary excitations of the surface (surface plasmons, *e–h* pair excitation, transitions between surface state levels etc.). In this respect results of electron energy loss spectroscopy are similar to those obtained by absorption or reflection of light, discussed in chapter 8.1. A dielectric response theory can be formulated in both cases, the optical absorption being proportional to $-\mathrm{Im}\,\varepsilon$ and the electron losses to $\mathrm{Im}\, 1/(\varepsilon + 1)$, ε being the complex dielectric function of the surface [75F].

In impact scattering the interaction is between the primary electron and the atomic cores: the elementary excitations are then the surface phonons. The very high resolution necessary to detect losses caused by surface phonons (of the order of 10 meV) was reached through a considerable improvement of the electron spectrometer (the double monochromator-analyzer system) introduced by H. Ibach and coworkers [71I, 91I].

In the article the experimental data are divided according to the material (metals and semi-metals, semiconductors, insulators and oxides) and to the type of excitation (surface phonons, surface plasmons, electron-hole pair and Stoner excitations, surface state transitions etc.).

6.3 Elastic and inelastic scattering of ions, by P.F.A. Alkemade.

Ion beams interact strongly with surfaces, the scattering cross-section being in general larger than the area per surface atom. This means that the process of scattering is very complex because of multiple scattering, ion neutralization (and reionization) etc.. On the other hand surface sensitivity, at least at energies not too high, is very great.

Because of its complexity, the field of ion scattering developed slowly through the work of H.D. Hagstrum, D.P. Smith, M. Aono and others [54H, 67S, 81A].

The interest for surface physics stems from the possibility of exploiting the strong angular dependence of the scattered beam to obtain information on the structure of the surface. The determination of the models for surface reconstruction is within the reach of this technique [83T].

Various processes contribute to the scattering:

(i) For a given direction of the impinging ion there is a region behind each surface atom (shadow cone) into which ions cannot penetrate. Every atom within the shadow cone cannot then scatter the probing ion. This phenomenon is called shadowing.

(ii) Conversely, scattered ions cannot be detected if other atoms are located along the trajectory towards the analyzer. This phenomenon is called blocking. Through a combination of shadowing and blocking it is possible to obtain a map of intensities that are related to the structure of the surface. Because of multiple scattering detailed calculations are necessary, starting from a given surface model.

(iii) If the ion beam runs parallel to a major crystallographic direction all atoms below the surface are in the shadow cones of other atoms: the ions are compelled to travel in the open channels between the atomic rows. This phenomenon, especially important at high energies, is called channeling.

Since the scattering processes depend strongly upon the energy of the ions (commonly rare-gas ions, alkalis, H and a few simple molecules), it has become customary to divide the subject according to the energy of the projectile into: low energy ion scattering (LEIS, 40 eV $< E <$ 50 keV); medium energy ion scattering (MEIS, 50 keV $< E <$ 300 keV); and high energy ion scattering (HEIS, 0.3 MeV $< E <$ 3MeV).

In the article there are two types of tables containing, for the various energies, a) surface structure data and b) topics related to the various aspects of the scattering process.

The tables of the first type present data on: the change in first to second and second to third layer distances; the lateral displacement of atoms in the first layer; the amplitude of thermal vibrations of surface atoms with respect to the bulk; the surface Debye temperature (when available); various comments on the surface structures.

The tables of the second type summarize the various information that can be obtained in the literature on the scattering process itself.

Sputtering has been deliberately excluded from the compilation.

7 Interaction of atoms with surfaces, by V. Celli.

Elastic and inelastic scattering of atoms (mainly He) with thermal energies has become a major tool in surface physics, both for structure and phonon spectra determination. The technique has developed slowly over the past 60 years and, as in the case of electron diffraction, the first experiments were aimed at proving the wave character of slowly moving particles. Only after 1975, through the work of G. Boato, P. Toennies, and collaborators [76B, 81B] and others, high quality spectra were obtained that paved the way to extensive surface applications. Among the various techniques used in surface physics, atomic scattering is the one that shows the highest surface sensitivity.

In many cases simple kinematic theories are sufficient to describe with reasonable approximation the scattering process. This is mainly due to the form of the gas-surface interaction potential in which the hard-wall repulsive part predominates. Three (or many)-body interactions and multiple scattering are very often negligible.

A peculiar effect due to the combination of the strong repulsive and the weak van der Waals attractive interactions is the existence of bound states near the surface into which the slowest atoms (or the atoms slowed down by the creation of a surface phonon) fall down giving rise to resonance scattering. Kinematic conditions alone allow a precise determination of the bound state energies.

The article presents in its introduction a thorough discussion of the various aspects of atomic scattering theories with special attention to the problem of the gas-surface interaction potential. In the data section the results for the various surfaces (ordered alphabetically) include the parameters that characterize the surface potential well (well depths, van der Waals coefficients, bound state energies, characteristic distances, effective surface corrugation etc.) as well as the surface phonon spectra. A critical comment accompanies each phonon spectrum. In this respect some kind of superposition with the spectra reported in chapter 4.1 could not be avoided. In some case reference is made to the pertinent figure of chapter 4.1. In other cases, particularly when comments are extended and especially valuable, figures have been repeated. In such cases an asterisk ($\ast$) follows the figure number to remind the reader that additional information can be found in chapter 4.1.

8 Interaction of electromagnetic radiation with surfaces.

This part of the volume deals with: 8.1 Optical properties of surfaces, 8.2 Photoemission and inverse photoemission, 8.3 X-ray diffraction of surface structures.

Though electromagnetic radiation does not interact strongly with solids, surface sensitivity is attained through differential methods (8.1), by exploiting the short escape depth of the photoemitted electron (8.2) and by X-ray scattering in directions off the Bragg peaks of the bulk (8.3).

8.1 Optical properties of surfaces, by P. Chiaradia.

Light penetrates into a solid for a depth of the order of α^{-1}, α being the absorption coefficient of the bulk. Even for strongly absorbing solids $\alpha^{-1} \simeq 10^{-5}$ cm, a length much larger than the thickness of the layer where electronic surface states are localized.

Nevertheless, optical methods were the first to give a direct evidence of the existence of surface states on the "clean" semiconductor surfaces [68C, 71C], i.e. of states that in some way are connected with Tamm or Shockley states. Differential techniques and multiple reflections were used to improve considerably the surface sensitivity. The use of polarized light allows, on the other hand, the observation of anisotropies associated to surface reconstruction and revealed an ideal tool for checking reconstruction models [84C, 84O].

The dielectric theory of light reflection and absorption by a surface can be done on classical grounds [71M]; in semiconductors for energies below the bulk gap and in conditions of near-normal incidence the surface reflectivity is simply proportional to the imaginary part of the surface dielectric function and can be directly correlated to the electronic structure of the surface. For energies above the gap a more complex situation occurs.

The article presents in a succinct way the various methods by which the interaction of light with surfaces has been studied, i.e. differential reflectivity, ellipsometry, surface photoconductivity, photoluminescence, photovoltage spectroscopy, photothermal displacement, second harmonic generation etc.. Data concern mainly semiconductors and are arranged according to the periodic table.

The reader may consult chapter 6.2 for comparison with results on excitations observed by electron energy losses.

8.2 Photoemission and inverse photoemission, by A.M. Bradshaw, R. Hemmen, Th. Schedel-Niedrig, and D. Ricken.

The determination of the energy and momentum distribution of electrons emitted from a solid by an incident beam of UV (or X-) radiation revealed a very powerful method for investigating the electronic surface structure of solids. Its surface sensitivity descends from the very short escape length of low energy electrons, a property common to other electron spectroscopies and already mentioned for EELS and LEED.

The first experiments that showed the possibility of detecting surface states through photoemission were done in semiconductors at the beginning of the 70s [72E, 72W]. A great advancement was introduced when it was realized that the component of the ***k***-vector parallel to the surface is not changed in the emission process [64K] and that angularly resolved photoemission could allow the determination of the dispersion of the surface states in the 2D surface Brillouin zone [73F, 74S, 83H]. For this reason a strong connection exists between photoemission results and band structure calculations. As a consequence there is a certain contiguity between the present chapter and those (3.1 and 3.2) on the electronic surface structure. As in other cases, a few figures have been duplicated in order to facilitate the discussion of the results and to give an internal authonomy to the various subvolumes.

Photoemission probes occupied states. Conversely, inverse photoemission – a technique in which an electron of a well specified energy is shot at the sample and the intensity of the outcoming radiation is analyzed as a function of the energy of the incoming electron – probes the empty states of the surface structure.

Comparison of the energies obtained by photoemission and inverse photoemission with those of the optical transitions (reported in chapters 6.2 and 8.1) on the basis of the one-electron approximation should be done, however, with great caution since it neglects many-body effects, in particular the influence of relaxation and the change of electron correlation, as well as exciton binding energies.

In chapter 8.2 the results are ordered according to the groups of the periodic table and divided into metals and semiconductors.

8.3 X-ray diffraction of surface structures, by R. Colella.

X-rays interact only weakly with the atoms of solids so that surface sensitivity in X-ray diffraction is very poor. Nevertheless, surface reconstruction, because of change of symmetry, causes the appearance of a new set of Bragg peaks at angles where the bulk does not diffract. The problem of attaining suitable surface sensitivity is then similar to that of uncovering weak lines buried in an unwanted background. If such a problem is solved, X-ray diffraction has a remarkable advantage over, say, electron diffraction since multiple scattering is negligible and kinematic theories are in general adequate for the determination of atomic positions.

Surface X-ray diffraction developed only recently [81E], exploiting the high intensity of synchrotron radiation sources.

The article gives a detailed description of the principles of X-ray diffraction and presents the still scarce results mainly in a discursive way, with comments and suggestions. Some of the results (especially those on surface waves) refer to surfaces deliberately contaminated and have been included to give a more general view of a field in rapid evolution.

9 Electron tunneling at surfaces

The direct imaging of atoms is of the utmost importance for the study of surfaces. Several microscopies with atomic resolution have been developed throughout the years. The most commonly used are field emission and field ionization microscopy, atomic probe and scanning tunneling microscopy. They are described in chapters 9.1 and 9.2.

9.1 Field emission, field ionization and field desorption, by G.L. Kellogg.

Field emission, field ionization and field desorption are phenomena of the greatest interest for surface physics "per se" i.e. independent of their use in the various microscopies. Nevertheless, the development of microscopes with atomic resolution, based on the above techniques, has greatly enhanced their practical interest for surface physics. Such an important development was made possible through the work of E.W. Müller and a few others [36M, 51M, 68M].

The three techniques make use of strong electric fields applied, with a specific polarity, to a sample ordinarily in the shape of a tip.

In field electron emission, electrons tunnel through the surface barrier made thinner by the application of an external field. The data are interpreted in terms of the well known Fowler-Nordheim equation which relates the current density to the electric field and the work-function. In the field emission microscope, electrons ejected from the tip are imaged on a fluorescent screen. The contrast arises mainly from the local variations of the work function along the surface of the sample. The resolution (20 ··· 30 Å) is not sufficient, however, to image the individual atoms. Nevertheless, the field emission microscope gives the local value of the work function and can be used to follow the contamination of the surface exposed to gases or molecular beams.

In atomically clean surfaces the energy distribution of the emitted electrons has given evidence of the presence of surface states in metals. The work of Plummer and Gadzuk on W provided one of the earliest evidences of surface states in metals [70P].

When a gas is introduced at low pressure in the microscope operated in reverse polarity, gas atoms are ionized in the immediate vicinity of the surface, and accelerated towards the screen, giving rise to an image of the tip with atomic resolution. In this case contrast arises because gas atoms are ionized preferentially above the protruding surface atoms.

Field desorption consists in the removal of an atom from the surface by an intense electric field. The desorbed atom is focussed into a mass spectrometer that allows the determination of the chemical composition of the surface at a local level.

The article emphasizes both the techniques and the underlying principles. The data are presented in the form of tables that give: the work function values for various metals; the activation energies for surface self-diffusion and diffusion of foreign atoms; various crystallographic relations for epitaxially grown surfaces; best fields for imaging gases; evaporation fields for various ions; relative abundance of clusters observed in laser-stimulated field evaporation; depth profiles in alloy segregation.

Because of the nature of the subject, the article is not strictly limited to clean surfaces but includes data on gas-surface interaction, impurity segregation, effects of contamination etc..

9.2 Scanning tunneling microscopy, by R.J. Hamers.

The discovery of scanning tunneling microscopy (STM) by G. Binnig, H. Rohrer and coll. in 1982 [82B] opened the way to direct observation of surfaces with atomic resolution. Since then STM experienced an explosive growth and became the ideal probe for surface structure determination at a local level.

The technique has many variants and operation modes: its basic principle is the quantum-mechanical tunneling of an electron from a metallic tip (that has necessarily an atomic termination, i.e. is very sharp) to the surface of the sample. The tip is scanned along the sample at distances of the order of a few angstroms. When a voltage is applied, the tip drives a tunneling current that depends exponentially upon the distance and linearly upon the local density of surface states of the sample at energies close to the Fermi energy. Reversing the polarity electrons tunnel from filled states or into empty states of the surface. STM is then essentially a spectroscopic device that probes the electronic structure of the sample. It is a powerful tool for investigating reconstruction or relaxation, adsorption of impurities, epitaxial growth, surface defects etc..

The article presents, in the introduction, the principle of the method and some experimental details including the description of the various modes of operation.

The data section consists mainly of STM pictures of the various surfaces with an extended comment and information on models and processes pertaining to that surface.

1.3 How to consult the volume

The way a scientist consults a data-handbook is certainly a matter of personal experience and taste. So we shall not indulge in outlining a standard procedure, that would be in any way arbitrary.

Nevertheless, it should be noticed that scientists approaching a Landolt-Börnstein volume are aimed at different purposes and belong to different categories. As an example (certainly not exhaustive) we can think to the following classes of "users":

(i) scientists looking for data specific to their own field of research,
(ii) scientists looking for all what is known on a specific surface,
(iii) scientists working in a specific field of surface physics and wanting to know what can be obtained with a different technique or in a different area,
(iv) scientists approaching surface physics from a different field of research or for the first time,
(v) scientists looking for bibliographical references on a given subject.

The following suggestions can tentatively be given to the readers of the various classes:

For class (i): look at the list of contents at the beginning of each subvolume and go directly to the data section of the chapter of interest. The data may be arranged alphabetically or according to the periodic table. If doubts arise on the meaning of quantities reported therein, consult the introduction of the chapter.

For class (ii): consult the "general index" (at the end of the last subvolume) and then look at the data-section of chapters quoted therein.

For class (iii): look at the list of contents and then read the introduction of the chapter(s) of interest. Look at the data section(s) at a second moment.

For class (iv): read first the general introduction and then the introductions of chapters of interest.

For class (v): the same as for (ii).

Other information included in the "general introduction" for quick reference is a table of crystal structures and bulk lattice constants of crystals quoted in the volume; a list of symbols and acronyms used in the volume. The last information may be of special value of readers of articles in surface physics, especially those of class (iv).

1.4 List of frequently used symbols and abbreviations

The list is not intended to contain all of the symbols used in the following sections, but in any case each symbol is explained at the place where it is used. Some symbols are listed here in their general form; in some sections they are made more specific by adding superscripts or subscripts or by using both lower and upper case forms.

Symbols

Symbols	Units	Property
A ... Z		points or lines of high symmetry in bulk BZ
$\overline{\mathrm{A}} \ldots \overline{\mathrm{Z}}$		points or lines of high symmetry in surface BZ
a_0	m	Bohr radius
$a, b, c,$	Å, m	lattice parameters
d	Å, m	distance, spacing (d_{12} : first to second layer distance)
$\boldsymbol{E}$	$\mathrm{Vm^{-1}}$	electric field strength, $\mid \boldsymbol{E} \mid = E$
E	eV	energy
E_B	eV	binding energy
E_C	eV	conduction band energy minimum
E_F	eV	Fermi energy
E_V	eV	valence band energy maximum
E_i	eV	initial energy, energy of incident particles
E_kin	eV	kinetic energy
E_loss	eV	energy loss
E_s	eV	energy of scattered particle
E_sp	eV	surface plasmon energy
E_vac	eV	vacuum energy
e	C	electron charge
$\boldsymbol{g}$	$\mathrm{Å^{-1}}$	reciprocal lattice vector
h	Js	Planck constant, $h = 2\pi\hbar$
$h\nu$	eV	photon energy
$\hbar\omega$	eV	phonon energy
h, k, l		Miller indices
h_pv	Å	peak-to-valley height of the surface corrugation
I	-	Intensity
I	A	electric current
$\boldsymbol{k}$	$\mathrm{Å^{-1}}$	wave vector, $\mid \boldsymbol{k} \mid = k$
$\boldsymbol{k}_\parallel$	$\mathrm{Å^{-1}}$	wave vector parallel to the surface, $\mid \boldsymbol{k}_\parallel \mid = k_\parallel$
$\boldsymbol{k}_\perp$	$\mathrm{Å^{-1}}$	wave vector perpendicular to the surface, $\mid \boldsymbol{k}_\perp \mid = k_\perp$
k_B	$\mathrm{JK^{-1}}$	Boltzmann constant
L	-	surface peak area
M	kg	atom mass
m_e	kg	electron restmass
m^*	kg	effective mass
n	$\mathrm{cm^{-3}}$	electron density in a semiconductor
P^+	-	survival probability
p	$\mathrm{cm^{-3}}$	hole density in a semiconductor
$\boldsymbol{q}$	$\mathrm{Å^{-1}}$	wave vector, $\mid \boldsymbol{q} \mid = q$
$\boldsymbol{q}_\parallel$	$\mathrm{Å^{-1}}$	wave vector parallel to the surface, $\mid \boldsymbol{q}_\parallel \mid = q_\parallel$

Symbols (cont.)

Symbols	Units	Property
$\boldsymbol{q}_\perp$	$Å^{-1}$	wave vector perpendicular to the surface, $\lvert \boldsymbol{q}_\perp \rvert = q_\perp$
R	-	reliability factor
R	$Å^{-1}$	radius, radial distance
R	Å	shadow cone radius
r	$Å^{-1}$	radius, radial distance
T	K, °C	temperature
T_C	K	Curie temperature
T_c	K, °C	critical temperature
T_m	K, °C	bulk melting temperature
T_N	K	Néel temperature
U	eV	potential
V	eV	potential
V_{rep}	eV	short-range repulsion potential
V_{att}	eV	long-range attraction potential
V	V	voltage
v	ms^{-1}	velocity
W	m	width of depletion layer
X, Y, Z	-	Cartesian coordinates
x, y, z		
z	-	coordinate perpendicular to the surface
α	-	absorptance, absorption coefficient
α	deg	exit angle
Γ	-	half width, FWHM (free width at half maximum)
$\Gamma, \Delta, \Lambda, \Sigma$	-	points or lines of high symmetry in bulk BZ
$\overline{\Gamma}, \overline{\Delta}, \overline{\Lambda}, \overline{\Sigma}$	-	points or lines of high symmetry in surface BZ
ΔX	-	difference in quantity X
δx	-	difference in quantity x
ε	-	dielectric constant,
		complex dielectric constant $\varepsilon = \varepsilon_1 - i\varepsilon_2$
ε_n	eV	surface-bound energy levels (n = 0, 1, 2, ...)
ζ	-	reduced wave vector coordinate
η	deg	azimuthal exit angle
Θ_D	K	Debye temperature
θ	deg	angle of incidence, scattering angle
λ	Å, m	wavelength
ν	Hz, s^{-1}	frequency, $\nu = \omega / 2\pi$
$\overline{\nu}$	cm^{-1}	wavenumber, $\overline{\nu} = \omega / 2\pi c$
ρ	Å	thermal vibration amplitude
Φ	eV	work function
ϕ	eV	potential
ϕ, φ	deg	azimuthal angle
Ψ	-	wave function
ψ	deg	angle of incidence
ω	rad s^{-1}	circular frequency, $\omega = 2\pi\nu$
ω_p	rad s^{-1}	plasma frequency

Symbol	Unit	Property
ω_c	rad s^{-1}	cyclotron frequency

[1]) In most cases k and q are given in units of the sides of the Brillouin zone.

Abbreviations

ABD	atom beam diffraction
AES	Auger electron spectroscopy
AFM	atomic force microscope
ALICISS	alkali impact collision ions scattering spectroscopy
ann	annealing
APS	appearance potential spectroscopy
ARCIS	angle resolved constant initial state spectroscopy
AREDC	angle resolved energy distribution curve
ARIPS	angle resolved inverse photoelectron spectroscopy
ARUPS	angle resolved ultraviolet photoemission spectroscopy, angle resolved UV photoelectron spectroscopy
ARXPS	angle resolved X-ray photoelectron spectroscopy, angle resolved X-ray photoemission spectroscopy
ATR	attenuated total reflection
a.u.	atomic units
BB	back bond
BBZ	bulk Brillouin zone
bc	body centered
bcc	body centered cubic
BCM	binary collision model
BCSOS	body centered solid on solid
bl	blocking
BLS	Brillouin light scattering
BZ	Brillouin zone
CAICISS	coaxial impact collision ion scattering spectroscopy
CBM	conduction band minimum
CCV	core-core valence Auger transition
CEMS	conversion electron Mössbauer spectroscopy
chan	channeling
CITS	current imaging tunneling spectroscopy
CMA	cylindrical mirror analyzer
CMTA	constant-moment-transfer average
cub	cubic
CVD	chemical vapor deposition
CVV	core valence-valence Auger transition
2D	2-dimensional
DAS	dimer-adatom-stacking fault
DAT	dimer adatom terraces
DB	dangling bond
2D BZ	2-dimensional Brillouin zone

DFT	density functional theory
DOS	density of states
DR	differential reflectivity
DWBA	distorted wave Born approximation
ECS	electron capture spectroscopy
ECT	elastic continuum theory
EDC	energy distribution curve
EELS	electron energy loss spectroscopy
ELEED	elastic low energy electron diffraction
EM	electron microscopy
ESA	electrostatic analyzer
ESD-TOF	electron stimulated desorption – time of flight
ESR	electron spin resonance
ETBM	empirical tight-binding method
fcc	face centered cubic
FEED	field emission energy distribution
FEM	field emission microscope, field emission microscopy
FERP	field emission retarding potential
FIM	field ion microscope, field ion microscopy
FK	Franz–Keldish effect
FLAPW	full potential linearized augmented plane wave
FLUC	current fluctuation measurement
FMR	ferromagnetic resonance
FN	Fowler–Nordheim (workfunctions obtained by FN plots).
FWHM	full width at half maximum
GFM	Green's function method
HAS	helium atom scattering
HATOF	helium atom time of flight spectrometry
hcp	hexagonal close packed
HEIC	high energy ion channeling
HEIS	high energy ion scattering, high energy ion scattering spectroscopy
HETED	high energy transmission electron diffraction
hex	hexagonal
HF	Hartree Fock
HREELS	high resolution electron energy loss spectroscopy
HR-LEED	high resolution – LEED
HR-RHEED	high resolution – reflection high energy electron diffraction
HREELS	high resolution electron energy loss spectroscopy
HREM	high resolution electron microscope
IAP	imaging atom probe
IB	ion bombardment
IBA	ion bombardment and annealing
ICISS	impact collision ion scattering spectroscopy
IIAES	ion induced Auger electron spectroscopy
IPD	in plane diffraction
IPE	inverse photoemission
IPES	inverse photoelectron spectroscopy, inverse photoemission spectroscopy
IR	infrared
IRAS	infrared reflection absorption spectroscopy
IS	image state
ISS	ion scattering spectroscopy
KRIPES	K-resolved inverse photoelectron spectroscopy

KRIPS	K-resolved inverse photoemission spectroscopy
LA	longitudinal acoustic
LAPW	linearized augmented plane wave method
LDA	local density approximation
LDM	lattice dynamical model
LDOS	local density of states
LEED	low energy electron diffraction
LEED-IV	low energy electron diffraction intensity-voltage measurements
LEEM	low energy electron diffraction (LEED) microscopy, low energy electron microscope
LEERM	low energy electron reflection microscopy
LEIS	low energy ion scattering, low energy ion scattering spectroscopy
LENIS	low energy neutral ionization spectroscopy
LEPD	low energy positron diffraction
LIP	low index pole
LMTO	linearized muffin-tin orbital
LPE	liquid phase epitaxy
LRO	long range order
MBE	molecular beam epitaxy
MC	Monte Carlo simulation
MD	molecular dynamics
MEED	medium energy electron diffraction
MEIS	medium energy ion scattering
MEIS-CB	medium energy ion scattering-combined with channeling and blocking
ML	mono layer
MLR	multi layer relaxation
MOKE	magneto-optical Kerr effect
mon	monoclinic
MR	missing row
MS	mixed surface and bulk mode
N-DIC-m.	Nomarski-differential interference contrast microscopy
NBISS	neutral beams ion scattering spectroscopy
NFE	nearly free electron
NICISS	neutral impact collision ion scattering spectroscopy
oop	out of plane
orth	orthorhombic
PAX	photoemission of adsorbed Xe
PBBS	projected bulk band structure
PDS	photothermal displacement spectroscopy
PE	photoelectron
PEEM	photoemission electron microscope
PES	photoemission spectroscopy, photoelectron spectroscopy
PICS	positive ion channeling spectroscopy
PLAP	pulsed laser atom probe
PLEED	polarized LEED
POSAP	position-sensitive atom probe
2PPES	2-photon photoemission spectroscopy
PPM	pseudopotential method
PRM	polarization modulated reflectance
QD	quasi-double scattering
QS	quasi-single scattering
QT	quasi-triple scattering
RBS	Rutherford backscattering spectrometry, Rutherford back scattering

REM	reflection electron microscopy, reflection electron microscope
RFS	renormalized forward scattering
rh	rhombohedral
RHEED	reflection high energy electron diffraction
RPA	random phase approximation
RQ	rapid quenching
RRM	rotation relaxation model
RT	room temperature
SAM	scanning Auger microscope
SBZ	surface Brillouin zone
SC	simple cubic
SCLO	self consistent local orbital
SCLS	surface core-level shift
SCP	self consistent pseudopotential
SEM	scanning electron microscope
SEXAFS	surface-sensitive extended X-ray absorption fine structure
SH	shear horizontal surface mode
sh	shadowing
SHG	second harmonic generation
SIMS	secondary ions mass spectroscopy
SOS	solid on solid
SP	(surface mode polarized in the) saggital plane
SPA	spot profile analysis
SPARPES	spin polarized angle-resolved photoemission spectroscopy
SPIPES	spin polarized inverse photoemission spectroscopy
SPLEED	spin polarized LEED
SPT	surface phase transition
SPV	surface photovoltage spectroscopy
SR	surface resonance
SRO	short range order
SS	surface state
STA	scattering theoretical approach
STEM	scanning transmission electron microscope
STM	scanning tunneling microscopy, scanning tunneling microscope
STS	scanning tunneling spectroscopy
SXRD	surface X-ray diffraction
TE	transverse electric
TEAS	thermal energy atom scattering
TED	transmission electron diffraction
TEM	transmission electron microscopy, transmission electron microscope
tetr	tetragonal
TF	thermal field
TFM	time of flight model
TFM	Thomas–Fermi–Molière potential
THEED	transmission high energy electron diffraction
TLK	terrace-ledge-kink model
TM	transverse magnetic
TOF	time of flight
TOF-SARS	time of flight-scattering and recoiling ion spectrometry
TOM	torsion oscillation magnetometry
TPD	thermal programmed desorption
tric	triclinic

trig	trigonal
TRS	truncation rod scattering
UHV	ultra high vacuum
UPS	ultraviolet photoemission spectroscopy
VB	valence band
VBM	valence band maximum
VLEED	very low energy electron diffraction
XPS	X-ray photoelectron spectroscopy
ZBL	universal potential of Ziegler Biersack and Littmark

1.5 Conversion tables

Table 1. Atomic units

Quantity	Atomic unit [1]	SIU	Miscellaneous
Length l	1 a.u. $= a_0$	$0.52918 \cdot 10^{-10}$ m	0.52918 Å
Reciprocal length l^{-1}	1 a.u. $= a_0^{-1}$	$1.88973 \cdot 10^{10}$ m^{-1}	1.88973 Å^{-1}
Energy E	1 a.u. $= \hbar^2/m_e a_0^2$	$4.35975 \cdot 10^{-18}$ J	1 hartree = 2 Ry (Rydberg) = 27.21140 eV
DOS n	1 a.u = states/atom hartree	$2.29371 \cdot 10^{17}$ states/atom J	1/2 states/atom Ry = $3.675 \cdot 10^{-2}$ states/atom eV

[1]) a_0: Bohr radius, m_e: electron rest mass, $\hbar$: Planck constant.

Table 2. Energy conversion

	J	m^{-1}	Hz[1])	eV	hartree (a.u.)[2])
1 J	≙ 1	$5.03411 \cdot 10^{24}$	$1.50919 \cdot 10^{33}$	$6.24151 \cdot 10^{18}$	$2.29371 \cdot 10^{17}$
1 m^{-1}	≙ $1.98645 \cdot 10^{-25}$	1	$2.99792 \cdot 10^{8}$	$1.23984 \cdot 10^{-6}$	$4.55634 \cdot 10^{-8}$
1 Hz	≙ $6.62608 \cdot 10^{-34}$	$3.33564 \cdot 10^{-9}$	1	$4.13567 \cdot 10^{-15}$	$1.51983 \cdot 10^{-16}$
1 eV	≙ $1.60218 \cdot 10^{-19}$	$8.06554 \cdot 10^{5}$	$2.41799 \cdot 10^{14}$	1	$3.67493 \cdot 10^{-2}$
1 hartree	≙ $4.35975 \cdot 10^{-18}$	$2.19475 \cdot 10^{7}$	$6.57968 \cdot 10^{15}$	27.21140	1

[1]) 1 Hz = 6.2832 rad s^{-1}.
[2]) 1 hartree = 2 Ry; 1 Ry = 13.60570 eV.

(1 Langmuir: 1L = 10^{-6} Torr · s = $1.33 \cdot 10^{-4}$ Pa · s = $1.33 \cdot 10^{-6}$ mbar · s (a measure for the exposure of a surface to a gas atmosphere)).

Error notation: abc (xy)
(xy) applies to the last digits of the value abc: e.g.
3.478(21) Å = (3.478 ± 0.021) Å   9.0(2) eV = (9.0 ± 0.2) eV
13.4(21) Å = (13.4 ± 2.1) Å   9.0(25) eV = (9.0 ± 2.5) eV

1.6 Crystal structures and bulk lattice parameters of materials quoted in the volume

Table 3. Bulk structures and lattice parameters ($a, b, c, \alpha, \beta, \gamma$) of crystals quoted in the volume (value at RT if not otherwise stated) [58H, 58P, 65E, 65W, 69A, 71H, 73H, 82G, 82W].

Crystal	Structure[1])	a [Å]	Comments
Ag	fcc	4.0862	
Al	fcc	4.04964	
Al_2O_3	rh	$a = 5.140$; $\alpha = 55°16'$	
AlAs	zincblende	5.6622	
$AlFe_3$	fcc (DO3)	5.792	
AlP	zincblende	5.451	
As	rh (A7)	$a = 4.159$; $\alpha = 53°49'$	
Au	fcc	4.0786	
B	tetr	$a = 8.75$; $c = 5.06$	
$BaTiO_3$	trig-rh	$a = 5.772$; $c = 6.995$	$T < -80°C$
	orth	$a = 5.682$; $b = 5.669$; $c = 3.990$	$-80°C < T < 0°C$
	tetr	$a = 3.9945$; $c = 4.0335$	$0°C < T < 120°C$
	cub	4.009	$120°C < T < 1330°C$
	hex	$a = 5.74$; $c = 14$	$T > 1330°C$
Be	hcp	$a = 2.2854$; $c = 3.5807$	
BeO	wurtzite	$a = 2.698$; $c = 4.3772$	
Bi	rh	$a = 4.5459$; $c = 11.8622$	
$Bi_2CaSr_2Cu_2O_8$	tetr	$a = 5.41$; $c = 30.7$	
Bi_2Te_3	rh	$a = 10.47$; $\alpha = 24°8'$	
C (diamond)	fcc	3.5670	
C (graphite)	hex	$a = 2.4624$; $c = 6.7014$	
$C_{14}H_{10}$ (anthracene)	mon	$a = 8.561$; $b = 6.036$, $c = 11.163$ $\beta = 124°42'$	
CaF_2	fcc	5.46398	
CaO	fcc	4.8096	
CaO_2	bc tetr	$a = 5.03$; $c = 5.95 \cdots 5.97$	different values reported in literature
Cd	hcp	$a = 2.9794$; $c = 5.6186$	
CdF_2	fcc	5.388	
CdS	zincblende	5.818	
	wurtzite	$a = 4.1348$; $c = 6.7490$	metastable; stable for $T > 700°C$

(continued)

Table 3. (continued)

Crystal	Structure[1])	a [Å]	Comments
CdSe	zincblende	6.05	
	wurtzite	$a = 4.299$; $c = 7.010$	metastable
CdTe	zincblende	6.482	
	wurtzite	$a = 4.57$; $c = 7.47$	metastable
Ce	fcc	5.1612	
Co	hcp	$a = 2.5074$; $c = 4.0699$	
CoO	mon	$a = 5.183$; $b = 3.015$; $c = 3.017$ $\beta = 125°33.8°$	$T < 280$K
	fcc	4.2603	280 K $< T <$ 1223 K
	fcc	8.083	$T >$ 1223 K
CoTi	bcc	2.994	
Cr	bcc	2.8845	
Cs	bcc	6.045	measured at 5 K
Cu	fcc	3.61509	
Cu_3Au	cub ($L1_2$)	3.7477	
CuAu (I)	tetr ($L1_0$)	$a = 3.96$; $c = 3.67$	
CuAu (II)	orth	$a = 3.96$; $b = 39.66$; $c = 3.69$	
CuPd	cub	2.973	measured with 45.5% Pd atoms
Dy	hcp	$a = 3.5903$; $c = 5.6475$	
Er	hcp	$a = 3.5588$; $c = 5.5874$	
Eu	bcc	4.606	
EuO	fcc	5.143	
EuS	fcc	5.968	
Fe	bcc	2.86645	
Fe_2O_3	hex	$a = 5.0345$; $c = 13.7492$	α-phase
	rh	$a = 5.4271$; $\alpha = 55°15.8'$	
FeCo	cub (B2)	2.8570	alloy with 49.70% Fe atoms
FeCr	tetr ($D8_b$)	$a = 8.794$; $c = 4.552$	
FeNi	fcc	3.5698	alloy with 62.66% Ni atoms
FeTi	bcc	2.978	
$Ga_{0.7}Al_{0.3}As$	zincblende	5.6556	
GaAs	zincblende	5.65315	
GaP	zincblende	5.4505	
GaSb	zincblende	6.0954	
GaSe	hex	$a = 3.74$; $c = 15.92$	
	rh	$a = 3.746$; $c = 23.910$	
Gd	hcp	$a = 3.6360$; $c = 5.7826$	
$Gd_2(MoO_4)_3$	mon	$a = 7.58$; $b = 11.44$; $c = 11.42$ $\beta = 109.28°$	
	tetr	$a = 7.39$; $c = 10.67$	$T > 857$°C
Ge	diamond (fcc)	5.657764	
HfC	fcc	4.6365	

Table 3. (continued)

Crystal	Structure[1])	a [Å]	Comments
$Hg_xCd_{1-x}Te$	zincblende	6.462 (x = 0) < a < 6.482 (x = 1)	
Ho	hcp	a = 3.5773; c = 5.6158	
In	tetr (A6)	a = 4.5979; c = 4.9467	
InAs	zincblende	6.05838	
InP	zincblende	5.86875	
InSb	zincblende	6.47877	
Ir	fcc	3.8394	
K	bcc	5.225	measured at 5 K
KBr	fcc	6.5982	
KCl	fcc	6.29294	
$KTaO_3$	cub	3.989	
La	hcp	a = 3.770; c = 6.075	hex structure with $c' = 2c$ also reported
LaB_6	cub ($D2_1$)	4.156	
Li	bcc	3.5092	measured at 5 K
LiCl	fcc	5.13988	
LiF	fcc	4.0262	
LiH	fcc	4.093	
$LiNbO_3$	trig-rh	a = 5.1483; c = 13.863	
Lu	hcp	a = 3.5031; c = 5.5509	
Mg	hcp	a = 3.20939; c = 5.21053	
MgF_2	tetr	a = 4.63; c = 3.05	
MgO	fcc	4.212	
MnO	fcc	4.449	
	hex-rh		T < 120 K
Mo	bcc	3.1472	
MoO_3	orth	a = 13.825; b = 3.694; c = 3.954	
MoS_2	hex	a = 3.15; c = 12.32	
$MoSe_2$	hex	a = 3.29; c = 12.9	
Na	bcc	4.225	measured at 5 K
Na_2O	fcc	5.56	
NaCl	fcc	5.6401	
$NaClO_3$	cub (G3)	6.57584	
NaF	fcc	4.62344	
NaI	fcc	6.47	
Na_xWO_3			
$0 \leq x \leq 0.01$	mon	$a \cong 7.31$; $b \cong 7.48$; $c \cong 3.83$	measured for x = 0.010
$0.01 \leq x \leq 0.03$	mon + orth		
$0.03 \leq x \leq 0.04$	orth		
$0.04 \leq x \leq 0.07$	orth + tetr		

(continued)

Table 3. (continued)

Crystal	Structure[1])	a [Å]	Comments
Na_xWO_3 (cont.)			
$0.07 \leq x \leq 0.28$	tetr	$a = 5.248$; $c = 3.895$	measured for x = 0.10
$0.28 \leq x \leq 0.38$	tetr	$a = 12.094$; $c = 3.748$	measured for x = 0.28
$0.38 \leq x \leq 0.40$	tetr + cub		
$0.40 \leq x \leq 0.70$	cub	3.8305	measured for x = 0.56
		3.846	measured for x = 0.67
Nb	bcc	3.3008	
Nb_3Ir	cub (A15)	5.131	
NbC	fcc	4.470	
$NbSe_2$	hex		
Nd	hcp	$a = 3.657$; $c = 5.902$	hex structure with $c' = 2c$ also reported
Ni	fcc	3.5238	
$Ni_{0.76}Fe_{0.24}$	fcc	3.5522	measured with 75% Ni atoms
Ni_3Al	cub ($L1_2$)	3.589	
NiAl	cub	2.887	measured with 49.8% Ni atoms
NiO	rh	$a = 2.9518$; $\alpha = 60°4'$	
	fcc	4.1946	$T > 210°C$
NiS_2	pyrite (C2)	5.676	
Os	hcp	$a = 2.735$; $c = 4.319$	
Pb	fcc	4.9505	
PbTe	fcc	6.454	
Pd	fcc	3.8898	
Pr	hcp	$a = 3.669$; $c = 5.920$	hex structure with $c' = 2c$ also reported
Pt	fcc	3.9231	
Pt_3Sn	cub ($L1_2$)	4.01	
Pt_3Ti	cub ($L1_2$)	3.898	
PtNi	tetr ($L1_0$)	$a = 3.8227$; $c = 3.589$	
Rb	bcc	5.585	measured at 5 K
RbBr	fcc	6.889	
RbCl	fcc	6.590	
RbI	fcc	7.342	
Re	hcp	$a = 2.7608$; $c = 4.4582$	
Rh	fcc	3.8043	
Ru	hcp	$a = 2.70389$; $c = 4.28168$	
RuO_2	tetr	$a = 4.491$; $c = 3.106$;	
Sb	rh (A7)	$a = 4.5066$; $\alpha = 57°6.27'$	
Sc	fcc	4.541	dimorphic
	hcp	$a = 3.3090$; $c = 5.2733$	
Se	trig (A8)	$a = 4.3640$; $c = 4.9588$	
Si	diamond (fcc)	5.43072	

Table 3. (continued)

Crystal	Structure[1])	a [Å]	Comments
SiC	hex	$a = 3.082$; $c = 2.52$	α-phase. Values of c multiple of 2.52 also reported
	zincblende	4.358	β-phase
SiFe	cub (B20)	4.489	
SiO_2	trig (A8)	$a = 4.91343$; $c = 5.40506$	α-phase ($T < 573$°C), measured at 25°C
	hex	$a = 5.038$; $c = 5.460$	β-phase ($T > 573$°C), measured at 600°C; several other structures reported
Sm	rh	$a = 8.996$; $\alpha = 23°13'$	
SmB_6	cub	4.1333	
SnO_2	tetr (C4)	$a = 4.738$; $c = 3.188$	
SnTe	fcc	6.313	
$SrTiO_3$	cub	3.9050	$T > 110$ K
	bc tetr		65 K $< T <$ 110 K
	orth		35 K $< T <$ 65 K
	rh		4.2 K $< T <$ 35 K
Ta	bcc	3.3058	
TaC	fcc	4.4564	
TaS_2	trig (C6)	$a = 3.319$; $c = 6.275$	α-phase
$TaSe_2$	hex	$a = 3.4362$; $c = 25.399$	4H-$TaSe_2$
Tb	hcp	$a = 3.6010$; $c = 5.6936$	
Te	hex (A8)	$a = 4.4559$; $c = 5.9268$	
TeCd	zincblende	6.477	
Th	fcc	5.084	α-Th
ThO_2	fcc	5.597	
Ti	hcp	$a = 2.950$; $c = 4.686$	
Ti_2O_3	rh ($D5_1$)	$a = 5.454$; $\alpha = 59°6'$	
TiC	fcc	4.310	
TiO_2	tetr (C4)	$a = 4.5937$; $c = 2.95868$	(rutile); orthorhombic and bc tetragonal structures also reported
TiS_2	trig (C6)	$a = 3.41$; $c = 5.69$	
$TiSe_2$	trig (C6)	$a = 3.548$; $c = 5.988$	
Tm	hcp	$a = 3.5375$; $c = 5.55146$	
UO_2	fcc	5.4691	
	fcc	5.4616	$T < 31$ K
V	bcc	3.0399	
V_2O_5	orth	$a = 11.519$; $b = 4.373$; $c = 3.564$	
$VC_{0.75 \cdots 0.96}$	cub	4.136 ··· 4.182	
W	bcc	3.16517	

(continued)

Table 3. (continued)

Crystal	Structure[1])	a [Å]	Comments
WO_3	mon	$a = 5.27$; $b = 5.16$; $c = 7.67$ $\beta = 91°43'$	$T < -40°C$
	tric	$a = 7.30$; $b = 7.52$; $c = 7.69$ $\alpha = 88°50'$; $\beta = 90°55'$; $\gamma = 90°56'$	$-40°C < T < 17°C$
	mon	$a = 7.31$; $b = 7.54$; $c = 7.69$ $\beta = 90.88°$	$17°C < T < 310°C$
	orth	$a = 7.34$; $b = 7.55$; $c = 3.86$	$310°C < T < 730°C$
	tetr	$a = 15.27$; $c = 3.92$	$T > 730°C$
Y	hcp	$a = 3.6515$; $c = 5.7474$	
$Y_3Fe_5O_{12}$	bcc	12.376	
Yb	fcc	5.4862	
$YBa_2Cu_3O_7$	orth	$a = 3.8196$; $b = 3.8798$; $c = 11.6719$	
$YFeO_3$	orth	$a = 5.28$; $b = 5.59$; $c = 7.60$	
$YMnO_3$	hex	$a = 6.13$; $c = 11.4$	
	hex	$a = 6.233$; $c = 11.450$	$T > 660°C$
Zn	hcp	$a = 2.664$; $c = 4.945$	
ZnO	wurtzite	$a = 3.24950$; $c = 5.2069$	
	zincblende	4.63	metastable
ZnS	zincblende	5.4093	
	wurtzite	$a = 3.8140$; $c = 6.2576$	metastable, stable for $T > 1020°C$
ZnSe	zincblende	5.6676	
	wurtzite	$a = 3.996$; $c = 6.626$	metastable
ZnTe	zincblende	6.1037	
	wurtzite	$a = 4.27$; $c = 6.99$	metastable
Zr	hcp	$a = 3.2312$; $c = 5.1477$	α = phase
	bcc	3.6090	β-phase, $T > 865°C$
ZrC	fcc	4.678 ··· 4.695	different values reported in literature

[1]) bc body centered,
bcc body centered cubic,
cub cubic,
fcc face centered cubic,
hcp hexagonal close packed,
hex hexagonal,
mon monoclinic,
orth orthorhombic,
rh rhombohedral,
tetr tetragonal,
tric triclinic,
trig trigonal.

In case of ambiguities, Strukturbericht notations have been added, as defined in [58P] p. 85 ··· 121.

1.7 References for 1

32T Tamm, I.: Z. Phys. **76** (1932) 849.
36M Müller, E.W.: Phys. Z. **36** (1936) 838.
39S Shockley, W.: Phys. Rev. **56** (1939) 317.
51M Müller, E.W.: Z. Phys. **131** (1951) 136.
54H Hagstrum, H.D.: Phys. Rev. **96** (1954) 336.
58H Hansen, M., Anderko, K.: Constitution of binary alloys. New York: McGraw-Hill Book Company inc. 1958.
58P Pearson, W.B.: A handbook of lattice spacings and structures of metals and alloys. Oxford: Pergamon Press 1958.
59S Schlier, R.E., Farnsworth, H.E.: J. Chem. Phys. **30** (1959) 917.
63L1 Lander, J.J., Morrison, J.: J. Appl. Phys. **34** (1963) 1403.
63L2 Lander, J.J., Gobeli, G.W., Morrison, J.: J. Appl. Phys. **34** (1963) 2298.
64K Kane, E.O.: Phys. Rev. Lett. **12** (1964) 97.
65E Elliott, R.P.: Constitution of binary alloys. First supplement, New York: McGraw-Hill Book Company inc. 1965.
65M Many, A., Goldstein, Y., Grover, N.B.: Semiconductor Surfaces. Amsterdam: North-Holland 1965.
65W Wyckoff, R.W.G.: Crystal structures. Vol. 1–2, 2nd ed., Interscience Publishers 1965–67.
67F Frankl, D.R.: Electric Properties of Semiconductor Surfaces. Oxford: Pergamon Press 1967.
67S Smith, D.P.: J. Appl. Phys. **38** (1967) 340.
68C Chiarotti, G., Del Signore, G., Nannarone, S.: Phys. Rev. Lett. **21** (1968) 1170.
68M Müller, E.W., Panitz, J.A., McLane, S.B.: Rev. Sci. Instrum. **39** (1968) 83.
69A Abrikosov, N.K., Bankina, V.F., Poretskaya, L.V., Shelimova, L.E., Skudnova, E.V.: Semiconducting II-VI, IV-VI and V-VI compounds. New York: Plenum Press 1969.
70H Henzler, M.: Surf. Sci. **19** (1970) 159.
70L Lang, N.D., Kohn, W.: Phys. Rev. B **1** (1970) 4555.
70P Plummer, E.W., Gadzuk, J.W.: Phys. Rev. Lett. **25** (1970) 1493.
71C Chiarotti, G., Nannarone, S., Pastore, R., Chiaradia, P.: Phys. Rev. **4** B (1971) 3398.
71H Hellwege, K.-H., Hellwege, A.M. (eds.): Landolt-Börnstein, Vol. III/5a, b. Structure Data of Organic Crystals. Berlin: Springer Verlag 1971.
71I Ibach, H.: Phys. Rev. Lett. **27** (1971) 253.
71M McIntyre, J.D.E., Aspnes, D.E.: Surf. Sci. **24** (1971) 417.
72E Eastman, D.E., Grobman, W.D.: Phys. Rev. Lett. **28** (1972) 1378.
72W Wagner, L.F., Spicer, W.E.: Phys. Rev. Lett. **28** (1972) 1381.
73F Feuerbacher, B., Fitton, B.: Phys. Rev. Lett. **30** (1973) 923.
73H Hellwege, K.-H., Hellwege, A.M. (eds.): Landolt-Börnstein, Vol. III/7. Crystal Structure Data of Inorganic Compounds. Berlin: Springer Verlag 1973–1986.
74S Smith, N.V., Traum, M.M., Di Salvo, F.J.: Solid State Commun. **15** (1974) 211.
75F Froitzheim, H., Ibach, H., Mills, D.L.: Phys. Rev. B **11** (1975) 4980.
76B Boato, G., Cantini, P., Mattera, L.: Surf. Sci. **55** (1976) 14.
76E Eib, W., Alvarado, S.F.: Phys. Rev. Lett. **37** (1976) 444.
79M1 McRae, E.G.: Rev. Mod. Phys. **51** (1979) 541.
79M2 Moritz, W., Wolf, D.: Surf. Sci. **88** (1979) L29.
80K Kisker, E., Goudat, W., Kuhlmann, E., Klauberg, R., Campagna, M.: Phys. Rev. Lett. **45** (1980) 2053.

81A Aono, M., Oshima, C., Zaima, S., Otani, S., Ishizawa, Y.: Jpn. J. Appl. Phys. **20** (1981) L829.

81B Brusdeylins, G., Doak, R.B., Toennies, J.P.: Phys. Rev. Lett. **46** (1981) 437.

81E Eisenberger, P., Marra, W.C.: Phys. Rev. Lett. **46** (1981) 1081.

81F Feder, R.: J. Phys. C: Solid State Phys. **14** (1981) 2049.

81P Pandey, K.C.: Phys. Rev. Lett. **47** (1981) 1913.

82G Gray, D.E. (ed): American Institute of Physics Handbook, 3rd ed., New York: McGraw-Hill 1982.

82W Weast, R.C. (ed.): CRC Handbook of Chemistry and Physics. 63rd ed., Boca Raton, Florida: CRC Press inc. 1982-83.

83B1 Binnig, G., Rohrer, H., Gerber, C., Weibel, E.: Phys. Rev. Lett. **50** (1983) 120.

83B2 Binnig, G., Rohrer, H., Gerber, C., Weibel, E.: Surf. Sci. **131** (1983) L379.

83H Himpsel, F.J.: Adv. Phys. **32** (1983) 1.

83J Johnson, P.D., Smith, N.V.: Phys, Rev. B **27** (1983) 2527.

83T Tromp, R.M., Smit, L., van der Veen, J.F.: Phys. Rev. Lett. **51** (1983) 1672.

84C Chiaradia, P., Cricenti, A., Selci S., Chiarotti, G.: Phys. Rev. Lett. **52** (1984) 1145.

84O Olmstead, M.A., Amer, N.M.: Phys. Rev. Lett. **52** (1984) 1148.

84T Takayanagi, K.: J. Microsc. (Oxford) **136** (1984) 287.

87F Feenstra, R.M., Stroscio, J.A., Fein, A.P.: Surf. Sci **181** (1987) 295.

90A Ancillotto, F., Andreoni, W., Selloni, A., Car, R., Parrinello, M.: Phys. Rev. Lett. **65** (1990) 3148.

91I Ibach, H.: High Resolution Electron Energy Loss Spectrometers. Springer Series in Optical Science **63**, Berlin: Springer Verlag 1991.

6 Interaction of charged particles with surfaces

6.1 Elastic scattering and diffraction of electrons and positrons

6.1.1 Introduction

6.1.1.1 Definitions and historical layout

Of the various techniques devoted to the study of crystalline surfaces, those dealing with the scattering of charged particles are the most widely used. In particular, the elastic scattering of electrons and positrons have for many years been accepted as one of the principal tools for obtaining crystallographic data, i.e. for finding out the exact positions occupied by atoms in the first atomic layers. This is due to the fact that the wavelength is appropriate for the interatomic distances, and to the limited penetration depth of the particles.
Various acronyms are in use depending on the type of incident beam and on its energy:

- LEED or ELEED (for low-energy electron diffraction or elastic low-energy electron diffraction, in the energy range 30 ... 400 eV);
- VLEED (for very low-energy electron diffraction, up to 30 ... 40 eV);
- MEED (for medium-energy electron diffraction, up to thousands of eV);
- RHEED (for reflection high-energy electron diffraction, up to tens of thousands of eV);
- LEPD (for low-energy positron diffraction, in the LEED typical energy range. It is also to be noted that LEPD is sometimes used for low-energy photoelectron diffraction);
- PLEED and SPLEED (for polarized and spin polarized LEED).

The above mentioned techniques are all based on similar mechanisms of interactions of the particle with the solid, but their specific use covers different, and often complementary, aspects of the physics of surfaces.

The diffraction phenomenon of charged particles from a diffracting grating such as the regular arrays of atoms in a crystal was discovered in the twenties by Davisson and Germer [21D, 27D1, 27D2, 27D3], and Thompson and Reid [27T] with their historic experiments. The aim of those pioneering works was mainly to demonstrate wave-particle dualism, and not to obtain information about the physics of surfaces. Only H.E. Farnsworth, at the Brown University, and L.H. Germer, at the Bell Telephone Laboratories, used electrons as a probe to characterize solid surfaces from the thirties on. However, until the end of the fifties and the start of the sixties, the lack of valid vacuum technologies and of adequate means to obtain and maintain well reproducible clean surfaces rendered these works of historical interest only.

With the development of the UHV technology, able to keep surfaces clean for a long enough period to allow experimental observations, diffraction techniques emerged again as the most promising means to study the solid surfaces and the gas-solid interaction, soon supported by other techniques such as Auger spectroscopy, etc. The first commercial diffractometers quickly spread all over the world, and very soon researchers felt the need of an appropriate theory to obtain quantitative interpretations from data.

Various features of diffraction were early recognized as characteristic of a strong multiple scattering, and the theorists tried to work out approximate dynamical theories to explain the discrepancies from the simple Bragg laws [27B, 28B2], but only in the second half of the sixties the first more complete theories for LEED and MEED were set up, the first quantitative structural determinations being performed in the early seventies. By the mid-seventies the most important ideas and computing procedures have already been expressed and the scientists devoted their efforts more and more to the assessment of reliability as well as to the attainment of important refinements and to the solution of more complicated surface structures.

The influence of the electron spin in LEED was also investigated and polarized electrons were used to investigate the magnetic properties of the surfaces.

At the same time, important experiments were performed in the region of very low energy, and theoretical efforts were made to obtain accurate information about the surface barrier of potential.

The RHEED technique was used for a long time mainly for interface rather than surface studies, because of the major experimental difficulties it involved, and only in recent years a concerted effort to obtain a quantitative interpretation of experimental data has been made.

Finally, in the last decade LEPD has been attempted, and very recently some aspects of this technique showed it to be an important rival of LEED for surface crystallography.

Because of the wide interest and diffusion of electron diffraction several books are now available on this subject [74P, 79V, 85C, 85K2, 86V1] and a large number of review articles can be found in the literature [88J4].

In the following sections, the physical principles, the existing theories, instrumentation and experimental particulars and the reliability and precision of the above mentioned techniques will be briefly and schematically reviewed with the aim of providing a wide bibliographical basis for each topic.

In section 6.1.2, the structure of clean surfaces so far studied using the above diffraction techniques are reported in some detail with the pertinent bibliographical references. In view of the scope of this volume, no mention will be made of results on surface structures from adsorbed gases, contaminants, evaporated layers and epitaxy. The effects of surface roughness and studies on related topics will be dealt with in chapter 2.3 of subvolume III/24a.

6.1.1.2 The diffraction of electrons and positrons

When a charged particle impinges on a surface, the wave field in the vacuum is the superimposition of the incident plane wave and of a function representing the scattered wave. The total wave field must fulfil the Schrödinger equation and continuously match the solution inside the solid. Supposing the surface to be a perfect two-dimensional array of scatterers, one can express the wave field of the scattered particle as a Bloch wave, developed in a Fourier series:

$$\psi_S(\boldsymbol{r}) = \exp(\mathrm{i}\,\boldsymbol{k}_{\parallel}\,\boldsymbol{\rho}) \sum_{\boldsymbol{q}} \alpha_q(z) \exp(\mathrm{i}\,\boldsymbol{q}\,\boldsymbol{\rho})\,, \tag{1}$$

where $\boldsymbol{k}_{\parallel}$ and $\boldsymbol{\rho}$ are the surface components of the wave vector $\boldsymbol{k}$ and of the position vector $\boldsymbol{r}$, respectively, $\boldsymbol{q}$ is the two-dimensional reciprocal lattice vector and $\alpha_q(z)$ is a function of the z coordinate alone, taken to be normal to the surface and positive towards the external space. Substitution in the Schrödinger equation in the absence of a potential leads to

$$\alpha_q(z) = A_q \exp\{-\mathrm{i}\,[\,2E - (\boldsymbol{k}_{0,\parallel} + \boldsymbol{q})^2\,]^{1/2}\}\,, \tag{2}$$

i.e., the external wave is a set of plane waves of amplitude A_q which propagate in the directions specified by the reciprocal lattice vectors and which form a diffraction pattern on a detector at great distance, with beams of intensity given by $|A_q|^2$. The directions of the diffracted beams are defined by the conservation of the energy and of the momentum parallel to the surface:

$$|\boldsymbol{k}|^2 = |\boldsymbol{k}_0|^2\,; \quad \boldsymbol{k}_{\parallel} = \boldsymbol{k}_{0,\parallel} + \boldsymbol{q}\,. \tag{3}$$

The above exposition demonstrates that the diffraction pattern is characteristic of the translation symmetry of the surface lattice. The other symmetry properties (such as rotation axes, glide planes, etc.), the scattering properties of the single scatterers and their position inside the unit surface cell are responsible for the beam intensities, which also depend on the energy of the incident beam and on the angles of incidence and of azimuth.

The mere aspect of the diffraction diagram is not in itself characteristic of a single surface structure, as all structures having the same translation symmetry give the same pattern [67B1]. Nevertheless, in many cases of complicated structures, the only information obtainable from a diffraction pattern is this general symmetry, from which models of the structure can be formulated. This procedure was the one used for early surface studies using diffraction techniques. On the other hand, the exact structure can be derived from the quantitative analysis of beam intensities.

A useful way to represent the scattering mechanism is the Ewald construction in two dimensions. The Ewald sphere has a radius equal to the length of the incident wave vector $\boldsymbol{k}_0$. The tip of the $\boldsymbol{k}_0$ vector is at the origin of the reciprocal net, and the origin in the center of the sphere (Fig. 1a). A diffracted beam, characterized by the vector $\boldsymbol{k}_q$, arises where a reciprocal lattice rod intersects the sphere. When the energy of the incident beam is varied, the sphere radius varies, so that the reciprocal net rods are explored. In the case of RHEED, because the incident beam is almost parallel to the surface, to ensure high sensitivity at the surface region due to the deep penetration of high-energy electrons, and because the radius of the Ewald sphere is wide, only the rods which are in a particular azimuth are found on the sphere near the incident beam direction (Fig. 1b).

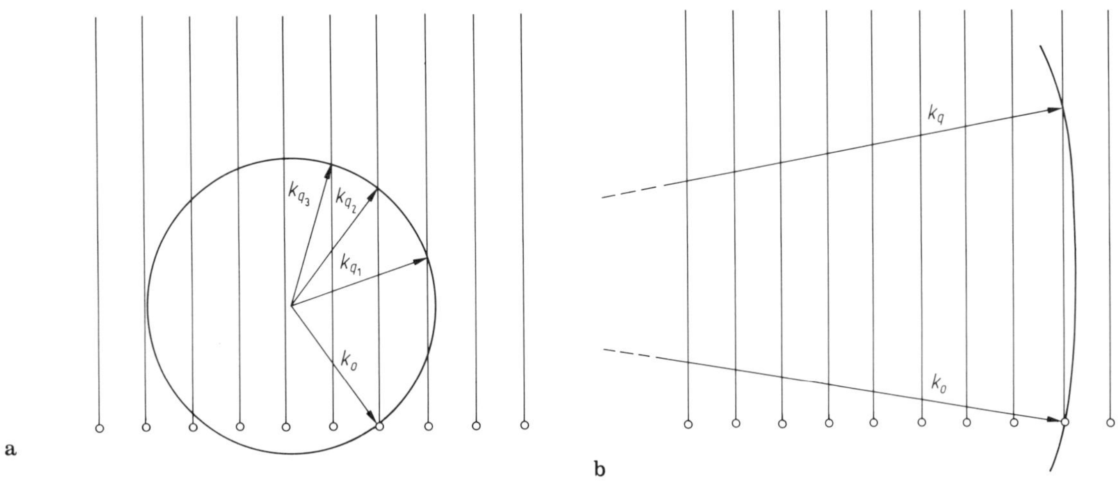

Fig. 1a, b. Ewald construction for LEED **a**, and RHEED **b**.

6.1.1.3 Experimental considerations

In principle, a diffractometer is a UHV chamber equipped with a source of particles, a sample manipulator and a detector, in addition to a number of accessories to clean the sample (e.g. ion bombardment), to verify the purity of the surface (e.g. Auger spectroscopy), and to other techniques for surface studies. The actual physical arrangement depends on the characteristics of the incident particles and on the purposes of the entire apparatus.

The early apparatuses were very similar to the present ones in their main features [33S, 34E, 62L2, 64P1, 65C]. The scheme of a typical LEED apparatus is reported in Fig. 2a. In general, a system of filtering grids is used to separate elastic from anelastic electrons. A fluorescent screen is used to see the diffraction pattern directly through a glass window. The screen, a hemispherical bowl with the sample in the center, simulates the Ewald sphere, and the observed pattern is the image of the reciprocal lattice. When varying the electron energy, instead of varying the Ewald sphere radius, one observes expansion or contraction of the reciprocal lattice, i.e. movement of the reflected beams on the specular beam.

In some cases, principally to save the flange used for the viewport for other applications, the fluorescent screen is transparent and the pattern is viewed from the rear.

In principle the RHEED apparatus is very similar to the LEED one: the main difference lies in the position of the gun which is placed on the opposite side of the sample in relation to the screen, in order to obtain a wide angle of incidence and to observe the pattern formed around the specular beam (Fig. 2b).

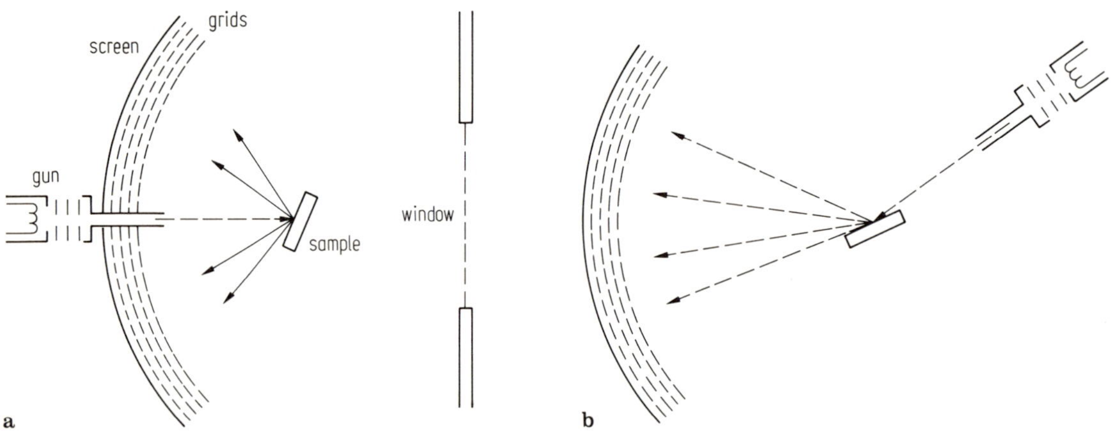

Fig. 2a, b. Scheme of the LEED **a**, and RHEED **b** apparatuses.

In general, no particular experimental sophistication is necessary to obtain a pattern of medium quality, such as those used when LEED is present as an ancillary tool in surface studies, i.e. when the only information required from the observation of the pattern is the surface unit mesh, its dimensions and its symmetry. Instead, when beam intensity measurements are performed, e.g. to obtain information on surface imperfections from beam profile measurements [83L1] and chapter 2.3, subvolume III/24a or to perform quantitative crystallography, particular devices and care are required.

The electron gun consists in general of a thermionic cathode, with a series of electrostatic lenses for extraction and focusing of the beam. The typical beam current is of the order of 1 μA, its angular divergence of the order of 0.5°, the diameter of about 0.5 mm, and the energy spread of about 0.5 eV. Better characteristics, such as those necessary for spot profiling, are achieved with particular designs or methods [83L1, 83M1]. A method to measure electron beam characteristics (e.g. divergence) is found in [81M2]. Particular requirements are also needed for measurements of fine structure in the low energy region (VLEED). For these purposes a miniature, movable electron gun and a gun with a focusing system using a 127° analyser are used [75R1, 80D4, 85M3, 85T, 88H1]; in this way, a resolution of 10 meV can be achieved, so that as many as four very narrow peaks in the intensity vs. energy curves are resolved in a range of a couple of eV. No particular problems are found with RHEED, mainly because the beam has very little angular divergence, due to the high energy. The RHEED technique is often performed in TEM apparatuses, improved for UHV, taking advantage of the already existing electron gun, sample manipulator, etc.

The sample manipulator is in general a mechanical device with some degrees of freedom to permit variations of the incidence and the azimuthal angles, and of other translational movements to vary the position of the sample. The manipulator permits heating of the sample and sometimes cooling down to liquid nitrogen temperature. In general the temperature is measured by a thermocouple. Particular variations have been designed to permit heating and electronic check of the temperature [79M10, 79T2], cooling at liquid helium temperature [71S1, 73T1] and mounting of various samples at the same time or rapid introduction of different samples [76C2, 80P2, 82S1, 82O, 83F1, 84K4, 84N1]. A simple method for rapidly mounting very small samples pressed on indium spheres on a grating support is described in [84G2].

The evaluation of θ and ϕ, the incidence and azimuthal angles, is not immediate, particularly when the intensities for quantitative analysis are measured. In general, the angle θ is measured directly on the sample manipulator, after having determined the normal incidence conditions. This can be done after appropriate calibration of the manipulator outside the chamber [73B3] or more easily by checking the correct degeneracy of the beams which must be equal because of symmetry, and the immobility of the specular beam when varying the energy of the incident beam. The angle ϕ is evaluated by measuring the angle between the trace described by the specular beam when varying θ, and the directions between some beams and the specular beam in normal incidence conditions. More sophisticated methods are also used, involving a single photograph of the fluorescent screen and a subsequent processing of the data from the angles measured on it [78C3]. A computer program for such elaboration is reported in [86V1]. If a large number of spots is used, an accuracy better of 0.1° can be achieved. Variations of this method are also reported to take into account the fact that the incident beam may be not perfectly aligned with the center of curvature of the screen and possible perturbations, such as residual magnetic field [82S4, 83M6].

For VLEED, measurements of high values of θ and ϕ (45° ... 70°) are usual. To obtain a good resolution in this case a method was described which uses a double modulation technique [80P3]. In one case a method to focus the beam directly in a Faraday cage is used, obtaining a quasi-stable position of the specular reflection as a function of energy [83D7].

In the early apparatuses the reflected beams were observed by using a movable Faraday cage. This method is precise, particularly in the case of measurements of the intensity of each beam, but it is not very practical. Display systems using a fluorescent screen on which the pattern as a whole becomes immediately visible are more convenient. The screen is a hemispherical bowl, with the sample in the center. The gun is in general centered in front of the screen in the case of the LEED, and on the opposite side in the case of the RHEED (Figs. 2a and 2b). Recently, an apparatus in which low-energy and high-energy electrons can be used together was built [85S2]. In this apparatus, the gun is at 90° from the screen, so that the experiment could also be carried out on surfaces which are not completely flat, as in normal RHEED.

In the conventional apparatuses a series of centered hemispherical grids (in general four) are put in front of the screen: the first one is at the same potential as the chamber in order to provide a space with no field for the diffracted electrons. The second and the third are filters for the inelastic electrons. The fourth is a shield for the post-acceleration electric field (in general of a few kilovolts) of the screen.

The measurements of the beam intensities for quantitative analysis are in general made beam by beam. Normally, the intensities as a function of energy of the incident electrons at fixed values of θ and ϕ are collected (the so called *I*/*V* curves or spectra). In other cases, measurements at fixed energy are made by varying the angles, in general ϕ (polar curves). In the case of RHEED, the most accessible data are the intensity curves as a function of θ (rocking curves).

The measurements require normalization with respect to the primary current and background subtraction.

An alternative method to collecting LEED data using a CMA (cylindrical mirror analyzer) without the screen, is reported in [75D1].

To measure the beam intensities in the case of a fluorescent screen, spot photometer was often used, this being operated from outside the chamber and through a glass window. The measurement of the intensity of the light spots on the screen requires accurate calibration of the response of the phosphor. In certain cases, corrections were made to take into account the transparency of the grids as a function of their curvature [74L4]. This method is cumbersome and time-consuming: to obtain a set of intensity data which will be sufficient for quantitative analysis normally requires several hours. In many cases this time is unacceptable, because of residual gas action on the surface or because of the instability of the surface structure under electron beam bombardment. Therefore, automatic methods are adopted, among these a method for recording the *I*/*V* curves by imaging the LEED spots into a photometer tube with a mirror which is rotatable around two orthogonal axes [82B1]. A number of photographic methods have also been proposed, in which a photographic film of the diffraction pattern is exposed at each energy and subsequently analyzed by a microdensitometer or by a scanning video analyzer [75S2, 76F4, 79T3].

The most useful and recent methods are based on a vidicon camera and a processing computer [76H2, 79L, 79W3, 80H1, 80L3, 85H2, 86O1, 86A3]. Today at least two commercial apparatuses of this kind exist. The method was also adapted for the measurement of intensity of rocking curves in RHEED [86B4].

Finally, it should be mentioned the fact that for some studies, mainly when structures of adsorbed species which alter under electron bombardment are present on the surface, a very low intensity of the incident beam is required. In these cases, a special amplified photometer [82S2] or an image intensifier (such as a channel plate multiplier array) is used, and currents of 10^{-6} less than in the traditional systems are reached [79W3, 79W4, 80S3, 87B3].

A different apparatus was constructed by the Grenoble group, in which a hemispherical transparent screen permits observation of the entire half space of the reflected beams. This system enables measurements to be made at very high angles of incidence [72G3, 74A1, 74D1], and very precise and reproducible conditions of incidence are reached by means of an external two-circle goniometer. A similar apparatus was also built in York [84D2]. For this kind of system a special miniaturized gun was designed [82C1].

The use of polarized electrons for surface studies requires special sources and/or detectors. When a primary polarized beam is necessary, sources such as those used in atomic physics are not very convenient as they involve atomic beams and they are consequently difficult to use under UHV conditions [76K1, 80C1]. More widely used are solid state (GaAs) sources [76R1]. In these, the polarized electrons are photoemitted by circular polarized infrared light and the polarization of the emitted electrons is due to the different rates of the electron-induced transitions. The polarized electrons are extracted from the GaAs surface if the latter is covered by Cs. Other sources have also been tried and discussed [85K2].

Early experiences in SPLEED were performed using a Mott detector. Of late, the most frequently used detector of polarized electrons is the so-called LEED detector. This works on the principle of the Mott detector, i.e. on the asymmetry due to spin-orbit coupling in electron-atom scattering. In practice, when the beam to be analyzed is normally impinging on the surface of a single crystal, two beams (degenerate in the case of non-polarization) are collected by two channeltrons oriented along the exit angle of the beams [83L1]. If the electrons are polarized, the intensities are unequal, and their difference is a measure of the polarization. Such a detector needs accurate calibration [79K1, 85K1, 85K2], but its use is convenient because of its small size, which means that it can also be used in a standard LEED apparatus. A particular arrangement in which a standard LEED has a movable gun and diffracted beams can be extracted from a hole in the screen, was described in [77F4].

A different type of detector is the absorbed current one. It is based on the observation that the zero crossing of the secondary electrons emitted by electron bombardment (i.e. the point at which the net current to a sample is zero) also depends on the orientation of the spin of the primary electrons relative to the magnetization. This effect was attributed to exchange interaction and to the spin dependence of the inelastic scattering, even if the exact nature of the phenomenon is not clear [81E1, 81E2, 81P2, 85K2].

In the case of LEPD, the main problem is to build a source of sufficient brightness. In general, a radioactive source (such as ^{58}Co or ^{22}Na) is used to irradiate a metal single-crystal (often Cu (111)) as a moderator. A fraction of 10^{-3} of the incident high-energy particle thermalizes within a positron diffusion length and is reemitted from the surface of the moderator, because of the negative work function of most metals for positron. The positron emitted at energies of < 3 eV are accelerated by the gun, focused, decelerated and focused again [79M8, 80M10, 80R3].

With a radioactive source of about 0.5 mCi, an intensity of (1 ... 3)$\cdot 10^6$ positrons/s is reached. The brightness can be enhanced by activating the surface of the moderator by adsorption of a gas (S or H_2S on Cu) [79M8, 80M12] or by using a second moderator. In the remoderator, the particles only penetrate distances of the order of 10^2 Å below the surface, and since the positron thermal diffusion is of the order of 10^3 Å nearly all the positrons are reemitted [80M12, 82C2, 85F]. A 1mm beam diameter and an energy dispersion of 0.07 eV at 100 eV can be obtained.

6.1.1.4 Diffraction theories and methods

The strong scattering power of matter for electrons and positrons is responsible for the complexity of the description of diffraction from a crystal, particularly in the case of low energy. This is the main difference between LEED and X-ray diffraction. In the latter, the incident wave is considered to undergo just one scattering event inside the solid. This is justified by the fact that the atomic X-ray cross sections are so small that the scattered intensity is a very low fraction of the incident intensity. Within this limit, the coherent combination of single scattering events from each atom is in general sufficient to describe photon interaction with the crystal (kinematic theory). In the case of electrons and positrons, on the other hand, the cross sections are various orders of magnitude greater than those of X-rays. The scattered intensity is very high and a substantial amount of multiple scattering is present. This leads to the need for dynamical theories.

6.1.1.4.1 LEED

Right from the early stages of LEED, the lack of simple kinematic interpretation was established by observing the shape of the *I*/*V* curves [64G, 64H, 64P2, 67G, 68G2]. Indeed, if the scattering was kinematic, only the Bragg peaks should be seen. Instead more highly-structured curves were obtained than in early, simplified dynamical theories (two or three beams) [66M3, 69F6].

In general, the scattering potential is considered in the Harthree-Fock framework, i.e. it is assumed that the scattering potential is the attractive nuclear potential and a Coulomb and exchange potential is averaged over the core states. In this way the Schrödinger equation is reduced to a one-electron equation. The most important contribution to elastic scattering is in the core potential, due to the nuclei and to the bound electrons. A complete analysis of the electron-solid interaction can be found in [71P2, 71P3, 74P]. The scattering properties of each atom in the solid are generally represented by appropriate phase shifts.

Various approaches were tried to solve the scattering problem inside the solid with the appropriate contour conditions at the surface, and to represent the discrete set of beams in the vacuum (Eq.(2)). The first elaborations were substantially extensions or generalizations of earlier dynamical X-rays or multiple scattering theories, such as those of Darwin or Lax [14D, 51L], or self-consistent formalisms to describe the backscattering of electrons from the crystal. This was done by McRae using the $\boldsymbol{k}$-space [66M4, 67M2, 68M2, 68M3], and by Beeby using a layer structural propagator in $\boldsymbol{l}$-space representation [68B1, 68W, 69W]. An important step forward was taken with the introduction of the approximation of the muffin-tin scattering potential, i.e. a spherically symmetrical potential around each scatterer up to a characteristic radius and constant potential between the atoms (as in the band structure calculation known as the KKR method). The muffin-tin potential is justified by the strong interaction of the electrons with the nucleus and the core electrons and by the small effect of the external electrons, due to the energy which is well above the Fermi level. Using this approximation and taking advantage of two-dimensional symmetry, a complete formal theory and important sum rules were established. In such a way the scattering from a periodic plane of atoms was expressed with a Green function [68B1]:

$$G^{ji}_{LL'} = -4\,\pi\,\mathrm{i}\,(2m/\hbar^2)\,k \sum_{L''}\sum_{r} i^{l'}\, C(L, L', L'')\, h^{(1)}_{l''}\,(k\,|\boldsymbol{r}_j-\boldsymbol{r}_i+\boldsymbol{r}|)\, Y_{L''}(\boldsymbol{r}_j-\boldsymbol{r}_i+\boldsymbol{r}) \\ \exp\,[-\mathrm{i}\,\boldsymbol{k}_0\,(\boldsymbol{r}_j-\boldsymbol{r}_i+\boldsymbol{r})] \tag{4}$$

in which $k = |\boldsymbol{k}^{\pm}_{\boldsymbol{q}}| = [\,|\boldsymbol{k}_{0,\parallel}+\boldsymbol{q}\,|^2 + (k^{\pm}_z)^2\,]^{1/2}$, $k^{\pm}_z$ being the component of the $\boldsymbol{k}_q$ vector normal to the surface ; $C(L, L', L'') = \int Y^*_L(\Omega)\, Y_{L'}(\Omega)\, Y^*_{L''}(\Omega)\, \mathrm{d}\Omega$ are the Clebsch-Gordan coefficient (Y_L are spherical harmonics, and L stands for the quantum numbers l and m; L'' has to satisfy the conditions $|l-l'| < l'' < l+l'$, and $m'' = m + m'$); $\boldsymbol{r}_i$ and $\boldsymbol{r}_j$ are the atomic sites in two different planes i and j; $\boldsymbol{r}$ is the position of a lattice point of any of the planes, excluding $\boldsymbol{r} = \boldsymbol{r}_i - \boldsymbol{r}_j$; $h^{(1)}_{l''}$ denotes the Hankel function of the first kind.

Stacking a set of n layers of atoms, we just have to take into account the difference of phase due to the distance of each plane from the surface, before adding the amplitudes together.

To evaluate the lattice sum in Eq.(4), a method using an Ewald scheme was proposed by Kambe [67K1, 67K2], in which the sum was carried out in a direct summation for the near lattice points and with a Fourier summation for the far ones. In such a way, Kambe formulated a theory for diffraction from a slab of planes [68K]. A generalization of the Kambe theory was proposed to take into account also the potential jump at the surface [70K] and the Ewald theory was recently reformulated [82S3].

The muffin-tin approximation was also used in the generalized Darwin theory [70J2], and it is now generally accepted.

Another approach leads again to band-structure theories. It expresses the wave inside the solid as a combination of Bloch functions, because of the two dimensional periodicity of the planes parallel to the surface. In such a way the problem is formulated in an exact formalism [71S2], even if its interest remains theoretical because of the difficulties in the matching with the external field. The interest of this formulation lies in the interpretation in terms of real and complex band structure of the semi-infinite solid. Relations between the *I*/*V* curves and the band structure have been recognized [71A4].

Since the regions between the planes have constant potential, the most appropriate formulation for planar scattering is obtained by transforming the above spherical wave expressions into plane waves in the space between planes. This is done with a diffraction matrix which connects an input plane wave, incident on a plane, and an output plane wave [71T4, 72J2, 74P, 77S2]. The final form for a layer is

$$M_{qq'}^{\pm\pm} = -(8\pi^2 m\mathrm{i})/(Akk_z^{+}\hbar^2)\cdot \sum_{lml'm'} \mathrm{i}^l(-1)^m Y_{l-m}(\boldsymbol{k}_q^{\pm})(1-X_{lml'm'})^{-1}\left[Y_{l'm'}(\boldsymbol{k}_{q'}^{\pm})(\mathrm{i})^{-l'}\right]\cdot$$
$$\exp(\mathrm{i}\,\delta_{l'})\sin\delta_{l'} + \delta_{qq'}\delta_{\pm\pm}\,; \tag{5}$$

where A is the area of the unit surface cell, δ_l are the phase shifts and

$$X_{lml'm'} = \sum_{l'+m'=\mathrm{even}} C^{l'}(l'm', l''m'')_{l'm'} \exp(\mathrm{i}\delta_{l'})\sin\delta_{l'}\,;$$

with

$$C(l'm', l''m'') = 4\pi(-1)^{(l-l'-l'')/2}(-1)^{m'+m''}\cdot Y_{l'-m'}(\pi/2, 0)\cdot \int Y_{lm}(\Omega)Y_{l'm'}(\Omega)Y_{l''m''}(\Omega)\mathrm{d}\Omega.$$

Other somewhat different forms were found by summing over the reciprocal lattice rather than over the direct space [75T1].

Assuming a constant potential between the layers, the wave function in this gap can be expressed as a set of waves moving towards the bulk of the crystal and coming back towards the surface [71J4]:

$$\psi_i = \sum_{\boldsymbol{k}}\left[\alpha_i(\boldsymbol{k}^+)\exp(\mathrm{i}k_z z) + \alpha_i(\boldsymbol{k}^-)\exp(-\mathrm{i}k_z z)\exp\left[\mathrm{i}(\boldsymbol{k}_0 + \boldsymbol{k}_{0,\parallel})\cdot \boldsymbol{r}\right]\right]. \tag{6}$$

The coefficients of the wave from one side of the layer to the other are linearly related by a transfer matrix [68J1] partitioned as

$$\mathbf{Q} = \begin{pmatrix} (P^+)^{-1} & -(P^+)^{-1}R^+ \\ R^-(P^+)^{-1} & P^- - R^-(P^+)^{-1}R^+ \end{pmatrix}, \tag{7}$$

where P^+ and P^- are forward and backward propagators, and R^+ and R^- are reflection matrices for the layer (the symbols are taken from another formulation of intralayer scattering [71P4, 71P5]). In this way, taking advantage of the periodicity in the direction perpendicular to the layers, we can use the Bloch theorem, so that the eigenvectors of the **Q** matrix are Bloch waves and the eigenvalues are exp(i *k d*) (*d* being the interlayer spacing): diagonalization of **Q** gives the appropriate Bloch waves inside the semi-infinite crystal. Moreover, a surface slab of several different layers, each with a different **Q** matrix, has a wave function obtained by simple multiplication of the **Q** matrices. The crystal wave function is finally matched to the wave in the vacuum. Only the non-evanescent waves are retained at great distance. Nevertheless, all the Bloch waves, both the propagating and the evanescent ones, contribute to the entire wave [68J2].

Various other approaches, often based on older theories and sometimes equivalent to them, or generalizations, have also been published [71H1, 71S3, 72L10, 74L5, 77M4, 76F2].

Finally it is worth mentioning one method based on the scattering from clusters of atoms on the surface rather than from planes [89R4], and another one which is suitable for the treatment of incommensurate overlayers on surfaces [80A].

Some important factors have to be taken into account in order to obtain realistic calculated *I*/*V* curves: inner potential, inelastic damping of the electrons and thermal atom vibrations.

The inner potential has its origin in the surface potential jump undergone by the electrons passing from the vacuum into the solid, and corresponds to the zero of the muffin-tin. It is the sum of the Fermi level energy and the work function of the sample (the work function of the cathode has also to be considered). It produces refraction effects of the beams. The inner potential is in general taken as constant, or as an empirical function which is slightly variable with the energy in the LEED typical range [80C3, 81J1], and added to the primary energy as a parameter V. This results in an almost rigid translation of the entire *I*/*V* curve along the energy axis [75V3].

Inelastic interactions are very important in the low-energy range, and are mainly due to energy loss arising from plasmon excitation. The resulting effect is a general attenuation of the electron beams, thus improving the convergence properties of the calculations. A phenomenological way is used to take into account inelastic damping, i.e. an imaginary part of the potential V_i is added to the inner potential (optical theorem) [37S, 57Y, 66H3, 68H2, 68O, 71T5]. An imaginary part in the potential affects the Hamiltonian, which becomes non-Hermitian and there is no longer conservation of the electron flux, thus simulating the attenuation of the beam. This term can be constant or an empirically-adjusted function of the energy. The effect on the curves is a general smoothing, whereas the width of the peaks is of the order of 2V [69P6, 71J5, 71T6, 74P, 86V1]. The importance of inelastic scattering was recognized in a different formulation of the entire theory, in which the propagators were renormalized to account for the various inelastic loss mechanisms [69D2, 71D2, 71D3, 71D4, 71L3, 71T5, 72D4].

In the case of molecular crystals, a theory including the effect of the creation of Frenkel excitons was elaborated [78N3, 78N4, 78N5, 78N6].

6.1.1.4.1.1 Alternative methods using LEED

Many simplifications can be introduced to reduce the computer time and/or the core memory required for the numerical calculations, mainly when large matrices are present because of the high number of atoms in the unit cell. The simplest and one of the earliest ones used is the kinematic theory, which is by no means sufficient to explain the features of the intensity curves. Nevertheless, it can be useful for understanding the mechanism of scattering [75V2] as well as for obtaining some qualitative information on the structure, mainly from the direct observation of the diffraction pattern. Moreover, the kinematic theory was often used to try alternative methods to the full dynamical one. In some cases, an optical simulation was used by irradiating a two-dimensional grating, such as a reduced photographic transparency of a model drawing, with a laser light [68C1, 68F1, 75L3, 81V1, 81V2, 82E1]. This method can be useful to create an interference pattern which resembles the electron diffraction pattern, not only in the obvious reproduction of the symmetry, but also in a tentative reproduction of the relative intensities of the various spots, which can be used to interpret large unit cell patterns.

An interesting attempt to take advantage of the kinematic theory was made by trying to reduce the dynamical features of the intensity curves by appropriate averages. The idea was anticipated by Lander and Morrison [63L] and stated by Lagally et al. [71L1] in the constant-moment-transfer average (CMTA) method: the intensity curves taken at various incidence angles were plotted as a function of the momentum transfer and averaged, so that only the kinematic features (depending directly on this transfer) remain, whereas the dynamical contributions do not produce systematic effects. Even if this procedure was demonstrated to be doubtful [72P], the method was developed and used [73N, 73W1, 75L1, 76U, 79A4]. It was also applied to the study of overlayers, but dynamical contributions due to the different scattering powers of the atoms of the substrate and of the overlayer renders it useless in this case [75M3, 80O4]. The method was extended to the average over the azimuthal angles [72A1, 76A1] and the curves compared with a two-beam theory derived from the Darwin theory [73A].

An intensity average over the energy to be compared with an effective kinematic atomic scattering factor was tried by Duke et al. [70T3, 72D3, 72T4].

An important feature of kinematic theory is in the possible use of direct methods such as those used in X-ray crystallography, mainly the Patterson function. This kind of application is by no means obvious, mainly because of the multiple scattering, of the effect of inner potential and of the data truncation. Nevertheless, a generalization of the Patterson function, which is suitable for LEED was introduced [64T, 66T, 72C1, 72C2, 73C1, 74W1, 79A1]. Problems connected with data truncation are minimized by window functions and using the quasi-kinematic data obtained by averaging methods [74B2, 76C1, 82M3]. Better data processing is to be found in generalizations called transform-deconvolution and convolution transform methods. The idea is that if the specular beam intensity (expressed as a kinematic intensity) is analytically Fourier-transformed, a set of delta functions positioned at the interplanar distances (or their multiple), convoluted with the transform of a window function, is obtained. Thus, the numerical transformation of the experimental intensity can be deconvoluted to obtain a set of "experimental" delta functions [74A2, 74L1, 75A2, 75L2, 77A1, 79A2, 79M2, 79M3] or one can directly try to obtain directly agreement between the experimental transform and a trial kinematic transform by minimizing an appropriate mean distance between the two curves (called residual) [77C6, 78C1, 78C2]. In any case, the above methods are to be taken as simply indicative, and they may be useful permitting the possible exclusion of a model.

6.1.1.4.1.2 Simplifications and calculations

As the main method for obtaining information about atom positions remains full dynamical calculation, many simplifications were introduced to shorten the computer time and core requirements. Sometimes several approximation schemes or simplifications are used together.

One of the most efficient and widely-used approximation is called "renormalized forward scattering" (RFS), worked out by Pendry [71P4, 71P6]. It works very well as long as the scattering between any pair of successive layers is weak, and the layer spacing is not too small. It is a perturbative scheme, in which the wave transmitted through a layer is considered only modified by forward scattering and by the other waves transmitted with various scattering angles. The reflectivity of each layer is calculated from the impinging wave. The forward and backward components of the wave are then iterated at various orders of approximation. The convergence is normally reached in few iterations and it is good at energies above 30 eV [75Z3]. A similar treatment, which is useful when subplanes in a layer are present [75Z2], and another, which is substantially equivalent and based on the iterative Gauss-Seidel-Aitken scheme [81A1], have also been formulated. A further approximation has been obtained by calculating the intralayer scattering in a kinematic manner [79C4].

In the approximations, perturbative schemes are often used to avoid inversion of the very large matrix involved in the theory by developing it in an infinite series and keeping just a few terms. A lot of different perturbative treatments have been tried [68M3, 70S1, 70S2, 74J2], but the necessity of using higher order terms was soon recognized [70D2, 71J4] and schemes to the third order were formulated [71T5, 72T1, 73H2, 73T3, 73T4, 73T5, 73T6, 74A3, 74H4, 75T2, 76H1]. A third order perturbative scheme was also proposed to calculate iso-intensity maps [75G2, 75G3]. Beyond the third order, the perturbative scheme becomes inoperative, due to the rapid growth of the number of terms to handle.

Another approximation often used is called "layer doubling". In this method, the forward and backward scattering by a couple of layers is solved exactly and from this a new set of scattering matrices are derived, describing the scattering of this double layer. Iteration of this procedure enables one to treat the scattering of two such double layers and so on for eight, sixteen, etc., layers, thus obtaining a net saving in computer time [75V4, 83A1].

An important saving in computing time is obtained by applying symmetry considerations. These can be of either structural type (i.e. depending on the arrangement of atoms at the surface and in the substrate) or of wave field symmetries (depending on the particular direction of the incident beam) [72F2, 72T3, 74R1, 74R2, 75R2, 76R2, 84M4]. In particular cases, systematic absence of expected beams in the diffraction pattern leads to formulation of plausible structural models [73H5, 78J1, 84W]. Finally, it may be interesting to note the presence in LEED of time-reversal symmetry, which had already been recognised in the early days of the technique [36F, 70W1, 71F1, 71L2]. Because of the time reversal, 1-fold and 2-fold symmetry are indistinguishable, as well as 3-fold and 6-fold symmetry, and so on.

A type of approximation used when very large unit cells with a large number of atoms are present consists in neglecting the multiple scattering in some phases of the process. Thus one has the quasi-dynamical method, in which the multiple scattering within each layer is neglected, while the multiple scattering between the layers is included [77T, 80H2, 84K5, 86T2]. A similar method with a single reflection approximation for the scattering within a sandwich of low spacing layers was described, this being suitable for layered compounds, such as metal dichalcogenides [79H2].

An approximation used in some cases is called the " beam set neglect" because it considers as negligible the effect of all other beams on each emerging beam except the others in its own beam set and the integral order beams [83V].

A very promising method was recently proposed, in which the intensities are considered as coming from unperturbed intensities obtained from a reference structure on which a full dynamical calculation has been performed. The scattered amplitudes from a trial structure are developed to the first order in the variation of the atomic coordinates from the unperturbed structure, thus obtaining a linear system which can be explicitly resolved with low computational effort. The method can be iterated up to the best convergence [86R1, 88P2, 89R3, 89R4]. This method is the first direct method in LEED; in it a structure can be obtained without previous detailed structural models. The method can also be used to evaluate the diffuse intensity in the case of disordered structures, and it is applied to the determination of adsorption sites .

The cluster approach was also re-elaborated using a kinematic approximation with renormalized atomic scattering factors in a form suitable for large unit cells [85S1].

A new formulation was also performed in the transfer-matrix method in the case of small interplanar distances, when a very large number of beams was necessary [80J2].

A simplification in the calculation of the structure constants of Kambe's theory was obtained by using the $3j$ symbols [69J].

Finally, a completely different formalism was used to express the scattering which has cylindrical instead of spherical waves as a set of base functions. The method is called the "chain method" because the multiple scattering within a single layer is constructed from the scattering of a linear chain of atoms. The idea comes from the fact that the scattering wave field from an atom in the crystal is strongly peaked in the forward direction, which results in an intense multiple scattering in a narrow cone around the forward direction. The method is especially designed for higher energies than those which are typical of LEED, and so it is mainly suitable for MEED [75P, 77M5, 80M3]. The calculations are in some cases performed using self-made programs. On the other hand, standard packages are often available, such as that one of Van Hove and Tong [79V] or the CAVLEED package by Pendry, Titterington, Kinniburgh, et al. [74H3, 74P]. Other programs are described in the literature [74R1, 75R2, 80J2, 80T2, 89R1, 89R2], and these also contain approximations and/or other ancillary programs. It is important to note that all versions give very similar results, and it is interesting to compare them [84S2].

6.1.1.4.1.3 The scattering potential in LEED

One of the most important ingredients in the calculations is the scattering potential. Ultimately, structural parameter accuracies depend on the availability of adequate scattering potentials. Due to the assumed spherical symmetry of the atomic potential, which is inherent to the muffin-tin picture, the scattering potential enters LEED calculations through the phase shifts, which are calculated using appropriate computer programs performing the numerical integration of the radial component of the Schrödinger equation. Various approximations are used to take into account mainly the conduction electrons of the solid [53M, 64M5, 65W, 67L1, 69P4, 74P]. The most delicate part of this calculation is the evaluation of the exchange and correlation terms. This is in principle a non-local potential, which is in general approximated by a local one for the sake of simplicity. These approximations are the Slater term, which is proportional to the 1/3 power of the charge density [51S], or other approximations usual in band structure calculations [65K]. Many authors tested the effect of different choices of the potential, and in particular of the exchange and correlation [75S1, 76E, 76T, 77E, 83N1]. The general conclusion is that the effect on the determination of structural parameters is not very important. The physical reason lies in the fact that the scattering is mainly due to the atomic core, particularly because of the relatively high energy of the electrons. The muffin-tin potential is also adequate in the case of semiconductors [76S3], where different exchange potentials also have little effect [79M6, 82F4].

A non-structural parameter entering the calculations is the inner potential. Its effect has also been investigated [69S2, 80H3, 80M7], but in general it is assumed that small variations of the inner potential do not change the shape of the intensity curves. It is less evident how the variation of the inner potential in the surface layers has to be taken into account. In general a surface inner potential which is equal to the bulk counterpart is assumed, because the effect of its variations is small [74L3]. Connected to the existence of the inner potential is the jump at the surface. Its detailed profile does not affect the general shape of the intensity curves much, at least when the energy is over approximately 30 eV and the shape of the step continuously matches the inner potential to the vacuum level [76G1, 78E2]. On the other hand, the surface barrier is fundamental in explaining some features of the fine structure at low energy, because of resonance, trapping of the electrons inside the potential step, and surface states [69M3, 77R1, 78E2, 81M3] (see also section 6.1.1.4.2).

In general, the electron spin is not considered, at least in the case of light scatterers. However, for high atomic number scatterers relativistic effects and the spin interaction become important and have to be taken into account. The effect can be summarized in effective phase shifts, averaged from the spin up and spin down phase shifts, and using a non relativistic theory [70J1, 75J1]. Spin-orbit interaction has to be taken into account and separate calculations for spin up and spin down have to be performed in the case of spin polarized electrons (see also section 6.1.1.4.3).

6.1.1.4.1.4 The effect of temperature

One of the most important factors affecting LEED intensities is temperature. The electron-phonon interaction is typically inelastic. However, the energy loss necessary for creating a phonon is so small that one can again treat as elastic the scattering by a rigid lattice, averaged over the different momentary distortions. Because of the multiple scattering the effect cannot be simply described as a Debye-Waller factor, as in the X-ray diffraction, and an appropriate treatment is necessary. The generally adopted method is to consider effective temperature-dependent phase shifts. These are obtained from the zero temperature ones and become complex, but the formal theory of the scattering by a rigid lattice is maintained [70D3, 70D4, 71H4, 72L3, 74P]:

$$\exp\left[\mathrm{i}\delta_l(T)\right]\sin\delta_l(T) = \sum_{m,n}\mathrm{i}^m \exp\left[-4\alpha\cdot(E+V)\right] j_m\left[-\mathrm{i}4\alpha\cdot(E+V)\right]\cdot$$
$$\exp(\mathrm{i}\,\delta_n)\sin(\delta_n)\left[4\pi(2m+1)(2n+1)/(2l+1)\right]^{1/2}\cdot\int Y_{n0}(\Omega)Y_{m0}(\Omega)Y_{l0}(\Omega)\,\mathrm{d}\Omega \qquad (8)$$

where $\alpha = 3T/(4\ M\ k_\mathrm{B}\ \Theta^2)$, and M is the mass, k_B the Boltzmann constant and Θ the Debye temperature (with $\hbar = 1$, M in electron mass unit, k_B in hartree/K, Θ in K).

It is also possible to include the anisotropy of the atomic vibrations in the calculations [78U1].

From the shift of the intensity peaks at various temperatures the surface thermal expansion may be measured [71W3].

The overall effect of temperature on the intensity curves is a general decrease in intensity, without any peak spreading [69D1, 70T2, 71I, 71T3]; the amount of this decrease depends on the energy. Mainly in the first few years of the LEED technique, the effect was described by means of an effective Debye-Waller factor, obtained in a kinematic approximation.

The Debye surface temperature was obtained from the slope of an approximate straight line by plotting the logarithm of the intensity as a function of temperature [64M1, 66J, 66L2, 68G3, 68S, 69E, 69M4, 69R, 70R1, 70R2, 70W2, 71K1, 71R2, 71T2, 72Q]. In such a way, many surface Debye temperatures were obtained, as well as the ratio of the components of the mean square amplitude of vibration, normal and parallel to the surface, to the bulk value. In general, this ratio is near 2 for the normal component. When the energy (i.e. the penetration of the electrons) is varied, the contribution of the first planes is also obtained. Better results were obtained in considering multiple scattering [75M5, 76P, 74J4]. The interest in these results lies in their usefulness for checking of the theories of surface vibrations, so that this kind of measurements was also recently performed.

6.1.1.4.1.5 Structure determination

The normal procedure for structure determination is that of the trial and error method, i.e. a model of the structure is hypothesized and the calculated intensity curves are compared to the experimental ones. The model is varied until good agreement between theory and experiment is reached. It is in general accepted that a better agreement corresponds to a more reliable result. The quality of the agreement, and the achievement of the solution of the structure, were visually judged in early studies. Since the absolute intensity of the experimental curves is never reproduced in the theory (the experimental ones always being lower by a factor of some units to some hundreds, differing from one beam to the other), the most important features of the curves to be compared are the overall shape, the existence of maxima and minima and their positions, the relative peak heights, the existence of shoulders, etc. The visual inspection being subjective, very soon scientists felt the necessity of using some systematic and objective means of comparison. This is now made by numerical factors obtained from algorithms called reliability factors (or R factors); these generally yield a number which is indicative of the closeness of the agreement, in the sense that the smaller they are, the better the agreement is.

Even in the early days of the quantitative use of LEED some simple forms of R factors were used, like the mean absolute deviation (or the mean square deviation) of the position of the theoretical peaks as compared to the experimental ones [73D1, 73M1], or the fraction of energy range in which experimental and theoretical curves have slopes of opposite signs [77V].

A more complete and complicated R factor is the R_{ZJ} factor [77Z1]. Its definition for a single beam comparison follows:

$$r = N \frac{\int_{E_1}^{E_2} w(E) \left| c I'_{th}(E) - I'_{\exp}(E) \right| \mathrm{d}E}{\int_{E_1}^{E_2} I_{\exp}(E) \mathrm{d}E}, \tag{9}$$

where

$$c = \frac{\int_{E_1}^{E_2} I_{\exp}(E) \mathrm{d}E}{\int_{E_1}^{E_2} I_{\mathrm{th}}(E) \mathrm{d}E} \tag{10}$$

is a factor to scale the compared curves to the same mean intensity and

$$w(E) = \frac{\left| c I''_{\mathrm{th}}(E) - I''_{\exp}(E) \right|}{\left| I'_{\exp}(E) \right| + \left| I'_{\exp} \right|_{\max}} \tag{11}$$

is a weight factor to enhance some features of the curves. Primes indicate first and second derivatives in respect to the energy. N is a normalization factor (equal to 1/0.027), which allows a value of 1 to be reached when random curves were compared.

To make an overall comparison of a theoretical and an experimental structure, a mean value of the above r for all the beams, weighted by the energy range $E_2 - E_1$, is used:

$$R_{\mathrm{ZJ}} = \frac{\sum_i r_i (E_{2,i} - E_{1,i})}{\sum_i (E_{2,i} - E_{1,i})}. \tag{12}$$

On an experimental basis, values of R_{ZJ} of 0.2 or less, of 0.3, and of 0.5 or more, were respectively indicated as characteristic of a very probable, doubtful or wrong solution of the structure.

A slight modification can be introduced to avoid the use of the N factor, by substituting it with $1 / \int \left| I''_{\exp} \right| \mathrm{d}E$ [77V].

This R_{ZJ} factor has been widely used.

Another important factor and now also widely used, is the R_p factor [80P1]. This exclusively emphasizes the peak positions, disregarding the absolute intensities:

$$R_{\mathrm{p}} = \frac{\sum_i \int_{E_1}^{E_2} \left| Y_{i,\mathrm{th}}(E) - Y_{i,\exp}(E) \right|^2 \mathrm{d}E}{\sum_i \int_{E_1}^{E_2} \left\{ \left[Y_{i,th}(E) \right]^2 + \left[Y_{i,\exp}(E) \right]^2 \right\} \mathrm{d}E}, \tag{13}$$

where $Y(E) = L^{-1}(E)/(L^{-2}(E)+2V_i)$; with $L(E) = I'(E)/I(E)$; and V_i being the imaginary part of the potential. The sums are over the number of beams utilized.

Other R factors used were the simple X-ray ones [78C2, 77Z1, 79A3], like

$$R = \frac{\int_{E_1}^{E_2} \left| cI_{\text{th}}(E) - I_{\text{exp}}(E) \right| \mathrm{d}E}{\int_{E_1}^{E_2} I_{\text{exp}}(E)\mathrm{d}E}, \tag{14}$$

$$R = \frac{\int_{E_1}^{E_2} \left[cI_{\text{th}}(E) - I_{\text{exp}}(E) \right]^2 \mathrm{d}E}{\int_{E_1}^{E_2} \left[I_{\text{exp}}(E) \right]^2 \mathrm{d}E}, \tag{15}$$

and analogues obtained by replacing $I(E)$ with $I'(E)$ and $I''(E)$.

Finally a totally different R factor was defined as metric distance, and was derived from already established topological considerations [84P1]. Transforming the energy range into a variable y ranging from 0 ... 1, and normalizing both theoretical and experimental intensities so that their integral is 1, the metric distance is defined as

$$R_2 = \int \left| F(x) - G(x) \right| \mathrm{d}x, \tag{16}$$

where $F(x) = \int_0^x I_{\text{exp}}(y)\mathrm{d}y$, and $G(x) = \int_0^x I_{\text{th}}(y)\mathrm{d}y$.

It is difficult to make a general assessment about the best factor, mainly because of the lack of a well-known structure to be used as a standard. In fact, the LEED technique is currently considered the most reliable technique for surface crystallography, and the most accurate structure determinations have been made by using the R factors. Comparison of various R factors (mainly R_{ZJ} and R_p) have been made by several authors [81T2, 82G2, 82M8, 82M6, 83H1]. The choice ultimately rests on the user's personal feeling, but the use of anyone of them leads to almost the same structure determination. Often an average of several R factors is used to avoid possible local minima of the single ones, which would give an incorrect structure. A general discussion of the philosophy of R factor use can be found in [86C3].

Different strategies have been proposed, for minimizing the number of trial models, mainly when many parameters have to be optimized. These are strategies for finding the minimum of R factors [79J2, 87C2], or a different use of the data base, i.e. experimental collection and theoretical calculation at large intervals of energy so that just 5 ... 10 values of intensity can be used [89K3]. Finally a somewhat different method, called $I(g)$, has been proposed [79C5, 79C6, 80C4, 81W4, 82C4]; in this the relative intensities of all the beams at fixed energy and incidence angles are simultaneously measured and compared with the calculations. In other cases, the use of polar curves (rotation diagrams) has been proposed, mainly for experiments at high angles of incidence [78F1, 79K2].

6.1.1.4.1.6 Precision and reliability

From the early years of the LEED technique, a large amount of work was devoted to the assessment of data reproducibility and to the analysis of accuracy in surface determinations. The first question regarded the cleanness and characterization of the surfaces. These were achieved with the ion bombardment and annealing technique [55F, 58F], a method generally also used today, and the effect of the damage from the sputtering on the diffraction pattern and on the intensities were also soon investigated [66P, 67F1, 69H1].

The general effect of disorder on the pattern is broadening of the spots, and the decrease of the contrast, as far as the disorder remains random. The presence of regularly distributed steps results in the splitting of the spots into particular symmetry directions [66M1, 68E2, 70H1]. Similar splitting can arise from the formation of equivalent surface zones separated by an even number of half wavelengths (antiphase domains). These points are reviewed in chapter 2.3 of subvolume III/24a and their effects on the intensities, and in particular the analysis of the spot profiles can be used to obtain information on the type of disorder and the height and distribution of steps.

Following the evolution of the pattern and of the intensities as a function of temperature, it is also possible to study phase transitions [85B5], even if the kinematic analysis does not seem always adequate [86B2, 86M2, 86M3].

The effect of disorder on the intensity curves was also investigated [66M1, 66P, 67J3, 69P3, 70H2, 71H7]. The discrepancy between the absolute intensity of the theoretical and experimental *I*/*V* curves is attributed to the surface disorder, but a general dynamical theory is not available. Only semi-qualitative evaluations, based on the interference of the wave field coming from surface zones separated by statistically-distributed steps, were tried [73H7, 73L2, 77Z2], whereas a more general approach was formulated to take into account the effect of the disorder [74D3, 74D4, 78J2, 78M11, 78W4, 78W5].

The accuracy of structural determinations also owes something to the evaluation of instrument response function, i.e. to the effect introduced by angular divergence and energy spreading of the incident beam on the coherence length of the electrons, and on the response of the intensity measurement apparatus. The coherent length for the electron has been variously estimated from 100 ... 500 Å [63K, 67C2] and the instrument response function has been investigated by several authors [71P1, 79C7, 79C8, 79F4, 79W2, 80L4, 81L2, 83L1]. The instrument response function seems more important for the application of LEED to the study of defects than for the structural parameters derived from the intensity analysis, because the overall effect, both instrumental and from disorder, results in a scale factor of the *I*/*V* curves, without any important changes in their shape [66P, 80Z1, 84Z1].

However, there are different causes of errors when comparing theoretical and experimental curves, some of these being errors in the determination of the angles of incidence and azimuth, or due to the residual magnetic field. To minimize their effect, an appropriate average of the intensity of beams which are symmetrically equivalent can be used [80N2, 82D1].

Experimental reproducibility has also been investigated [73W2, 74W2, 81S3], and a comparison of the results from different laboratories performed and analyzed [87J1, 87J2].

The influence of the choices of the non-structural parameters used in the calculations has been extensively tested (see above sections, and [82N1, 83O2, 84D1]).

It is difficult to make an overall evaluation of the degree of precision that can be reached, because of the lack of different methods which are absolutely sure and of well known structural determinations. As a general rule, it may be accepted that the interplanar distances are determined to better than 0.03 Å, even if there are more pessimistic evaluations [80U, 84U].

6.1.1.4.2 VLEED

Whereas LEED measurements are in general performed above about 30 eV, the spot intensity up to this energy level (normally only the specular beam is present) shows some very narrow peaks, which represent the fine structure of the *I*/*V* curves [88J4]. To observe the narrow intensity oscillations, high resolution in both angle and energy is needed [77A2, 81B3, 81B4, 81B5, 81J1, 83B1, 85B1, 86B5, 88H1]. Observations by means of spin polarized electrons have also been performed [81M4, 81P1].

Some peaks at very low energies (a few eV) are band structure effects [63K, 71L5, 72M1, 73B2, 75A3, 81H, 82J2]. Narrow peaks arising from narrow gaps in the band structure tend to be cancelled due to inelastic scattering, unless this is very weak, as at very low energies. The relation with surface states has been outlined in [73B1, 74H1, 84L4, 86L3, 87B2, 87L3] and a Krönig-Penney mono-dimensional model has been used to study the features of electronic surface resonance [81B2]. These peaks can be reproduced in the standard LEED theory at low energy, including relativistic effects [86T1] and also taking into account the non-sphericity of the potential [89K4].

The effect due to the surface barrier is more important . This is also called "threshold" effect, because it is connected with the emergence of diffracted beams. It is also known as "resonance" [64M2, 69M1,

79M5]. The basic mechanism arises from the interference of the components of the wave field which can propagate in the crystal, without having sufficient energy to escape, and which are thus reflected between the bulk and the surface barrier [68G1, 68H1, 70A1, 71M, 72G1, 72S1, 73Z, 75H, 77M6, 80S4]. While the exact shape of the barrier does not greatly influence LEED intensity over 30 ... 40 eV, the induced oscillations at lower energies are very sensitive to the details of the surface potential step [82L4, 83G1, 84G1, 85L1, 87D2] and in turn to the atomic structure [73J1, 78M6, 78M7]. Therefore, an accurate description of the potential barrier is important.

The scattering by the surface barrier can be determined by numerical integration of the Schrödinger equation, and the reflection and transmission coefficients can also be calculated [71J1, 80M1, 83J3]. The general shape of the barrier derives from modifications of the simple rectangular step, and can be calculated using the method of the images in electrostatic [78R1, 80D4, 80J1, 80M13, 81B5, 82T2, 85J2, 85R1].

The standard LEED theory has been adjusted to calculate the fine structure correctly [67M3, 68M1, 68M2, 72T2, 74J2, 74M2, 75M4, 76F3, 81H, 81M3, 82J3, 83L5, 83T]. Effects due to the surface conditions, in particular the changes induced by the adsorption [83B2, 85J2, 86J1, 87L3] and by the ion sputtering damage [83T] have been investigated. Calculations have also been performed in the SPLEED formalism [72J1, 78J4].

Various models have been formulated for the potential barrier at the surface [78W6, 82W7, 82W8, 82W9, 84L2, 84L3]. With the more sophisticated models, containing some adjustable parameters, the intensity curves are successfully simulated and particular features are reproduced, for example the fact that the intensity oscillations are grouped in a Rydberg-like series, with a constant of 1/16 [77A3, 77R1, 77R2, 80D4, 80M13, 82L3, 84J1]. The Rydberg series has also been found for spin-polarized electrons [81M1].

The comparison between theoretical calculations and experimental data leads to the definition of the barrier shape [78M5, 78J3, 79J1, 79P1, 86J2]. The comparison can be carried out by means of standard *R* factors or via metric-distances [84L1].

The agreement between theory and experiment leads to the evaluation of the parameters, among which are the inner potential, the image plane location, the point at which the inner potential is reached, and others. An imaginary component, damping towards the vacuum, is also used to simulate the inelastic scattering [72G1, 74J3, 74M3, 76M2, 76M3, 76M4, 78J3, 78R1, 84L1, 85B2, 85B3, 85L2, 87B2, 88J3]. Sometimes, agreement was reached with different sets of parameters [88J3].

The barrier effect was also observed in polycrystalline materials [86G].

6.1.1.4.3 SPLEED

If one is not interested in polarization studies, a conventional theory, with effective phase shifts, is adequate. On the other hand, to describe electron scattering taking into account electron spin, one has to introduce an additional scattering potential, the sign of which depends on the side where the electrons with spin-up and spin-down travel in relation to the scattering centers. This potential arises from the spin-orbit interaction, which is stronger in the vicinity of the nucleus and larger for heavy elements. As a result of the spin-orbit interaction, a non-polarized electron becomes polarized.

A further effect due to spin, which follows from the Pauli principle, is the exchange interaction. A spin-dependent exchange potential can be introduced which, according to the relative orientation of the spins, can either increase or decrease the effective scattering potential. These two kinds of interactions may finally interfere with each other [68V, 71J2, 74J1, 75F, 76F1, 80D5, 80F1, 81F1, 82R, 85K2].

When an unpolarized incident beam is used, the polarization of the emergent electrons is measured by the asymmetry in the number of scattered electrons, i.e. the normalized difference of spin-up and spin-down electrons. When a source of polarized electrons is used, the asymmetry in the scattered intensity with spin-up and spin-down is measured. The two quantities coincide in particular symmetry conditions, i.e. when the scattering plane coincides with a mirror plane of the crystal [79W1, 81W3, 85K2].

It is easy to show that in kinematic approximation the spin-dependent intensity scattered by a periodic set of atoms is equal to that scattered by a single atom [85K2]. In this approximation no crystallographic information is obtained from polarized electrons. On the other hand, strong characteristic dynamical effects, which are particularly apparent in rotation diagrams, show that surface information can be obtained through the analysis of polarized diffracted electrons [69F2, 79K1, 82E2].

To take the electron spin into account, a two-component spinor has to be used in the vacuum . A two-component spinor is also used inside the crystal in the form of a Bloch wave. This scheme should indeed also be used when one is not interested in the polarization studies. Using the two component spinor, the remaining diffraction theory is very similar to that of conventional LEED. Coulomb singularities at the ion sites can also be removed [86S7].

Because of the spin and the consequent doubling of the components, the matrix dimensions in the calculation increase so that particularly efficient and fast schemes have to be used. The final output of the calculations can be either the beam intensities and the polarization, or the asymmetry.

As in conventional LEED, an imaginary part in the inner potential is used to take the absorption into account .

Particular difficulties arise from the consideration of the exchange interaction, as it is needed for scattering from ferromagnetic materials [83F4, 83F5, 84A1]. This is simpler if the interference between the spin-orbit and the exchange scattering is weak and can be considered separately.

The time-reversal symmetry is no longer valid, as it was for LEED, thus removing the degeneracy of the 1-fold and 2-fold and of the 3-fold and 6-fold symmetries. Other effects due to spatial symmetry, in conjunction with time-reversal, have been discussed [80B2, 80F1, 81F3, 83B4]. This is particularly important for scattering from magnetic materials.

The comparison between theory and experiment is able to yield crystallographic information, because of the sensitivity to the structural parameters of the surface [75O, 76F3, 77F2, 77F3, 78K4, 80B3, 80R2, 81M9, 81R2, 82L2, 86L2], mainly because of the interplanar distances, whereas complicated reconstructions have not so far been investigated. For these purposes, as things stand SPLEED does not appear to be anymore convenient than conventional LEED.

VLEED would seem to be sensitive to the spin polarization of the electrons. The image potential does not cause significant spin-orbit interaction, but in the reflection from the substrate barrier polarization effects arise even for relatively light elements. With spin-dependent calculations some features, like double peaks, are correctly reproduced, thus providing the depth of the barrier, its position and the other relevant characteristics [70J1, 71F2, 71J3, 72F1, 72J1, 73F, 74F, 76F3, 77F1, 78J4, 81M1, 86T1].

Particular effects appear as a function of temperature. Polarization may increase, decrease or remain constant, depending on the energy [77C1, 78R2, 78U2, 79R1, 80R1, 81F2, 81K]. At high temperatures, as the scattering approaches the kinematic limit, the spin-dependent features approach the atomic limit. The polarization is also sensitive to the thermal lattice expansion and to phase transitions induced by temperature variations.

An important application of SPLEED is the study of magnetic materials. In 3d ferromagnets the spin-orbit coupling is weak and becomes of the same order as the exchange interaction. The effects can be separated exploiting particular symmetry conditions.

The magnetization in the surface region is conveniently investigated with low-energy electrons [79C1, 81W2, 82A3, 82A5, 82F1, 82T1, 83F3, 83G2, 84K3, 88M3]. The question of the existence of a zero magnetization layer (the so-called "dead" layer) has been resolved, in the sense that the top layer shows a larger magnetization than the bulk one. This is probably due to the presence of a high density of surface states at the Fermi level.

The temperature effects of both non-magnetic and magnetic origin have been investigated using temperature ranges around the Curie temperature. It has also been possible to obtain the critical exponent of the relationship between exchange magnetization and temperature [70L, 81A4, 82A4, 82A6, 83F3].

6.1.1.4.4 RHEED

The use of higher-energy electrons for surface studies is older than LEED; however, only by using LEED very detailed knowledge of surface crystallography has been obtained. This is due to the major experimental difficulties of RHEED and because very flat surfaces have to be studied in high angle reflection conditions. MEED requires lower flatness than RHEED and was in the past occasionally preferred for this reason [72M4]. Nevertheless, the potentiality of the use of grazing incidence was soon recognized [65M, 65S, 67S1] and the method was applied to obtain surface unit meshes [73H3]. More recently, a method to obtain lattice parameter and surface composition (to be used in alloy studies) has been reported [78V1].

At present, the main applications are to the study of interfaces and in particular to the study of growth by molecular beam epitaxy via the intensity oscillations as a function of deposed depth on a surface (see for example [87C1]). Nevertheless, a notable recent effort to obtain quantitative analysis of the rocking curves for surface crystallographic purposes can be found in the literature.

The need for a dynamical approach to the problem of electron diffraction was soon recognized also in the case of high energy electrons. Many dynamical theories, often based on Bethe's early studies [28B1], were formulated [54M, 57C, 58C, 59C1, 59C2, 66M3]. More elaborated n-beam theories were later formulated, also to reproduce the rocking curves, but with poor results in comparison with experiments [72C3, 72C4, 72M3, 72M4]. On the other hand, simple kinematic analysis is used in some cases, mainly in the extraction of semiqualitative information from the diffraction pattern [89I1].

When the electron energy is in the medium range (MEED) the most appropriate theory is based on the chain method [75P, 77M5, 80M3]. The same method was also tried for RHEED [76M1]. A different MEED theory, used to calculate the rocking curves for electrons up to 10 keV was formulated and compared with the chain method [84M1].

More recently, the problem of reproduction of the rocking curves was investigated in a more complete manner [78B3]. A somewhat general theory was formulated by Beeby [79B1] and used by other authors [79H4].

The effect of the incidence angle on the rocking curves was also investigated [82M1].

A similar approach to that of LEED, with a periodic potential in the planes which is parallel to the surface and not periodic in the third dimension was followed using a transfer matrix method [81I, 82I, 83I, 85I1], for the energy range of some keV. Since the electrons are mainly scattered from the nuclei and from the core electrons (in LEED also from the valence electrons), an inner potential different from the LEED one is found.

Other dynamical theories have been formulated. Of these, a theory based on the plane-wave expansion of Bloch waves [86L4, 88K1], inspired by older works [61L2, 68L]. The appropriate boundary conditions for Bloch waves have been investigated [89M1, 89P].

The reflectivity matrix can also be directly calculated [89M2].

Another kind of multislice theory was also worked out, in which the crystal is cut into almost perpendicular planes to the incident beam. The scattering of each slice is characterized by a grating function, associated with the potential distribution of the atoms in this slice [77S1, 86P2, 87W1, 88P4, 88P5, 88W1, 88W2, 89W1, 89W4].

A promising recent approach is the calculation of the reflectivity matrix based on the numerical integration, using a stable invariant-embedding method [88J1, 88Z, 88T2].

Of the investigated effects, intensity enhancement (resonance), when a beam is travelling parallel to the surface, is one of the most interesting [54M, 62K, 85M2, 88P5]. This was explained using surface geometrical arguments [80I2, 86L1] and also introducing tridimensional effects [87P2].

The appearance of streaking in the diffraction pattern from flat surfaces was interpreted in terms of thermal diffuse scattering [78H] and involves scattering by phonons [84A2].

As in LEED, inelastic scattering was introduced by an imaginary part of the inner potential. The relation between the width of the peaks and the beam attenuation has been investigated [75E1].

The effect of multiple inelastic scattering was also introduced in the multislice method. Multiple plasmon excitation results in a significant perturbation of the electron distribution near the surface, thus obtaining an enhancement of the reflectance, according to the experimental reflectance which is twice the theoretical one, based on the pure elastic scattering theory. Introduction of different energy-losses also produces different RHEED patterns [88W3, 89W2, 89W3].

To evaluate the agreement between theoretical and experimental results an R factor has been defined [87I2].

The variations of diffracted intensities as a function of the orientation of the incident beam were used to characterize the surface symmetry [84S1]. More generally, the symmetries were introduced into the calculations to minimize the size of the effective diffraction matrices [88G1].

From the intensity variations with temperature, the surface atom vibrations were obtained [73C4, 80B4].

The intensity from stepped surfaces may be qualitatively understood in terms of thermal diffuse scattering [79H3]. The effect of the steps on the intensities and on the rocking curves was also introduced in the theories [84K1, 87I1, 88I, 89I2]. Also amorphous surfaces were studied, introducing both a refractive correction and an effective mean inner potential, and using a kinematic theory [78M4]. This is justified by the ineffectiveness of dynamical effects from amorphous materials. The average short-range order of the atomic arrangement is obtained from the radial distribution function [86R2].

6.1.1.4.5 LEPD

The use of the positron as a structural probe for surfaces is recent. The first experiences date back about ten years. This is mainly due to the experimental difficulties in obtaining well collimated and sufficiently intense incident beams [79M8, 80M10, 80M11, 80R3, 82C2, 85F]. The first calculations using a modified LEED theory, and the comparison with the experiment, were also tried in the same period [80J3, 81R1, 83W2].

The first observations outlined the differences between electron and positron diffraction. The main differences reside in the different interaction between the incident particle and the solid due to the positive charge of the positrons. For electrons, the potential from the ion-core is attractive, reduced by the electron-electron repulsion and enhanced by both exchange and correlation. For positrons, the repulsion of positrons by the ion-core is reduced by positron-electron interaction. The exchange is zero and the correlation is doubtful. The different interactions result in very different phase shifts [90D], which for positrons are largely independent of the atomic species of the scatterer, so that the positron scattering factors are similar for all the elements.

The inner potential is equal to the positron work function. The imaginary part of the potential is of the same order as in LEED [80O2, 81R1]. The resulting free path is from about 0.7 ... 1 times the free path of the electrons and varies more slowly with the energy [83W2].

The calculation of the diffracted positron intensities has been performed in some cases [84C1, 85M4, 87M, 89H2]. The fact that, in one of the most recent theory to experiment comparison, the agreement reached using positrons was better than that using electrons is of particular interest [89H2]. It may perhaps be ascribed to the lower sensitivity of positrons to inner layer distortion.

One of the typical results using positrons is the observation of Bragg reflections at very low energy [80M12, 87G2, 88G2]. This peak shows the typical "Darwin top hat" structure [14D], indicating the existence of a gap in the energy bands, and it permits exact measurements of the inner potential.

Measurements at very low energy have also been performed to obtain information on the surface barrier for positrons, as in the VLEED [83R2]. However, the characteristics of the fine structure are different, because the new beams are evanescent waves until their emergence, due to the very low inner potential value [88J2].

Theoretical studies of LEPD show polarization effects at energies over 100 eV for heavy materials, like W [80F2], whereas relativistic effects are computed for a periodic potential and they are similar to electron ones, if one reverses the electric charge sign [82G3].

6.1.1.5 Clean surfaces

Diffraction techniques have been widely applied to the studies of surfaces, which may be clean or have adsorbates or epitaxial layers. Their use ranges from the simple qualitative observation of the diffraction pattern to the qualitative determination of the surface structure from the pattern symmetry, to the semi-quantitative determination of surface vibrations, to the quantitative determination of the atom coordinates in the first atomic layers, to the investigation of the surface barrier and of surface magnetic properties.

One of the main characteristics, noted by the very first observers, is the presence of reconstructions of most semiconductor surfaces, and of a number of metallic surfaces.

In the case of reconstruction, only by coordinating many techniques it is possible to determine the correct structural model: e.g. HREELS, ISS and STM.

When the surface is unreconstructed, i.e. its structure corresponds to an ideal plane of the bulk parallel to the surface, a common result is the oscillatory feature of the interplanar distances of the first planes, mainly for the "open" surfaces, like the (110) of fcc. These features have only been obtained in the structural determinations of the last decade, because the necessary precision for this kind of determination has only recently been achieved. These results are very important, in particular because they constitute a sure basis for testing full energy calculations to obtain the equilibrium positions of the surface atoms theoretically. Up to now, this kind of calculations has only produced qualitative agreement with the experiment.

Recent developments in surface structural investigations include studies of alloys and of the formation kinetics of the various phases and phase transitions.

6.1.2 Data *)

In this section the various clean surfaces studied by means of electron and positron diffraction are considered. Each element or compound is presented in alphabetical order. For compounds, the order is determined by the first element in the chemical formula.

For each case the main results obtained using the above techniques are matched with the appropriate bibliographical references. Other bibliographical references can be found in the references section of each paper cited in the present article. Moreover, it is worth mentioning that many detailed crystallographic data on surfaces are found in [88M1] and in the associated software for personal computer which is able to display the structure of each surface.

The data reported in the following refer to clean and regular surfaces. We have omitted the cases where a structure is known to be stabilized by a contaminant, considering such cases as pertinent to adsorption. The surfaces generated by epitaxial growth are also excluded from this directory, because of the fact that there is usually no relation between the film and the bulk phase. The form of the sample and its preparation may result in the growth of crystal structure which is unique to the substrate and conditions of growth. Moreover, the distinction between a film and an adsorbed layer is questionable.

Standard nomenclature has been used [64W, 80V2].

If not otherwise stated, the distance between the first and second atomic layer is indicated by d_{12} , the distance between the second and the third layer by d_{23} , and so on. The symbol Δd_{ij} indicates the variation of such distances from the bulk value. These distances are to be understood as being from planes passing through the center of the atoms of each layer. In some complex cases other distances (like first and second neighbor distances), or other geometrical quantities, are often indicated with reference to a figure.

*) The results reported in this section include data appeared in the literature before early 1990.
The most relevant papers which appeared afterwards until 1993 have been included in the Bibliography, see section 6.1.4. The elements and compounds which are concerned are listed in section 6.1.3 (Appendix).
The reader should consult the appropriate articles for a possible up-dating of the results for these elements and compounds.

Ag (Silver)

Ag (100)

Regular 1x1 structure, with first interplanar distance almost equal to the bulk value [73J2, 77E].
Observed a c(2x2) structure at 600 ... 750 °C, probably due to impurities [67M1].
Experimental intensity LEED curves in [73H6].
Theoretical intensity LEED curves in [73J2, 77E].
Calculated RHEED intensities in [79H4, 88Z].
Surface Debye temperature 104 K [66J, 69M4].
Studied the effect of incidence angle on the RHEED intensities in[82M1].
Absorption potential studied by VLEED in [76M2, 76M3].
Surface barrier studied by LEPD in [83R2].

Ag (110)

Regular 1x1 structure. Surface multilayer relaxation determinated with $\Delta d_{12} = -7(2)\%$, $\Delta d_{23} = +1(2)\%$, $\Delta d_{34} = -2(2)\%$ [89L].
Experimental and theoretical *I*/*V* LEED curves in [77M1, 77Z2, 81N, 82N2, 82N3, 82N4, 89L].
Semiqualitative effect of randomly distributed surface steps on the *I*/*V* LEED curves in [77Z2].
Effect of non structural parameters on *I*/*V* LEED curves in [77E, 82N3, 82N4].
Structure studied by CMTA in [79A4].
Structure studied by convolution-deconvolution method in [77C2].
Surface Debye temperature 150 ... 190 K [66J, 69M4].

Ag (111)

Regular 1x1 structure.
LEED intensity curves in [69S1].
Studied by CMTA in [73N].
Surface Debye temperature, normal to the surface 155 K, parallel to the surface 226 K [66J, 68C4, 74J4].

Ag (103)

Stepped surface observed by RHEED [79H3].

Ag (331)

Vicinal surface observed by LEED and found stable up to 900 K [78M3].

Ag spherical

Observed by LEED various reflected beams due to various faces [73K2].

Al (Aluminum)

Al (100)

Regular 1x1 structure. No relaxation of the first interplanar distance.
LEED observations in [67J1, 72G3, 73Q].
Many LEED intensity calculations, mainly to test early theories [69F1, 69M2, 71D1, 71D4, 71H2, 71L4, 72J3, 72L4, 72L5, 72L6, 72L7, 72T1, 73J3, 73T4, 75A1, 76V2, 77G].
MEED intensity calculations exhibit little expansion of d_{12} [77M5, 83M2].
Inner potential 13.2 eV determined by RHEED [83O1].
Average of rotation diagrams in [76A1].
Transform-deconvolution in [75L2].
Surface Debye temperature 200 K [72Q, 82V].
Surface resonances studied by VLEED in [71L5, 75H, 85L1].
Bragg peak observed at very low energy using LEPD [80M12].

Al (110)

Regular 1x1 surface, with oscillatory relaxation of the first interplanar distance: $\Delta d_{12} = -8.5(10)\%$, $\Delta d_{23} = +5.0(11)\%$, $\Delta d_{34} = -1.6(12)\%$ [84A3].
The above relaxation is almost equal in the first two distances to another determination [84N2], which reports $\Delta d_{12} = +2.2(13)\%$ (note reversed sign) and $\Delta d_{23} = +1.6(16)\%$, but with presence of misoriented domains.
LEED observations in [67J1].
Many LEED intensity calculations, mainly to test the early theories [71D4, 71H3, 71L4, 72J3, 72J5, 72L5, 72L6, 72L7, 72T1, 73J3, 73M2, 73T4, 75A1, 77G].
MEED intensity calculations [77M5, 84M1].
Inner potential 13.2 eV determined by RHEED [83O1].
Average of rotation diagrams in [76A1].
Convolution-transform in [77C2].
Surface Debye temperature 140 K [72Q, 82V].
Surface resonances studied by VLEED in [75H].
Bragg peak observed at very low energy using LEPD [80M12]

Al (111)

Regular 1x1 structure, with expansion of the first interplanar distance by 2.2%, from many LEED intensity calculations, mainly to test early theories [71H3, 71L4, 72J3, 72L5, 72L6, 72L7, 73J3, 73M2, 80J4, 82N1].
LEED observations in [67J1].
Inner potential 13.2 eV determined by RHEED [83O1].
Surface Debye temperature 200 K [82V].
Surface resonances studied by VLEED in [75H].
Bragg peak observed at very low energy using LEPD in [80M12].

Al (210)

Stepped surfaces with relaxation $\Delta d_{12} = -16(2)\%$, $\Delta d_{23} = -1(3)\%$, $\Delta d_{34} = +9(3)\%$, $\Delta d_{45} = -4(4)\%$, and lateral displacement in <120> direction $\Delta r_1 = 0(3)\%$, $\Delta r_2 = -3(3)\%$, $\Delta r_3 = +2(3)\%$, $\Delta r_4 = -2(4)\%$, $\Delta r_5 = -1(5)\%$ [88A1].

Al (311)

Stepped surface, with relaxation $\Delta d_{12} = -13\%$, $\Delta d_{23} = +9\%$. Lateral displacement of the first plane [85N].

Al (331)

Stepped surface, with relaxation $\Delta d_{12} = -11.7(23)\%$, $\Delta d_{23} = -4.1(31)\%$, $\Delta d_{34} = +10.3(27)\%$,
$\Delta d_{45} = -4.8(41)\%$, and lateral displacement of the first plane of -2.8% [86A1].
The sequence of signs in the surface relaxation is consistent with theoretical prediction [83B3].

Al_2O_3 Aluminum oxide (alumina)

Al_2O_3 (0001)

α-alumina surface exhibiting 1x1 structure up to 1250 °C, $(\sqrt{3}\times\sqrt{3})R30°$, $(3\sqrt{3}\times3\sqrt{3})R30°$ and $(\sqrt{31}\times\sqrt{31})R9°$ (two domains) structure up to 1700 °C. These structures remain stable when brought back to room temperature. Observed by LEED, using a second electron gun at grazing incidence to avoid charging of the surface [67C1, 68C3, 70F2, 71C2].
Measured LEED *I*/*V* curves of the specular beam from the 1x1 structure at various incidence angles [72W].
Absorption potential determined by VLEED [76M2].

Al_2O_3 ($\bar{1}012$)
α-alumina surface exhibiting 1x1 structure up to 900 °C, and 2x1 structure at higher temperatures [72W].
Absorption potential determined by VLEED [76M2].

Al_2O_3 ($1\bar{1}\bar{2}3$)
α-alumina surface exhibiting 1x1 structure up to 900 °C, and 4x5 structure at higher temperatures [72W].
Absorption potential determined by VLEED [76M2].

$Al_xGa_{1-x}As$ (Aluminum gallium arsenide)

$Al_{0.3}Ga_{0.7}As$
Unreconstructed surface showing different compositions depending on the annealing temperature [88D1].
The bulk composition can be obtained only in a narrow temperature range (400 ... 450 °C).

As (Arsenic)

As (0001)
Cleaved surface. Shows facets <1 0 $\bar{1}$ 15> and <1 0 1 23> [74E1, 74T].
The LEED pattern exhibits evidence of energy loss before scattering [75E2].

Au (Gold)

Au (100)
Surface usually reconstructed 5x20. This structure is stable up to about 400 °C. A 6x6 is seen up to 550 °C. A metastable 1x1 probably simple bulk truncation is also observed.
Early LEED observations interpreted the structure as 5x1 (two orthogonal domains), attributed to impurities. A more correct interpretation was soon performed as a 5x20, determined by a hexagonal incommensurate layer [66F, 67F2, 67F3, 67P, 67S2, 68P2, 69F3, 69P2].
The stability range of the 5x20 and the formation of 6x6 is described in [67M1, 67S2].
Other observations of 5x20, its interpretation as a hexagonal layer and discussion of the validity of the model in [72D2, 72G2, 73D2, 73H4, 73R1, 74Z1, 75Z1, 87D1].
Observation of the structure by RHEED and interpretation as hexagonal layer in [72G2].
Possible interpretation as superimpositions of domains (001) and (211) is also presented [74Z1].
Non-exhaustive confirmation of the formation of a non-buckled layer from LEED calculations in [81V1, 81V2].
Finer LEED observation shows that the 5x20 structure is actually a c(26± 1x68± 2.5) [81V1, 81V2].
The simple 1x1 structure was observed as a metastable one [78W3] and determined by LEED intensity calculations using a quasi-dynamical theory [82L1]. Other calculations in the SPLEED formalism are in [77F2].
A method for stabilizing the 1x1 structure (with formation of steps and disorder) is to use laser pulses [83M5].
Surface Debye temperature is 82 K [73K1].
Absorption potential determined via VLEED [76M2].

Au (110)
Reconstructed 1x2 structure, exhibiting a reversible transition at about 600 °C to a 1x1 structure [67F2, 67F3]. The 1x2 was determined as a missing-row structure, with $\Delta d_{12} = -20.0(35)\%$,
$\Delta d_{23} = -6.25\%$, $\Delta d_{34} = -6.25\%$. In the second layer, a row-pairing is present with lateral displacement of 0.01 Å. In the third layer there is buckling of 0.24 Å [85M5].
Other LEED intensity calculations using various models in [79M11, 79N, 82B3].

The case of disorder, with diffuse spots in the LEED pattern in direction <010>, is used for LEED calculation with random disorder [78W4, 78W5].
Surface Debye temperature 83 K [73K1].
Symmetry considerations using SPLEED lead to exclusion of some models (among which the missing-row is one) [80R2].
Other SPLEED measurements in [81R2, 82E2]. The 1x1 structure (at high temperature) is used for SPLEED measurements in [77F4].

Au (111)
Regular 1x1 structure [67F3]. Other phases are observed: a 3x22 phase, stable, which becomes 1x1 probably by adsorption of impurities [87D1]; also a phase $\sqrt{3}$x22 (three domains rotated by 120°) [81V1].
LEED intensity calculations using relativistic phase-shifts (without comparison with the experience) in [78F2].
Surface Debye temperature 83 K [73K1].

Au (311), (410), (510), (511), (10 1 1), (11 1 1)
Stable stepped surfaces [89S1].

Au (711), (911), (10 1 0), (14 1 1)
Stepped surfaces formed of stable (511) and (100) surfaces [89S1].

B (Boron)

B (111)
Surface of the β-rhombohedral B prepared and observed in [73R2]. The surface does not exhibit any reconstructions.

$BaTiO_3$ (Barium titanate)

$BaTiO_3$ (100)
The surface structure is 1x1 at temperatures between 20 °C and 650 °C [71A2]. Various other phases observed: 2x2, $3\sqrt{2}$x$3\sqrt{2}$, 2x2n (partially disordered), 3x3n (partially disordered). By heating at 850 °C a 1x1 structure is formed. At temperatures over 1000 °C a $\sqrt{5}$x$\sqrt{5}$ is formed [71A1]. The phases 2x1 (two domains) and 2x2 are also observed after Ar ion bombardment [80C6].
LEED intensity curves for the 1x1 phase are recorded up to 500 eV [71A2].

Be (Beryllium)

Be (0001)
Simple bulk truncation structure.
($\sqrt{3}$x$\sqrt{3}$)R30° is also observed [69A2].
LEED intensity curves and comparison with kinematic theory in [72B2].
LEED intensity calculations using the inelastic-collision model and s-wave scattering in [71D5].
Other approximate theoretical studies using perturbation schemes in [70S2, 71S2, 72J4, 75S3].
Structural values to be refined.
Other observations in [78G2].

Bi (Bismuth)

Bi (0001)
Ideal 1x1 surface observed in [67J2].
Surface Debye temperature 1.87 times lower than the bulk value [69G1].
Absorption potential determined via VLEED [76M2].

Bi ($01\bar{1}2$)
Ideal 1x1 surface observed in [67J2].
Surface Debye temperature 2.19 times lower than the bulk value [69G1].
Absorption potential determined via VLEED [76M2].

Bi ($11\bar{2}0$)
Reconstructed 2x10 surface observed in [67J2].
Absorption potential determined via VLEED [76M2].

Bi_2Te_3 (Bismuth telluride)

Bi_2Te_3 (0001)
Ideal 1x1 surface observed in [60H2].

$Bi_2CaSr_2Cu_2O_8$

$Bi_2CaSr_2Cu_2O_8$ (100)
Preparation of cleaved surface and observation of LEED pattern [89S2].

$Bi_2CaSr_2Cu_2O_{8+\delta}$ (100)
Cleaved surface exhibits the superstructure observed in the bulk [89C], and an incommensurate surface structure with a periodicity of 27.0(5)Å [88L2].

C (Carbon)

C (0001)
Basal plane of graphite, observed in [64L1]. The surface is also prepared by cleaving [80S1].
LEED intensity calculations show $\Delta d_{12} = -1.5\%$ [82W11].
LEED intensity calculations, showing surface resonance, in [67M3].
Using LEPD a d_{12} indicating low contraction is obtained [87G2].
Surface Debye temperature 350 K [71A3].
Absorption potential determined via VLEED [76M2].

C ($10\bar{1}0$)
A c(2x2/3) structure arising from the edge or prism surface of graphite is observed in [84K2].

C (100)
Diamond surface, observed with fractional order beams (0 1/2) in [64M4].
The structure is found to be a superimposition of 1x2 and 2x1 or 1x1 and 2x2, depending on the preparation procedure [77L6].
Kinematic analysis of the LEED intensities indicates a d_{12} equal to the bulk one within $\pm 5\%$ [77L6].

C (110)
Diamond surface, unreconstructed [77L6].

C (111)
Diamond surface, observed with fractional order beams (0 1/2) in [64M4, 66L1].
The structure is found to be a superimposition of 1x2 and 2x1 or 1x1 and 2x2 depending on the preparation procedure [77L6], or more likely a superimposition of two domains 2x1 rotated by 90° [88S1].
For the 1x1 phase LEED calculations show a simple bulk truncation with a $\Delta d_{12} < -20\%$ [82Y].

CaO (Calcium oxide)

CaO (100)
Simple truncated bulk, with $\Delta d_{12} = -1.3\%$, without rumpling of the first plane [79P3, 79P4].

Cd (Cadmium)

Cd (0001)
Simple truncated bulk [76S1].

CdS (Cadmium sulfide)

CdS $(10\bar{1}0)$
Observation in [68C2, 73B2]. First interplanar distance determined as equal to the bulk value using kinematic LEED theory [72B1].

CdS $(11\bar{2}0)$
Observation in [75M2]. First interplanar distance determined as equal to the bulk value using kinematic LEED theory [72B1].

CdSe (Cadmium selenide)

CdSe $(10\bar{1}0)$
Surface distortion with Se atoms tilted outward, involving bond-length-conserving rotation. The values, as determined by LEED [88D2, 89D1, 89D2, 89H2], are reported in Table 1, and the symbols refer to Fig. 3. Using LEPD, somewhat different values are obtained (i.e. ω =15(5)°, $\Delta_{1,\perp}$ = 0.68 Å, $d_{12,\perp}$= 0.65 Å), probably due to different sensitivity to inner planes [89H2].

Table 1.

$\omega = 21.3°$	
$\Delta_{1,\perp} = 1.03$ Å	$d_{12,\perp} = 0.45$ Å
$\Delta_{2,\perp} = 0.0$ Å	$d_{12,y} = 4.10$ Å
$\Delta_{1,y} = 4.60$ Å	$d_{23,\perp} = 2.48$ Å

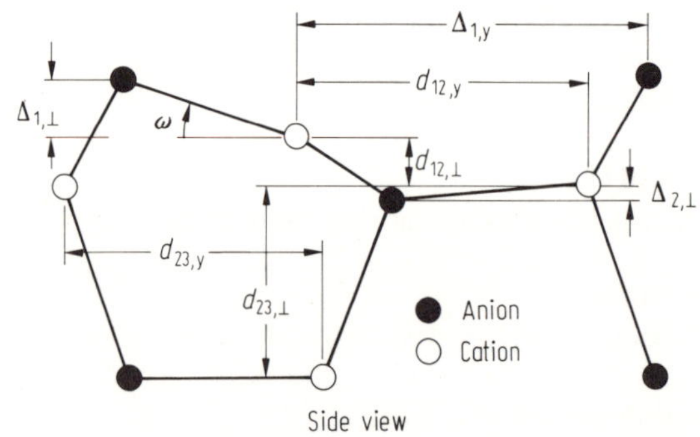

Fig. 3.

CdSe ($11\bar{2}0$)
Surface distortion involving bond-length-conserving rotation ω = 33.5(30)°, $d_{12,\perp}$ = 1.49Å , $\Delta_{1,\perp}$ = 0.745 Å (for symbols, see Fig. 3) as determined using LEED [89H2]. Using LEPD, somewhat different values are obtained (i.e. ω = 27(5)° , $d_{12,\perp}$ = 1.62 Å, $\Delta_{1,\perp}$ = 0.61 Å), probably due to different sensitivity to inner planes [89H2].

CdTe (Cadmium telluride)

CdTe (110)
Surface preparation by cleavage in [65F].
Surface distortion involving bond-length-conserving rotation ω = 30.5(15)°, relaxation of the rotated top layer toward the substrate 0.05(5) Å. In the second layer, the Cd atoms are displaced upwards by 0.09(5) Å, and the Te atoms are displaced inwards to same extent [81D4, 82D4, 87C2, 88C].

Co (Cobalt)

hexagonal phase

Co (0001)
Surface of the room temperature phase, prepared and observed in [76B2, 77B2, 81B7]. Structure determination by LEED reveals no relaxation of the first interplanar distance [78L2].

Co ($10\bar{1}2$)
Surface of the room temperature phase, prepared and observed in [78P2, 79P2]. The phase transition to cubic structure is also observed. A LEED kinematic analysis shows the surface termination with the d_{12} = 1/3 and d_{23} = 2/3 of the reticular parameter [79P2].

Co ($11\bar{2}0$)
Surface of the room temperature phase, prepared and observed in [83W3]. The phase transition to the cubic structure is also observed. Structure determination by LEED reveals $\Delta d_{12} = -8.5(30)\%$.

cubic phase

Co (100)
1x1 surface structure of the cubic phase, stable at high temperatures, (over 450 °C), prepared and observed at room temperature by quenching [78M1]. Structure determination obtained by LEED with $\Delta d_{12} = -4\%$ [78M2].

Co (111)
1x1 surface structure of the cubic phase at high temperatures (at about the transition temperature). The LEED intensity analysis shows a pure fcc termination without relaxation of the interplanar distances [78L2].

CoO (Cobalt oxide)

CoO (100)
Surface obtained by cleavage. The analysis shows no relaxation within 3% [79F3, 86T2].

CoTi (Cobalt titanium)

CoTi (110)
1x1 surface observed after ion bombardment and annealing at 740 °C for 10 sec [85R2].

Cr (Chromium)

Cr (100)
Surface observed in [69G3]. A ($\sqrt{2}$x$\sqrt{2}$)R45° phase is also observed [79G].
Expansion coefficient was measured by LEED, using Bragg peak shifts with temperature variation. The coefficient is two to three times the bulk value [71W3].
Surface Debye temperature 175 K [71T2].

Cr (110)
The surface is not stable and facets forming (001) faces at temperatures over 1000 °C [69G3].
Surface Debye temperature 238 K [71K1].

Cu (Copper)

Cu (100)
Simple 1x1 structure, exhibiting an oscillatory relaxation of the first layers. Δd_{12} = −1.1(4)%, Δd_{23} = +1.7(6)%, Δd_{34} within 2% [78N1]. These values were confirmed by measurements of asymmetry using SPLEED [86L2, 87L2].
LEED intensity curves in [69A3, 72R1, 85H2].
Other theoretical LEED intensity curves in [71C1, 71L4, 74L2].
The effect of different exchange potentials is tested in [77E].
The inner potential is 14.2 eV, determined by LEED [80H3, 81J1] and 19.1 eV by RHEED [83O1].
The structure is also investigated using CMTA [74B3, 75K4].
The effect of the temperature on the LEED intensity curves is in [72R2].
Surface Debye temperature 230 K [70R1].
Dependence on the incidence of the primary beam is studied in [71R1].
Resonances, potential barrier and inner potential are studied using VLEED [70A1, 76M4, 78R1, 79J1, 80D4, 82T2, 83G1, 85R1, 88H1].
Absorption potential studied by VLEED in [76M2].
This surface was also studied by LEPD [83W2, 87M].

Cu (110)
Simple 1x1 structure, showing oscillatory relaxation of the first layers. Δd_{12} = −8.5(6)%, Δd_{23} = +2.3(8)% [78N2, 79D1, 82A2, 83A2, 83D1]. The LEED values are compared with the ones obtained by HEIS in [82A2]. Other comparisons between experiment and theory in [72R1].
The effect of the temperature on the LEED intensity curves in [72R2].
The potential barrier and the inner potential are studied by VLEED in [82T2].

Cu (111)
Simple 1x1 structure with little interplanar contraction: Δd_{12} = −0.6(10)% [81T2, 83N1], confirmed by using VLEED [84L4].
Other LEED intensity calculations in [72R1, 74L2, 78W1].
Surface Debye temperature 245 K [70W2].
The effect of the temperature on the LEED intensity curves is in [70W2, 72R2].
The potential barrier and the inner potential are studied by VLEED in [82T2, 83L3].
The surface is also investigated by LEPD [80J3, 80R3, 81R1, 83W2].

Cu (211)
Stepped surface studied by RHEED, which indicates monoatomic steps [82M4].

Cu (310), (410)
Stepped surface observed and used in [71P7].

Cu (311)
Stepped surface showing multilayer relaxation, with $\Delta d_{12} = -7.3(30)\%$, $\Delta d_{23} = +3.7(10)\%$, $\Delta d_{34} = 0(1)\%$ [78S4, 88W4].
A LEED kinematic analysis, via Fourier transform is in [77M10].

Cu (530), (531)
Stepped surfaces studied by RHEED, which indicates monoatomic steps [82M4].

Cu spherical
Observed as a solidification of a drop of Cu melt [67M4].

$Cu_{1-x}Al_x$ (Copper aluminum)

$Cu_{1-x}Al_x$ (100)
α-phase, 12.5 at% Al, and 5.7 at% Al, observed in [81B1]. The LEED pattern is similar to that of Cu.

$Cu_{1-x}Al_x$ (111)
α-phase, 5.7 at% Al, observed in [81B1]. The LEED pattern is similar to that of Cu.
At 300 °C the reversible transition from 1x1 to $(\sqrt{3}\mathrm{x}\sqrt{3})$R30° is observed [81B1]. This structure is already observed at room temperature in an alloy 10 at% Al [77W3].
Alloy with 16 at% Al also is a $(\sqrt{3}\mathrm{x}\sqrt{3})$R30°, giving a 1/3 monolayer substitutional occupancy in the first layer, without distortion. The deeper layers have random distribution of Al and Cu atoms [86B1].

Cu–Au (Copper gold)

CuAu (100)
Observed structures c(2x2) and 2x2 [79F5].

CuAu (110)
Observed structures 1x2, 2x2 and c(3x1) [79F5].

CuAu $(1\bar{1}1)$
Observed structure $(\frac{2}{3}\sqrt{3}\,\mathrm{x}\,\frac{2}{3}\sqrt{3})$R30° [79F5].

Cu_3Au (100)
Observed as a function of the temperature, using LEED intensity curves, the surface exhibits a continuous order-disorder transition about 390 °C [73S, 74S, 84M3].
Surface Debye temperatures, both normal and parallel to the surface, between 164 and 225 K [76S4].

Cu_3Au (110)
Observed as a function of the temperature, using SPLEED, the surface exhibits a continuous order-disorder transition at about 390 °C [85J1].

CuPd (Copper palladium)

CuPd (100)
Observed structures c(2x2) and 2x2 [79F5].

CuPd (110)
Observed structure 2x1 [79F5].

CuPd $(1\bar{1}1)$
Observed structure 1x1 [79F5].

EuO (Europium oxid)

EuO (100)
Magnetic surface disordered after cleavage. When ordered, it produces a LEED pattern with half order beam, indicating scattering from ordered antiferromagnetic structure [73T7, 76D1, 79F2].
The surface is unstable, probably because of an O or Eu enrichment on the surface [74B1].

Fe (Iron)

Fe (100)
1x1 structure with surface relaxation $\Delta d_{12} = -1.4(30)\%$ [77L3].
Other LEED intensity calculations in [82F1].
This surface was also investigated with LEED kinematic theory [71K2].
A c(2x2) reconstruction was observed, perhaps due to impurities [64P3, 69K].

Fe (110)
1x1 structure with $\Delta d_{12} = +0.5(20)\%$ [80S2].
Other LEED intensity calculations in [76F2].
Surface magnetization of 30% determined by SPLEED [84K3].
Also investigated with LEED kinematic theory [71K2].
K-ion bombardment produces the formation of terraces. The perfect surface is again obtained by annealing at 500 °C [85B6].

Fe (111)
1x1 structure with $\Delta d_{12} = -16.9(30)\%$, $\Delta d_{23} = -9.8(30)\%$, $\Delta d_{34} = +4.2(36)\%$, $\Delta d_{45} = -2.2(36)\%$ [81S2, 86S3].
Also investigated with LEED kinematic theory [71K2].

Fe (210)
Stepped surface with $\Delta d_{12} = -22.0(47)\%$, $\Delta d_{23} = -11.1(47)\%$, $\Delta d_{34} = +17.0(47)\%$, $\Delta d_{45} = +4.8(47)\%$, and with lateral displacement in <210> direction $\Delta a_{12} = +7.1(16)\%$, $\Delta a_{23} = +1.4(26)\%$, $\Delta a_{34} = 0.0(26)\%$, $\Delta a_{45} = +4.0(26)\%$ [84S4, 85S3].
Comparative discussion of the relaxation found for (210), (211), and (310) surfaces in [84S4].

Fe (211)
Stepped surface with $\Delta d_{12} = -10.4(26)\%$, $\Delta d_{23} = +5.4(26)\%$, $\Delta d_{34} = -1.8(34)\%$, and with lateral displacement in <111> direction $\Delta a_{12} = -14.5\%$, $\Delta a_{23} = +2.4\%$ [83S, 84S5].
Comparative discussion of the relaxation found for (210), (211), and (310) surfaces in [84S4].

Fe (310)
Stepped surface with Δd_{12} = −16.1(33)%, Δd_{23} = +12.5(33)%, Δd_{34} = −4.0(45)%, with lateral displacement in <310> direction Δa_{12} = 0.13 Å , Δa_{23} = 0.03 Å [84S3].
Comparative discussion of the relaxation found for (210), (211), and (310) surfaces in [84S4].

FeCo (Iron cobalt)

FeCo (111)
Preparation and observation in [81A3]. The surface is not affected by the transition at 730 °C, where the bcc lattice splits into two simple cubic lattices, each of them occupied by a preferential type of atom.

$Fe_{1-x}Cr_x$ (Iron chromium)

$Fe_{1-x}Cr_x$ (110)
Alloy prepared and observed with 5, 10, 16, and 20 at% Cr. At a Cr content of 12 at% a transition at 1100 K is observed. Segregation of Cr at the surface is present so that the above percentage becomes 20, 25, 35, and 55, respectively. Facets are present on the surface and an increase of the lattice constant with an increase of the Cr content is observed [76L1].
A c(2x2) reconstruction is observed, probably due to oxygen or sulphur impurities.

$FeNi_3$ (Iron nickel)

$FeNi_3$ (100)
Observed in [86V2]. A surface Debye temperature of 230 K was measured at temperatures <370 °C (transition temperature). At higher temperatures a 10 ... 15% higher value is observed, probably due to Fe segregation.

$FeNi_3$ (110)
Observed in [86V2]. A surface Debye temperature of 250 K is measured at temperatures <370 °C (transition temperature). At higher temperatures a 10 ... 15% higher value is observed, probably due to Fe segregation.

$FeNi_3$ (111)
Observed in [86V2]. A surface Debye temperature of 190 K is measured at temperatures <370 °C (transition temperature). At higher temperatures a 10 ... 15% higher value is observed, probably due to Fe segregation.

Fe_2O_3 (Iron oxide)

Fe_2O_3 (100)
Surface showing a good LEED pattern at 700 °C. It shows a phase transition at 820 °C, with 3-fold symmetry and a 2x2 structure [83K]. A ($\sqrt{3}$x$\sqrt{3}$)R30° structure is also observed [83K, 75J2].
Above 820 °C the pattern of an incommensurate overlayer appears [83K].

Fe_2O_3 (0001)
Surface prepared and observed in [85H4, 88L1]. After ion bombardment a Fe_3O_4 (111) phase is also present. After annealing in oxygen atmosphere at 900 °C only Fe_2O_3 (0001) is observed.

Fe_2O_3 $(10\bar{1}2)$
Surface prepared and observed in [85H4, 88L1]. The surface is probably formed of 1x1 and 1x2 islands. After annealing at 900 °C streaking of the beams along the $\langle 02\bar{2}1 \rangle$ direction is present.

FeTi (Iron titanium)

FeTi (100)
Prepared and observed in [82F2]. At temperatures of about 930 °C the cubic phase appears, showing the (001) surface. At temperatures above 1130 °C, the hexagonal phase appears. The lattice parameters observed are slightly greater than those determined by X-rays.

GaAs (Gallium arsenide)

GaAs (100)
Reconstructed 2x4 surface terminating with As atoms [84J2, 88K2], studied by RHEED. A stable 1x1 structure is obtained by irradiation with laser pulses and observed by LEED [83M5].
A c(2x8) phase is also observed [84J2], and interpreted as being simply related to the 2x4 structure by random positioning in the direction of two-fold periodicity, with a different sequencing of dimer chains in the <110> direction.

GaAs (110)
1x1 structure exhibiting rumpling with the As atoms tilted outwards and the Ga atoms inwards, and atomic displacements propagating in the inner planes [84D4]. The scheme of the interatomic distances is given in Fig. 3, see p. 54. The values are reported in Table 2. Other structure determinations in [76L3, 77T, 78T3, 79C4, 79D3, 79M6, 79M12, 80M4, 82M2, 83D6, 84T1, 85D5, 85P2].
Calculations using kinematic LEED theories and CMTA method in [77M2, 78K1, 78K2, 78K3].
Conditions of preparation and experimental observations in [64M3, 77M3, 78S2, 79M1]. One of these is the observation that at temperatures >350 °C a dissociation occurs, inducing a local reconstruction on the surfaces with Ga vacancies. At temperatures >580 °C there is a dissociation with formation of As. At temperatures >700 °C a faceting occurs with formation of (111)2x2 faces [83P2].
The structure was also investigated using calculations of RHEED rocking curves [88J1, 88T2].
Stepped surfaces were also observed by RHEED [85M2].

Table 2.

$\omega = 27.4°$, $\Delta_{1,\perp} = 0.65$ Å, $d_{12,\perp} = 1.43$ Å, $\Delta_{1,y} = 4.39$ Å, $d_{12,y} = 3.31$ Å, $\Delta_{2,\perp} = -0.12$ Å, $d_{23,\perp} = 2.06$ Å

GaAs (111)
Reconstructed 2x2 surface, terminating with Ga atoms [60H1, 61H, 66M1, 71C3, 75G1].
Depending on the surface treatment 1x1, 3x3 and $(\sqrt{19} \times \sqrt{19})$ structures are also observed [85A3].
On heating at 600 °C the surface shows facets in the <110> direction.
The structure of the (2x2) phase is determined in [84T2]. The structure corresponds to a vacancy-buckling model, with vacancy of Ga atoms on the first atomic layer and relaxation of the other atoms.

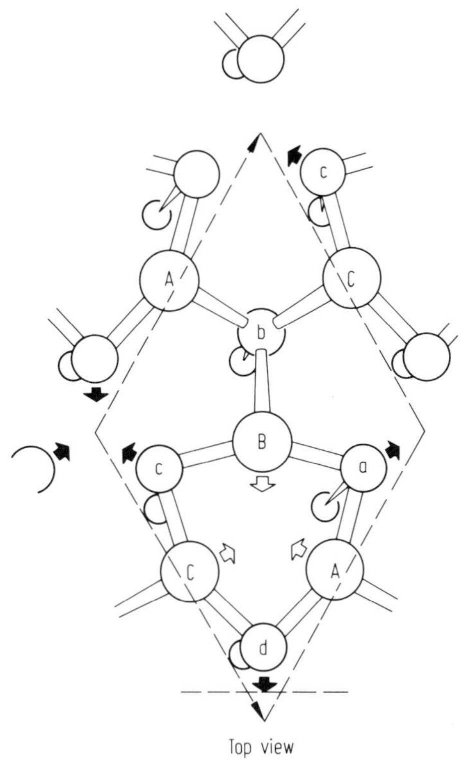

Fig. 4. Scheme of the interatomic distances. A, B, C denote surface layer Ga atoms and a, b, c, d As atoms in the layer immediately below. The smallest circles denote Ga atoms in the third layer, and the arrows indicate displacements. The A, B, C atoms are displaced by –0.706 Å vertically and have a lateral displacement of 0.10 Å. The a, c, d atoms are displaced by +0.04 Å vertically and have a lateral displacement of 0.28 Å. The b atoms are displaced by –0.08 Å vertically. The Ga atoms of the second layer bound to a, c, d atoms have a vertical displacement of +0.01 Å. The other Ga atoms of the second layer have a vertical displacement of –0.08 Å.

GaAs ($\bar{1}\bar{1}\bar{1}$)

Surface terminating with As atoms and showing 3x3 structure [60H1], faceting [61H] and 1x1 structure [66M1].

On heating at 600 °C the surface forms facets in the <110> direction.

GaAs (211)

Surface prepared and observed in [84H3] shows faceting in the <110> direction, with facets forming 30°, –30°, and 54° angles to the (211) plane.

GaAs (311)

1x1 surface studied by LEED intensity analysis [86D2], but has to be considered unresolved, probably because of a 1x1 reconstruction.

GaAs vicinal

Vicinal surfaces with misorientations within 5% and azimuth angles within 10% observed by RHEED [84P2]. Streaking due to interference between steps is observed.

GaP (Gallium phosphide)

GaP (110)

1x1 structure showing rumpling with the P atoms tilted outwards and the Ga atoms inwards, without reconstruction of the second layer, and conserving bond-lengths [81D3, 81L1]. The scheme of the distances is indicated in Fig. 3, see p. 54. The values are reported in Table 3.

Table 3.

ω =27.5°, $\Delta_{1,\perp}$ = 0.63 Å, $d_{12,\perp}$ = 1.44 Å, $\Delta_{2,\perp}$ = 0.0 Å, $d_{12,y}$ = 3.20 Å, $\Delta_{1,y}$ = 4.24 Å, $d_{23,\perp}$ = 1.93 Å

GaP (111)
Surface exhibiting 1x1 structure [73G1, 75D3, 78M9] after ion bombardment and annealing, or stabilized by laser irradiation [87K].
Other structures, 2x2, 3x3, and 17x17, are observed [73G1, 80V1, 87K]. The 2x2 is also investigated by LEED intensity analysis and gives a structure with missing Ga atoms [85X1].

The structure is reported in Fig. 4, see p. 61. In this figure, A, B, C denote surface layer Ga atoms and a, b, c, d denote P atoms in the layer immediately below. The smallest circles denote Ga atoms in the third layer. The arrows indicate displacements. The A, B, C atoms are displaced by −0.746 Å vertically and have a lateral displacement of 0.136 Å. The a, c, d atoms are displaced by +0.04 A vertically and have a lateral displacement of 0.23 Å. The b atoms are displaced by −0.05 Å vertically. The Ga atoms of the second layer bound to a, c, d atoms have a vertical displacement of +0.02 Å. The other Ga atoms of the second layer have vertical displacement of −0.1 Å.
Above 600 °C a faceting of (110) surfaces is formed [81L1].

GaP $(\bar{1}\bar{1}\bar{1})$
Surface exhibiting 1x1 structure [73G1, 75D3, 78M9] after ion bombardment and annealing.

GaSb (Gallium antimonide)

GaSb (110)
Surface 1x1 produced by cleavage [64M3], rumpled with the Sb atoms tilted outwards and the Ga atoms inwards [83D3, 83D4].
Interatomic distances are given in Table 4, with reference to Fig. 3, see p. 54.

Table 4.

$\omega = 30°$, $\Delta_{1,\perp} = 0.77$ Å, $d_{12,\perp} = 1.61$ Å, $\Delta_{1,y} = 4.79$ Å, $d_{12,y} = 3.27$ Å, $\Delta_{2,\perp} = 0.0$ Å, $d_{23,\perp} = 2.16$ Å

GaSb (111)
Structure exhibiting 2x2 and 3x3 structures, observed by LEED and RHEED [60H1, 76R3].
Above 550 °C surface facets along the $<110>$ directions [60H1].

GaSb $(\bar{1}\bar{1}\bar{1})$
Structure exhibiting 2x2 and 3x3 structures, observed by LEED and RHEED [60H1, 76R3].
Above 550 °C surface facets along the $<110>$ direction [60H1].

Ge (Germanium)

Ge (100)
Surface with 2x1 and 4x2 reconstructions [59S, 77O1, 77O2, 81O2]. Vicinal surfaces, cut at 3° 20' and 7° 20' in the $<011>$ zone and at 7° 10' in the $<010>$ zone, form (001) terraces [77O1].
LEED calculations and structural models in [81F4, 84W].

Ge (110)
Surface with c(8x10) reconstructions at temperatures <380 °C and >430 °C [59S, 77O1, 77O2, 81O2]. Vicinal surfaces cut at 2° and 3° 50' in the $<\bar{1}11>$ zone form facets (17 15 1), $(15\,17\,\bar{1})$, $(17\,15\,\bar{1})$, and (15 17 1). If cut at 3° 30' in the $<\bar{1}10>$ zone, they facet (17 15 1) and (15 17 1) up to 480 °C. At higher temperatures the surfaces become stepped with (110) planes.

Ge (111)
Surface obtained by cleavage and annealing [55F, 58F, 59S, 61L1]. A variety of superstructures is observed: $\sqrt{3}$x1 at higher temperatures, which transforms into 12x12 at 300 °C, and again into 8x8 at 200 °C [61L1]; 2x8 and c(2x8) [77O1, 85P1], also observed by RHEED [79I]; 1x1 at room temperature if cleaved at −195 °C [68P1, 82H1, 84Z2]; again 1x1 from 2x8 by heating at 200 ... 400 °C [68P1]; the c(2x8) phase transforms into 2x1 at 300 °C, which may be interpreted as three different orientations of a quasi-periodical 2x1 or as four domains of a quasi-periodical 2x2 [85P1]; the 2x1 may be coexistent with the 1x1, and 2x2 observed by RHEED [80I1]; ($\sqrt{3}$x$\sqrt{3}$)R30° observed in [89F1], where the LEED intensity curves were also measured.
2x8 structure investigated by LEED kinematic analysis in [64H, 78T1].
Vicinal surfaces cut at 5°, 7° and 10° are investigated in [77O2, 81O2].

Ge (210)
Surface with 2x2 reconstruction [81O2], which at 500 °C presents an order-disorder transition.

Ge (211)
Surface with 1x2 reconstruction [81O2].

Ge (311)
Surface with 3x1 reconstruction [81O2].

Ge (331)
Surface with 5x1 reconstruction [81O2], which at 630 °C presents an order-disorder transition.

Ge (510)
Surface with 1x2 reconstruction [81O2], which at 520 °C presents an order-disorder transition.

Ge (511)
Surface with 3x1 reconstruction [81O2], which at 600 °C presents an order-disorder transition.

Ge (551)
Surface with 5x2 reconstruction [81O2], which at 630 °C presents an order-disorder transition.

InAs (Indium arsenide)

InAs (110)
Surface prepared by cleavage [64M3] and analyzed with LEED dynamical intensity analysis [83D5]. The scheme of the structure is shown in Fig. 3, see p. 54. The distances indicated in the figure are given in Table 5.

Table 5.

$\omega = 36.5°$, $\Delta_{1,\perp} = 0.78$ Å, $d_{12,\perp} = 1.50$ Å, $\Delta_{1,y} = 4.98$ Å, $d_{12,y} = 3.60$ Å, $\Delta_{2,\perp} = -0.15$ Å, $d_{23,\perp} = 2.21$ Å

InAs (111)
Surface with 2x2 reconstruction observed in [71G].

InAs($\overline{1}\overline{1}\overline{1}$)
Surface observed in [71G] and exhibiting faint streaks between the integral order spots in the LEED pattern and faceting when heating.

InP (Indium phosphide)

InP (100)
Surface with 4x2 reconstruction [87H2], terminating both with In and P atoms. The structure becomes 1x1 by laser irradiation [82M5, 83M5]. The (4x1)R30° structure is also observed [77C4].

InP (110)
Surface obtained by cleavage and ion sputtering and annealing [75W, 77C4, 80T3].
The surface structure is investigated in [80M6]. The geometry is reported in Table 6, with reference to Fig. 3, see p. 54.
Other LEED calculations in [80T1, 82F4].
Investigation via RHEED shows that the annealing of the ion bombarded surface produces agglomeration of In atoms and formation of randomly oriented crystallites [83P1].

Table 6.

ω =28°, $\Delta_{1,\perp}$ = 0.69 Å, $d_{12,\perp}$ = 1.59 Å, $\Delta_{1,y}$ = 4.57 Å, $d_{12,y}$ = 3.45 Å, $\Delta_{2,\perp}$ = 0 Å, , $d_{23,\perp}$ = 2.07 Å

InP (111)
Surface prepared in [77C4].

InSb (Indium antimonide)

InSb (110)
Surface obtained by cleavage [64M3].
The geometry of the structure is reported in Table 7, with reference to Fig. 3, see p. 54 [80M5, 87D3].
Other LEED intensity calculations in [82F4, 80D6, 86C3].

Table 7.

ω = 28.8°, $\Delta_{1,\perp}$ = 0.78 Å, $\Delta_{1,y}$ = 5.06 Å, $\Delta_{2,\perp}$ = −0.18 Å, $d_{12,\perp}$ = 1.60 Å, $d_{12,y}$ = 3.82 Å, $d_{23,\perp}$ = 2.38 Å

InSb (111)
Surface with 2x2 reconstruction, observed in [60H1, 61H].

InSb ($\bar{1}\bar{1}\bar{1}$)
Structure exhibiting 2x2 and 3x3 structures, observed by RHEED [76R3].

Ir (Iridium)

Ir (100)
Surface with reconstruction 1x5 stable up to 1800 °C [69G2, 79H5, 81V1]. The surface structure is probably formed of a hexhagonal layer with buckling and a two-bridge registry of the top layer [81V1, 81V2, 83L2, 85B7].
A 1x1 structure is also observed; this is both metastable and stabilized [79K3, 83B5]. The kinetic of the transition from 1x1 to 1x5 at about 500 °C is investigated in [85H3]. The metastable phase is investigated by LEED intensity analysis in [82L1]. The stable 1x1 structure exhibits a relaxation of the first layer of Δd_{12} = −2(2)% [83B5].

Ir (110)
Surface with reconstruction 1x2 [73C2]. Its structure is a missing row with paired rows in the second layer and buckling in the third [80C2, 86C1].
Other models are investigated in [79C2, 82B3].
A 1x1 phase is also observed and determined as a relaxed surface with $\Delta d_{12} = -4\%$ [83H2].
Measurements of surface Debye temperature (150 K) in [79C3].

Ir (111)
Unreconstructed surface, investigated by the CMTA method [77C3].
Measurements of surface Debye temperature (170 K) in [79C3].

KBr (Potassium bromide)

KBr (100)
Observed and analyzed with LEED kinematic theory [68T].

KCl (Potassium chloride)

KCl (100)
Observed and analyzed with LEED kinematic theory [68T].

LaB_6 (Lanthanum boride)

LaB_6 (100)
Unreconstructed surface, probably with the La atoms in the first layer, prepared and observed by LEED in [67B2, 67B3, 69B1, 69B2, 78A1, 78G1, 79A5, 79A6, 80N1].

LaB_6 (110)
Reconstructed c(2x2) surface, which transforms into 1x1 at temperatures > 850 °C [80N1].

LaB_6 (111)
Unreconstructed surface, probably with the La atoms in the first layer, prepared and observed by LEED in [80N1].

$LaCuO_4$ (Lanthanum copper oxide)

$LaCuO_4$ (0001)
Unreconstructed surface prepared and observed in [88B3]. By annealing a 1x8 structure is formed.

LiF (Lithium fluoride)

LiF (100)
1x1 surface formed by cleavage, observed by LEED at temperatures > 300 °C to avoid charging [64M2, 71W1]. The surface is probably buckled with Li and F atoms, displaced inwards and outwards the surface, respectively [67M3, 71H5, 73L3].

LiH (Lithium hydride)

LiH (100)

Surface whose LEED pattern is observed during cooling in the range 400 ... 120 °C, at incident beam energy above 70eV. Probably there is an excess of Li atoms fixed on the surface during the heating [72H].

MgO (Magnesium oxide)

MgO (100)

Surface unrelaxed with rumpling of 2(2)%, as determined by LEED intensity analysis [82W5]. The rumpling appears after annealing at 300 °C, but is not present after cleavage [81G2].
Other authors do not find any differences among preparation methods [83U].
Other LEED observation and *I*/*V* curve measurements in [74H2, 74L4, 75K3, 76K2, 78W2, 79P4, 80B5].
Measurements and analysis of RHEED rocking curves in [85M1].

MgO (110)

Surface exhibiting thermal faceting into (100) planes [76H3].

MgO (111)

Surface exhibiting thermal faceting into (100) planes [76H3].

MnO (Manganese oxide)

MnO (100)

Surface which reconstructs 2x2 at 800 °C and 6x6 at 1000 °C [87L1]. The reconstructions are irreversible and stoichiometry-preserving. The transition temperatures are affected by the impurities.

Mo (Molybdenum)

Mo (100)

Unreconstructed surface with $\Delta d_{12} = -9.5(20)\%$, and $\Delta d_{23} \approx -1.5\%$ [80C3]. Other LEED calculations in [75I, 79C6].
A phase transition is observed at low temperatures (< 0°C); this appears to be incommensurate [77F5, 80B1]. Its structure can be described in terms of a periodic lattice distortion.
A surface Debye temperature of 239 K [71T1, 71T2] and an expansion coefficient of 2 ... 3 times the bulk value is determined [71W3].

Mo (110)

Unreconstructed surface prepared and observed in [66H2], and with $\Delta d_{12} = -1.6(20)\%$ [81M7].

MoS_2 (Molybdenum sulfide)

MoS_2 (0001)

Cleaved surface prepared and observed in [68I, 71W2]. LEED intensity calculations show a relaxation of $\Delta d_{12} = -4.7(3)\%$. The sandwiches of three layers are separated by a gap whose distance at the surface varies by $\Delta d_{gap} = -3.0(5)\%$ [77M8, 77M9, 77V].

$MoSe_2$ (Molybdenum selenide)

$MoSe_2$ (0001)
Surface of the layered compound, resulting in a quasi-ideal stable surface [77M8, 84S6].

NaCl (Sodium chloride)

NaCl (100)
Surface prepared in [67M3], but appearing unstable under electron bombardment [70G1, 76B1]. LEED intensity curves are analyzed using kinematic theory [68T] and dynamical theory [66M2, 72L8], also in the chain method formalism [79M4]. On the surface, rumpling is probably present [68B2, 72L8].

NaF Sodium fluoride

NaF (100)
1x1 surface formed by cleavage, observed by LEED at high temperatures to avoid charging [68T, 71W1]. Surface resonances observed and calculated in [67M3].

$Na_x(WO_3)_{1-x}$ (Sodiumn tungsten oxide)

$Na_x(WO_3)_{1-x}$ (100)
Observed various compositions (x = 0.47, 0.72, 0.80, 0.82, 0.91).
The LEED pattern shows the WO_3 square symmetry plus fractional order features corresponding to ordered Na row structures, which depend on the x value [77L1, 80L1]. When x = 0.47 at 500 °C a 3x1 structure is formed. The other compositions occasionally give a c(2x2) at 550 °C. When x = 0.80, a 2x1 structure is observed [81S1].

$Na_{2/3}(WO_3)_{1/3}$ (110)
Reconstructed 3x1 surface, which is unstable under electron bombardment [85E]. The superstructure is interpreted as being due to ordering of the Na atoms.

Nb (Niobium)

Nb (100)
Stable unreconstructed surface in the range –260 ... 700 °C [81M5].
Surface Debye temperature is measured and is 2.66 times lower than the bulk value [70T1].

Nb (110)
Unreconstructed surface observed in [66H1].

$NbSe_2$ (Niobium selenide)

$NbSe_2$ (0001)
Unreconstructed surface of the layered compound. LEED intensity calculations show a relaxation of $\Delta d_{12} = -1.4(14)\%$. The sandwiches of three layers are separated by a gap whose distance at the surface varies by $\Delta d_{12} = -0.6(6)\%$ [77M8, 77M9, 77V].

Nb_3Ir (Niobium iridium)

Nb_3Ir (100)
Unreconstructed surface [86S2]. At T > 795 °C a 2x2 structure is formed. At 940 °C and 1210 °C a 3x3 structure is formed. The superstructures are perhaps due to impurities.

Nb_3Ir (110)
Unreconstructed surface [86S2]. A 1x3 structure is also observed.

Nb_3Ir (111)
Unreconstructed surface [86S2]. A 3x3 structure is also observed.

Ni (Nickel)

Ni (100)
Unreconstructed, with $\Delta d_{12} = +1.1(11)\%$ [75D2].
Preparations and experimental observations in [55F, 58F, 66P, 73C3, 74D2].
Studies using kinematic LEED and CMTA in [71A5, 72S2, 76U, 77C2].
Other dynamical LEED calculations in [71A4, 71T4, 72D1, 72T1, 73T2, 73L1, 76T].
Surface Debye temperature in normal direction to the surface measurements in [64M1, 70A2, 85G3].
SPLEED investigations in [81A4, 82F1, 82L2, 83F3] and surface magnetization studies in [82A4, 82A6].
VLEED studies for surface barrier and surface resonances in [72S1, 74M3, 78M5, 78R1, 79J1, 83L5].
Absorption potential determined by VLEED in [76M3].

Ni (110)
Unreconstructed surface with oscillatory relaxation of the $\Delta d_{12} = -8.5(15)\%$, $\Delta d_{23} = +3.5(15)\%$, $\Delta d_{34} = +1.0(15)\%$ [87R].
Preparations and experimental observations in [61G, 64P2, 74D2].
Study using the convolution transform method in [77C2].
Other dynamical LEED calculations in [71H6, 72T1, 73T2, 75D2, 82C4, 82G1, 82G2, 85A1, 85X2].
Surface Debye temperature 216 K (normal to the surface) and 344 K (parallel to the surface) [81M8, 85G3]. Other Debye temperature measurements in [62M, 64M1, 69R].
SPLEED investigations for magnetization studies in [79C1] and to find the dependence of polarization by the temperature layer by layer in [81W2].

Ni (111)
Unreconstructed surface with $\Delta d_{12} = -1.2(12)\%$ [75D2].
Experimental observations in [63G, 64P2, 66P].
Other dynamical LEED calculations in [69P5, 71H6, 73L1, 74D2].
Investigations using kinematic LEED theory and CMTA method in [64G, 70F1, 72L1, 73N].
Surface Debye temperature 208 K (normal to the surface) and 348 K (parallel to the surface) [85G3]. Other Debye temperature measurements in [64M1, 75M5, 82M7].
SPLEED studies in [77F1, 88M3].

Ni (311)
Vicinal surface with oscillatory relaxation $\Delta d_{12} = -15.9(10)\%$, $\Delta d_{23} = +4.1(15)\%$, $\Delta d_{34} = -1.6(16)\%$, and with displacement parallel to the surface in the $<\bar{2}33>$ direction; $r_1 = -0.8(19)\%$, $r_2 = -1.4(19)\%$, $r_3 = -0.5(32)\%$ [82M7, 85A2].

Ni ($51\bar{1}$)
Stepped surface observed and investigated in [83C].

Ni (911)
Stepped surface observed and investigated in [83C].

Ni–Al (Nickel aluminum)

NiAl (100)
Alloy surface, terminating with a layer of Al, with $\Delta d_{12}(\text{Al}) = -8.33\%$, $\Delta d_{12}(\text{Ni}) = +4.17\%$ [87D2].

NiAl (110)
Alloy surface exhibiting rippling on the first layer with Al atoms displaced outwards by $\Delta d_{12}(\text{Al}) = +4.6\%$ and the Ni atoms displaced inwards by $\Delta d_{12}(\text{Ni}) = -6.0\%$ [85D2, 85D3].

NiAl (111)
Alloy surface unreconstructed, which can terminate with Ni atoms or Al atoms. Both terminations are present in different domains with $\Delta d_{12}(\text{Al}) = -5\%$, $\Delta d_{23}(\text{Al}) = +5\%$, and $\Delta d_{12}(\text{Ni}) = -50\%$, $\Delta d_{23}(\text{Ni}) = +15\%$ [88N].

Ni_3Al (100)
Surface with a mixed layer of Ni and Al atoms with $\Delta d_{12} = -2.8(16)\%$, and buckling with the Al atoms outwards of +1.1(16)% [85S4, 86S4].

Ni_3Al (110)
Surface with a mixed layer of Ni and Al atoms, exhibiting buckling with the Al atoms outwards of +1.2(25)%. Oscillatory relaxation is also present, with $\Delta d_{12} = -12\%$, $\Delta d_{23} = +3.17\%$ [86S6].

Ni_3Al (111)
Surface with small relaxation ($\Delta d_{12} = -0.5\%$) and with rippling because of displacement of the Al atoms outwards by $\Delta d_{12}(\text{Al}) = +2.9(15)\%$ and of the Ni atoms inwards by $\Delta d_{23} = -0.5(15)\%$ [86S5].

$Ni_{0.992}Au_{0.008}$ (Nickel gold)

$Ni_{0.992}Au_{0.008}$ (110)
Surface obtained by enrichment of Au, prepared by segregation and observed in [86M1].

$Ni_{0.76}Fe_{0.24}$ (Nickel iron)

$Ni_{0.76}Fe_{0.24}$ (100)
Surface of the alloy obtained and observed in [79B2].

NiO (Nickel oxide)

NiO (100)
Antiferromagnetic material showing a spin superlattice at temperatures below the Néel temperature [68P3, 69P1, 75N].
LEED intensity analysis shows $0 > \Delta d_{12} > -3\%$, with no rumpling [77K2, 77W1, 80W2].

Pb (Lead)

Pb (110)
Surface with phase transition at 140 °C, observed in [89Y].

Pb (111)
Surface Debye temperature 55 K, measurement in [68G3].

PbTe (Lead telluride)

PbTe (111)
Surface prepared and observed by RHEED in [80B4]. The anisotropy of the atomic vibrations is observed in normal and parallel directions to the surface.

$Pb_{1-x}SnTe_x$ (Lead tin telluride)

$Pb_{1-x}SnTe_x$

Polycrystalline blocks with different surface orientations and with various compositions (x = 0.21(2), 0.23(1), 0.29(20), 0.295(5), 0.32(2), 0.36(1), 0.43(3), 0.44(5)), prepared by ion bombardment and annealing at 900 °C [86C2].

Pd (Palladium)

Pd (100)
Surface Debye temperature 140 K, measured in [68G3].

Pd (110)
Surface with oscillatory relaxation of $\Delta d_{12} = -5.1(15)\%$, $\Delta d_{23} = +2.9(15)\%$ [87S]. Other LEED intensity analyses in [85B4, 85D4].
Problems connected with impurity contamination in [85B4, 87W3, 88B2].

Pd (111)
LEED intensity measurements in [73C3].
Surface Debye temperature 140 K, measured in [68G3].

Pt Platinum

Pt (100)
Surface in general reconstructed as 1x5, stable between −196 °C and > 1600 °C [67L2, 73L1, 73T2]. The 1x5 structure is really a hexagonal layer rotated by 0.7° resulting a

$$\begin{pmatrix} 14 & 1 \\ -1 & 5 \end{pmatrix} \text{or} \begin{pmatrix} 14 & 1 \\ -3 & 10 \end{pmatrix} \text{or} \begin{pmatrix} 12 & 1 \\ -3 & 5 \end{pmatrix} \text{or} \begin{pmatrix} 27 & 2 \\ -3 & 4 \end{pmatrix}$$

structure [77H1, 81V1].
Other superstructures are observed: 1x1, metastable [78B2, 82H2], which evolves into a 20x5 or 25x5 at 125 °C and into the hexagonal R0.7° at 835 ... 885 °C [77H1, 81V1, 82H3], and 2x1 at 300 ... 500 °C [67L2].

LEED intensity calculations of the 1x1 [82L1].
SPLEED calculations of the unreconstructed surface in [77F2].
Surface Debye temperature of 118 K, measured in [66L2].

Pt (110)
Reconstructed surface 1x2, resulting in a missing-row, with multilayer relaxation of Δd_{12} = −18.4%, Δd_{23} = −12.6%, Δd_{34} = −8.7%, Δd_{45} = −0.9%, and row-pairing in the second layer with variation of the distance between rows of 0.130 Å, and buckling of the third plane of 0.320 Å [88S2, 86C1].
Other structural models and LEED intensity analysis in [80C2, 82B3, 83H2, 87F, 88F].
The formation and stability of the structure is investigated in [67L2, 77B3].
A 1x1 phase is also observed; this is reversible at 830 °C [87H1].
The same 1x1 structure is formed by ion bombardment at 25 °C and transforms into 1x2 at 250 °C.
A 1x3 phase is formed and analyzed by LEED intensity calculations, resulting in a missing-row with Δd_{12} = −20%, buckling of the third layer of 0.17 Å, and a lateral shift of the second layer by 0.04 Å [88F].
SPLEED calculations of the unreconstructed surface in [77F2].
Surface Debye temperature of 107 K, measured in [66L2].

Pt (111)
Unreconstructed surface with multilayer relaxation by Δd_{12} = 0.0(10)%, Δd_{23} = −1.0(10)%, Δd_{34} = 0.0(20)% [86O2].
Other LEED dynamical calculations in [73L1, 73T2, 75K2, 77K1, 79A3, 85H1].
Kinematic LEED intensity analysis in [73C3].
The stability of the structure was investigated in [77B3].
2x2 and 3x3 phases forming at 800 ... 1000 °C are reported in [67L2].
SPLEED calculations are performed in [77F2, 80B2, 81F3, 83B4].
RHEED rocking curves calculated in [88Z].
Surface Debye temperature of 111 K, measured in [66L2].

Pt (012)
Stepped surface stable up to 1200 °C, observed in [77B3].

Pt (112)
Stable stepped surface, observed in [77B3].

Pt (113)
Stable stepped surface, observed in [77B3].

Pt (122)
Stepped surface stable up to 1200 °C, observed in [77B3].
Pt (015)
Stepped surface, evolving into (100)1x5 and (210), observed in [77B3].

Pt (133)
Stepped surface stable up to 1200 °C, observed in [77B3].

Pt (233)
Stepped surface stable up to 1200 °C, observed in [77B3].

Pt (045)
Stepped surface stable up to 1200 °C with faceting, observed in [77B3].

Pt (155)
Stepped surface which facets into (110), (331), and (221), observed in [77B3].

Pt (119)
Stepped surface stable up to 1200 °C, observed in [72L2, 77B3].

Pt (445)
Stepped surface stable up to 1200 °C [77B3].

Pt (456)
Stepped surface stable up to 1200 °C [77B3].

Pt (655)
Stepped surface stable up to 1200 °C, observed in [77B3].

Pt (1 1 15)
Stepped surface stable up to 1100 °C, observed in [77B3].

Pt (557)
Stepped surface stable up to 1600 °C [72L2, 77B3].

Pt (559)
Stepped surface stable until 900 °C with facets (533). At higher temperatures facets into (211) and (553) [77B3].

Pt (679)
Stepped surface stable up to 1200 °C [77B3].

Pt (799)
Stepped surface stable up to 1400 °C [72L2, 77B3].

Pt (997)
Stepped surface, with faceting, observed in [80C5].

Pt (11 12 16)
Stepped surface stable up to 1200 °C [77B3].

Pt (12 12 13)
Stepped surface stable up to 1400 °C [77B3].

Pt (33 34 47)
Stepped surface stable up to 1200 °C [77B3].

Pt (69 72 72)
Stepped surface stable up to 1400 °C [77B3].

Pt polycrystalline
Observed by VLEED in [86G].

$Pt_{1-x}Co_x$ (Platinum cobalt)

$Pt_{1-x}Co_x$ (100)
Alloy with x = 0.2, exhibiting a 5x1 or quasi-hexagonal reconstruction, similar to those of Pt (100) [89B1].

$Pt_{1-x}Ni_x$ (Platinum nickel)

$Pt_{1-x}Ni_x$ (110)

Random substitutional alloy with x = 0.5, studied by LEED intensity analysis, using averaged **t**-matrix approximation to obtain the scattering amplitude layer by layer [87G1]. The result is $\Delta d_{12} = -19.0(7)\%$, $\Delta d_{23} = +11.0(15)\%$, $\Delta d_{34} = -1.0(15)\%$, with 0(+5−3) at% Pt, 95(4) at% Ni and 3(+7−3) at% vacancies in the first layer, and 17(7) at% Pt, 48(13) at% Ni in the second layer.

$Pt_{1-x}Ni_x$ (111)

Random substitutional alloy with x = 0.5, 0.22 and 0.9, studied by LEED intensity analysis, using averaged **t**-matrix approximation to obtain the scattering amplitude layer by layer [85G1, 85G2, 86B3]. For x = 0.5, the result is $\Delta d_{12} = -2(1)\%$, $\Delta d_{23} = -2(1)\%$, and with Pt concentration of 88(2) at% in the first layer, 9(5) at% in the second layer, 65(10) at% in the third layer. For x = 0.22, $\Delta d_{12} = -1.8(9)\%$, $\Delta d_{23} = -1.8(9)\%$, and with Pt concentration of 99(1) at% in the first layer, 30(5) at% in the second, 87(10) at% in the third, 78 at% in the fourth. For x = 0.9, $\Delta d_{12} = 0.0(2)\%$, $\Delta d_{23} = -0.8(9)\%$, and the Pt concentration is 30(4) at% in the first layer, 5(3) at% in the second.
The case x = 0.9 was also observed in [82B2].

Pt_3Ti (Platinum titanium)

Pt_3Ti (100)

Surface prepared and observed in [84B], giving a regular surface, with periodicity determined by the mixed PtTi layer, i.e. c(2x2) with respect to the Pt(100).

Pt_3Ti (111)

Surface prepared and observed in [84B], giving a regular surface, with periodicity determined by the mixed PtTi layer, i.e. 2x2 with respect to the Pt(111).

Pt_3Ti (510)

Stepped surface, showing terraces corresponding to a bulk truncation model [88B1].

Re (Rhenium)

Re (0001)

1x1 surface prepared and observed in [69F4, 70D1, 72Z, 77H2], often to be used in adsorption studies. On heating at 1600 °C two domains of $(11\bar{1}n)$ planes are formed. By electropolishing, planes $(11\bar{2}n)$ (n = 2, 3, 4) are formed.
LEED intensity calculations using approximate theory are performed in [74Z2].

Re $(10\bar{1}0)$

Surface prepared and observed in [72Z]. The termination is with the second layer located directly below the centers of the geometric rectangle of the first layer, with $\Delta d_{12} = -17\%$ and $\Delta d_{23} = +2\%$ [80D1].

Rh (Rhodium)

Rh (100)

Unreconstructed surface prepared in [66T], analyzed by LEED intensity analysis and having $\Delta d_{12} = +0.5(20)\%$, $\Delta d_{23} = 0.0(15)\%$ [88O]. Other LEED intensity calculations in [77M7, 78W1, 80H3].

Rh (110)
Surface analyzed in [78F3, 87N], with $\Delta d_{12} = -6.9(10)\%$, $\Delta d_{23} = -1.9(10)\%$.

Rh (111)
Surface observed in [70G2]. LEED intensity calculations produce Δd_{12} = 0.0(20)% or −2.0(20)%, depending on the potential used [78S1].
The surface is also studied by convolution-transform method in [78C1].
The surface Debye temperature is 197 K [78C1].

Ru (Ruthenium)

Ru (100)
Surface with $\Delta d_{12} = -2\%$ [83M4]. The experimental curves are in [82W6, 83M3].

Ru (0001)
Unreconstructed surface observed in [70G2].

RuO_2 (Ruthenium oxide)

RuO_2 (100)
Surface observed in [88A2]. Annealed at 450 °C gives a diffuse LEED pattern. If annealed in presence of oxygen the pattern becomes sharp. By further annealing in a vacuum a 1x2 structure is formed over 450 °C and a c(2x2) at 550 °C. All these structures show an oxygen enrichment, probably by formation of RuO_3.

RuO_2 (110)
Surface with Ru enrichment [88A2]. During heating up to 450 °C a c(2x2) is formed. At 450 °C the structure becomes hexagonal.

Sb (Antimony)

Sb (0001)
Unreconstructed surface observed in [67J2].

Sb ($01\bar{1}2$)
Unreconstructed surface observed in [67J2].

Sb ($11\bar{2}0$)
Surface observed in [67J2], showing a 6x3 reconstruction.

Sc (Scandium)

Sc (0001)
Surface with hcp stacking and $\Delta d_{12} = -2.0(8)\%$ [82T4].

Sc (100)
Unreconstructed surface, with the first interplanar distance unrelaxed [82J1].

Se (Selenium)

Se (1$\bar{2}$10)

Surface obtained by cleavage, which forms two inequivalent faces, one having the ideal bulk structure, the other showing faceting at 120 °C, with facets (10$\bar{1}$0) and (01$\bar{1}$0) [78A2].

Si (Silicon)

Si (100)

Surface generally reconstructed as 2x1, first observed in [59S, 62L1].
LEED intensity calculations show a distortion of the ideal structure with formation of dimers, as indicated in Fig. 5. The values and definitions are reported in Table 8 and [83J2, 83Y, 84H2].
Other LEED intensity calculations in [77J, 78M8, 78T4, 79J3, 82W4].
Studies using the CMTA method in [77W2].
RHEED investigations in [79I, 83N2, 86A4].
Comparison between various experimental sets of data of LEED intensity in [77I].
The stability of the 2x1 structure is studied in [81O1, 88M2].
Other structures are observed. A 2x2 structure reported in [74I, 82W10]. A c(2x2) structure reported in [83J1]. A c(4x2) structure formed at 900 °C and 1250 °C and studied with the CMTA method [78P1], also observed in [82W10, 83J1, 84M5]. This structure transforms into 2x1 at −73 °C with a second order transition [87T].
The c(4x4), c(8x8) and 2x8 structures are reported in [84M5].
The 16x2 structure is reported in [87I3, 89N].
The surface Debye temperature is measured in [71N, 73C4, 77B4] and for the 2x2 structure in [74I].
Vicinal surfaces cut at 7° 20' in <011> direction and 6°10' in <001> direction are observed to form facets (47 35 $\bar{7}$) and (35 47 7) [80O3]. Surfaces cut at 7°, 3.5°, and 2.5°, are observed in [80K, 88A3] and gave in two domains 2x1 and 1x2.

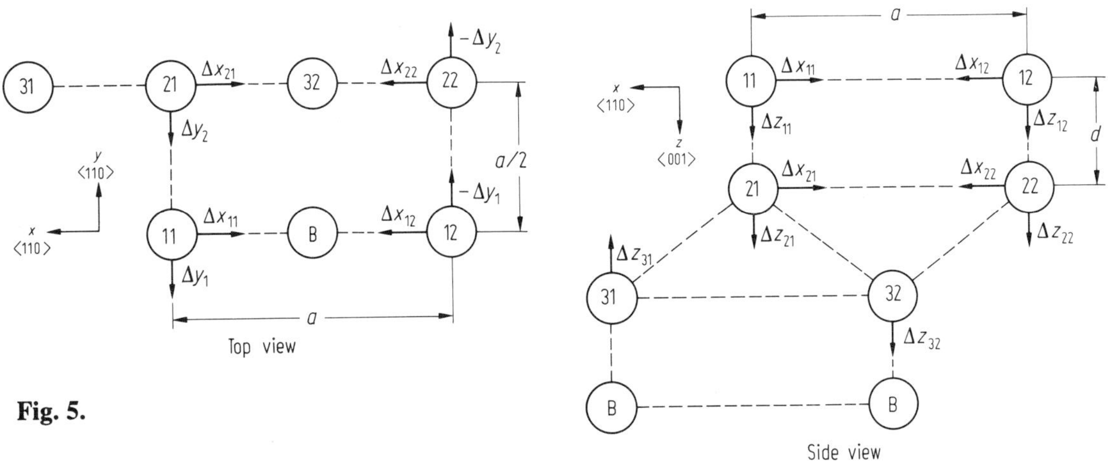

Fig. 5.

Table 8. Distortion of the ideal structure in Å, see Fig. 5.

$\Delta x_{11} = -0.65$, $\Delta x_{12} = 0.75$, $\Delta x_{21} = -0.06$, $\Delta x_{22} = 0.12$, $\Delta y_1 = -0.30$, $\Delta y_2 = -0.10$
$\Delta z_{11} = 0.045$, $\Delta z_{12} = 0.445$, $\Delta z_{21} = 0.136$, $\Delta z_{22} = 0.136$, $\Delta z_{31} = -0.152$, $\Delta z_{32} = 0.208$

Si (110)

Surface exhibiting various reconstructions. The 2x1, 5x1, and 4x5 reported in [77O3, 86N]. The 2x1 and 5x1 structures transform into each other at 600 ... 750 °C, and can be frozen at low temperatures. During the transition 7x1 and 9x1 are also observed. A 16x2 (and sometimes 32x2) structure seems to be the real reconstruction, the above ones probably being due to impurities [85I2, 86A2, 86Y].

Vicinal surfaces cut at 3° 30' in $\langle 1\bar{1}0\rangle$ direction, at 4° and 9° 30' in <001> direction and at 6° 50' in $\langle 1\bar{1}1\rangle$ direction were observed to form facets ($7\,\bar{7}\,12$).

Si (111)

Surface generally reconstructed as 7x7, first observed in [59S, 61L1, 64L2, 65L, 82T4]. The structure is indicated in Fig. 6 and the atomic coordinates are reported in Table 9 [88H2, 88T1].

In Fig.6, the unit cell (a) is indicated, as made up of triangles (b). In the panels (b), (c), and (d) the two base triangles A and B are sketched, in the first, second and third layer, and the atoms are indicated and numbered. Panels (e), (f), (g), and (h) are side views. In Table 9 the origin is placed on atom 103.

Other LEED intensity calculations [83W1].

LEED intensity analysis using kinematic considerations in [77L4, 77L5, 79M7, 80M9, 81M6].

The structure is also investigated by RHEED in [69H2, 72M2, 83H4, 87I2, 89H1].

Studies on the stability of this structure and on its transitions into other structures in [78M10, 81B6, 81O1, 87P1]. The 7x7 structure transforms into 1x1 at 870 °C and during the annealing in the range 290 ... 380 °C.

The 1x1 structure is also obtained using stabilization by laser pulses or by quenching [73H1, 80Z2, 80Z3, 81Z, 82C3, 83M5, 86J3] and it is a relaxed bulk, with $\Delta d_{12} = -25.5\%$ and $\Delta d_{23} = +3.2\%$.

Another important structure is 2x1 [78M10, 83A3, 87P1]. The structure is reported in Fig. 7 and the atomic coordinates in Table 10 [84H1, 86S1]. The table gives the atomic coordinates, using x, y and z axes parallel to $\langle\bar{1}\bar{1}1\rangle$, $\langle 1\bar{1}0\rangle$, and $\langle\bar{1}\bar{1}2\rangle$ directions, respectively. Other LEED intensity calculations in [79F1, 83F2, 84F, 83L4].

The structure ($\sqrt{3}$ x$\sqrt{3}$)R30° was also obtained in [64L2, 65L, 89F2] and gives the structure reported in Fig. 8. It is a vacancy model in which $\Delta d_{12} = -64\%$, $\Delta d_{23} = -3\%$, $\Delta d_{34} = -18\%$. The atoms of the second layer are laterally displaced along the arrows of Fig. 8 by 0.64 Å. The atoms of the third layer are displaced along the arrows by 0.24 Å. Other structures observed are: (2$\sqrt{3}$ x2$\sqrt{3}$)R30°, 4x1, 24x24 [64L2, 65L], 2x2 [64L2, 65L, 87P1], $\sqrt{19}$ x$\sqrt{19}$ [72M2], 5x5 [89U].

Vicinal surfaces, when treated with laser pulses, become 7x7 [80Z4]. When cut at 6° along $\langle 1\bar{1}0\rangle$ direction and 4° along $\langle\bar{4}\bar{3}1\rangle$ direction and treated with laser quenching the surface becomes 2x2 [88P1].

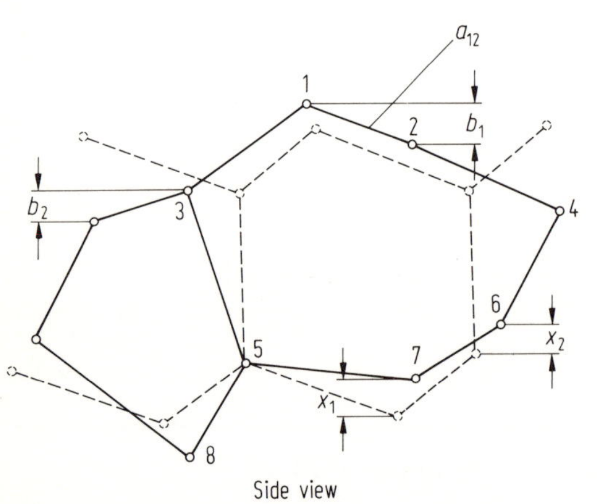

Fig. 7. Si (111)1x1.Further information, see text. The atomic coordinates are given in Table 10.

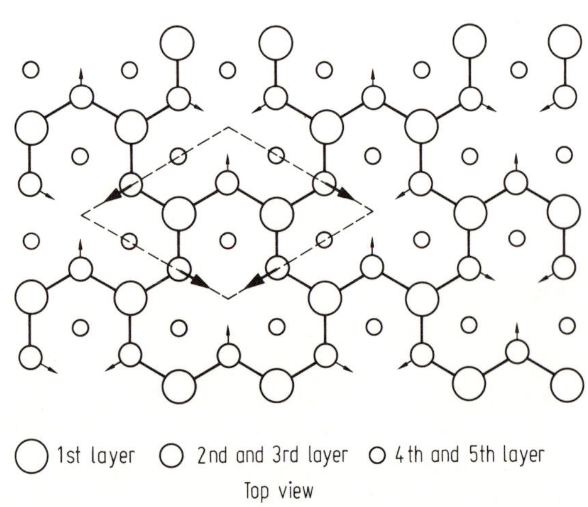

Fig. 8. ($\sqrt{3}$ x$\sqrt{3}$)R30° structure for Si (111).

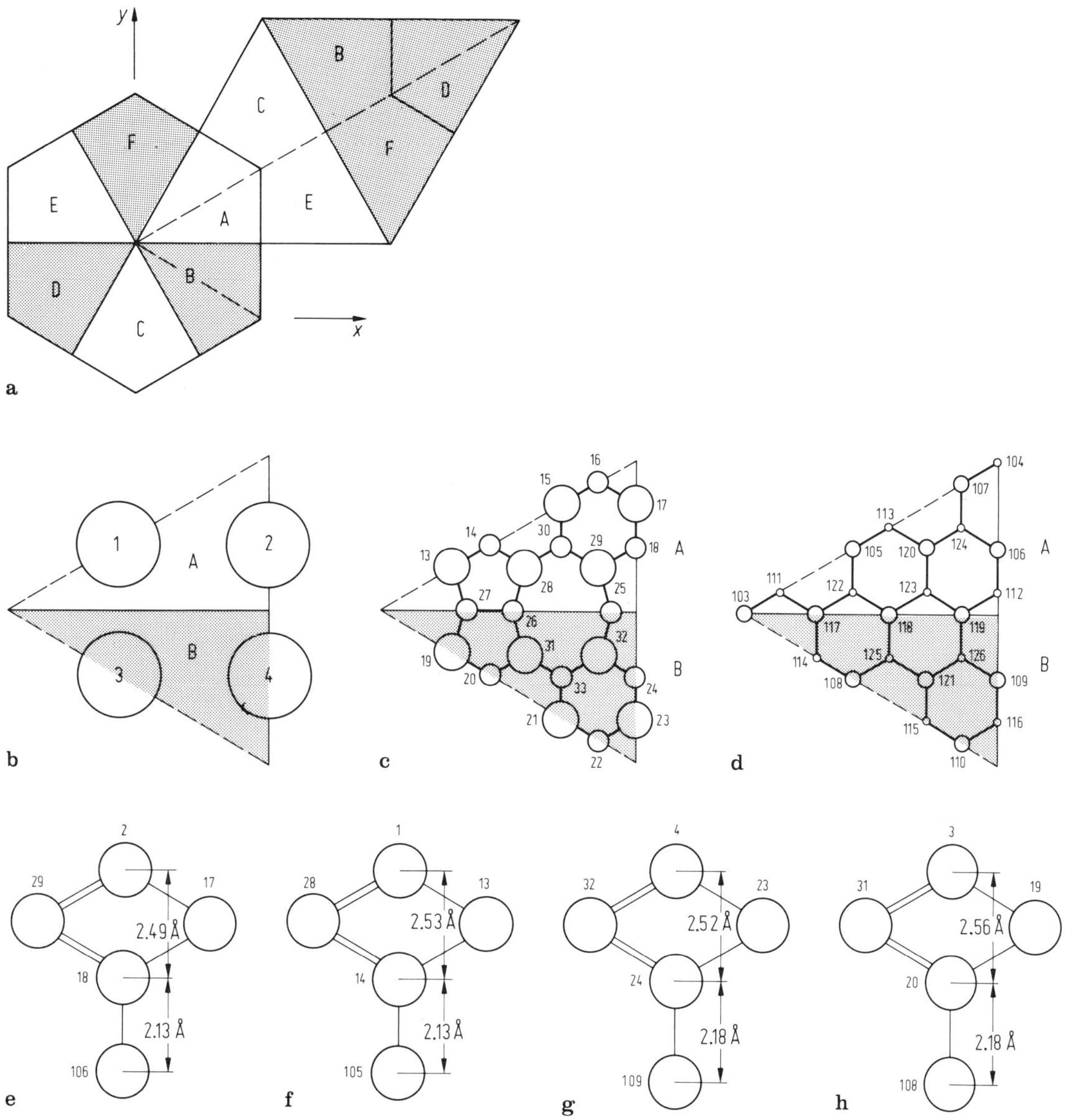

Fig. 6a-h. Si (111)7x7. Further information see text.The atomic coordinates are given in Table 9.

Table 9. Atomic coordinates. The origin is placed on atom number 103, see Fig.6.

Atom no.	x [Å]	y [Å]	z [Å]
1	5.76	3.33	−4.21
2	13.44	3.33	−4.17
3	5.76	−3.33	−4.29
4	13.44	−3.33	−4.25
13	4.01	2.32	−2.94
14	5.76	3.33	−1.68
15	9.60	5.54	−3.11
16	11.55	6.67	−2.41
17	13.44	5.32	−2.98
18	13.44	3.33	−1.68
19	4.01	−2.32	−2.99
20	5.76	−3.33	−1.73
21	9.60	−5.54	−3.16
22	11.55	−6.67	−2.46
23	13.44	−5.32	−3.03
24	13.44	−3.33	−1.73
25	12.21	0.00	−2.24
26	6.98	0.00	−2.24
27	4.53	0.00	−2.24
28	7.49	2.33	−2.98
29	11.71	2.33	−2.98
30	9.60	3.29	−2.41
31	7.49	−2.33	−3.03
32	11.71	−2.33	−3.03
33	9.60	−3.29	−2.46

Atom no.	x [Å]	y [Å]	z [Å]
103	0.00	0.00	−0.05
104	13.44	7.76	0.73
105	5.76	3.33	0.45
106	13.44	3.33	0.45
107	11.52	6.65	−0.07
108	5.76	−3.33	0.45
109	13.44	−3.33	0.45
110	11.52	−6.65	−0.07
111	1.92	1.11	0.74
112	13.44	1.11	0.87
113	7.68	4.43	0.80
114	3.84	−2.22	0.85
115	9.60	−5.54	0.73
116	13.44	−5.54	0.80
117	3.84	0.00	0.00
118	7.68	0.00	0.00
119	11.52	0.00	0.00
120	9.60	3.33	−0.07
121	9.60	−3.33	−0.07
122	5.76	1.11	0.87
123	9.60	1.11	0.72
124	11.52	4.43	0.80
125	7.68	−2.22	0.83
126	11.52	−2.22	0.83

Table 10. Atomic coordinates of first and second overlayer, see Fig.7.
$a_{12} = 2.25$ Å, $b_1 = 0.35$ Å, $b_2 = -0.05$ Å, $x_1 = 0.25$ Å, $x_2 = 0.15$ Å

Atom	First overlayer			Second overlayer		
no.	x [Å]	y [Å]	z [Å]	x [Å]	y [Å]	z [Å]
1	0.00	0.00	0.00	0.00	0.00	0.00
2	0.35	1.92	1.12	0.38	1.92	1.12
3	1.26	0.00	−1.98	1.26	0.00	−2.00
4	1.21	1.92	3.31	1.19	1.92	3.26
5	3.46	0.00	−1.26	3.46	0.00	−1.10
6	3.31	1.92	2.20	3.39	1.92	2.40
7	3.99	0.00	1.03	4.06	0.00	1.20
8	4.31	1.92	−2.32	4.26	1.92	−2.12

Si (210)
Stepped surface reconstructed 2x2, with transition temperature 850 °C, observed in [80O3].

Si (211)
Stepped surface reconstructed 4x2, observed in [80O3].

Si (320)
Stepped surface reconstructed 1x2 at temperatures < 680 °C and 1x1 at temperatures > 850 °C. At intermediate temperatures the surface shows facets (23 15 3) [80O3].

Si (311)
Stepped surface reconstructed 3x2, with transition temperature 750 °C, observed in [80O3, 81O1, 89M5].

Si (510)
Stepped surface reconstructed 1x2, with transition temperature 700 °C, observed in [80O3].

Si (331)
Stepped surface reconstructed 13x1, with transition temperature 800 °C, observed in [80O3].

Si (511)
Stepped surface reconstructed 3x1, with transition temperature 620 °C, observed in [80O3].

Si (711)
Stepped surface observed in [80O3].

Si (533)
Stepped surface observed in [80O3].

SiC (Silicon carbide)

SiC (100)
Surface of β-SiC, which exhibits the superstructures 3x2 at 850 °C, c(4x2) and 2x1 at 1100 °C, c(2x2) at 1175 °C, 1x1 above 1200 °C [85D1, 86D1, 89K1].

SiC (111)
Surface of β-SiC, which exhibits the superstructures 3x3 at 850 °C, ($\sqrt{3}$ x $\sqrt{3}$)R30° at 950 °C, 1x1 above 1000 °C [89K1].

SiC (0001)
Surface of β-SiC observed in [59D]. The fractional order beam 2/3 is sometimes observed in the LEED pattern.
The two basal faces have different terminations: the Si face reconstructs $\sqrt{3}$ x $\sqrt{3}$ at 250 °C, and (6$\sqrt{3}$ x 6$\sqrt{3}$)R30° at 800 °C. The C surface is 1x1 and becomes 2x2 and 3x3 at 800 °C and 4x4 at 1000 °C [75V1].
The surface of 6H-SiC exhibits the superstructures 3x3 at 850 °C, ($\sqrt{3}$ x $\sqrt{3}$)R30° at 950 °C, 1x1 above 1000 °C [89K1].

SiFe (Silicon iron)

SiFe ($1\bar{1}0$)
Reconstructed c(7x1), probably deriving from two antiphase domains, or ($\sqrt{51/3} \times \sqrt{8/3}$)R43.3° and R35.3° [78V2].

SmB_6 (Samarium boride)

SmB_6 (100)
Surface reconstructed 2x2 and 3x3, observed in [79A6].

SnO_2 (Tin oxide)

SnO_2 (110)
Amorphous surface, becomes 1x1 by annealing, up to 250 °C. In the range 250 ... 370 °C it reconstructs 4x1 and c(1x1) (or 4x2); in the range 370 ... 470 °C, 4x1, at temperatures above 470 °C 1x2 and diffuse 1x1. The transition temperature shifts towards higher values after various cycles of annealing [80D3, 81D1, 82D3, 83D2].

$SrTiO_3$ (Strontium titanate)

$SrTiO_3$ (100)
This surface has no LEED pattern if sputtered. After annealing it is 1x1 [80C6]. At low temperatures (–250 °C) it has a mixed termination of Sr and Ti atoms, with the O atoms outwards. For the Ti termination, Δd_{12} = +2(2)%, Δd_{23} = –2(2)%, oxygen-titanium distance = +0.08(8) Å. For the Sr termination, Δd_{12} = –10(2)%, Δd_{23} = +4(2)%, oxygen-strontium distance = +0.16(8) Å [89B2].

$SrTiO_3$ (111)
Surface stable if annealed at temperatures below 600 °C after ion bombardment at room temperature. If ion bombardment is performed at 600 °C, a complicated faceting forms [78L3].

Ta (Tantalum)

Ta (100)
Surface cleaned at 2000 °C, and observed between –260 °C and 400 °C as 1x1 [80O1, 84C2]. The structure shows relaxation Δd_{12} = –11% and Δd_{23} = +1% [82T3].

Ta (110)
Surface which under ion bombardment facets (130) and (310). At 700 °C it becomes 5x5 up to 1400 °C [82T3].

Te (Tellurium)

Te (0001)
Surface reconstructed 2x1, observed in [68A].

Te ($10\bar{1}0$)
After ion bombardment, the 1x1 structure is observed with facets (h 1 h+1 0) [68A].

Unreconstructed structure indicated in Fig. 9, with parameter values in Table 11 [80M8].

Table 11.

$\omega = 10.5°$	
$\alpha = 3.5°$	$\beta = 11.4°$
$\Delta x_1 = -0.40$ Å	$\Delta x_2 = -0.16$ Å
$\Delta y_1 = -0.16$ Å	$\Delta y_2 = -0.17$ Å
$\Delta z_1 = -0.21$ Å	$\Delta z_2 = +0.46$ Å

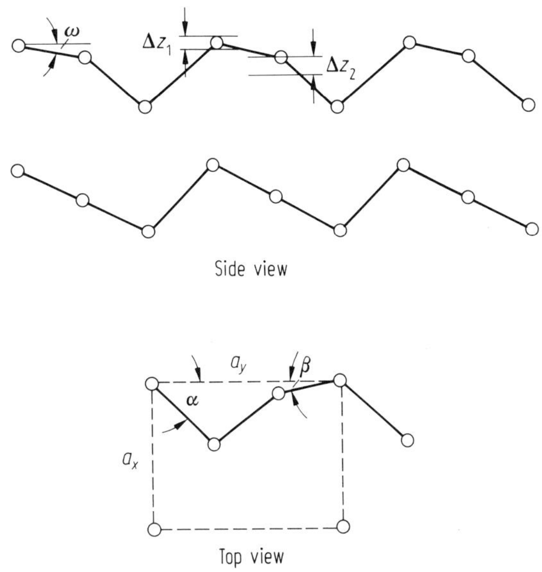

Fig. 9.

Te $(1\bar{2}10)$

This surface can have two different terminations. One is a $(1\bar{2}10)$ plane, the other a superimposition of $(1\bar{1}00)$ and $(0\bar{1}10)$, rotated by 120° [76A2].

Th (Thorium)

Th (111)

At temperature < 1350 °C the surface is 1x1. After annealing at 730 °C it is 9x9 [77B1].

ThO_2 (Thorium oxide)

ThO_2 (111)

Surface observed in [78T2].

Ti (Titanium)

Ti (100)

Unreconstructed surface terminating with $d_{12} = 0.800$ Å, observed in [89M4].

Ti (1000)

Unreconstructed surface cleaned and observed in [55F, 58F, 59G] and with $\Delta d_{12} = -2(2)\%$ [76S2]. Surface Debye temperatures of 240 ... 280 K are measured as a function of incident electron energy [87W2].

Ti (10$\bar{1}$0)
Surface observed by LEED without reconstructions [89M3]. The structures 4x4 at temperatures 25 ... 250 °C, 3x3 at 250 ... 450 °C, 2x2 at 450 ... 600 °C, and 1x1 at 600 ... 800 °C, were observed by RHEED [75K1].

TiC (Titanium carbide)

TiC (100)
Stable 1x1 observed in [59G, 80W1].

TiC (110)
Surface observed in [59G]. At high temperatures the surface facets.

TiC (111)
Stable 1x1 observed in [80W1].

Ti–O (Titanium oxide)

TiO_2 (100)
Stable surface which reconstructs 1x3 at 600 °C, 1x5 at 800 °C, 1x7 at 1200 °C [77C5, 82F3]. Also reported as unstable, which facets and at 950 °C reconstructs $(n/\sqrt{2}\,\text{x}1)$R45° (n = 4, 5, 8) [79T1].

TiO_2 (011)
Stable surface with 2x1 reconstruction [82F3].

TiO_2 (110)
Prepared by annealing at 820 °C, with evidence of oxygen vacancy production [76H4].

Ti_2O_3 (047)
Surface prepared by cleaving [83H3] and by ion bombardment and annealing [84M2].

$TiSe_2$ (Titanium selenide)

$TiSe_2$ (100)
1T–$TiSe_2$, exhibiting 1x1 structure with $\Delta d_{12} = -5\%$ [86P1, 89K2].

UO_2 (Uranium oxide)

UO_2 (100)
Surface unreconstructed, stable at temperatures < 1800 °C. From the systematic extinction of spots in the LEED pattern it is possible to argue that zig-zag rows of oxygen exist on the surface. At temperatures > 800 °C c(2x2), and c(4x4) structures and faceting form [68C1, 68E1, 78T2, 81T1].

UO_2 (110)
Surface observed, with many reconstructions perhaps due to contaminants.
At high temperature it facets: at 620 °C in <311> direction, at 730 °C in <331> direction, at 1080 °C in <n, n, n–2> direction, and at 1400 °C in <111> direction [68E1, 78T2].

UO_2 (111)
Unreconstructed surface observed in [68E1, 68E2], and interpreted using LEED kinematic theory. It also exhibits faceting in <311> and <553> directions [74E2].

UO_2 (311)
Stepped surface observed in [78T2].

UO_2 (410)
Stepped surface observed in [78T2].

UO_2 (510)
Stepped surface observed in [78T2].

UO_2 (711)
Stepped surface observed in [78T2].

UO_2 (553)
Stepped surface observed in [78T2].

V (Vanadium)

V (100)
Unreconstructed surface, with $\Delta d_{12} = -7.0(7)\%$, $\Delta d_{23} = +1.0(7)\%$ [82J4].
The 1x1 structure is also reported in [71F4, 76S5].
A 5x1 structure is observed [80D2], but is probably due to oxygen absorption or CO adsorption [81J2].
A surface Debye temperature of 250 K is reported in [75C].

V (110)
Unreconstructed surface with $\Delta d_{12} = -0.5(5)\%$ [81A2, 82A1], also prepared and observed in [67H].

V_6C_5 (Vanadium carbide)

V_6C_5 (100)
Surface exhibiting two-fold symmetry and faceting in <711> and <311> directions [87U].

V_2O_5 (Vanadium oxide)

V_2O_5 (010)
1x1 surface, unstable: under the electron beam it becomes V_6O_{13} (010), with structures 4x1 and 1x2 [67F4, 68F2, 69F5, 73G2].

Y Yttrium

Y (11$\bar{2}$0)
At temperatures > 340 °C a 1x1 structure is observed. At lower temperatures, two sets of LEED diffraction spots are observed, each with 6-fold symmetry, rotated by 30° to each other.
Hypothesized formation of terraces of close-packed atoms, with 60% increase in atom density [87B1].

$YBa_2Cu_3O_{7-x}$ (Yttrium barium copper oxide)

$YBa_2Cu_3O_{7-x}$ (100)
1x1 surface, prepared in a vacuum by cleavage [88S3].

$YFeO_3$ (Yttrium iron oxide)

$YFeO_3$
Surface prepared by etching at 100 °C in orthophosphoric acid to remove the amorphous layer present after polishing. Observed by LEED and RHEED in [72A2].

$YMnO_3$ (Yttrium manganese oxide)

$YMnO_3$ (0001)
2x2 structure after heating and annealing, probably with a different surface composition to the bulk one [69A1].

W (Tungsten)

W (100)
Surface presenting a reconstruction c(2x2) at temperatures < 130 °C. The high temperature 1x1 phase results in a simple bulk truncation with $\Delta d_{12} = -8.0(15)\%$ [80M2, 88P3].
Other LEED intensity calculations in [70R2, 75J1, 76V1, 77D3, 77F3, 77L2, 79H1, 80C4].
The structure is also attempted by the study of the rotation diagrams [78F1, 79K2], by the transfer-convolution method [79A2, 79M3] and by the Fourier transform method [78C2].
The precision of the measurements and the data reproducibility are investigated in [81S3, 82D2].
The c(2x2) structure produces a LEED pattern with characteristic p2mg symmetry [77D2, 81W1]. The structure corresponds to a zig-zag model, with $0 > \Delta d_{12} > -6.3\%$, and with a lateral shift of the displaced surface atoms of 0.15 ... 0.30 Å [78B1].
The transition from c(2x2) to 1x1 is investigated in [77D1, 79D2, 81S3].
Studies using relativistic theories and polarized electrons are in [71J2, 71J3, 73F, 74F, 75F, 76F1, 76F3, 77C1, 78J4, 78R2, 79K1, 79K2, 79W1, 81F2, 81P1, 81W3, 82W1, 82W3, 83J3].
Using VLEED, the surface effective potential is obtained and from this the lateral structure, which is not in agreement with the Debe and King model [82S5]. Other VLEED investigations in [72M1, 77A2, 77A3, 78J3, 81B4, 84L1, 85B1, 85B2, 85L2, 86J2].
Surface Debye temperature is 210(40) K for 1x1 and 400(200) K for c(2x2) [79H1]. A lower value is reported in [69E].
A stepped surface, with c(2x2) reconstruction, is studied in [82W2].

W (110)
Unreconstructed surface, first observed in [63K], found to be unrelaxed [76V1].
Investigations by kinematic LEED theory in [69S2, 72S2].
Investigations using CMTA and Fourier transform methods in [73W1, 75B2, 77C6].
Potential barrier calculations in [86J2].
Surface Debye temperature 190 ... 235 K [65A, 68S].

W (120)
Surface observed in [63K].

W (650)
Stepped surface, vicinal of W (110), investigated in [75B1].

W (750)
Stepped surface, vicinal of W (110), investigated in [75B1].

W (40 37 1)
Stepped surface, vicinal of W(110), investigated in [75B1].

W cylindrical
A cylinder with axis parallel to <110> direction, exhibiting low index planes and their stepped vicinals, observed in [81G1].

WO_3 (Tungsten oxide)

WO_3 (100)
Surface showing distorted triclinic symmetry, with evidence of twin domain formation and with normal vectors at 3.5° to each other observed in [80L2].

Zn (Zinc)

Zn (0001)
Surface analyzed by CMTA method in [75U].

ZnO (Zinc oxide)

ZnO (0001)
Surface prepared in [70C, 71F3, 72V, 72L9, 74M1, 75M1], and observed by LEED (in [72V] also by RHEED) with 6-fold symmetric pattern. The surface is 1x1 at temperatures below 600 °C, and ($\sqrt{3}$ x $\sqrt{3}$) at higher temperatures [75M1]. LEED intensity calculations show $\Delta d_{12} = -24\%$ [76L2].
Other calculations in [75D4].
Kinematic theory used in [72B2].

ZnO ($000\bar{1}$)
Surface prepared and observed by LEED in [72L9, 72V] and by RHEED [72V].
LEED intensity calculations show probable contraction of the first interplanar distance [75D4, 76L2].

ZnO ($10\bar{1}0$)
Surface prepared and observed in [72L9, 74C1, 78D, 78L1]. When produced by cleavage the surface is 1x1, with steps and faceting. If cut and polished it is 1x1 at temperatures < 600 °C and 2x2 at higher temperatures, for the Zn termination, and ($\sqrt{3}$ x $\sqrt{3}$)R30° for the O termination [74C1].
LEED intensity calculations for the 1x1 structure indicate a reconstruction relative to the truncated bulk [78D], with Δd_{12}(O) = −5.3(100)%, Δd_{12}(Zn) = −47.8(100)%, and with lateral displacement of Zn atoms by Δd = 3.1(60)%.
Other LEED intensity calculations in [76L2, 77D5]. LEED kinematic calculations in [75M2].
Surface Debye temperature 1.8 times lower than the bulk value [76G2].

ZnO ($11\bar{2}0$)
Surface prepared in [72L9, 74C2]. Analyzed using LEED kinematic theory in [75M2], and dynamical theory in [76D2, 76L2], giving no surface relaxation.

ZnS (Zinc sulfide)

ZnS (110)
Surface analyzed by LEED intensity analysis [81D2, 84D3]. The scheme of the surface structure is shown in Fig. 3, see p. 54 , and the values are given in Table 12.

Table 12.

$\omega = 28°$, $\Delta_{1,\perp} = 0.59$ Å, $d_{12,\perp} = 1.40$ Å, $\Delta_{1,y} = 4.3$ Å, $d_{12,y} = 3.15$ Å, $\Delta_{2,\perp} = 0$ Å, $d_{23,\perp} = 1.91$ Å

ZnSe (Zinc selenide)

ZnSe (110)
Unreconstructed surface. The structure is determined in [77D4, 84D5] but the exact values of the interatomic distances remain uncertain.

ZnTe (Zinc telluride)

ZnTe (110)
1x1 surface observed in [78E1, 79D4]. The structure obtained by LEED intensity analysis is reported in Fig. 3,see p. 54, with the geometric values given in Table 13 [83D3]. Other calculations in [80M7, 82F4].

Table 13.

$\omega = 28°$, $\Delta_{1,\perp} = 0.71$ Å, $\Delta_{1,y} = 4.75$ Å, $\Delta_{2,\perp} = 0.05$ Å, $d_{12,\perp} = 1.57$ Å, $d_{12,y} = 3.58$ Å, $d_{23,\perp} = 2.18$ Å

ZnTe (111)
1x1 surface, which becomes 2x2 after heating to various temperatures [78E1].

ZnTe($\bar{1}\bar{1}\bar{1}$)
1x1 surface, which facets in <110> and <331> directions [78E1].

Zr (Zirconium)

Zr (0001)
Unreconstructed surface with hcp termination and $\Delta d_{12} = -1(2)\%$ [79M9].

6.1.3 Appendix

The results reported for the diffraction methods in section 6.1.1, and the results for the various elements and compounds given in section 6.1.2 include data which appeared in the literature before early 1990. The most relevant papers which appeared afterwards up to 1993 have been included in the Bibliography of section 6.1.4.
The reader should consult the appropriate papers for possible up-dating of theory and methods and of results concerning the elements and compounds, which are listed in the following table :

Theory and method developments:

LEED [90A, 90D2, 90P2, 90R2, 90P5, 90H3, 90H4, 90K2, 90L2, 90L6, 90M1, 90M3, 90P3, 90R1, 90S1, 90Z1, 90Z2, 91C2, 92D2, 92J, 92L1, 92O, 92W1, 92W2, 93F1, 93F2, 93h2, 93M2, 93M3, 93N1, 93O, 93R2, 93T, 93W2, 93Z1, 93Z2];
RHEED [90I, 90K3, 90K4, 90N1, 90P4, 90S2, 91S, 92M2, 92P1, 93Z2];
LEPD [93T];

Elements and compounds

Ag (0 0 1) [91L2];
Ag_{62} Au_{38} (1 1 1) [93D4];
Al (1 1 1) [90N2];
α-Al_2 O_3 $(1 1 \bar{2} 0)$ [92S];
Au (0 0 1) [90L4];
Au (4 3 0) [90B4];
Au (9 9 1) [91B5];
Be (0 0 0 1) [92D1];
Be $(1 1 \bar{2} 0)$ [92H1];
CdS $(1 1 \bar{2} 0)$ [90K1, 91K];
CdSe $(1 0 \bar{1} 0)$ [92H2];
CdSe $(1 1 \bar{2} 0)$ [92H2];
Co $(1 0 \bar{1} 0)$ [91O];
$CoSi_2$ (1 1 1) [93Z3];
Cu (0 0 1) [91H4, 92G1];
Cu (1 1 1) [91L1];
CuBr (0 0 1) [93D5];
CuCl (1 1 0) [92K1];
Cu_3Au (1 1 0) [90M2, 93H3];
Cu_{85} Pd_{15} (1 1 0) [90H5];
Er $(1 1 \bar{2} 0)$ [91B1];
Fe (1 1 0) [90O];
Fe_{85} Cr_{15} (0 0 1) [90U1];
Fe-3%Si-0.04%V-C (0 0 1) [90U2];
GaAs (0 0 1) [90L1, 91M, 92M1, 93M1];
GaAs (1 1 0) [92L3, 92Z2, 93C1];
GaP $(\bar{1}\,\bar{1}\,\bar{1})$ [90H6];
Gd (0 0 0 1) [92Q];
Gd $(1 1 \bar{2} 0)$ [93Q];
Ho $(1 1 \bar{2} 0)$ [91B1];
InP (1 1 0) [93C1];
Ir (1 1 0) [91H3];
$LiNbO_3$ (0 0 0 1) [93R1];
Mg (0 0 0 1) [93S];
MgO (0 0 1) [91B4];
Mo (0 0 1) [93D1];
Na_xNO_3 (0 0 1) [93P];
Ni (1 1 1) [93K];
Ni (1 1 1) 5° $(1 1 \bar{2})$ [91C1];
Ni (1 1 1) 5° $(1 \bar{1} 0)$ [91C1];
Ni_{20} Au_{80} (1 1 0) [93D2];
Pb (0 0 1) [90L5, 90P1, 93H1];
Pb (1 1 0) [90B5];
Pb (1 1 1) [91L3];
Pb (3 1 1) [91L4];
Pd (0 0 1) [90Q];
Pd (1 1 0) [93W1];
Pt (1 1 0) [90S3, 90Z3, 92K2, 92K3];
Pt (2 1 0) [91Z];
Pt_{80} Co_{20} (0 0 1) [90B1];
Pt_{25} Co_{75} (1 1 0) [92G2];
Pt_{80} Co_{20} (1 1 1) [92G2];
Pt_{25} Co_{75} (1 1 1) [93A1, 93G];
Pt_{80} Fe_{20} (1 1 0) [91B3];
Pt_{80} Fe_{20} (1 1 1) [90B2, 92G2];
Pt_{10} Ni_{90} (0 0 1) [90G2];
Pt_{80} Ni_{20} (1 1 1) [92G2];
Pt_3 S (0 0 1) [91H1, 92A1];
Pt_3 S (1 1 0) [90H1];
Pt_3 S (1 1 1) [91H1, 92A1, 93A2];
Pt_3 Sn (1 1 1) [93A1];
Pt_3 Ti (0 0 1) [92A2];
Pt_3 Ti (1 1 1) [93C2];
Rh (0 0 1) [93C1];
Rh (1 1 0) [93D3];
Rh (3 1 1) [90L3];
Si (1 1 1) [90G1];
Si (1 1 3) [90X];
β-SiC (0 0 1) [90H2, 91P, 92P2];
$(SnS)_{1.16}$ TaS_2 [92E];
$SrTiO_3$ (0 0 1) [90B3, 93H4]
Tb (0 0 0 1) [91Q];
Tb $(1 1 \bar{2} 0)$ [92L4];
TiO_2 (0 0 1) [92P2];
V (1 1 0) [93N2];
V (1 1 10 [93N2];
$VNi_{0.89}$ (0 0 1) [90G3];
$VC_{0.80}$ (1 1 1) [92R];
W (0 0 1) [93V];
Y $(1 1 \bar{2} 0)$ [91B2];

6.1.4 References for 6.1

14D Darwin, C.G.: Philos. Mag. **27** (1914) 675.
21D Davisson, C.J., Kunsman, C.H.: Science **54** (1921) 522.
27B Bethe, H.: Naturwissenschaften **15** (1927) 786.
27D1 Davisson, C.J., Germer, L.H.: Nature (London) **119** (1927) 558.
27D2 Davisson, C.J., Germer, L.H.: Phys. Rev. **29** (1927) 908.
27D3 Davisson, C.J., Germer, L.H.: Phys. Rev. **30** (1927) 705.
27T Thompson, C.P., Reid, A.: Nature (London) **119** (1927) 890.
28B1 Bethe, H.: Ann. Phys. (Leipzig) **87** (1928) 55.
28B2 Bethe, H.: Naturwissenschaften **16** (1928) 333.
33S Sproull, W.T.: Rev. Sci. Instrum. **4** (1933) 193.
34E Ehremberg, W.: Philos. Mag. **18** (1934) 878.
36F Farnsworth, H.E.: Phys. Rev. **49** (1936) 402.
37S Slater, J.C.: Phys. Rev. **51** (1937) 840.
51L Lax, M.: Rev. Mod. Phys. **23** (1951) 287.
51S Slater, J.C.: Phys. Rev. **81** (1951) 385.
53M Moruzzi, V.L., Feshbach, H.: Method of Theoretical Physics. New York: McGraw-Hill 1953, p. 1574.
54M Miyake, S., Kohra, K., Takagi, M.: Acta Crystallogr. **7** (1954) 393.
55F Farnsworth, H.E., Schlier, R.E., George, T.H., Burger, R.M.: J. Appl. Phys. **26** (1955) 252.
57C Cowley, J.M., Moodie, A.F.: Acta Crystallogr. **10** (1957) 609.
57Y Yoshioka, H.: J. Phys. Soc. Jpn. **12** (1957) 618.
58C Cowley, J.M., Moodie, A.F.: Proc. Phys. Soc. **71** (1958) 533.
58F Farnsworth, H.E., Schlier, R.E., George, T.H., Burger, R.M.: J. Appl. Phys. **29** (1958) 1150.
59C1 Cowley, J.M., Moodie, A.F.: Acta Crystallogr. **12** (1959) 353.
59C2 Cowley, J.M., Moodie, A.F.: Acta Crystallogr. **12** (1959) 360.
59D Dillon, J.A., Schlier, R.E., Farnsworth, H.E.: J. Appl. Phys. **30** (1959) 675.
59G George, T.H., Farnsworth, H.E., Schlier, R.E.: J. Chem. Phys. **31** (1959) 89.
59S Schlier, R.E., Farnsworth, H.E.: J. Chem. Phys. **30** (1959) 917.
60H1 Haneman, D.: Phys. Chem. Solids **14** (1960) 162.
60H2 Haneman, D.: Phys. Rev. **119** (1960) 567.
61G Germer, L.H., MacRae, A.U., Hartman, C.D.: J. Appl. Phys. **32** (1961) 2432.
61H Haneman, D.: Phys. Rev. **121** (1961) 1093.
61L1 Lander, J.J., Gobeli, G.W., Morrison, J.: J. Appl. Phys. **34** (1961) 2298.
61L2 Lehmpfuhl, G., Molière, K.: Z. Phys. **164** (1961) 389.
62K Kohra, K., Molière, K., Nakano, S., Ariyama, M.: J. Phys. Soc. Jpn. Suppl. B **2** (1962) 82.
62L1 Lander, J.J., Morrison, J.: J. Chem. Phys. **37** (1962) 729.
62L2 Lander, J.J., Morrison, J., Unterwald, F.: Rev. Sci. Instrum. **33** (1962) 782.
62M MacRae, A.U., Germer, L.H.: Phys. Rev. Lett. **8** (1962) 489.
63G Germer, L.H., MacRae, A.U.: Ann. N.Y. Acad. Sci. **101** (1963) 605.
63K Khan, I.H., Hobson, J.P., Amstrong, R.A.: Phys. Rev. **129** (1963) 1513.
63L Lander, J.J., Morrison, J.: J. Appl. Phys. **34** (1963) 3517.
64G Gafner, G.: Surf. Sci. **2** (1964) 534.
64H Hansen, N.R., Haneman, D.: Surf. Sci. **2** (1964) 566.
64L1 Lander, J.J., Morrison, J.: J. Appl. Phys. **35** (1964) 3593.
64L2 Lander, J.J., Morrison, J.: Surf. Sci. **2** (1964) 553.
64M1 MacRae, A.U.: Surf. Sci. **2** (1964) 522.
64M2 MacRae, A.U., Gobeli, G.W.: J. Appl. Phys. **35** (1964) 1629.
64M3 Marsh, J.B., Farnsworth, H.E.: Surf. Sci. **1** (1964) 3.
64M4 Mattheiss, L.F.: Phys. Rev. **133** (1964) A 1399.
64M5 McRae, E.G., Caldwell, C.W.: Surf. Sci. **2** (1964) 509.
64P1 Park, R.L., Farnsworth, H.E.: Rev. Sci. Instrum. **35** (1964) 1592.
64P2 Park, R.L., Farnsworth, H.E.: Surf. Sci. **2** (1964) 527.
64P3 Pignocco, A.J., Pellissier, G.E.: J. Electrochem. Soc. **162** (1964) 1188.
64T Tucker jr., C.W.: Surf. Sci. **2** (1964) 516.
64W Wood, E.A.: J. Appl. Phys. **35** (1964) 1306.
65A Aldag, J., Stern, R.M.: Phys. Rev. Lett. **14** (1965) 857.
65C Caldwell, C.W.: Rev. Sci. Instrum. **35** (1965) 1500.
65F Feinstein, L.G., Schoemaker, D.P.: Surf. Sci. **3** (1965) 294.
65K Kohn, W., Sham, L.J.: Phys. Rev. **140** (1965) A 1133.

65L Lander, J.J., Morrison, J.: J. Appl. Phys. **36** (1965) 1706.
65M Mitchell, D.F., Simmons, G.W., Lawless, K.R.: Appl. Phys. Lett. **7** (1965) 32.
65S Sewell, P.B., Cohen, M.: Appl. Phys. Lett. **7** (1965) 32.
65W Wakoh, S.: J. Phys. Soc. Jpn. **20** (1965) 1894.
66F Fedak, D.G., Gjostein, N.A.: Phys. Rev. Lett. **16** (1966) 171.
66H1 Haas, T.W.: Surf. Sci. **5** (1966) 345.
66H2 Haas, T.W., Jackson, A.G.: J. Chem. Phys. **44** (1966) 2921.
66H3 Hirabayashi, H., Takeishi, Y.: Surf. Sci. **4** (1966) 150.
66J Jones, E.R., McKinney, J.T., Webb, M.B.: Phys. Rev. **151** (1966) 476.
66L1 Lander, J.J., Morrison, J.: Surf. Sci. **4** (1966) 241.
66L2 Lyon, H.B., Somorjai, G.A.: J. Chem. Phys. **44** (1966) 3707.
66M1 MacRae, A.U.: Surf. Sci. **4** (1966) 247.
66M2 Marklund, I., Andersson, S.: Surf. Sci. **5** (1966) 197.
66M3 Miyake, S., Hayakawa, K.: J. Phys. Soc. Jpn. **21** (1966) 363.
66M4 McRae, E.G.: J. Chem. Phys. **45** (1966) 3258.
66P Park, R.L.: J. Appl. Phys. **37** (1966) 295.
66T Tucker jr., C.W.: J. Appl. Phys. **37** (1966) 3013.
67B1 Bauer, E.: Surf. Sci. **7** (1967) 351.
67B2 Broers, A.N.: J. Appl. Phys. **38** (1967) 1991.
67B3 Broers, A.N.: J. Appl. Phys. **38** (1967) 3040.
67C1 Charig, J.M.: Appl. Phys. Lett. **10** (1967) 139.
67C2 Chutjian, A.: Phys. Lett. A **24** (1967) 615.
67F1 Farnsworth, H.E., Hayek, K.: Surf. Sci. **8** (1967) 35.
67F2 Fedak, D.G., Gjostein, N.A.: Acta Metall. **15** (1967) 827.
67F3 Fedak, D.G., Gjostein, N.A.: Surf. Sci. **8** (1967) 77.
67F4 Fiermans, L., Vennik, J.: Phys. Lett. A **25** (1967) 687.
67G Gerlach, R.L., Rhodin, T.N.: Surf. Sci. **8** (1967) 1.
67H Haas, T., Jackson, A.G., Hooker, P.M.: J. Chem. Phys. **46** (1967) 3025.
67J1 Jona, F.: J. Chem. Phys. Solids **28** (1967) 2155.
67J2 Jona, F.: Surf. Sci. **8** (1967) 57.
67J3 Jona, F.: Surf. Sci. **8** (1967) 478.
67K1 Kambe, K.: Z. Naturforsch. A **22** (1967) 322.
67K2 Kambe, K.: Z. Naturforsch. A **22** (1967) 422.
67L1 Loucks, T.: Augmented Plane Wave Method. New York: Benjamin 1967.
67L2 Lyon, H.B., Somorjai, G.A.: J. Chem. Phys. **46** (1967) 2539.
67M1 Mattera, A.M., Goodman, R.M., Somorjai, G.A.: Surf. Sci. **7** (1967) 26.
67M2 McRae, E.G.: Surf. Sci. **8** (1967) 14.
67M3 McRae, E.G., Caldwell, C.W.: Surf. Sci. **7** (1967) 41.
67M4 Menzel, E., Schott, O.: Surf. Sci. **8** (1967) 217.
67P Palmberg, P.W., Rhodin, T.N.: Phys. Rev. **161** (1967) 586.
67S1 Siegel, B.M., Menadue, J.F.: Surf. Sci. **8** (1967) 206.
67S2 Somorjai, G.A.: Surf. Sci. **8** (1967) 98.
68A Andersson, S., Andersson, D., Marklund, I.: Surf. Sci. **12** (1968) 284.
68B1 Beeby, J.I.: J. Phys. C **1** (1968) 82.
68B2 Benson, G.C., Claxton, T.A.: J. Chem. Phys. **48** (1968) 1356.
68C1 Campbell, B.D., Ellis, W.P.: Surf. Sci. **10** (1968) 124.
68C2 Campbell, B.D., Farnsworth, H.E.: Surf. Sci. **10** (1968) 197.
68C3 Chang, C.C.: J. Appl. Phys. **38** (1968) 5570.
68C4 Corotte, C., Ducros, P., Mascall, A.: Compt. Rend. (Paris) B **267** (1968) 544.
68E1 Ellis, W.P.: J. Chem. Phys. **48** (1968) 5695.
68E2 Ellis, W.P., Schwoebel, R.L.: Surf. Sci. **11** (1968) 82.
68F1 Fedak, D.G., Fischer, T.E., Robertson, W.D.: J. Appl. Phys. **39** (1968) 5658.
68F2 Fiermans, L., Vennik, J.: Surf. Sci. **9** (1968) 187.
68G1 Gerstner, J., Cutler, P.H.: Surf. Sci. **9** (1968) 198.
68G2 Gervais, A., Stern, R.M., Menes, M.: Acta Crystallogr. A **24** (1968) 191.
68G3 Goodman, R.M., Farrell, H.H., Somorjai, G.A.: J. Chem. Phys. **48** (1968) 1046.
68H1 Hirabayashi, K.: J. Phys. Soc. Jpn. **25** (1968) 856.
68H2 Hirabayashi, K.: J. Phys. Soc. Jpn. **26** (1968) 846.
68I Ino, S., Ogawa, S.: J. Appl. Phys. **7** (1968) 308.
68J1 Jepsen, D.W., Marcus, P.M.: Phys. Rev. Lett. **20** (1968) 925.
68J2 Jona, F., Lever, R.F., Gunn, J.B.: Surf. Sci. **9** (1968) 468.
68K Kambe, K.: Z. Naturforsch. A **23** (1968) 1280.

68L Lehmpfuhl, G., Reissland, A.: Z. Phys. A **23** (1968) 544.
68M1 McRae, E.G.: Surf. Sci. **11** (1968) 429.
68M2 McRae, E.G.: Surf. Sci. **11** (1968) 479.
68M3 McRae, E.G.: Surf. Sci. **11** (1968) 492.
68O Ohtsuki, Y.H.: J. Phys. Soc. Jpn. **24** (1968) 1116.
68P1 Palmberg, P.W.: Surf. Sci. **11** (1968) 153.
68P2 Palmberg, P.W., Rhodin, T.N.: J. Chem. Phys. **49** (1968) 134.
68P3 Palmberg, P.W., De Wames, R.E., Vredevoe, L.A.: Phys. Rev. Lett. **21** (1968) 682.
68S Stern, R.M.: Trans. Am. Crystallogr. Assoc. **4** (1968) 14.
68T Tokutaka, H., Prutton, M.: Surf. Sci. **11** (1968) 216.
68V Vredevoe, L.A., De Wames, R.E.: Phys. Rev. **176** (1968) 684.
68W Watts, C.M.K.: J. Phys. C **1** (1968) 1237.
69A1 Aberdam, D., Bouchet, G., Ducros, P., Daval, J., Grunberg, G.: Surf. Sci. **14** (1969) 121.
69A2 Adams, R.D., in: The Structure and Chemistry of Solid Surfaces. Somorjai, G.A.(ed.), New York : Wiley 1969, p. 70.
69A3 Andersson, S.: Surf. Sci. **18** (1969) 325.
69B1 Broers, A.N.: J. Phys. E **2** (1969) 273.
69B2 Broers, A.N.: Rev. Sci. Instrum. **40** (1969) 1040.
69D1 Duke, C.B., Tucker jr., C.W.: Surf. Sci. **15** (1969) 231.
69D2 Duke, C.B., Tucker jr., C.W.: Surf. Sci. **15** (1969) 273.
69E Estrup, P.J., in: The Structure and Chemistry of Solid Surfaces. Somorjai, G.A. (ed.), New York: Wiley 1969, p. 19.
69F1 Farrell, H.H., Somorjai, G.A.: Phys. Rev. B **182** (1969) 751.
69F2 Feder, F., Meister, H.J.: Z. Phys. **229** (1969) 309.
69F3 Fedak, D.G., Florio, J.V., Robertson, W.D., in: The Structure and Chemistry of Solid Surfaces. Somorjai, G.A. (ed.), New York: Wiley 1969, p. 74.
69F4 Feinstein, L.G., Macrakis, M.S.: Surf. Sci. **18** (1969) 277.
69F5 Fiermans, L., Vennik, J.: Surf. Sci. **18** (1969) 317.
69F6 Friedman, S., Stern, R.M.: Surf. Sci. **17** (1969) 214.
69G1 Goodman, R.M.: Ph. D. Thesis, Univ. of California, 1969.
69G2 Grant, J.T.: Surf. Sci. **18** (1969) 228.
69G3 Grant, J.T., Haas, T.W.: Surf. Sci. **17** (1969) 484.
69H1 Heckingbottom, R.: Surf. Sci. **17** (1969) 394.
69H2 Henderson, R.C., Polito, W.J.: Surf. Sci. **14** (1969) 473.
69J Jackson, A.G.: Surf. Sci. **17** (1969) 482.
69K Kobayashi, H., Kato, S.: Surf. Sci. **12** (1969) 398.
69M1 McRae, E.G., Jennings, P.J.: Surf. Sci. **15** (1969) 345.
69M2 Marcus, P.M., Jepsen, D.W., Jona, F.: Surf. Sci. **17** (1969) 442.
69M3 Miyake, S., Hayakawa, K.: Acta Crystallogr. A **26** (1969) 60.
69M4 Morabito, J.M., Steiger, R.F., Somorjai, G.A.: Phys. Rev. B **179** (1969) 638.
69P1 Palmberg, P.W., De Wames, R.E., Vredevoe, L.A., Wolfram, T.: J. Appl. Phys. **40** (1969) 1158.
69P2 Palmberg, P.W., Rhodin, T.N., Plummer, E.W., in: The Structure and Chemistry of Solid Surfaces. Somorjai, G.A. (ed.), New York: Wiley 1969, p. 22.
69P3 Park, R.L., Houston, J.E.: Surf. Sci. **18** (1969) 213.
69P4 Pendry, J.B.: J. Phys. C **2** (1969) 1215.
69P5 Pendry, J.B.: J. Phys. C **2** (1969) 2273.
69P6 Pendry, J.B.: J. Phys. C **2** (1969) 2283.
69R Rovida, G., Torrini, M., Zanazzi, E.: Chem. Phys. Lett. **3** (1969) 201.
69S1 Seah, M.P.: Surf. Sci. **17** (1969) 181.
69S2 Stern, R.M., Gervais, A.: Surf. Sci. **17** (1969) 273.
69W Watts, C.M.K.: J. Phys. C **2** (1969) 966.
70A1 Andersson, S.: Surf. Sci. **19** (1970) 21.
70A2 Andersson, S., Kasemo, B.: Solid State Commun. **8** (1970) 1885.
70C Chung, M.F., Farnsworth, H.E.: Surf. Sci. **22** (1970) 93.
70D1 Dooley III, G.C., Haas, T.W.: Surf. Sci. **19** (1970) 1.
70D2 Duke, C.B., Anderson, J.R., Tucker jr., C.W.: Surf. Sci. **19** (1970) 117.
70D3 Duke, C.B., Laramore, G.E.: Phys. Rev. B **2** (1970) 4765.
70D4 Duke, C.B., Laramore, G.E.: Phys. Rev. B **2** (1970) 4783.
70F1 Feinstein, L.G.: Surf. Sci. **19** (1970) 366.
70F2 French, T.M., Somorjai, G.A.: J. Chem. Phys. **74** (1970) 2489.
70G1 Gallon, T.E., Kigginbotham, I.G., Prutton, M., Tokutaka, H.: Surf. Sci. **21** (1970) 224.
70G2 Grant, J.T., Haas, T.W.: Surf. Sci. **21** (1970) 76.

70H1 Henzler, M.: Surf. Sci. **19** (1970) 159.
70H2 Houston, J.E., Park, R.L.: Surf. Sci. **21** (1970) 209.
70J1 Jennings, P.J.: Surf. Sci. **20** (1970) 18.
70J2 Jennings, P.J., McRae, E.G.: Surf. Sci. **23** (1970) 363.
70K Kerre, E., Phariseau, P.: Phys. Status Solidi (b) **41** (1970) 759.
70L Liebermann, L.N., Clinton, J., Edwards, D.M., Mathon, J.: Phys. Rev. Lett. **25** (1970) 232.
70R1 Reid, R.J.: Phys. Status Solidi (a) **2** (1970) K109.
70R2 Rovida, G., Torrini, M., Zanazzi, E.: Accad. Naz. Lincei Cl. Sci. Fis. Mat. Nat. Atti Rend. **48** (1970) 242.
70S1 Strozier jr., J.A., Jones, R.O.: Phys. Rev. Lett. **23** (1970) 1163.
70S2 Strozier jr., J.A., Jones, R.O.: Phys. Rev. Lett. **25** (1970) 516.
70T1 Tabor, D., Wilson, J.: Surf. Sci. **20** (1970) 203.
70T2 Theeten, J.B., Domange, J.L., Bonnerot, J.: Solid State Commun. **8** (1970) 643.
70T3 Tucker jr., C.W., Duke, C.B.: Surf. Sci. **23** (1970) 411.
70W1 Woodruff, D.P., Holland, B.W.: Phys. Lett. A **31** (1970) 207.
70W2 Woodruff, D.P., Seah, M.P.: Phys. Status Solidi (a) **1** (1970) 429.
71A1 Aberdam, D., Bouchet, G., Ducros, P.: Surf. Sci. **27** (1971) 559.
71A2 Aberdam, D., Gaubert, G.: Surf. Sci. **27** (1971) 571.
71A3 Albinet, A., Biberian, J.P., Bienfait, M.: Phys. Rev. B **3** (1971) 2015.
71A4 Andersson, S., Kasemo, B.: Phys. Status Solidi (a) **4** (1971) 373.
71A5 Andersson, S., Kasemo, B.: Surf. Sci. **25** (1971) 273.
71C1 Capart, G.: Surf. Sci. **26** (1971) 429.
71C2 Chang, C.C.: J. Vac. Sci. Technol. **8** (1971) 500.
71C3 Chen, J.M.: Surf. Sci. **25** (1971) 305.
71D1 De Bersuder, L., Hoffstein, V., Lauzier, J.: Surf. Sci. **27** (1971) 338.
71D2 Duke, C.B., Howsmon, A.J., Laramore, G.E.: J. Vac. Sci. Technol. **8** (1971) 10.
71D3 Duke, C.B., Laramore, G.E.: Phys. Rev. B **3** (1971) 3183.
71D4 Duke, C.B., Laramore, G.E., Holland, B.W., Gibbons, A.M.: Surf. Sci. **27** (1971) 523.
71D5 Duke, C.B., Tucker jr., C.W.: Phys. Rev. B **3** (1971) 3561.
71F1 Farnsworth, H.E.: Phys. Lett. A **36** (1971) 56.
71F2 Feder, R.: Phys. Status Solidi (b) **46** (1971) K31.
71F3 Fiermans, L., Arijs, E., Vennik, J., Maenhont-Van der Vorst, W.: Phys. Status Solidi (a) **6** (1971) K1.
71F4 Fiermans, L., Vennik, J.: Surf. Sci. **24** (1971) 541.
71G Grant, J.T., Haas, T.W.: Surf. Sci. **26** (1971) 669.
71H1 Hirabayashi, K.: J. Phys. Soc. Jpn. **30** (1971) 211.
71H2 Hirabayashi, K.: Surf. Sci. **28** (1971) 621.
71H3 Hoffstein, V., Boudreaux, D.S.: Phys. Rev. B **3** (1971) 2447.
71H4 Holland, B.H.: Surf. Sci. **28** (1971) 258.
71H5 Holland, B.W., Hannum, R.W., Gibbons, A.M.: Surf. Sci. **25** (1971) 561.
71H6 Holland, B.W., Hannum, R.W., Gibbons, A.M., Woodruff, D.P.: Surf. Sci. **25** (1971) 576.
71H7 Houston, J.E., Park, R.L.: Surf. Sci. **26** (1971) 269.
71I Ignatiev, A., Rhodin, T.N., Tong, S.Y., Lundqvist, B.I., Pendry, J.B.: Solid State Commun. **9** (1971) 1851.
71J1 Jennings, P.J.: Surf. Sci. **25** (1971) 513.
71J2 Jennings, P.J.: Surf. Sci. **26** (1971) 509.
71J3 Jennings, P.J.: Surf. Sci. **27** (1971) 221.
71J4 Jepsen, D.W., Marcus, P.M., in: Computational Methods in Band Theory. Marcus, P.M., Janak, J.F., Williams, A.R. (eds.), New York: Plenum Press 1971, p. 416.
71J5 Jepsen, D.W., Marcus, P.M., Jona, F.: Phys. Rev. Lett. **26** (1971) 1365.
71K1 Kaplan, R., Somorjai, G.A.: Solid State Commun. **9** (1971) 505.
71K2 Kato, S., Kobayashi, H.: Surf. Sci. **27** (1971) 625.
71L1 Lagally, M.G., Ngoc, T.C., Webb, M.B.: Phys. Rev. Lett. **25** (1971) 1557.
71L2 Lagally, M.G., Ngoc, T.C., Webb, M.B.: Surf. Sci. **25** (1971) 444.
71L3 Laramore, G.E., Duke, C.B.: Phys. Rev. B **3** (1971) 3198.
71L4 Laramore, G.E., Duke, C.B., Bagchi, A., Kunz, A.B.: Phys. Rev. B **4** (1971) 2058.
71L5 Lauzier, J., De Bersuder, L., Hoffstein, V.: Phys. Rev. Lett. **27** (1971) 735.
71M McRae, E.G.: Surf. Sci. **25** (1971) 491.
71N Nesterenko, B.A., Borodkin, A.D.: Sov. Phys. Solid State (English Transl.) **12** (1971) 1621.
71P1 Park, R.L., Houston, J.E., Schreiner, D.G.: Rev. Sci. Instrum. **42** (1971) 60.
71P2 Pendry, J.B.: J. Phys. C **4** (1971) 2501.
71P3 Pendry, J.B.: J. Phys. C **4** (1971) 2515.

71P4 Pendry, J.B.: J. Phys. C **4** (1971) 3095.
71P5 Pendry, J.B.: Phys. Rev. Lett. **27** (1971) 356.
71P6 Pendry, J.B.: Phys. Rev. Lett. **27** (1971) 856.
71P7 Perdereau, J., Rhead, G.E.: Surf. Sci. **24** (1971) 555.
71R1 Reid, R.J.: Phys. Status Solidi (a) **4** (1971) K211.
71R2 Rovida, G., Torrini, M., Zanazzi, E.: Nuovo Cimento B **4** (1971) 97.
71S1 Strongin, M., Dickey, J.M., Farrell, H.H., Arns, T.F., Hrabak, G.: Rev. Sci. Instrum. **42** (1971) 311.
71S2 Strozier jr., J.A., Jones, R.O.: Phys. Rev. B **3** (1971) 3228.
71S3 Shen, A.P.: Phys. Rev. B **4** (1971) 382.
71T1 Tabor, D., Wilson, J.M.: J. Vac. Sci. Technol. **9** (1971) 2.
71T2 Tabor, D., Wilson, J.M., Bastow, T.J.: Surf. Sci. **26** (1971) 471.
71T3 Theeten, J.B., Bonnerot, J., Domange, J.L., Hurault, J.P.: Solid State Commun. **9** (1971) 1121.
71T4 Tong, S.Y.: Solid State Commun. **16** (1971) 75.
71T5 Tong, S.Y., Rhodin, T.N.: Phys. Rev. Lett. **26** (1971) 711.
71T6 Tucker jr., C.W., Duke, C.B.: Surf. Sci. **24** (1971) 31.
71W1 Wei, P.S.P.: Surf. Sci. **24** (1971) 219.
71W2 Williams, R.H., McMevoy, A.J.: J. Phys. D **4** (1971) 456.
71W3 Wilson, J.M., Bastow, T.J.: Surf. Sci. **26** (1971) 461.
72A1 Aberdam, A., Baudoing, R.: Solid State Commun. **10** (1972) 1199.
72A2 Augustus, P.D., Fairholme, R.J., Gill, G.P., Ogden, R., Maguire, H.G.: Phys. Status Solidi (a) **10** (1972) K1.
72B1 Baidyaroy, S., Bottoms, W.R., Mark, P.: Surf. Sci. **29** (1972) 165.
72B2 Baker, J.M., Blakely, J.M.: Surf. Sci. **32** (1972) 45.
72C1 Clark, T.A., Mason, R., Tescari, M.: Proc. R. Soc. London A **331** (1972) 321.
72C2 Clark, T.A., Mason, R., Tescari, M.: Surf. Sci. **30** (1972) 553.
72C3 Colella, R.: Acta Crystallogr. A **28** (1972) 11.
72C4 Colella, R., Menadue, J.F.: Acta Crystallogr. A **28** (1972) 16.
72D1 Demuth, J., Tong, S.Y., Webb, M.B.: J. Vac. Sci. Technol. **9** (1972) 639.
72D2 Dennis, P.N.J., Dobson, P.J.: Surf. Sci. **33** (1972) 187.
72D3 Duke, C.B., Smith, D.L.: Phys. Rev. B **5** (1972) 4730.
72D4 Duke, C.B., Smith, D.L., Holland, B.W.: Phys. Rev. B **5** (1972) 3358.
72F1 Feder, R.: Phys. Status Solidi (b) **49** (1972) 699.
72F2 Fingerland, A.: Surf. Sci. **32** (1972) 639.
72G1 Gersten, J.I., McRae, E.G.: Surf. Sci. **29** (1972) 483.
72G2 Grønlund, F., Højlund Nielsen, P.E.: J. Appl. Phys. **43** (1972) 3919.
72G3 Groupe d'Etudes des Surfaces: Surf. Sci. **32** (1972) 297.
72H Holcombe jr., C.E., Powell, G.L., Clausing, R.E.: Surf. Sci. **30** (1972) 561.
72J1 Jennings, P.J., Sim, B.K.: Surf. Sci. **33** (1972) 1.
72J2 Jepsen, D.W., Marcus, P.M., Jona, F.: Phys. Rev. B **5** (1972) 3933.
72J3 Jepsen, D.W., Marcus, P.M., Jona, F.: Phys. Rev. B **6** (1972) 3684.
72J4 Jona, F., Strozier jr., J.A., Kumar, J., Jones, R.O.: Phys. Rev. B **6** (1972) 407.
72J5 Jona, F., Strozier jr., J.A., Wong, C.: Surf. Sci. **30** (1972) 225.
72L1 Lagally, M.G., Ngoc, T.C., Webb, M.B.: J. Vac. Sci. Technol. **9** (1972) 645.
72L2 Lang, B., Joyner, R.W., Somorjai, G.A.: Surf. Sci. **30** (1972) 440.
72L3 Laramore, G.E.: Phys. Rev. B **6** (1972) 1097.
72L4 Laramore, G.E.: Phys. Rev. B **6** (1972) 2950.
72L5 Laramore, G.E., Duke, C.B.: Phys. Rev. B **5** (1972) 267.
72L6 Laramore, G.E., Duke, C.B.: Phys. Rev. B **6** (1972) 1609.
72L7 Laramore, G.E., Duke, C.B., Bagchi, A., Kunz, A.B.: Phys. Rev. B **6** (1972) 3163.
72L8 Laramore, G.E., Switendick, A.C.: Phys. Rev. B **7** (1972) 3615.
72L9 Levine, J.D., Willis, A., Bottoms, W.R., Mark, P.: Surf. Sci. **29** (1972) 144.
72L10 Lynch, D.F., Moodie, A.F.: Surf. Sci. **32** (1972) 422.
72M1 McRae, E.G., Wheatley, G.H.: Surf. Sci. **29** (1972) 342.
72M2 Menadue, J.F.: Acta Crystallogr. A **28** (1972) 1.
72M3 Moon, A.R.: Z. Naturforsch. A **27** (1972) 390.
72M4 Moon, A.R., Cowley, J.M.: J. Vac. Sci. Technol. **9** (1972) 649.
72P Pendry, J.B.: J. Phys. C **5** (1972) 2567.
72Q Quinto, D.T., Holland, B.W., Robertson, D.W.: Surf. Sci. **32** (1972) 139.
72R1 Reid, R.J.: Surf. Sci. **29** (1972) 603.
72R2 Reid, R.J.: Surf. Sci. **29** (1972) 623.
72S1 Sinharoy, S., Stern, R.M., Goldstone, P.D.: Surf. Sci. **30** (1972) 207.

72S2 Stern, R.M., Sinharoy, S.: Surf. Sci. **33** (1972) 131.
72T1 Tait, R.H., Tong, S.Y., Rhodin, T.N.: Phys. Rev. Lett. **28** (1972) 553.
72T2 Taub, H.: Surf. Sci. **30** (1972) 161.
72T3 Torrini, M., Zanazzi, E.: J. Chem. Phys. **57** (1972) 2380.
72T4 Tucker jr., C.W., Duke, C.B.: Surf. Sci. **29** (1972) 237.
72V Van Hove, H., Leysen, R.: Phys. Status Solidi (a) **9** (1972) 361.
72W Wei, P.S., Smith, A.W.: J. Vac. Sci. Technol. **9** (1972) 1209.
72Z Zenher, D.M., Farnsworth, H.E.: Surf. Sci. **30** (1972) 335.
73A Aberdam, A., Baudoing, R., Gaubert, C.: Surf. Sci. **52** (1973) 125.
73B1 Bartos, I.: Surf. Sci. **34** (1973) 791.
73B2 Bedell, L.R., Farnsworth, H.E.: Surf. Sci. **41** (1973) 165.
73B3 Burkstrand, J.M.: Rev. Sci. Instrum. **44** (1973) 774.
73C1 Clark, T.A., Mason, R., Tescari, M.: Surf. Sci. **40** (1973) 15.
73C2 Christmann, K., Ertl, G.: Z. Naturforsch. A **28** (1973) 1144.
73C3 Christmann, K., Ertl, G., Schober, O.: Surf. Sci. **40** (1973) 61.
73C4 Colella, R., Batterman, B.W., Menadue, J.F.: Acta Crystallogr. A **29** (1973) 151.
73D1 Demuth, J.E., Jepsen, D.W., Marcus, P.M.: Solid State Commun. **13** (1973) 311.
73D2 Dennis, P.N.J., Dobson, P.J.: Surf. Sci. **36** (1973) 781.
73F Feder, R.: Phys. Status Solidi (b) **56** (1973) K43.
73G1 Glachant, A., Derrien, J., Bienfait, M.: Surf. Sci. **40** (1973) 683.
73G2 Grymonprez, G., Fiermans, L., Vennik, J.: Surf. Sci. **36** (1973) 370.
73H1 Hagstrom, H.D., Becker, G.E.: Phys. Rev. B **8** (1973) 1580.
73H2 Hoffstein, V., Albinet, G.: Surf. Sci. **38** (1973) 506.
73H3 Højlund Nielsen, P.E.: Surf. Sci. **35** (1973) 194.
73H4 Højlund Nielsen, P.E.: Surf. Sci. **36** (1973) 778.
73H5 Holland, B.W., Woodruff, D.P.: Surf. Sci. **36** (1973) 488.
73H6 Hu, E.W., Goodman, R.M., Jona, F.: Phys. Rev. B **8** (1973) 5519.
73H7 Houston, J.E, Laramore, G.E, Park, R.L.: Surf. Sci. **34** (1973) 477.
73J1 Jennings, P.J.: Surf. Sci. **34** (1973) 668.
73J2 Jepsen, D.W., Marcus, P.M., Jona, F.: Phys. Rev. B **8** (1973) 5523.
73J3 Jepsen, D.W., Marcus, P.M., Jona, F.: Surf. Sci. **38** (1973) 525.
73K1 Kosterlitz, M., Domange, J.L.: Solid State Commun. **13** (1973) 241.
73K2 Kreutz, E.W., Sotnik, N., Waidelich, W.: Surf. Sci. **38** (1973) 210.
73L1 Laramore, G.E.: Phys. Rev. B **8** (1973) 515.
73L2 Laramore, G.E., Houston, J.E., Park, R.L.: J. Vac. Sci. Technol. **10** (1973) 196.
73L3 Laramore, G.E., Switendick, A.C.: Phys. Rev. B **7** (1973) 3615.
73M1 Marcus, P.M., Demuth, J.E., Jepsen, D.W.: Surf. Sci. **53** (1973) 501.
73M2 Martin, M.R., Somorjai, G.A.: Phys. Rev. B **7** (1973) 3607.
73N Ngoc, T.C., Lagally, M.G., Webb, M.B.: Surf. Sci. **35** (1973) 117.
73Q Quinto, D.T., Robertson, W.D.: Surf. Sci. **34** (1973) 501.
73R1 Rhead, G.C.: J. Phys. F **3** (1973) L53.
73R2 Rovida, G., Maglietta, M.: J. Appl. Phys. **44** (1973) 3801.
73S Sundaran, V.S., Farrell, B., Alben, R.S., Robertson, W.D.: Phys. Rev. Lett. **31** (1973) 1136.
73T1 Theeten, J.B., Domange, J.L., Hurault, J.P.: Surf. Sci. **35** (1973) 145.
73T2 Tong, S.Y., Kesmodel, L.L.: Phys. Rev. B **8** (1973) 3753.
73T3 Tong, S.Y., Rhodin, T.N., Tait, R.H.: Phys. Rev. B **8** (1973) 421.
73T4 Tong, S.Y., Rhodin, T.N., Tait, R.H.: Phys. Rev. B **8** (1973) 430.
73T5 Tong, S.Y., Rhodin, T.N., Tait, R.H.: Phys. Rev. B **8** (1973) 5361.
73T6 Tong, S.Y., Rhodin, T.N., Tait, R.H.: Surf. Sci. **34** (1973) 457.
73T7 Trullinger, S.E., Mills, D.L.: Solid State Commun. **12** (1973) 819.
73W1 Webb, M.B., Lagally, M.G.: Solid State Phys. **28** (1973) 301.
73W2 Woodruff, D.P., McDonnell, L.: Surf. Sci. **40** (1973) 200.
73Z Zimmer, R.S.: Surf. Sci. **36** (1973) 430.
74A1 Aberdam, A., Baudoing, R., De Bersuder, L.: Rev. Sci. Instrum. **45** (1974) 1573.
74A2 Adams, D.L., Landman, U.: Phys. Rev. Lett.. **33** (1974) 585.
74A3 Albinet, G., Baudoing, R., Aberdam, D., Hoffstein, V.: Surf. Sci. **42** (1974) 467.
74B1 Bas, E., Bäninger, U., Mülethaler, H.: Jpn. J. Ápl. Phys., Suppl. **2**, Pt. 2 (1974) 671.
74B2 Buchholtz, J.C., Lagally, M.G., Webb, M.B.: Surf. Sci. **41** (1974) 248.
74B3 Burkstrand, J.M., Kleiman, G.G., Arlinghaus, F.J.: Surf. Sci. **46** (1974) 43.
74C1 Chang, S.C., Mark, P.: Surf. Sci. **46** (1974) 293.
74C2 Chang, S.C., Mark, P.: Surf. Sci. **45** (1974) 721.
74D1 De Bersuder, L.: Rev. Sci. Instrum. **45** (1974) 1569.

74D2 Demuth, J., Rhodin, T.N.: Surf. Sci. **42** (1974) 261.
74D3 Duke, C.B., Liebsch, A.: Phys. Rev. B **9** (1974) 1126.
74D4 Duke, C.B., Liebsch, A.: Phys. Rev. B **9** (1974) 1150.
74E1 Ellis, W.P.: Surf. Sci. **41** (1974) 125.
74E2 Ellis, W.P.: Surf. Sci. **45** (1974) 569.
74F Feder, R.: Phys. Status Solidi (b) **62** (1974) 135.
74H1 Harris, J., Jones, R.O.: J. Phys. C **7** (1974) 3751.
74H2 Hayakawa, K., Miyake, S.: Acta Crystallogr. A **30** (1974) 374.
74H3 Hoffstein, V.: Comput. Phys. Commun. **7** (1974) 50.
74H4 Hoffstein, V., Albinet, G.: Surf. Sci. **43** (1974) 385.
74I Ignatiev, A., Jona, F.: Surf. Sci. **42** (1974) 605.
74J1 Jennings, P.J.: Jpn. J. Appl. Phys., Suppl. **2**, Pt 2 (1974) 661.
74J2 Jennings, P.J.: Surf. Sci. **41** (1974) 67.
74J3 Jennings, P.J., Read, M.N.: Surf. Sci. **41** (1974) 113.
74J4 Jepsen, D.W., Marcus, P.M., Jona, F.: Surf. Sci. **41** (1974) 223.
74L1 Landman, U., Adams, D.L.: J. Vac. Sci. Technol. **11** (1974) 195.
74L2 Laramore, G.E.: Phys. Rev. B **9** (1974) 1204.
74L3 Laramore, G.E., Duke, C.B., Lipari, N.O.: Phys. Rev. B **10** (1974) 2246.
74L4 Legg, K.O., Prutton, M., Kinniburgh, C.: J. Phys. C **7** (1974) 4236.
74L5 Liebsch, A.: J. Vac. Sci. Technol. **11** (1974) 201.
74M1 Margoninski, Y., Kalman, Z.H.: Phys. Lett. A **47** (1974) 201.
74M2 McRae, E.G.: Surf. Sci. **42** (1974) 413.
74M3 McRae, E.G.: Surf. Sci. **42** (1974) 427.
74P Pendry, J.B.: Low Energy Electron Diffraction. London and New York: Academic Press 1974.
74R1 Rundgren, J., Salwèn, A.: Comput. Phys. Commun. **7** (1974) 369.
74R2 Rundgren, J., Salwèn, A.: J. Phys. C **7** (1974) 4247.
74S Sundaran, V.S., Alben, R.S., Robertson, W.D.: Surf. Sci. **46** (1974) 653.
74T Tester, J.W., Herrick, C.C., Ellis, W.P.: Surf. Sci. **41** (1974) 619.
74W1 Woodruff, D.P., Mitchell, K.A.R., McDonnell, L.: Surf. Sci. **42** (1974) 355.
74W2 Woodruff, D.P., McDonnell, L.: Surf. Sci. **42** (1974) 616.
74Z1 Zehner, D.M., Noggle, T.S., Jenkins, L.H.: Surf. Sci. **41** (1974) 601.
74Z2 Zimmer, R.S., Robertson, W.D.: Surf. Sci. **43** (1974) 61.
75A1 Aberdam, A., Baudoing, R., Gaubert, C.: Surf. Sci. **52** (1975) 125.
75A2 Adams, D.L., Landman, U., Hamilton, J.C.: J. Vac. Sci. Technol. **12** (1975) 260.
75A3 Amstrong, R.A.: Surf. Sci. **47** (1975) 666.
75B1 Besock, K., Wagner, H.: Surf. Sci. **52** (1975) 653.
75B2 Buchholtz, J.C., Wang, G.-C., Lagally, M.G.: Surf. Sci. **49** (1975) 508.
75C Cheng, D.J., Wallis, R.F., Megerle, C., Somorjai, G.A.: Phys. Rev. B **12** (1975) 5599.
75D1 David, J.M., Fain jr., S.C.: Surf. Sci. **52** (1975) 670.
75D2 Demuth, J., Marcus, P.M., Jepsen, D.W.: Phys. Rev. B **11** (1975) 1460.
75D3 Derrien, J., Arnaud D'Avitaya, F., Glachant, A.: Surf. Sci. **47** (1975) 162.
75D4 Duke, C.B., Lubinsky, A.R.: Surf. Sci. **50** (1975) 605.
75E1 Echenique, P.M., Pendry, J.B.: J. Phys. C **8** (1975) 2936.
75E2 Ellis, W.P.: Surf. Sci. **50** (1975) 178.
75F Feder, R.: Surf. Sci. **51** (1975) 297.
75G1 Goldstein, B.: Surf. Sci. **47** (1975) 143.
75G2 Groupe d'Etude des Surfaces: Surf. Sci. **48** (1975) 497.
75G3 Groupe d'Etude des Surfaces: Surf. Sci. **48** (1975) 509.
75H Henrich, V.E.: Surf. Sci. **49** (1975) 675.
75I Ignatiev, A., Jona, F., Shih, H.D., Jepsen, D.W., Marcus, P.M.: Phys. Rev. B **11** (1975) 4787.
75J1 Jennings, P.J., Read, M.N.: J. Phys. C **8** (1975) L285.
75J2 Jona, F.: Faraday Discuss. Chem. Soc. **60** (1975) 210.
75K1 Khan, I.H.: Surf. Sci. **48** (1975) 537.
75K2 Kesmodel, L.L., Somorjai, G.A.: Phys. Rev. B **11** (1975) 630.
75K3 Kinniburgh, C.G.: J. Phys. C **8** (1975) 2382.
75K4 Kleiman, G.G., Burkstrand, J.M.: Surf. Sci. **50** (1975) 493.
75L1 Lagally, M.G., Buchholtz, J.C., Wang, G.-C.: J. Vac. Sci. Technol. **12** (1975) 213.
75L2 Landman, U., Adams, D.L.: Surf. Sci. **51** (1975) 149.
75L3 Lee, B.W., Mark, P.: Surf. Sci. **52** (1975) 285.
75M1 Margoninski, Y., Kirby, R.E.: J. Phys. C **8** (1975) 1516.
75M2 Mark, P., Chang, S.C., Creighton, W.F., Lee, B.W.: Crit. Rev. Solid State Sci. **5** (1975) 189.
75M3 McDonnell, L., Woodruff, D.P., Mitchell, K.A.R.: Surf. Sci. **45** (1975) 1.

75M4 McRae, E.G.: Surf. Sci. **47** (1975) 167.
75M5 Mróz, S.: Surf. Sci. **51** (1975) 365.
75N Netzer, F.P., Prutton, M.: J. Phys. C **8** (1975) 2401.
75O O'Neill, M.R., Kalisvaart, M., Dunning, F.B., Walters, G.K.: Phys. Rev. Lett. **34** (1975) 1167.
75P Pendry, J.B., Gard, P.: J. Phys. C **8** (1975) 2048.
75R1 Roy, D., Delage, A., Carette, J.D.: J. Phys. E **8** (1975) 109.
75R2 Rundgren, J., Salwèn, A.: Comput. Phys. Commun. **9** (1975) 312.
75S1 Salwèn, A., Rundgren, J.: Surf. Sci. **53** (1975) 523.
75S2 Stair, P.C., Kaninska, T.J., Kesmodel, L.L., Somorjai, G.A.: Phys. Rev. B **11** (1975) 623.
75S3 Strozier jr., J.A., Jepsen, D.W., Jona, F., in: Surface Physics of Crystalline Materials. Blakely, J.M. (ed.), New York: Academic Press 1975, p. 1.
75T1 Tong, S.Y.: Progr. Surf. Sci. **7** (1975) 1.
75T2 Tong, S.Y., Van Hove, M.A.: Phys. Rev. B **16** (1977) 1459.
75U Unertl, W.N., Thapliyal, H.V.: J. Vac. Sci. Technol. **12** (1975) 263.
75V1 Van Bommel, A.J., Crombeen, J.E., Van Tooren, A.: Surf. Sci. **48** (1975) 463.
75V2 Van Hove, M.A.: Surf. Sci. **49** (1975) 181.
75V3 Van Hove, M.A.: Surf. Sci. **48** (1975) 406.
75V4 Van Hove, M.A., Pendry, J.B.: J. Phys. C **8** (1975) 1362.
75W Williams, R.H., McGovern, I.T.: Surf. Sci. **51** (1975) 14.
75Z1 Zehner, D.M., Appleton, B.R., Noggle, T.S., Miller, J.W., Barret, J.H., Jenkins, L.H., Schow III, O.E.: J. Vac. Sci. Technol. **12** (1975) 454.
75Z2 Zimmer, R.S., Holland, B.W.: J. Phys. C **8** (1975) 2395.
75Z3 Zimmer, R.S., Holland, B.W.: Surf. Sci. **47** (1975) 717.
76A1 Aberdam, A., Baudoing, R., Gaubert, C., McRae, E.G.: Surf. Sci. **57** (1976) 715.
76A2 Auboiroux, M., Marchand, D., Tougne, P., Legrand, A.P.: Surf. Sci. **59** (1976) 218.
76B1 Bandet, J., Malavaud, A., Oustry, A.: Surf. Sci. **60** (1976) 147.
76B2 Berning, G.L.P.: Surf. Sci. **61** (1976) 673.
76C1 Cohen, P.I., Unguris, J., Webb, M.B.: Surf. Sci. **58** (1976) 429.
76C2 Crider, C.A., Cisneros, G., Mark, P., Levine, J.D.: J. Vac. Sci. Technol. **13** (1976) 1202.
76D1 Demongeat, C., Mills, D.L.: Phys. Rev. B **14** (1976) 4997.
76D2 Duke, C.B., Lubinsky, A.R., Lee, B.W., Mark, P.: J. Vac. Sci. Technol. **13** (1976) 761.
76E Echenique, P.M.: J. Phys. C **9** (1976) 3193.
76F1 Feder, R.: Phys. Rev. Lett. **36** (1976) 598.
76F2 Feder, R., Gafner, G.: Surf. Sci. **57** (1976) 45.
76F3 Feder, R., Jennings, P.J., Jones, R.O.: Surf. Sci. **61** (1976) 307.
76F4 Frost, D.C., Mitchell, K.A.R., Sheferd, F.R., Watson, P.R.: J. Vac. Sci. Technol. **13** (1976) 1196.
76G1 Garcia, N., Solara, J.: Physica B **81** (1976) 296.
76G2 Göpel, W., Neuenfeldt, G.: Surf. Sci. **55** (1976) 362.
76H1 Hamauzu, Y.: Phys. Lett. A **56** (1976) 417.
76H2 Heilmann, P., Lang, E., Heinz, K., Müller, K.:Appl. Phys. **9** (1976) 247.
76H3 Henrich, V.E.: Surf. Sci. **57** (1976) 385.
76H4 Henrich, V.E., Dresselhaus, G., Zeiger, H.J.: Phys. Rev. Lett. **36** (1976) 1335.
76K1 Kessler, J.: Polarized Electrons. Berlin: Springer-Verlag 1976.
76K2 Kinniburgh, C.G.: J. Phys. C **9** (1976) 2695.
76L1 Leygraf, G., Hultquist, G.: Surf. Sci. **61** (1976) 60.
76L2 Lubinsky, A.R., Duke, C.B., Chang, S.C., Lee, B.W., Mark, P.: J. Vac. Sci. Technol. **13** (1976) 189.
76L3 Lubinsky, A.R., Duke, C.B., Lee, P.W., Mark, P.: Phys. Rev. Lett. **36** (1976) 1058.
76M1 Masud, N., Pendry, J.B.: J. Phys. C **9** (1976) 1833.
76M2 McRae, E.G.: Surf. Sci. **57** (1976) 761.
76M3 McRae, E.G.: Surf. Sci. **57** (1976) 766.
76M4 McRae, E.G., Caldwell, C.W.: Surf. Sci. **57** (1976) 77.
76P Payling, R., Ramsey, J.A.: Surf. Sci. **58** (1976) 375.
76R1 Rierce, D.T., Meier, F.: Phys. Rev. B **13** (1976) 5484.
76R2 Rundgren, J., Salwèn, A.: J. Phys. C **9** (1976) 3701.
76R3 Russell, G.J.: Surf. Sci. **55** (1976) 380.
76S1 Shih, H.D., Jona, F., Jepsen, D.W., Marcus, P.M.: Commun. Phys. **1** (1976) 25.
76S2 Shih, H.D., Jona, F., Jepsen, D.W., Marcus, P.M.: J. Phys. C **9** (1976) 1405.
76S3 Shih, H.D., Jona, F., Jepsen, D.W., Marcus, P.M.: Phys. Rev. Lett. **24** (1976) 1622.
76S4 Sundaran, V.S., Robertson, W.D.: Surf. Sci. **55** (1976) 324.
76S5 Szalkowski, F.J., Somorjai, G.A.: J. Chem. Phys. **64** (1976) 2985.

76T Tong, S.Y., Pendry, J.B., Kesmodel, L.L.: Surf. Sci. **54** (1976) 21.
76U Unertl, W.N., Webb, M.B.: Surf. Sci. **59** (1976) 373.
76V1 Van Hove, M.A., Tong, S.Y.: Surf. Sci. **54** (1976) 91.
76V2 Van Hove, M.A., Tong, S.Y., Stoner, N.: Surf. Sci. **54** (1976) 259.
77A1 Adams, D.L., Landman, U.: Phys. Rev. B **15** (1977) 3775.
77A2 Adnot, A., Carette, J.-D.: Phys. Rev. Lett. **38** (1977) 1084.
77A3 Adnot, A., Carette, J.-D.: Phys. Rev. B **16** (1977) 4703.
77B1 Bastaz, R., Colmenares, C.A., Smith, R.L., Somorjai, G.A.: Surf. Sci. **67** (1977) 45.
77B2 Berning, G.L.P.: J. Vac. Sci. Technol. **14** (1977) 485.
77B3 Blakely, D.W., Somorjai, G.A.: Surf. Sci. **65** (1977) 419.
77B4 Britze, K., Meyer-Ehmsen, G.: Surf. Sci. **67** (1977) 358.
77C1 Calvert, R.L., Russell, G.J., Haneman, D.: Phys. Rev. Lett. **39** (1977) 1226.
77C2 Chan, C.-M., Cunningham, S.L., Van Hove, M.A., Weinberg, W.H.: Surf. Sci. **67** (1977) 1.
77C3 Chan, C.-M., Cunningham, S.L., Van Hove, M.A., Weinberg, W.H., Withrow, S.P.: Surf. Sci. **66** (1977) 394.
77C4 Chang, K.-H., Meijer, P.H.E.: J. Vac. Sci. Technol. **14** (1977) 789.
77C5 Chung, Y.W., Lo, W.J., Somorjai, G.A.: Surf. Sci. **64** (1977) 588.
77C6 Cunningham, S.L., Chan, C.-M., Weinberg, W.H.: J. Vac. Sci. Technol. **14** (1977) 312.
77D1 Debe, M.K., King, D.A.: J. Phys. C **10** (1977) L303.
77D2 Debe, M.K., King, D.A.: Phys. Rev. Lett. **39** (1977) 708.
77D3 Debe, M.K., King, D.A., Marsh, F.S.: Surf. Sci. **68** (1977) 437.
77D4 Duke, C.B., Lubinsky, A.R., Bonn, M., Cisneros, G., Mark, P.: J. Vac. Sci. Technol. **14** (1977) 294.
77D5 Duke, C.B., Lubinsky, A.R., Chang, S.C., Lee, B.W., Mark, P.: Phys. Rev. B **15** (1977) 4865.
77E Echenique, P.M., Titterington, D.J.: J. Phys. C **10** (1977) 625.
77F1 Feder, R.: Phys. Rev. B **15** (1977) 1751.
77F2 Feder, R.: Surf. Sci. **68** (1977) 229.
77F3 Feder, R.: Surf. Sci. **63** (1977) 283.
77F4 Feder, R., Müller, N., Wolf, D.: Z. Phys. B **28** (1977) 265.
77F5 Felter, T.E., Barker, R.A., Estrup, P.J.: Phys. Rev. Lett. **38** (1977) 1138.
77G Group d'Etudes des Surfaces: Surf. Sci. **62** (1977) 567.
77H1 Heinz, K., Heilmann, P., Müller, K.: Z. Naturforsch. A **32** (1977) 28.
77H2 Housley, M., Ducros, R., Piquard, G., Cassuto, A.: Surf. Sci. **68** (1977) 277.
77I Ignatiev, A., Jona, F., Debe, M., Johnson, D.E., White, S.J., Woodruff, D.P.: J. Phys. C **10** (1977) 1109.
77J Jona, F., Shih, H.D., Ignatiev, A., Jepsen, D.W., Marcus, P.M.: J. Phys. C **10** (1977) L67.
77K1 Kesmodel, L.L., Stair, P.C., Somorjai, G.A.: Surf. Sci. **64** (1977) 342.
77K2 Kinniburgh, C.G., Walker, J.A.: Surf. Sci. **63** (1977) 274.
77L1 Langell, M.A., Bernasek, S.L.: Surf. Sci. **69** (1977) 727.
77L2 Lee, B.W., Ignatiev, A., Tong, S.Y., Van Hove, M.A.: J. Vac. Sci. Technol. **14** (1977) 291.
77L3 Legg, K.O., Jona, F., Jepsen, D.W., Marcus, P.M.: J. Phys. C **10** (1977) 937.
77L4 Levine, J.D., Mark, P., McFarlane, S.H.: J. Vac. Sci. Technol. **14** (1977) 878.
77L5 Levine, J.D., McFarlane, S.H., Mark, P.: Phys. Rev. B **16** (1977) 5415.
77L6 Lurie, P.G., Wilson, J.M.: Surf. Sci. **65** (1977) 453.
77M1 Maglietta, M., Zanazzi, E., Jona, F., Jepsen, D.W., Marcus, P.M.: J. Phys. C **10** (1977) 3287.
77M2 Mark, P., Cisneros, G., Bonn, M., Kahn, A., Duke, C.B., Paton, A., Lubinski, A.R.: J. Vac. Sci. Technol. **14** (1977) 910.
77M3 Mark, P., Pianetta, P., Lindau, I., Spicer, W.E.: Surf. Sci. **69** (1977) 735.
77M4 Masel, R.I., Merrill, R.P.: J. Vac. Sci. Technol. **14** (1977) 483.
77M5 Masud, N., Kinniburgh, C.G., Pendry, J.B.: J. Phys. C **10** (1977) 1.
77M6 McRae, E.G., Landwehr, J.M, Caldwell, C.W.: Phys. Rev. Lett. **38** (1977) 1422.
77M7 Mitchell, K.A.R., Sheferd, F.R., Watson, P.R., Frost, D.C.: Surf. Sci. **64** (1977) 737.
77M8 Mrstik, B.J., Kaplan, R., Reinecke, T.L., Van Hove, M.A., Tong, S.Y.: Nuovo Cimento B **38** (1977) 387.
77M9 Mrstik, B.J., Kaplan, R., Reinecke, T.L., Van Hove, M.A., Tong, S.Y.: Phys. Rev. B **15** (1977) 897.
77M10 Moore, W.T., Streater, R.M., Frost, D.C., Mitchell, K.A.R.: Solid State Commun. **24** (1977) 139.
77O1 Olshanetsky, B.Z., Repinsky, S.M., Shklyaev, A.A.: Surf. Sci. **64** (1977) 225.
77O2 Olshanetsky, B.Z., Repinsky, S.M., Shklyaev, A.A.: Surf. Sci. **69** (1977) 205.
77O3 Olshanetsky, B.Z., Shklyaev, A.A.: Surf. Sci. **67** (1977) 581.
77R1 Rundgren, J., Malmström, G.: J. Phys. C **10** (1977) 4671.

77R2 Rundgren, J., Malmström, G.: Phys. Rev. Lett. **38** (1977) 836.
77S1 Schuman, H.: Ultramicroscopy **2** (1977) 261.
77S2 Stoner, N., Van Hove, M.A., Tong, S.Y., in: Characterization of Metal and Polymer Science. Vol. 1, Lee, L.-H. (ed.), New York: Academic Press 1977.
77T Tong, S.Y., Van Hove, M.A., Mrstik, B.J.: Proc. 7th Int. Vac. Congr., 3rd Int. Conf. Solid Surf., Vienna 1977, p. 2407.
77V Van Hove, M.A., Tong, S.Y., Elconin, M.H.: Surf. Sci. **64** (1977) 85.
77W1 Walker, J.A., Kinniburgh, C.G., Matthew, J.A.D.: Surf. Sci. **78** (1977) 221.
77W2 White, S.J., Woodruff, D.P.: Surf. Sci. **64** (1977) 131.
77W3 Wynblatt, P., Ku, R.C.: Surf. Sci. **65** (1977) 511.
77Z1 Zanazzi, E., Jona, F.: Surf. Sci. **62** (1977) 61.
77Z2 Zanazzi, E., Jona, F., Jepsen, D.W., Marcus, P.M.: J. Phys. C **10** (1977) 375.
78A1 Aono, M., Oshima, C., Tanaka, T., Bannai, E., Kawai, S.: J. Appl. Phys. **49** (1978) 2761.
78A2 Auboiroux, M., Marchaud, D., Legrand, A.P., Thuillier, J.C.: Surf. Sci. **78** (1978) 104.
78B1 Barker, R.A., Estrup, P.J., Jona, F., Marcus, P.M.: Solid State Commun. **25** (1978) 375.
78B2 Bonzel, H.P., Brodén, G., Pirug, G.: J. Catal. **53** (1978) 96.
78B3 Britze, K., Meyer-Ehmsen, G.: Surf. Sci. **77** (1978) 131.
78C1 Chan, C.-M., Thiel, P.A., Yates jr., J.T., Weinberg, W.H.: Surf. Sci. **76** (1978) 296.
78C2 Cunningham, S.L., Chan, C.-M., Weinberg, W.H.: Phys. Rev. B **18** (1978) 1537.
78C3 Cunningham, S.L., Weinberg, W.H.: Rev. Sci. Instrum. **49** (1978) 752.
78D Duke, C.B., Meyer, R.J., Paton, A., Mark, P.: Phys. Rev. B **18** (1978) 4225.
78E1 Ebina, A., Asamo, K., Takahashi, T.: Phys. Rev. B **18** (1978) 4332.
78E2 Echenique, P.M., Pendry, J.B.: J. Phys. C **11** (1978) 2065.
78F1 Feder, R., Kirschner, J.: Phys. Status Solidi (a) **45** (1978) K117.
78F2 Feder, R., Moritz, W.: Surf. Sci. **77** (1978) 505.
78F3 Frost, D.C., Hengrasmee, S., Mitchell, K.A.R., Sheferd, F.R., Watson, P.R.: Surf. Sci. **76** (1978) L585.
78G1 Goldstein, B., Szostak, D.J.: Surf. Sci. **74** (1978) 461.
78G2 Green, A.K., Bauer, E.: Surf. Sci. **74** (1978) 676.
78H Holloway, S., Beeby, J.L.: J. Phys. C **11** (1978) L247.
78J1 Jagodzinski, H.: Phys. Status Solidi (a) **49** (1978) 119.
78J2 Jagodzinski, H., Moritz, W., Wolf, D.: Surf. Sci. **77** (1978) 233.
78J3 Jennings, P.J.: Surf. Sci. **75** (1978) L773.
78J4 Jennings, P.J., Jones, R.O.: Surf. Sci. **71** (1978) 101.
78K1 Kahn, A., Cisneros, G., Bonn, M., Mark, P., Duke, C.B.: Surf. Sci. **71** (1978) 387.
78K2 Kahn, A., So, E., Mark, P., Duke, C.B.: J. Vac. Sci. Technol. **15** (1978) 580.
78K3 Kahn, A., So, E., Mark, P., Duke, C.B., Meyer, R.J.: J. Vac. Sci. Technol. **15** (1978) 1223.
78K4 Kalisvaart, M., O'Neill, M.R., Riddle, T.W., Dunning, F.B., Walters, G.B.: Phys. Rev. B **17** (1978) 1570.
78L1 Lanks, I.R., Green, M.: Surf. Sci. **71** (1978) 735.
78L2 Lee, B.W., Alsenz, R., Ignatiev, A., Van Hove, M.A.: Phys. Rev. B **17** (1978) 1510.
78L3 Lo, W.L., Somorjai, G.A.: Phys. Rev. B **17** (1978) 4942.
78M1 Maglietta, M., Rovida, G.: Surf. Sci. **71** (1978) 495.
78M2 Maglietta, M., Zanazzi, E., Jona, F., Jepsen, D.W., Marcus, P.M.: Appl. Phys. **15** (1978) 409.
78M3 Marbrow, R.A., Lambert, R.M.: Surf. Sci. **71** (1978) 107.
78M4 Matysik, K.J.: J. Appl. Phys. **49** (1978) 3600.
78M5 McRae, E.G., Aberdam, G., Baudoing, R., Gauthier, Y.: Surf. Sci. **78** (1978) 518.
78M6 McRae, E.G., Caldwell, C.W.: J. Vac. Sci. Technol. **15** (1978) 574.
78M7 McRae, E.G., Caldwell, C.W.: Surf. Sci. **74** (1978) 285.
78M8 Mitchell, K.A.R., Van Hove, M.A.: Surf. Sci. **75** (1978) L147.
78M9 Miyao, M., Goto, R., Sukegawa, T., Hagino, M.: Surf. Sci. **71** (1978) 148.
78M10 Mönch, W., Auer, P.P.: J. Vac. Sci. Technol. **15** (1978) 1230.
78M11 Moritz, W., Jagodzinski, H., Wolf, D.: Surf. Sci. **77** (1978) 249.
78N1 Noonan, J.R., Davis, H.L., Jenkins, L.H.: Bull. Am.Phys. Soc. **23** (1978) 391.
78N2 Noonan, J.R., Davis, H.L., Jenkins, L.H.: J. Vac. Sci. Technol. **15** (1978) 619.
78N3 Nuyts, N., Phariseau, P.: Physica A **90** (1978) 260.
78N4 Nuyts, N., Phariseau, P.: Physica A **91** (1978) 113.
78N5 Nuyts, N., Phariseau, P.: Physica A **91** (1978) 575.
78N6 Nuyts, N., Phariseau, P.: Physica A **91** (1978) 585.
78P1 Poppendieck, T.D., Ngoc, T.C., Webb, M.B.: Surf. Sci. **75** (1978) 287.
78P2 Prior, K.A., Schwaha, K., Lambert, R.M.: Surf. Sci. **77** (1978) 193.
78R1 Read, M.N., Jennings, P.J.: Surf. Sci. **74** (1978) 54.

78R2 Riddle, T.W., Mahan, A.H., Dunning, F.D., Walters, G.K.: J. Vac. Sci. Technol. **15** (1978) 1686.
78S1 Sheferd, F.R., Watson, P.R., Frost, D.C., Mitchell, K.A.R.: J. Phys. C **11** (1978) 4591.
78S2 Skeath, P., Saperstein, W.A., Pianetta, P., Lindau, I., Spicer, W.E.: J. Vac. Sci. Technol. **15** (1978) 1219.
78S3 Stoner, N., Van Hove, M.A., Tong, S.Y., Webb, M.B.: Phys. Rev. Lett. **40** (1978) 243.
78S4 Streater, R.W., Moore, W.T., Watson, P.R., Frost, D.C., Mitchell, K.A.R.: Surf. Sci. **72** (1978) 744.
78T1 Taubenblatt, M., So, E., Sih, P., Kahn, A., Mark, P.: J. Vac. Sci. Technol. **15** (1978) 1143.
78T2 Taylor, T.N., Ellis, W.P.: Surf. Sci. **77** (1978) 321.
78T3 Tong, S.Y., Lubinsky, A.R., Mrstik, B.J., Van Hove, M.A.: Phys. Rev. B **17** (1978) 3303.
78T4 Tong, S.Y., Maldonado, A.L.: Surf. Sci. **78** (1978) 459.
78U1 Ulehla, M., Davis, H.L.: J. Vac. Sci. Technol. **15** (1978) 642.
78U2 Unertl, W.N.: J. Vac. Sci. Technol. **15** (1978) 591.
78V1 Vitali, G., Fainelli, E., Petrocco, G.: Appl. Phys. A **15** (1978) 315.
78V2 Viljoen, P.E., Roux, J.P., Boonstra, E.G.: J. Vac. Sci. Technol. **15** (1978) 626.
78W1 Watson, P.R., Shepherd, F.R., Frost, D.C., Mitchell, K.A.R.: Surf. Sci. **72** (1978) 562.
78W2 Welton-Cook, M.R., Prutton, M.: Surf. Sci. **74** (1978) 276.
78W3 Wendelke, J.F., Zehner, D.M.: Surf. Sci. **71** (1978) 178.
78W4 Wolf, D., Jagodzinski, H., Moritz, W.: Surf. Sci. **77** (1978) 265.
78W5 Wolf, D., Jagodzinski, H., Moritz, W.: Surf. Sci. **77** (1978) 283.
78W6 Wolf, H., Solbrig, H.: Phys. Status Solidi (b) **86** (1978) K15.
79A1 Aberdam, D.: Surf. Sci. **87** (1979) L701.
79A2 Adams, D.L.: Surf. Sci. **81** (1979) L337.
79A3 Adams, D.L., Nielsen, H.B., Van Hove, M.A.: Phys. Rev. B **20** (1979) 4789.
79A4 Alff, M., Moritz, W.: Surf. Sci. **80** (1979) 24.
79A5 Aono, M., Nishitani, R., Oshima, C., Tanaka, T., Bannai, E., Kawai, S.: J. Appl. Phys. **50** (1979) 4802.
79A6 Aono, M., Nishitani, R., Oshima, C., Tanaka, T., Bannai, E., Kawai, S.: Surf. Sci. **86** (1979) 631.
79B1 Beeby, J.L.: Surf. Sci. **80** (1979) 56.
79B2 Brundle, C.R.: J. Vac. Sci. Technol. **16** (1979) 474.
79C1 Celotta, R., Pierce, D.T., Wang, G.-C., Bader, S.D., Felcher, G.P.: Phys. Rev. Lett. **43** (1979) 728.
79C2 Chan, C.-M., Van Hove, M.A., Weinberg, W.H., Williams, E.D.: Solid State Commun. **30** (1979) 47.
79C3 Chan, C.-M., Williams, E.D., Weinberg, W.H.: Surf. Sci. **82** (1979) L577.
79C4 Cisneros, G.: J. Vac. Sci. Technol. **16** (1979) 584.
79C5 Clarke, L.J.: Vacuum **29** (1979) 405.
79C6 Clarke, L.J.: Surf. Sci. **80** (1979) 32.
79C7 Comsa, G.: Surf. Sci. **81** (1979) 57.
79C8 Comsa, G.: Surf. Sci. **84** (1979) L489.
79D1 Davis, H.L., Noonan, J.R., Jenkins, L.H.: Surf. Sci. **83** (1979) 559.
79D2 Debe, M.K., King, D.A.: Surf. Sci. **81** (1979) 193.
79D3 Duke, C.B., Meyer, R.J., Paton, A., Mark, P., Kahn, A., So, E., Yeh, J.L.: J. Vac. Sci. Technol. **16** (1979) 1252.
79D4 Duke, C.B., Meyer, R.J., Paton, A., Mark, P., So, E., Yeh, J.L.: J. Vac. Sci. Technol. **16** (1979) 647.
79F1 Feder, R., Mönch, W., Auer, P.P.: J. Phys. C **12** (1979) L179.
79F2 Felton, R.C., Prutton, M., Matthew, J.A.D., Zinn, W.: Surf. Sci. **79** (1979) 117.
79F3 Felton, R.C., Prutton, M., Tear, S.P., Welton-Cook, M.R.: Surf. Sci. **88** (1979) 474.
79F4 Frankl, D.R.: Surf. Sci. **84** (1979) L485.
79F5 Fujinaga, Y.: Surf. Sci. **86** (1979) 581.
79G Gewinner, G., Perucchetti, J.C., Jaegle, A., Riedinger, R.: Phys. Rev. Lett. **43** (1979) 935.
79H1 Heilmann, P., Heinz, K., Müller, K.: Surf. Sci. **89** (1979) 84.
79H2 Hjalmarson, H.P., Dow, J.D.: J. Vac. Sci. Technol. **16** (1979) 1262.
79H3 Holloway, S.: Phys. Lett. A **71** (1979) 476.
79H4 Holloway, S.: Surf. Sci. **80** (1979) 62.
79H5 Heilmann, P., Heinz, K., Müller, K.: Surf. Sci. **83** (1979) 487.
79I Ichikawa, T., Ino, S.: Surf. Sci. **85** (1979) 221.
79J1 Jennings, P.J.: Surf. Sci. **88** (1979) L25.
79J2 Jona, F., Shih, H.D.: J. Vac. Sci. Technol. **16** (1979) 1248.

79J3 Jona, F., Shih, H.D., Jepsen, D.W., Marcus, P.M.: J. Phys. C **12** (1979) L455.
79K1 Kirschner, J., Feder, R.: Phys. Rev. Lett. **42** (1979) 1008.
79K2 Kirschner, J., Feder, R.: Surf. Sci. **79** (1979) 176.
79K3 Küppers, J., Michel, H.: Appl. Surf. Sci. **3** (1979) 179.
79L Lang, E., Heilman, P., Hanke, G., Heinz, K., Müller, K.: Appl. Phys. **19** (1979) 287.
79M1 Mark, P., Chang, S.C., Creighton, W.F., Lee, B.W.: Crit. Rev. Solid State Sci. **5** (1979) 189.
79M2 Marsh, F.S., King, D.A.: Surf. Sci. **79** (1979) 445.
79M3 Marsh, F.S., King, D.A.: Surf. Sci. **81** (1979) L343.
79M4 Masud, N., Kinniburgh, C.G., Titterington, D.J.: J. Phys. C **12** (1979) 5263.
79M5 McRae, E.G.: Rev. Mod. Phys. **51** (1979) 541.
79M6 Meyer, R.J., Duke, C.B., Paton, A., Kahn, A., So, E., Yeh, J.L., Mark, P.: Phys. Rev. B **19** (1979) 5194.
79M7 Miller, D.J., Haneman, D.: J. Vac. Sci. Technol. **16** (1979) 1270.
79M8 Mills jr., A.P.: Appl. Phys. Lett. **35** (1979) 427.
79M9 Moore, W.T., Watson, P.R., Frost, D.C., Mitchell, K.A.R.: J. Phys. C **12** (1979) L887.
79M10 Morgen, P., Junl, T., Larsen, E.: J. Vac. Sci. Technol. **16** (1979) 89.
79M11 Moritz, W., Wolf, D.: Surf. Sci. **88** (1979) L29.
79M12 Mrstik, B.J., Tong, S.Y., Van Hove, M.A.: J. Vac. Sci. Technol. **16** (1979) 1258.
79N Noonan, J.R., Davis, H.L.: J. Vac. Sci. Technol. **16** (1979) 587.
79P1 Price, G.L., Jennings, P.J., Best, P.E., Cornish, J.C.L.: Surf. Sci. **89** (1979) 151.
79P2 Prior, K.A., Schwaha, K., Bridge, M.E., Lambert, R.M.: Chem. Phys. Lett. **65** (1979) 472.
79P3 Prutton, M., Ramsey, J.A., Walker, J.A., Welton-Cook, M.R.: J. Phys. C **12** (1979) 5271.
79P4 Prutton, M., Walker, J.A., Welton-Cook, M.R., Felton, R.C., Ramsey, J.A.: Surf. Sci. **89** (1979) 95.
79R1 Rasolt, M., Davis, H.L.: Phys. Rev. B **20** (1979) 5059.
79R2 Read, M.N., Russell, G.J.: Surf. Sci. **88** (1979) 95.
79T1 Tait, R.H., Kasowski, R.V.: Phys. Rev. B **20** (1979) 5178.
79T2 Thomas, G.E., Weinberg, W.H.: J. Vac. Sci. Technol. **16** (1979) 87.
79T3 Tommet, T.N., Olszewski, G.B., Chadwick, P.A., Bernasek, S.L.: Rev. Sci. Instrum. **50** (1979) 147.
79V Van Hove, M.A., Tong, S.Y.: Surface Crystallography by LEED. Berlin: Springer-Verlag 1979.
79W1 Wang, G.-C., Dunlap, B.I., Celotta, R.J., Pierce, D.T.: Phys. Rev. Lett. **42** (1979) 1349.
79W2 Wang, G.-C., Lagally, M.G.: Surf. Sci. **81** (1979) 69.
79W3 Weeks, S.P., Rowe, J.E., Christman, S.B., Chaban, E.E.: Rev. Sci. Instrum. **50** (1979) 1249.
79W4 Welkie, D.G., Lagally, M.G.: Appl. Surf. Sci. **3** (1979) 272.
80A Andersson, S., Pendry, J.B.: J. Phys. C **13** (1980) 3547.
80B1 Barker, R.A., Semancik, S., Estrup, P.J.: Surf. Sci. **94** (1980) L162.
80B2 Bauer, P., Feder, R., Müller, N.: Solid State Commun. **36** (1980) 249.
80B3 Bauer, P., Feder, R., Müller, N.: Surf. Sci. **99** (1980) L395.
80B4 Benz, V., Weil, K.G.: Surf. Sci. **91** (1980) 629.
80B5 Berndt, W.: Vide Couches Minces **201** (1980) 1133.
80C1 Celotta, R.J., Pierce, D.T., in: Advances in Atomic and Molecular Physics. Vol. 16, Bates, D.R., Bederson, R. (eds.), New York: Academic Press 1980.
80C2 Chan, C.-M., Van Hove, M.A., Weinberg, W.H., Williams, E.D.: Surf. Sci. **91** (1980) 440.
80C3 Clarke, L.J.: Surf. Sci. **91** (1980) 131.
80C4 Clarke, L.J., Morales de la Garza, M.: Surf. Sci. **99** (1980) 419.
80C5 Comsa, G., Mechtersheimer, G., Poelsema, B.: Surf. Sci. **97** (1980) L297.
80C6 Courths, R.: Phys. Status Solidi (b) **100** (1980) 135.
80D1 Davis, H.L., Zehner, D.M.: J. Vac. Sci. Technol. **17** (1980) 190.
80D2 Davis, P.H., Lambert, R.M.: Surf. Sci. **95** (1980) 571.
80D3 De Frésart, E., Darville, J., Gilles, J.M.: Vide Couches Minces **201** (1980) 431.
80D4 Dietz, R.E., McRae, E.G., Campbell, R.L.: Phys. Rev. Lett. **45** (1980) 1280.
80D5 Dunlap, B.I.: Solid State Commun. **35** (1980) 141.
80D6 Duke, C.B., Meyer, R.J., Paton, A., Yeh, J.L., Tsang, J.C., Kahn, A., Mark, P.: J. Vac. Sci. Technol. **17** (1980) 501.
80F1 Feder, R.: Phys. Lett. A **78** (1980) 103.
80F2 Feder, R.: Solid State Commun. **34** (1980) 541.
80H1 Heinz, K., Müller, K., in: Springer Tracts in Modern Physics. Vol. 91, Berlin: Springer-Verlag, 1980.
80H2 Hengrasmee, S., Frost, D.C., Mitchell, K.A.R.: Surf. Sci. **97** (1980) 503.
80H3 Hengrasmee, S., Mitchell, K.A.R., Watson, P.R., White, S.J.: Can. J. Phys. **58** (1980) 200.
80I1 Ichikawa, T., Ino, S.: Solid State Commun. **34** (1980) 349.

80I2 Ichimiya, A., Kambe, K., Lehmpfuhl, G.: J. Phys. Soc. Jpn. **49** (1980) 684.
80J1 Jennings, P.J., Price, G.L.: Surf. Sci. **93** (1980) L124.
80J2 Jepsen, D.W.: Phys. Rev. B **22** (1980) 5701.
80J3 Jona, F., Jepsen, D.W., Marcus, P.M., Rosenberg, I.J., Weiss, A.H., Canter, K.F.: Solid State Commun. **36** (1980) 957.
80J4 Jona, F., Sondericker, D., Marcus, P.M.: J. Phys. C **13** (1980) L155.
80K Kaplan, R.: Surf. Sci. **93** (1980) 145.
80L1 Langell, M.A., Bernasek, S.L.: J. Vac. Sci. Technol. **17** (1980) 1287.
80L2 Langell, M.A., Bernasek, S.L.: J. Vac. Sci. Technol. **17** (1980) 1296.
80L3 Leonhard, H., Gutmann, A., Hayek, K.: J. Phys. E **13** (1980) 298.
80L4 Lu, T.-M., Lagally, M.G.: Surf. Sci. **99** (1980) 695.
80M1 Malmström, G., Rundgren, J.: Comp. Phys. Commun. **19** (1980) 263.
80M2 Marsh, F.S., Debe, M.K., King, D.A.: J. Phys. C **13** (1980) 2799.
80M3 Masud, N.: J. Phys. C **13** (1980) 6359.
80M4 Meyer, R., Duke, C.B., Paton, A.: Surf. Sci. **97** (1980) 512.
80M5 Meyer, R.J., Duke, C.B., Paton, A., Yeh, J.L., Tsang, J.C., Kahn, A., Mark, P.: Phys. Rev. B **21** (1980) 4740.
80M6 Meyer, R.J., Duke, C.B., Paton, A., Tsang, J.C., Yeh, J.L., Kahn, A., Mark, P.: Phys. Rev. B **22** (1980) 6171.
80M7 Meyer, R.J., Duke, C.B., Paton, A., So, E., Yeh, J.L., Kahn, A., Mark, P.: Phys. Rev. B **22** (1980) 2875.
80M8 Meyer, R.J., Salaneck, W.R., Duke, C.B., Paton, A., Griffiths, C.H., Kovnat, L., Meyer, L.E.: Phys. Rev. B **21** (1980) 4542.
80M9 Miller, D.J., Haneman, D., Walker, L.H.: Surf. Sci. **94** (1980) 555.
80M10 Mills jr., A.P.: Appl. Phys. Lett. **37** (1980) 667.
80M11 Mills jr., A.P.: Appl. Phys. **23** (1980) 189.
80M12 Mills jr., A.P., Platzman, P.M.: Solid State Commun. **35** (1980) 321.
80M13 Malmstrøm, G., Rundgren, J.: J.Phys. C **13** (1980) L61.
80N1 Nishitani, R., Aono, M., Tanaka, T., Oshima, C., Kawai, S., Iwasaki, H., Nakamura, S.: Surf. Sci. **93** (1980) 535.
80N2 Noonan, J.R., Davis, H.L.: Surf. Sci. **99** (1980) L424.
80O1 O'Callaghan, P.: Vide Couches Minces **201** (1980) 225.
80O2 Oliva, J.: Phys. Rev. B **21** (1980) 4909.
80O3 Olshanetsky, B.Z., Shklyaev, A.A.: Surf. Sci. **82** (1979) 445.
80O4 Onuferko, J.H., Woodruff, D.P.: Surf. Sci. **91** (1980) 400.
80P1 Pendry, J.B.: J. Phys. C **13** (1980) 937.
80P2 Polizzotti, R.S., Schwarz, J.A.: J. Vac. Sci. Technol. **17** (1980) 655.
80P3 Price, G.L.: Rev. Sci. Instrum. **51** (1980) 605.
80R1 Rasolt, M., Davis, H.L.: Phys. Rev. B **21** (1980) 1445.
80R2 Rehil, B., Dunlap, B.D.: Appl. Phys. Lett. **37** (1980) 941.
80R3 Rosenberg, I.J., Weiss, A.H., Canter, K.F.: Phys. Rev. Lett. **44** (1980) 1139.
80S1 Schrader, M.E.: J. Phys. Chem. **84** (1980) 2774.
80S2 Shih, H.D., Jona, F., Bardi, U., Marcus, P.M.: J. Phys. C **13** (1980) 3801.
80S3 Stair, P.C.: Rev. Sci. Instrum. **51** (1980) 132.
80S4 Stevens, M.A., Russell, G.J.: Solid State Commun. **34** (1980) 785.
80T1 Tear, S.P., Welton-Cook, M.R., Prutton, M., Walker, J.A.: Surf. Sci. **99** (1980) 598.
80T2 Titterington, D.J., Kinniburgh, C.G.: Comput. Phys. Commun. **20** (1980) 230.
80T3 Tsang, J.C., Kahn, A., Mark, P.: Surf. Sci. **97** (1980) 119.
80U Unertl, W.N.: J. Vac. Sci. Technol. **17** (1980) 186.
80V1 Van Bommel, A.J., Crombeen, J.E.: Surf. Sci. **93** (1980) 383.
80V2 Van Hove, M.A., Somorjai, G.A.: Surf. Sci. **92** (1980) 489.
80W1 Weaver, J.H., Bradshaw, A.M., Van der Veen, J.F., Himpsel, F.J., Eastman, D.E., Politis, C.: Phys. Rev. B **22** (1980) 4921.
80W2 Welton-Cook, M.R., Prutton, M.: J. Phys. C **13** (1980) 3993.
80Z1 Zanazzi, E., Bardi, U., Maglietta, M.: J. Phys. C **13** (1980) 4001.
80Z2 Zehner, D.M., White, C.W., Ownby, G.W.: Appl. Phys. Lett. **36** (1980) 56.
80Z3 Zehner, D.M., White, C.W., Ownby, G.W.: Appl. Phys. Lett. **37** (1980) 456.
80Z4 Zehner, D.M., White, C.W., Ownby, G.W.: Surf. Sci. **92** (1980) L67.
81A1 Adams, D.L.: J. Phys. C **14** (1981) 789.
81A2 Adams, D.L., Nielsen, H.B.: Surf. Sci. **107** (1981) 305.
81A3 Allié, G., Lauroz, G., Villemain, P.: Surf. Sci. **104** (1981) 583.
81A4 Alvarado, S.F., Hopster, H., Feder, R., Pleyer, H.: Solid State Commun. **39** (1981) 1319.

81B1 Baird, R.J., Eberhardt, W.: J. Vac. Sci. Technol. **18** (1981) 538.
81B2 Baribeau, J.-M., Carette, J.-D.: Physica B **106** (1981) 221.
81B3 Baribeau, J.-M., Carette, J.-D.: Phys. Rev. B **23** (1981) 6201.
81B4 Baribeau, J.-M., Carette, J.-D.: Surf. Sci. **111** (1981) L641.
81B5 Baribeau, J.-M., Carette, J.-D.: Surf. Sci. **112** (1981) 241.
81B6 Bennett, P.A., Webb, M.W.: Surf. Sci. **104** (1981) 74.
81B7 Berning, G.L.P., Alldridge, G.P., Viljoen, P.E.: Surf. Sci. **104** (1981) L225.
81D1 De Frésart, E., Darville, J., Gilles, J.M.: Solid State Commun. **37** (1981) 13.
81D2 Duke, C.B., Meyer, R.J., Paton, A., Kahn, A., Carelli, J., Yeh, J.L.: J. Vac. Sci. Technol. **18** (1981) 866.
81D3 Duke, C.B., Paton, A., Ford, W.K., Kahn, A., Carelli, J.: Phys. Rev. B **24** (1981) 562.
81D4 Duke, C.B., Paton, A., Ford, W.K., Kahn, A., Scott, G.: Phys. Rev. B **24** (1981) 3310.
81E1 Erbudak, M., Müller, N.: Appl. Phys. Lett. **38** (1981) 575.
81E2 Erbudak, M., Ravano, G.: J. Appl. Phys. **52** (1981) 5032.
81F1 Feder, R.: J. Phys. C **14** (1981) 2049.
81F2 Feder, R., Kirschner, J.: Surf. Sci. **103** (1981) 75.
81F3 Feder, R., Pleyer, H., Bauer, P., Müller, N.: Surf. Sci. **109** (1981) 419.
81F4 Fernandez, J.C., Yang, W.S., Shih, H.D., Jona, F., Jepsen, D.W., Marcus, P.M.: J. Phys. C **14** (1981) L455.
81G1 Gardiner, T.M., Kramer, H.M., Bauer, E.: Surf. Sci. **112** (1981) 181.
81G2 Gotoh, T., Murakami, S., Kinosita, K., Murata, Y.: J. Phys. Soc. Jpn. **50** (1981) 2063.
81H Hertl, H.-J., Feder, R., Meister, G., Bauer, E.: Solid State Commun. **38** (1981) 973.
81I Ichimiya, A.: Surf. Sci. **110** (1981) 423.
81J1 Jennings, P.J., Thurgate, S.M.: Surf. Sci. **104** (1981) L210.
81J2 Jensen, V., Andersen, J.N., Nielsen, H.B., Adams, D.L.: Surf. Sci. **112** (1981) L785.
81K Kirschner, J., Feder, R.: Surf. Sci. **104** (1981) 448.
81L1 Lee, D.W., Ni, R.K., Masud, N., Wang, X.R., Wang, D.C., Rowe, M.: J. Vac. Sci. Technol. **19** (1981) 294.
81L2 Lu, T.-M., Lagally, M.G., Wang, G.-C.: Surf. Sci. **104** (1981) L229.
81M1 Malmstrøm, G., Rundgren, J.: J. Phys. C **14** (1981) 4937.
81M2 Martin, J.A., Lagally, M.G.: J. Vac. Sci. Technol. **18** (1981) 58.
81M3 McRae, E.G., Kane, M.L.: Surf. Sci. **108** (1981) 435.
81M4 McRae, E.G., Pierce, D.T., Wang, G.C., Celotta, R.J.: Phys. Rev. B **24** (1981) 4230.
81M5 Melmed, A.J., Ceyer, S.T., Tung, R.T., Graham, W.R.: Surf. Sci. **111** (1981) L701.
81M6 Miller, D.J., Haneman, D.: Surf. Sci. **104** (1981) L237.
81M7 Morales de la Garza, L., Clarke, L.J.: J. Phys. C **14** (1981) 5391.
81M8 Mróz, S., Mróz, A.: Surf. Sci. **109** (1981) 444.
81M9 Müller, N., Erbudak, E., Wolf, D.: Solid State Commun. **39** (1981) 1247.
81N Noonan, J.R., Davis, H.L.: Bull. Am. Phys. Soc. **26** (1981) 224.
81O1 Olshanetsky, B.Z., Mashanov, V.I.: Surf. Sci. **111** (1981) 414.
81O2 Olshanetsky, B.Z., Mashanov, V.I., Nikiforov, A.I.: Surf. Sci. **111** (1981) 429.
81P1 Pierce, D.T., Celotta, R.J., Wang, G.C., McRae, E.G.: Solid State Commun. **39** (1981) 1053.
81P2 Pierce, D.T., Girvin, S.M., Unguris, J., Celotta, R.J.: Rev. Sci. Instrum. **52** (1981) 1437.
81R1 Read, M.N., Lowy, D.N.: Surf. Sci. **107** (1981) L313.
81R2 Reihl, B.: Z. Phys. B **41** (1981) 21.
81S1 Schramm jr., C.J., Langell, M.A., Bernasek, S.L.: Surf. Sci. **110** (1981) 217.
81S2 Shih, H.D., Jona, F., Jepsen, D.W., Marcus, P.M.: Surf. Sci. **104** (1981) 39.
81S3 Stevens, M.A., Russell, G.J.: Surf. Sci. **104** (1981) 354.
81T1 Taylor, T.N., Ellis, W.P.: Surf. Sci. **107** (1981) 249.
81T2 Tear, S.P., Röll, K., Prutton, M.: J. Phys. C **14** (1981) 3297.
81V1 Van Hove, M.A., Koester, R.J., Stair, P.C., Biberian, J., Kesmodel, L.L., Bartos, I., Somorjai, G.A.: Surf. Sci. **103** (1981) 189.
81V2 Van Hove, M.A., Koester, R.J., Stair, P.C., Biberian, J., Kesmodel, L.L., Bartos, I., Somorjai, G.A.: Surf. Sci. **103** (1981) 218.
81W1 Walker, J.A., Debe, M.K., King, D.A.: Surf. Sci. **104** (1981) 405.
81W2 Wang, S.-W.: J. Appl. Phys. **52** (1981) 2499.
81W3 Wang, G.-C., Celotta, R.J., Pierce, D.T.: Phys. Rev. B **23** (1981) 1761.
81W4 White, S.J., Frost, D.C., Mitchell, K.A.R.: Surf. Sci. **108** (1981) L435.
81Z Zehner, D.M., Noonan, J.R., Davis, H.L., White, C.W.: J. Vac. Sci. Technol. **18** (1981) 852.
82A1 Adams, D.L., Nielsen, H.B.: Surf. Sci. **116** (1982) 598.
82A2 Adams, D.L., Nielsen, H.B., Andersen, J.N., Stensgaard, I., Feidenhansl, R., Søresen, J.E.: Phys. Rev. Lett. **49** (1982) 669.

82A3 Alvarado, S.F., Campagna, M., Ciccacci, F., Hopster, H.: J. Appl. Phys. **53** (1982) 7920.
82A4 Alvarado, S.F., Campagna, M., Hopster, H.: Phys. Rev. Lett. **48** (1982) 51.
82A5 Alvarado, S.F., Feder, R., Hopster, H., Ciccacci, F., Pleyer, H.: Z. Phys. B **49** (1982) 129.
82A6 Alvarado, S.F., Hopster, H., Campagna, M.: Surf. Sci. **117** (1982) 294.
82B1 Berndt, B.: Rev. Sci. Instrum. **53** (1982) 221.
82B2 Bertolini, J.C., Massardier, J., Delichere, P., Tardy, B., Imelic, B., Jugnet, Y., Duc, T.M., De Tennerman, L., Creemers, C., Van Hove, H., Neyens, A.: Surf. Sci. **119** (1982) 95.
82B3 Bonzel, H.P., Ferrer, S.: Surf. Sci. **118** (1982) L263.
82C1 Cowell, P.G.: J. Phys. E **15** (1982) 994.
82C2 Canter, K.F., Mills jr., A.P.: Can. J. Phys. **60** (1982) 551.
82C3 Chabal, Y.J., Rowe, J.E., Christman, S.B.: J. Vac. Sci. Technol. **20** (1982) 763.
82C4 Clarke, L.J., Baudoing, R., Gauthier, Y.: J. Phys. C **15** (1982) 3249.
82D1 Davis, H.L., Noonan, J.R.: Surf. Sci. **115** (1982) L75.
82D2 Debe, M.K., King, D.A.: J. Phys. C **15** (1982) 2257.
82D3 De Frésart, E., Darville, J., Gilles, J.M.: Appl. Surf. Sci. **11/12** (1982) 637.
82D4 Duke, C.B., Paton, A., Ford, W.K., Kahn, A., Scott, G.: J. Vac. Sci. Technol. **20** (1982) 778.
82E1 Ellis, W.P., Taylor, T.N.: Surf. Sci. **123** (1982) 77.
82E2 Erbudak, E., Ravano, G., Müller, N.: Phys. Lett. A **90** (1982) 62.
82F1 Feder, R., Pleyer, H.: Surf. Sci. **117** (1982) 285.
82F2 Felter, T.E., Steward, S.A., Uribe, F.S.: Surf. Sci. **122** (1982) 69.
82F3 Firment, L.E.: Surf. Sci. **116** (1982) 205.
82F4 Ford, W.K., Duke, C.B., Paton, A.: Surf. Sci. **115** (1982) 195.
82G1 Gauthier, Y., Baudoing, R., Clarke, L.J.: J. Phys. C **15** (1982) 3223.
82G2 Gauthier, Y., Baudoing, R., Clarke, L.J.: J. Phys. C **15** (1982) 3231.
82G3 Gevers, R., David, M.: Phys. Status Solidi (b) **113** (1982) 665.
82H1 Haneman, D., Bachrach, R.Z.: J. Vac. Sci. Technol. **21** (1982) 337.
82H2 Heinz, K., Lang, E., Strauss, K., Müller, K.: Appl. Surf. Sci. **11/12** (1982) 611.
82H3 Heinz, K., Lang, E., Strauss, K., Müller, K.: Surf. Sci. **120** (1982) L401.
82I Ishizuka, K.: Acta Crystallogr. A **38** (1982) 773.
82J1 Jagodzinski, J.: Z. Naturforsch. A **37** (1982) 1103.
82J2 Jaklevic, R.C., Davis, L.C.: Phys. Rev. B **26** (1982) 5391.
82J3 Jennings, P.J., Thurgate, S.M., Price, G.L.: Appl. Surf. Sci. **13** (1982) 180.
82J4 Jensen, V., Andersen, J.N., Nielsen, H.B., Adams, D.L.: Surf. Sci. **116** (1982) 66.
82L1 Lang, E., Grimm, W., Heinz, K.: Surf. Sci. **117** (1982) 169.
82L2 Lang, J.K., Jamison, K.D., Dunning, F.B., Walters, G.K., Passler, M.A., Ignatiev, A., Tamura, E., Feder, R.: Surf. Sci. **123** (1982) 247.
82L3 Le Bossé, J.C., Lopez, J., Gaubert, C., Gauthier, Y., Baudoing, R.: J. Phys. C **15** (1982) 3425.
82L4 Le Bossé, J.C., Lopez, J., Gaubert, C., Gauthier, Y., Baudoing, R.: J. Phys. C **15** (1982) 6087.
82M1 Maksym, P.A., Beeby, J.L.: Appl. Surf. Sci. **11/12** (1982) 663.
82M2 Masud, N.: J. Phys. C **15** (1982) 3209.
82M3 Miller, D.J.: Appl. Surf. Sci. **13** (1982) 104.
82M4 Milne, R.H.: Surf. Sci. **122** (1982) 474.
82M5 Moison, J.M., Bersoussan, M.: J. Vac. Sci. Technol. **21** (1982) 315.
82M6 Moore, W.T., Frost, D.C., Mitchell, K.A.R.: J. Phys. C **15** (1982) L5.
82M7 Moore, W.T., White, S.J., Frost, D.C., Mitchell, K.A.R.: Surf. Sci. **116** (1982) 253.
82M8 Moore, W.T., White, S.J., Frost, D.C., Mitchell, K.A.R.: Surf. Sci. **116** (1982) 261.
82N1 Nielsen, H.B., Adams, D.L.: J. Phys. C **15** (1982) 615.
82N2 Noonan, J.R., Davis, H.L.: Bull. Am. Phys. Soc. **27** (1982) 237.
82N3 Noonan, J.R., Davis, H.L.: Surf. Sci. **115** (1982) L75.
82N4 Noonan, J.R., Davis, H.L.: Vacuum **32** (1982) 107.
82O Outlaw, R.A., Baugh, B.T.: J. Vac. Sci. Technol. **21** (1982) 1037.
82R Ravano, G., Erbudak, M., Siegmann, H.C.: Phys. Rev. Lett. **49** (1982) 80.
82S1 Schlier, A.R.: J. Vac. Sci. Technol. **20** (1982) 96.
82S2 Schrott, A.G., Chinn, M.D., Shaw, C.G., Fain jr., S.C.: J. Vac. Sci. Technol. **21** (82) 101.
82S3 Solbrig, H.: Phys. Status Solidi (a) **72** (1982) 199.
82S4 Sobrero, A.C., Weinberg, W.H.: Rev. Sci. Instrum. **53** (1982) 1566.
82S5 Stevens-Kalceff, M.A., Russell, G.J.: Appl. Surf. Sci. **13** (1982) 94.
82T1 Tamura, E., Feder, R.: Solid State Commun. **44** (1982) 1101.
82T2 Thurgate, S.M., Jennings, P.J.: Surf. Sci. **114** (1982) 395.
82T3 Titov, A., Moritz, W.: Surf. Sci. **123** (1982) L709.
82T4 Tougaard, S., Ignatiev, A.: Surf. Sci. **115** (1982) 270.
82V Vilijoen, P.E., Wessels, B.J., Berning, G.L.P., Roux, J.P.: J. Vac. Sci. Technol. **20** (1982) 204.

82W1 Wang, G.-C., Celotta, R.J., Pierce, D.T.: Surf. Sci. **119** (1982) 479.
82W2 Wang, G.-C., Lu, T.-M.: Surf. Sci. **122** (1982) L635.
82W3 Wang, G.-C., Unguris, J., Pierce, D.T., Celotta, R.J.: Surf. Sci. **114** (1982) L35.
82W4 Wang, W.S., Jona, F., Marcus, P.M.: Solid State Commun. **43** (1982) 847.
82W5 Welton-Cook, M.R., Berndt, W.: J. Phys. C **15** (1982) 5691.
82W6 Williams, E.D., Weinberg, W.H., Sobrero, A.C.: J. Chem. Phys. **76** (1982) 1150.
82W7 Wolf, H., Solbrig, H.: Phys. Status Solidi (b) **110** (1982) 191.
82W8 Wolf, H., Solbrig, H.: Phys. Status Solidi (b) **110** (1982) 543.
82W9 Wolf, H., Solbrig, H.: Phys. Status Solidi (b) **111** (1982) 203.
82W10 White, S.J., Frost, D.C., Mitchell, K.A.R.: Solid State Commun. **42** (1982) 763.
82W11 Wu, N.J., Ignatiev, A.: Phys. Rev. B **25** (1982) 2983.
82Y Yang, W.S., Sokolov, J., Jona, F., Marcus, P.M.: Solid State Commun. **41** (1982) 191.
83A1 Adams, D.L.: J. Phys. C **16** (1983) 6101.
83A2 Adams, D.L., Andersen, J.N., Nielsen, H.B.: Surf. Sci. **128** (1983) 294.
83A3 Aristov, V.Y., Batov, I.E., Grazhulis, V.A.: Surf. Sci. **132** (1983) 73.
83B1 Baribeau, J.-M., Carette, J.-D.: J. Vac. Sci. Technol. A **1** (1983) 1174.
83B2 Baribeau, J.-M., Carette, J.-D.: Surf. Sci. **134** (1983) 886.
83B3 Barnett, R.N., Landman, U., Cleveland, C.L.: Phys. Rev. Lett. **51** (1983) 1359.
83B4 Bauer, P., Eckstein, W., Müller, N.: Z. Phys. B **52** (1983) 185.
83B5 Besold, G., Heinz, K., Lang, E., Müller, K.: J. Vac. Sci. Technol. A **1** (1983) 1473.
83C Cinti, R.C., Nguyen, T.T.A., Capiomont, Y., Kennon, S.: Surf. Sci. **134** (1983) 755.
83D1 Davis, H.L., Noonan, J.R.: Surf. Sci. **126** (1983) 245.
83D2 De Frésart, E., Darville, J., Gilles, J.M.: Surf. Sci. **126** (1983) 518.
83D3 Duke, C.B., Paton, A.: J. Vac. Sci. Technol. A **1** (1983) 672.
83D4 Duke, C.B., Paton, A., Kahn, A.: Phys. Rev. B **27** (1983) 3436.
83D5 Duke, C.B., Paton, A., Kahn, A., Bonapace, C.R.: Phys. Rev. B **27** (1983) 6189.
83D6 Duke, C.B., Richardson, S.L., Paton, A., Kahn, A.: Surf. Sci. **127** (1983) L135.
83D7 Dupuy, J.C., Laydevaut, L.: J. Phys. E **16** (1983) 217.
83F1 Fearon, D.H., Helbig, H.F., Steward, S.A., Uribe, F.S.: J. Vac. Sci. Technol. A **1** (1983) 1558.
83F2 Feder, R.: Solid State Commun. **45** (1983) 51.
83F3 Feder, R., Alvarado, S.F., Tamura, E., Kisker, E.: Surf. Sci. **127** (1983) 83.
83F4 Feder, R., Rosicky, F., Ackermann, B.: Z. Phys. B **52** (1983) 31.
83F5 Feder, R., Rosicky, F., Ackermann, B.: Z. Phys. B **53** (1983) 244.
83G1 Gaubert, G., Baudoing, R., Gauthier, Y., Le Bossé, J.C., Lopez, J.: J. Phys. C **16** (1983) 2625.
83G2 Gradmann, U., Waller, G., Feder, R., Tamura, E.: J. Magn. Magn. Mater. **31/34** (1983) 883.
83H1 Heinz, K., Besold, G.: J. Phys. C **16** (1983) 1299.
83H2 Heinz, K., Besold, G.: Surf. Sci. **125** (1983) 515.
83H3 Henrich, V.E.: Progr. Surf. Sci. **14** (1983) 175.
83H4 Horio, Y., Ichimiya, A.: Surf. Sci. **133** (1983) 393.
83I Ichimiya, A.: Jpn. J. Appl. Phys. **22** (1983) 176.
83J1 Jones, G.J.R., Holland, B.W.: Solid State Commun. **46** (1983) 651.
83J2 Jones, G.J.R., Holland, B.W.: Vacuum **33** (1983) 627.
83J3 Jones, R.O., Jennings, P.J.: Phys. Rev. B **27** (1983) 4702.
83K Kurtz, R.L., Henrich, V.E.: Surf. Sci. **129** (1983) 345.
83L1 Lagally, M.G., Martin, J.A.: Rev. Sci. Instrum. **54** (1983) 1273.
83L2 Lang, E., Müller, K., Heinz, K., Van Hove, M.A., Koestner, R.J., Somorjai, G.A.: Surf. Sci. **127** (1983) 347.
83L3 Lindgren, S.A., Walldén, L., Rundgren, J., Westrin, P.: Phys. Rev. Lett. **50** (1983) 368.
83L4 Liu, H., Cook, M.R., Jona, F., Marcus, P.M.: Phys. Rev. B **28** (1983) 6137.
83L5 Lopez, J., Le Bossé, J.C., Gaubert, C., Baudoing, R., Gauthier, Y.: Surf. Sci. **126** (1983) 286.
83M1 Martin, J.A., Lagally, M.G.: J. Vac. Sci. Technol. A **1** (1983) 1210.
83M2 Masud, N., Baudoing, R., Aberdam, D., Gaubert, C.: Surf. Sci. **133** (1983) 580.
83M3 Menzel, D., Pfnür, H., Feulner, P.: Surf. Sci. **126** (1983) 374.
83M4 Michalk, G., Moritz, W., Pfnür, H., Menzel, D.: Surf. Sci. **129** (1983) 92.
83M5 Moison, J.M., Bensoussan, M.: Surf. Sci. **126** (1983) 294.
83M6 Mróz, S., Gruniewski, A., Bak, E.: J. Phys. E **16** (1983) 859.
83N1 Neve, J., Westrin, P., Rundgren, J.: J. Phys. C **16** (1983) 1291.
83N2 Norton, N.G.: Vacuum **33** (1983) 621.
83O1 Ochando, M.A., Sanchèz, A., Serna, C.R.: J. Phys. C **16** (1983) L401.
83O2 Olsziewski, G.B., Bernasek, S.L.: J. Chem. Phys. **79** (1983) 3581.
83P1 Peacock, D.C.: Vacuum **33** (1983) 601.
83P2 Proix, F., Akremi, A., Zhong, Z.T.: J. Phys. C **16** (1983) 5449.

83R1 Ravano, G., Erbudak, M., Siegmann, H.C.: Phys. Rev. Lett. **50** (1983) 1022.
83R2 Read, M.N.: Solid State Commun. **47** (1983) 1.
83S Solokov, J., Shih, H.D., Bardi, U., Jona, F., Marcus, P.M.: Solid State Commun. **48** (1983) 739.
83T Thurgate, S.M., Jennings, P.J.: Appl. Surf. Sci. **22/23** (1983) 478.
83U Urano, T., Kanaji, T., Kaburagi, M.: Surf. Sci. **134** (1983) 109.
83V Van Hove, M.A., Lin, R., Somorjai, G.A.: Phys. Rev. Lett. **51** (1983) 778.
83W1 Wang, W.S., Jona, F.: Solid State Commun. **48** (1983) 377.
83W2 Weiss, A.H., Rosenberg, I.J., Canter, K.F., Duke, C.B., Paton, A.: Phys. Rev. B **27** (1983) 867.
83W3 Welz, M., Moritz, W., Wolf, D.: Surf. Sci. **125** (1983) 473.
83Y Yang, W.S., Jona, F., Marcus, P.M.: Phys. Rev. B **28** (1983) 2049.
84A1 Ackermann, B., Feder, R.: Solid State Commun. **49** (1984) 489.
84A2 Agrawal, B.K.: Phys. Rev. B **30** (1984) 4412.
84A3 Andersen, J.N., Nielsen, H.B., Petersen, L., Adams, D.L.: J. Phys. C **17** (1984) 173.
84B Bardi, U., Ross, P.N.: Surf. Sci. **146** (1984) L555.
84C1 Cook, D.R., Horsky, T.N., Coleman, P.G.: Appl. Phys. A **34** (1984) 237.
84C2 Ceyer, S.T., Melmed, A.J., Carroll, J.J., Graham, W.R.: Surf. Sci. **144** (1984) L444.
84D1 Davis, H.L., Noonan, J.R., in: Determination of Surface Structure by LEED. Marcus, P.M., Jona, F. (eds.), New York: Plenum Press 1984, p. 215.
84D2 De Carvalho, V.E., Cook, M.W., Cowell, P.G., Heavens, O.S., Prutton, M., Tear, S.P.: Vacuum **34** (1984) 893.
84D3 Duke, C.B., Paton, A.: J. Vac. Sci. Technol. A **2** (1984) 515.
84D4 Duke, C.B., Paton, A.: J. Vac. Sci. Technol. B **2** (1984) 327.
84D5 Duke, C.B., Paton, A., Kahn, A., Lu, D.W.: J. Vac. Sci. Technol. B **2** (1984) 366.
84F Feder, R., Mönch, W.: Solid State Commun. **50** (1984) 311.
84G1 Gaubert, C., Baudoing, R., Gauthier, Y., Rundgren, J.: Surf. Sci. **147** (1984) 162.
84G2 Gusthall, P.L.: J. Vac. Sci. Technol. B **2** (1984) 332.
84H1 Himpsel, F.J., Marcus, P.M., Tromp, R., Batra, I.P., Cook, M.R., Jona, F., Lin, H.: Phys. Rev. B **30** (1984) 2257.
84H2 Holland, B.W., Duke, C.B., Paton, A.: Surf. Sci. **140** (1984) L269.
84H3 Hren, P., Tu, D.W., Kahn, A.: Surf. Sci. **146** (1984) 69.
84J1 Jones, R.O., Jennings, P.J., Jepsen, O.: Phys. Rev. B **29** (1984) 6474.
84J2 Joyce, B.A., Neave, J.H., Dobson, P.J., Larsen, P.K.: Phys. Rev. B **29** (1984) 814.
84K1 Kawamura, T., Maksym, P.A., Iijima, T.: Surf. Sci. **148** (1984) L671.
84K2 Kelemen, S.R., Mims, C.A.: Surf. Sci. **136** (1984) L35.
84K3 Kirschner, J.: Surf. Sci. **138** (1984) 191.
84K4 Klebanoff, J., Ritz, V.H., Thomas, R.E.: J. Vac. Sci. Technol. A **2** (1984) 1396.
84K5 Koehler, H., Heinz, K.: Surf. Sci. **126** (1984) 294.
84L1 Le Bossé, J.C., Lopez, J., Baribeau, J.-M., Carette, J.-D.: Surf. Sci. **137** (1984) 361.
84L2 Lopez, J., Le Bossé, J.C., Baudoing, R.: Phys. Status Solidi (b) **124** (1984) K63.
84L3 Lopez, J., Le Bossé, J.C., Baudoing, R.: Phys. Status Solidi (b) **124** (1984) 381.
84L4 Lindgren, S.A., Walldén, L., Rundgren, J., Westrin, P.: Phys. Rev. B **29** (1984) 576.
84M1 Maksym, P.A., Beeby, J.L.: Surf. Sci. **140** (1984) 77.
84M2 McKay, J.M., Henrich, V.E.: Surf. Sci. **137** (1984) 463.
84M3 McRae, E.G., Malic, R.A.: Surf. Sci. **148** (1984) 551.
84M4 Moritz, W.: J. Phys. C **17** (1984) 353.
84M5 Müller, K., Land, E., Hammer, L., Grimm, W., Heilmann, P., Heinz, K., in: Determination of Surface Structure by LEED. Marcus, P.M., Jona, F. (eds.), New York: Plenum Press 1984, p. 483.
84N1 Nelson, G.C., Shapnek, L.R.: J. Vac. Sci. Technol. A **2** (1984) 1395.
84N2 Noonan, J.R., Davis, H.L.: Phys. Rev. B **29** (1984) 4349.
84P1 Philip, J., Rundgren, J., in: Determination of Surface Structure by LEED. Marcus, P.M., Jona, F. (eds.), New York: Plenum Press 1984, p. 409.
84P2 Pukite, P.R., Van Hove, J.M., Cohen, P.I.: Appl. Phys. Lett. **44** (1984) 456.
84S1 Shannon, M.D., Eades, J.A., Meichle, M.E., Turner, P.S., Buxton, B.F.: Phys. Rev. Lett. **53** (1984) 2125.
84S2 Shih, H.D., in: Determination of Surface Structure by LEED. Marcus, P.M., Jona, F. (eds.), New York: Plenum Press 1984, p. 67.
84S3 Sokolov, J., Jona, F., Marcus, P.M.: Phys. Rev. B **29** (1984) 5402.
84S4 Sokolov, J., Jona, F., Marcus, P.M.: Solid State Commun. **49** (1984) 307.
84S5 Sokolov, J., Shih, H.D., Bardi, U., Jona, F., Marcus, P.M.: J. Phys. C **17** (1984) 371.
84S6 Stickney, J.L., Rosasco, S.D., Schardt, B.C., Solomun, T., Hubbard, A.T., Parkinson, B.A.: Surf. Sci. **136** (1984) 15.

84T1 Tong, S.Y., Mei, W.N.: J. Vac. Sci. Technol. B **2** (1984) 393.
84T2 Tong, S.Y., Xu, G., Mei, W.N.: Phys. Rev. Lett. **52** (1984) 1693.
84U Unertl, W.N., McKay, S.R., in: Determination of Surface Structure by LEED. Marcus, P.M., Jona, F. (eds.), New York: Plenum Press 1984, p. 261.
84W Wang, W.S., Jona, F.: Phys. Rev. B **29** (1984) 899.
84Z1 Zanazzi, E., in: Determination of Surface Structure by LEED. Marcus, P.M., Jona, F. (eds.), New York: Plenum Press 1984, p. 401.
84Z2 Zharkikh, Y.S., Lysochenko, S.V.: Surf. Sci. **145** (1984) L513.
85A1 Adams, D.L., Peterson, L.E., Sørensen, C.S.: J. Phys. C **18** (1985) 1753.
85A2 Adams, D.L., Moore, W.T., Mitchell, K.A.R.: Surf. Sci. **149** (1985) 407.
85A3 Alonso, M., Soria, F., Sacedón, J.L.: J. Vac. Sci. Technol. A **3** (1985) 1598.
85B1 Baribeau, J.-M., Carette, J.-D., Jennings, P.J., Jones, R.O.: Phys. Rev. B **32** (1985) 6131.
85B2 Baribeau, J.-M., Le Bossé, J.C., Lopez, J.: J. Phys. C **18** (1985) L73.
85B3 Baribeau, J.-M., Lopez, J., Le Bossé, J.C.: J. Phys. C **18** (1985) 3083.
85B4 Barnes, C.J., Ding, M.Q., Lindroos, M., Diehl, R.D., King, D.A.: Surf. Sci. **162** (1985) 59.
85B5 Bartelt, N.C., Einstein, T.L., Roelofs, L.D.: Phys. Rev. Lett. **32** (1985) 2993.
85B6 Behner, H., Wedler, G.: Surf. Sci. **160** (1985) 271.
85B7 Bickel, N., Heinz, K.: Surf. Sci. **163** (1985) 435.
85C Clarke, L.J.: Surface Crystallography. Chichester: John Wiley and Sons 1985.
85D1 Dayan, M.: J. Vac. Sci. Technol. A **3** (1985) 361.
85D2 Davis, H.L., Noonan, J.R.: J. Vac. Sci. Technol. A **3** (1985) 1507.
85D3 Davis, H.L., Noonan, J.R.: Phys. Rev. Lett. **54** (1985) 566.
85D4 Diehl, R.D., Lindroos, M., Kearsley, A., Barnes, C.J., King, D.A.: J. Phys. C **18** (1985) 4069.
85D5 Duke, C.B., Paton, A.: Surf. Sci. **165** (1985) L797.
85E Egdell, R.G., Innes, H., Hill, M.D.: Surf. Sci. **149** (1985) 33.
85F Frieze, W.E., Gidley, D.W., Lynn, K.G.: Phys. Rev. B **31** (1985) 5628.
85G1 Gauthier, Y., Baudoing, R., Joly, Y., Rundgren, J., Bertolini, J.C., Massardier, J.: Surf. Sci. **162** (1985) 342.
85G2 Gauthier, Y., Joly, Y., Baudoing, R., Rundgren, J.: Phys. Rev. B **31** (1985) 6216.
85G3 Grudniewski, A., Mróz, S.: J. Phys. C **18** (1985) 3387.
85H1 Hayek, K., Glassl, H., Gutmann, A., Leonhard, H., Prutton, M., Tear, S.P., Welton-Cook, M.R.: Surf. Sci. **152** (1985) 419.
85H2 Heinz, K., Bickel, N., Besold, G., Müller, K.: J. Phys. C **18** (1985) 933.
85H3 Heinz, K., Schmidt, G., Hammer, L., Müller, K.: Phys. Rev. B **32** (1985) 6214.
85H4 Henrich, V.E.: Rep. Progr. Phys. **48** (1985) 1481.
85I1 Ichimiya, A.: Jpn. J. Appl. Phys. **24** (1985) 1365.
85I2 Ichinokawa, T., Ampo, H., Miura, S., Tamura, A.: Phys. Rev. B **31** (1985) 5183.
85J1 Jamison, K.D., Lind, D.M., Dunning, F.B., Walters, G.K.: Surf. Sci. **159** (1985) L451.
85J2 Jennings, P.J., Jones, R.O.: Appl. Surf. Sci. **22/23** (1985) 35.
85K1 Kirschner, J.: Appl. Phys. A **36** (1985) 121.
85K2 Kirschner, J.: Polarized Electrons at Surfaces. Berlin: Springer-Verlag 1985.
85L1 Lopez, J., Le Bossé, J.C.: Surf. Sci. **162** (1985) 945.
85L2 Lopez, J., Le Bossé, J.C., Baribeau, J.-M.: J. Phys. C **18** (1985) 2197.
85M1 Maksym, P.A.: Surf. Sci. **149** (1985) 157.
85M2 Marten, M., Meyer-Ehmsen, G.: Surf. Sci. **151** (1985) 570.
85M3 McRae, E.G., Malic, R.A., Kapilow, D.A.: Rev. Sci. Instrum. **56** (1985) 2077.
85M4 Mills jr., A.P., Crane, W.S.: Phys. Rev. B **31** (1985) 3988.
85M5 Moritz, W., Wolf, D.: Surf. Sci. **163** (1985) L655.
85N Noonan, J.R., Davis, H.L., Erley, W.: Surf. Sci. **152/153** (1985) 142.
85P1 Paneuf, R.J., Webb, M.B.: Surf. Sci. **164** (1985) 167.
85P2 Puga, M.W., Xu, G., Tong, S.Y.: Surf. Sci. **164** (1985) L789.
85R1 Read, M.N.: Appl. Surf. Sci. **22/23** (1985) 48.
85R2 Rivière, J.C.: Surf. Sci. **161** (1985) 521.
85S1 Saldin, D.K., Pendry, J.B.: Surf. Sci. **162** (1985) 941.
85S2 Shimaoka, G.: Appl. Surf. Sci. **22/23** (1985) 55.
85S3 Sokolov, J., Jona, F., Marcus, P.M.: Phys. Rev. B **31** (1985) 1929.
85S4 Sondericker, D., Jona, F., Marcus, P.M.: Solid State Commun. **53** (1985) 175.
85T Thurgate, S., Hitchen, G.: Appl. Surf. Sci. **24** (1985) 202.
85X1 Xu, G., Hu, W.Y., Puga, M.W., Tong, S.Y., Yeh, J.L., Wang, S.R., Lee, B.W.: Phys. Rev. B **32** (1985) 8473.
85X2 Xu, M.L., Tong, S.Y.: Phys. Rev. B **31** (1985) 6332.
86A1 Adams, D.L., Sørensen, C.S.: Surf. Sci. **166** (1986) 495.

86A2 Ampo, H., Miura, S., Kato, K., Ohkawa, Y., Tamura, A.: Phys. Rev. B **34** (1986) 2329.
86A3 Anderegg, J.W., Thiel, P.A.: J. Vac. Sci. Technol. A **4** (1986) 1367.
86A4 Ashby, J.V., Norton, N., Maksym, P.A.: Surf. Sci. **175** (1986) 604.
86B1 Baird, R.J., Ogletree, D.F., Van Hove, M.A., Somorjai, G.A.: Surf. Sci. **165** (1986) 345.
86B2 Bartelt, N.C., Einstein, T.L., Roelofs, L.D.: Phys. Rev. Lett. **56** (1986) 2881.
86B3 Baudoing, R., Gauthier, Y., Lundberg, M., Rundgren, J.: J.Phys. C **19** (1986) 2825.
86B4 Bölger, B., Larsen, P.K.: Rev. Sci. Instrum. **57** (1986) 1363.
86B5 Baribeau, J.-M., Roy, D.: Surf. Sci. **166** (1986) 234.
86C1 Chan, C.-M., Van Hove, M.A.: Surf. Sci. **171** (1986) 226.
86C2 Chrzanowski, E., Dobinski, W., Mróz, S., Worzniak, E.: Phys. Status Solidi (a) **98** (1986) 359.
86C3 Cowell, P.G., Prutton, M., Tear, S.P.: Surf. Sci. **177** (1986) L915.
86D1 Dayan, M.: J.Vac. Sci. Technol. A **4** (1986) 38.
86D2 Duke, C.B., Mailhiot, C., Paton, A., Kahn, A., Stiles, K.: J. Vac. Sci. Technol. A **4** (1986) 947.
86G Gaubert, C., Baudoing, R., Gauthier, Y., Michaud, M., Sanche, L.: Appl. Surf. Sci. **25** (1986) 195.
86J1 Jennings, P.J., Jones, R.O.: Surf. Sci. **176** (1986) 691.
86J2 Jennings, P.J., Jones, R.O.: Phys. Rev. B **34** (1986) 6699.
86J3 Jona, F., Marcus, P.M., Davis, H.L., Noonan, J.R.: Phys. Rev. B **33** (1986) 4005.
86L1 Lehmpfuhl, G., Dowell, W.C.T.: Acta Crystallogr. A **42** (1986) 569.
86L2 Lind, D.M., Jamison, K.D., Dunning, F.B., Waltres, G.K., Davis, H.L.: J. Vac. Sci. Technol. A **4** (1986) 1271.
86L3 Lindroos, M., Pfnür, H., Menzel, D.: Phys. Rev. B **33** (1986) 6684.
86L4 Litzman, O.: Acta Crystallogr. A **42** (1986) 552.
86M1 McRae, E.G., Malic, R.A.: Surf. Sci. **177** (1986) 53.
86M2 Moritz, W., Lagally, M.G.: Phys. Rev. Lett. **56** (1986) 865.
86M3 Moritz, W., Lagally, M.G.: Phys. Rev. Lett. **56** (1986) 2882.
86N Nesterenko, B.A., Brovii, A.V., Sorokovykh, A.I.: Surf. Sci. **171** (1986) 495.
86O1 Ogletree, D.F., Somorjai, G.A., Katz, J.E.: Rev. Sci. Instrum. **57** (1986) 3012.
86O2 Ogletree, D.F., Van Hove, M.A., Somorjai, G.A.: Surf. Sci. **173** (1986) 351.
86P1 Pehlke, E., Schattke, W.: Surf. Sci. **173** (1986) 20.
86P2 Peng, L.-M., Cowley, J.M.: Acta Crystallogr. A **42** (1986) 545.
86R1 Rous, P.J., Pendry, J.B., Saldin, D.K., Heinz, K., Müller, K., Bickel, N.: Phys. Rev. Lett. **57** (1986) 2951.
86R2 Rudakoff, G., Müller, B.: Phys. Status Solidi (a) **93** (1986) 471.
86S1 Sakana, H., Kawazu, A., Ueda, K.: Phys. Rev. B **34** (1986) 1367.
86S2 Sinharoy, S., Braginski, A.I., Talvacchio, J.: Surf. Sci. **167** (1986) 401.
86S3 Sokolov, J., Jona, F., Marcus, P.M.: Phys. Rev. B **33** (1986) 1397.
86S4 Sondericker, D., Jona, F., Marcus, P.M.: Phys. Rev. B **33** (1986) 900.
86S5 Sondericker, D., Jona, F., Marcus, P.M.: Phys. Rev. B **34** (1986) 6770.
86S6 Sondericker, D., Jona, F., Marcus, P.M.: Phys. Rev. B **34** (1986) 6775.
86S7 Stachulec, K.: Physica B **142** (1986) 332.
86T1 Tamura, E., Feder, R.: Solid State Commun. **58** (1986) 729.
86T2 Tear, S.P., Yousif, S.Y., Matthew, J.A.D.: Surf. Sci. **175** (1986) L715.
86V1 Van Hove, M.A., Weinberg, W.H., Chan, C.-M.: Low Energy Electron Diffraction. Berlin: Springer-Verlag 1986.
86V2 Vasiliev, M.A., Gorodetsky, S.D.: Surf. Sci. **171** (1986) 543.
86Y Yamamoto, Y., Ino, S., Ichikawa, T.: Jpn. J. Appl. Phys. **25** (1986) L331.
87B1 Barrett, S.D., Jordan, R.G., Begley, A.: J. Phys. F **17** (1987) L145.
87B2 Bartos, L., Koukal, J.: Surf. Sci. **183** (1987) 21.
87B3 Bermudez, V.M.: J. Vac. Sci. Technol. A **5** (1987) 2975.
87C1 Clarke, S., Wedensky, D.D.: Phys. Rev. Lett. **58** (1987) 2235.
87C2 Cowell, P.G., De Carvalho, V.E.: Surf. Sci. **187** (1987) 175.
87D1 D'Agostino, A.T., Ross jr., P.N.: Surf. Sci. **185** (1987) 88.
87D2 Davis, H.L., Noonan, J.R., in: The Structure of Surfaces II. Van der Veen, J.F., Van Hove, M.A. (eds.), Berlin: Springer-Verlag 1987, p. 152.
87D3 De Carvalho, V.E., Prutton, M., Tear, S.P.: Surf. Sci. **184** (1987) 198.
87F Fery, P., Moritz, W., Wolf, D.: Verh. Dtsch. Phys. Ges. **22** (1987) O 131.
87G1 Gauthier, Y., Joly, Y., Baudoing, R., Rundgren, J.: Phys. Rev. B **35** (1987) 7867.
87G2 Gullikson, E.M., Mills jr., A.P.: Phys. Rev. B **36** (1987) 8777.
87H1 Heinz, K., Barthel, A., Hammer, L., Müller, K.: Surf. Sci. **191** (1987) 174.
87H2 Hou, X., Dong, G., Ding, X., Wang, X.: J. Phys. C **20** (1987) L121.
87I1 Ichimiya, A.: Surf. Sci. **187** (1987) 194.

87I2 Ichimiya, A.: Surf. Sci. **192** (1987) L893.
87I3 Ishikawa, Y., Hosokawa, Y., Hamaguchi, I., Ichinokawa, T.: Surf. Sci. **187** (1987) L606.
87J1 Jona, F.: Surf. Sci. **192** (1987) 398.
87J2 Jona, F., Yang, J., Marcus, P.M.: Surf. Sci. **192** (1987) 414.
87K Kumazaki, Y., Nakai, Y., Itoh, N.: Surf. Sci. **184** (1987) L445.
87L1 Langell, M.A., Cameron, N.R.: Surf. Sci. **185** (1987) 105.
87L2 Lind, D.M., Dunning, F.B., Walters, G.K., Davis, H.L.: Phys. Rev. B **35** (1987) 9037.
87L3 Lindroos, M., Pfnür, H., Menzel, D.: Surf. Sci. **192** (1987) 421.
87M Mayer, R., Zhang, C.-S., Lynn, K.G., Frieze, W.E., Jona, F., Marcus, P.M.: Phys. Rev. B **35** (1987) 3102.
87N Nichtl, W., Bickel, N., Hammer, L., Heinz, K., Müller, K.: Surf. Sci. **188** (1987) L729.
87P1 Phaneuf, R.J., Williams, E.D.: Phys. Rev. B **35** (1987) 4155.
87P2 Peng, L.-M., Cowley, J.M.: J. Electron. Microsc. Tech. **6** (1987) 43.
87R Reiner, W., Penka, V., Skotte, M., Behm, R.J., Ertl, G., Moritz, W.: Surf. Sci. **186** (1987) 45.
87S Skottke, M., Behm, R.J., Ertl, G., Penka, V., Moritz, W.: J. Chem. Phys. **87** (1987) 6191.
87T Tabata, T., Aruga, T., Murata, Y.: Surf. Sci. **179** (1987) L63.
87U Usami, S., Nakai, H., Yaguchi, T., Kumashiro, Y., Fujimori, A.: J. Vac. Sci. Technol. A **5** (1987) 985.
87W1 Wang, Z.L., Lu, P., Cowley, J.M.: Ultramicroscopy **23** (1987) 205.
87W2 Watson, P.R., Mischenko III, J.: Surf. Sci. **186** (1987) 184.
87W3 Wolf, M., Gorschick, A., Loboda-Cackovic, J., Grunze, M., Unertl, W.N., Block, J.H.: Surf. Sci. **182** (1987) 489.
88A1 Adams, D.L., Jensen, V., Sun, X.F., Vollesen, J.H.: Phys. Rev. B **38** (1988) 7913.
88A2 Atanasoska, L., O'Grady, W.E., Atanasoski, R.T., Pollak, F.H.: Surf. Sci. **202** (1988) 142.
88A3 Aumann, C.E., Sauvage, D.E., Kariotis, R., Lagally, M.G.: J. Vac. Sci. Technol. A **6** (1988) 1963.
88B1 Bardi, U., Santucci, A., Rovida, G., Ross, P.N., in: The Structure of Surfaces II. Van der Veen, J.F., Van Hove, M.A. (eds.), Berlin: Springer-Verlag 1988, p. 147.
88B2 Barnes, C.J., Lindroos, M., King, D.A.: Surf. Sci. **210** (1988) 108.
88B3 Brookes, N., Viescas, A., Johnsons, P.D., Remeika, J.P., Cooper, A.S., Smith, N.V.: Surf. Sci. **203** (1988) L627.
88C Cowell, P.G., de Carvahlo, V.E.: J. Phys. C **21** (1988) 2983.
88D1 Dong, G., Ding, X., Huang, C., Chen, P., Madhukar, A.: Solid State Commun. **68** (1988) 411.
88D2 Duke, C..B., Paton, A., Wang, Y.R., Stiles, K., Kahn, A.: Surf. Sci. **197** (1988) 11.
88F Fery, P., Moritz, W., Wolf, D.: Phys. Rev. B **38** (1988) 7275.
88G1 Gratias, D., Cornier, M., Portier, R.: Acta Crystallogr. A **44** (1988) 789.
88G2 Gullikson, E.M., Mills jr., A.P., McRae, E.G.: Phys. Rev. B **37** (1988) 588.
88H1 Hitchen, G., Thurgate, S.: Surf. Sci. **197** (1988) 24.
88H2 Huang, H., Tong, S.Y., Packard, W.E., Webb, M.B.: Phys. Lett. **130** (1988) 166.
88I Ichimiya, A.: Acta Crystallogr. A **44** (1988) 1042.
88J1 Jamison, K.D., Zhon, D.N., Cohen, P.I., Zhao, T.C., Tong, S.Y.: J. Vac. Sci. Technol. A **6** (1988) 611.
88J2 Jennings, P.J.: Surf. Sci. **198** (1988) 180.
88J3 Jennings, P.J., Jones, R.O., Weinert, M.: Phys. Rev. B **37** (1988) 6113.
88J4 Jones, R.O., Jennings, P.J.: Surf. Sci. Rep. **9** (1988) 165.
88K1 Kambe, K.: Acta Crystallogr. A **44** (1988) 885.
88K2 Knibb, M. G., Maksym, P.A.: Surf. Sci. **195** (1988) 475.
88L1 Lad, R.J., Henrich, V.E.: Surf. Sci. **193** (1988) 81.
88L2 Lindberg, P.A.P., Shen, Z.-X., Wells, B.O., Mitzi, D.B., Lindau, I., Spicer, W.E., Kapitulnik, A.: Appl. Phys. Lett. **53** (1988) 2563.
88M1 MacLaren, J.M., Pendry, J.B., Rous, P.J., Saldin, D.K., Somorjai, G.A., Van Hove, M.A., Wedensky, D.D.: Surface Crystallographic Information Service. Dordrecht: D. Reidel Publishing Company 1988.
88M2 Men, F.K., Packard, W.E., Webb, M.B.: Phys. Rev. Lett. **61** (1988) 2469.
88M3 Mulhollan, G.A., Köymen, A.R., Lind, D.M., Dunning, F.D., Walters, G.K., Tamura, E., Feder, R.: Surf. Sci. **204** (1988) 503.
88N Noonan, J.R., Davis, H.L.: J. Vac. Sci. Technol. A **6** (1988) 722.
88O Oed, W., Dötsch, B., Hammer, L., Heinz, K., Müller, K.: Surf. Sci. **207** (1988) 55.
88P1 Phaneuf, R.J., Williams, E.D.: Surf. Sci. **195** (1988) 330.
88P2 Pendry, J.B., Heinz, K., Oed, W.: Phys. Rev. Lett. **61** (1988) 2953.
88P3 Pendry, J.B., Heinz, K., Oed, W., Landskron, H., Müller, K., Schmidtlein, G.: Surf. Sci. **193** (1988) L1.

88P4 Peng, L.-M., Cowley, J.M.: Surf. Sci. **199** (1988) 609.
88P5 Peng, L.-M., Cowley, J.M.: Surf. Sci. **201** (1988) 559.
88S1 Sowa, E.C., Kubiak, G.D., Stulen, R.H., Van Hove, M.A.: J. Vac. Sci. Technol. A **6** (1988) 832.
88S2 Sowa, E.C., Van Hove, M.A., Adams, D.L.: Surf. Sci. **199** (1988) 174.
88S3 Stoffel, N.G., Chang, Y., Kelly, M.K., Dottl, L., Onellion, M., Morris, P.A., Bonner, W.A., Margaritondo, G.: Phys. Rev. B **37** (1988) 7952.
88T1 Tong, S.Y., Huang, H., Wei, C.M., Packard, W.E., Men, F.K., Glander, G., Webb, M.W.: J. Vac. Sci. Technol. A **6** (1988) 615.
88T2 Tong, S.Y., Zhao, T.C., Poon, H.C., Jamison, K.D., Zhon, D.N., Cohen, P.I.: Phys. Lett. **128** (1988) 447.
88W1 Wang, Z.L.: Ultramicroscopy **24** (1988) 371.
88W2 Wang, Z.L., Lu, P.: Ultramicroscopy **26** (1988) 217.
88W3 Wang, Z.L., Cowley, J.M.: Surf. Sci. **193** (1988) 501.
88W4 Watson, P.R., Mitchell, K.A.R.: Surf. Sci. **203** (1988) 323.
88Z Zhao, T.C., Poon, H.C., Tong, S.Y.: Phys. Rev. B **38** (1988) 1172.
89B1 Bardi, U., Atrei, A., Ross, P.N., Zanazzi, E., Rovida, G.: Surf. Sci. **211/212** (1989) 441.
89B2 Bickel, N., Schmidt, G., Heinz, K., Müller, K.: Phys. Rev. Lett. **62** (1989) 2009.
89C Claessen, R., Manzke, R., Carstensen, H., Burandt, B., Buslaps, T., Skibowski, M., Fink, J.: Phys. Rev. B **39** (1989) 7316.
89D1 Duke, C.B., Paton, A., Wang, Y.R., Stiles, K., Kahn, A.: Surf. Sci. **214** (1989) 334.
89D2 Duke, C.B., Paton, A., Wang, Y.R., Stiles, K., Kahn, A.: Surf. Sci. **221** (1989) 619.
89F1 Fan, W.C., Ignatiev, A.: Phys. Rev. B **40** (1989) 5479.
89F2 Fan, W.C., Ignatiev, A., Huang, H., Tong, S.Y.: Phys. Rev. Lett. **62** (1989) 1516.
89H1 Horio, Y., Ichimiya, A.: Surf. Sci. **219** (1989) 128.
89H2 Horsky, T.N., Brandes, G.R., Canter, K.F., Duke, C.B., Horng, S.F., Kahn, A., Lessor, D.L., Mills jr., A.P., Paton, A., Steven, K., Stiles, K.: Phys. Rev. Lett. **62** (1989) 1876.
89I1 Ichimiya, A.: Surf. Sci. **219** (1989) 128.
89I2 Ichimiya, A.: Surf. Sci. **219** (1989) 352.
89K1 Kaplan, R.: Surf. Sci. **215** (1989) 111.
89K2 Kasch, M., Pehlke, E., Schattke, W., Kurberg, T., Barnscheidt, H.P., Manzke, R., Skibowski, M.: Surf. Sci. **214** (1989) 436.
89K3 Kleinle, G., Moritz, W., Adams, D.L., Ertl, G.: Surf. Sci. **219** (1989) L637.
89K4 Krewer, J.W., Feder, R.: Solid State Commun. **69** (1989) 87.
89L Lindroos, M., Barnes, C.-J., Valden, M., King, D.A.: Surf. Sci. **218** (1989) 269.
89M1 Ma, Y., Marks, L.D.: Acta Crystallogr. A **45** (1989) 174.
89M2 Meyer-Ehmsen, G.: Surf. Sci. **219** (1989) 177.
89M3 Mischenko III, J., Watson, P.R.: Surf. Sci. **209** (1989) L105.
89M4 Mischenko III, J., Watson, P.R.: Surf. Sci. **220** (1989) L667.
89M5 Myler, U., Jacobi, K.: Surf. Sci. **220** (1989) 353.
89N Nesterenko, B.A., Shkrebtii, A.I.: Surf. Sci. **213** (1989) 309.
89P Peng, L.-M.: Surf. Sci. **222** (1989) 296.
89R1 Rous, P.J., Pendry, J.B.: Comput. Phys. Commun. **54** (1989) 136.
89R2 Rous, P.J., Pendry, J.B.: Comput. Phys. Commun. **54** (1989) 157.
89R3 Rous, P.J., Pendry, J.B.: Surf. Sci. **219** (1989) 355.
89R4 Rous, P.J., Pendry, J.B.: Surf. Sci. **219** (1989) 373.
89S1 Sotto, M., Boulliard, J.C.: Surf. Sci. **214** (1989) 97.
89S2 Stoffel, N.G., Morris, P.A., Bonner, W.A., Chang, Y., Tang, M., Zanoni, R., Dottl, L., Chen, Q.B., Joyut, R., Huber, D.L., Onellion, M., Margaritondo, G.: Surf. Sci. **211/212** (1989) 1123.
89U Uhrberg, R.I.G., Landemark, E., Johansson, L.S.O.: Phys. Rev. B **39** (1989) 13525.
89W1 Wang, Z.L.: Surf. Sci. **214** (1989) 44.
89W2 Wang, Z.L.: Surf. Sci. **215** (1989) 201.
89W3 Wang, Z.L.: Surf. Sci. **215** (1989) 217.
89W4 Wang, Z.L., Lin, J., Lu, P., Cowley, J.M.: Ultramicroscopy **27** (1989) 101.
89Y Yang, H.-N., Lu, T.-M., Wang, G.-C.: Phys. Rev. Lett. **63** (1989) 1621.
90A Adams, D.L., Andersen, S.P., Buchhardt, J., in: The Structure of Surfaces III, Tong, S.Y., Van Hove, M.A., Takayanagi, K., Xie, X.D. (eds.), Berlin: Springer-Verlag 1990, p. 156.
90B1 Bardi, U., Atrei, A., Zanazzi, E., Rovida, G., Ross, P.N.: Vacuum **41** (1990) 437.
90B2 Beccat, P., Gauthier, Y., Baudoing-Savois, R., Bertolini, J.C.: Surf. Sci. **238** (1990) 105.
90B3 Bickel, N., Schmidt, G., Heinz, K., Mueller, K.: Vacuum **41** (1990) 46.
90B4 Borbonus, M., Haase, O., Koch, R., Rieder, K.H.: Vacuum **41** (1990) 321.

90B5 Breuer, U., Prince, K.C., Bonzel, H.P., Oed, W., Heinz, K., Schmidt, G., Mueller, K.: Surf. Sci. **239** (1990) L493.
90D1 Duke, C.B., Lessor, D.L.: Surf. Sci. **225** (1990) 81.
90D2 Duszak, R., Prince, R.H.: Surf. Sci. **234** (1990) L259.
90G1 Garni, B., Savage, D.E., Lagally, M.G.: Surf. Sci. **235** (1990) L324.
90G2 Gauthier, Y., Hoffmann, W., Wuttig, M.: Surf. Sci. **233** (1990) 239.
90G3 Gauthier, Y., Joly, Y., Rundgren, J., Johansson, L.I., Wincott, P.: Phys. Rev. B **42** (1990) 9328.
90H1 Haner, A.N., Ross, P.N., Bardi, U., in: The Structure of Surfaces III, Tong, S.Y., Van Hove, M.A., Takayanagi, K., Xie, X.D. (eds.), Berlin: Springer-Verlag 1990, p. 276.
90H2 Hara, S., Slijkerman, W.F.J., Van der Veen, J.F., Ohdomazi, I., Misawa, S., Sakuma, E., Yoshida, S.: Surf. Sci. **231** (1990) L196.
90H3 Heinz, K.: Vacuum **41** (1990) 328.
90H4 Heinz, K, Oed, W., Pendry, J.B., in: The Structure of Surfaces III, Tong, S.Y., Van Hove, M.A., Takayanagi, K., Xie, X.D. (eds.), Berlin: Springer-Verlag 1990, p. 139.
90H5 Holmes, D.J., King, D.A., Barnes, C.J.: Surf. Sci. **227** (1990) 179.
90H6 Hou, X.Y., Lu, X.K., Hao, P.H., Ding., X.M., Chen, P., Wang, X., in: The Structure of Surfaces III, Tong, S.Y., Van Hove, M.A., Takayanagi, K., Xie, X.D. (eds.), Berlin: Springer-Verlag 1990, p. 560.
90I Ichimiya, A., in: The Structure of Surfaces III, Tong, S.Y., Van Hove, M.A., Takayanagi, K., Xie, X.D. (eds.), Berlin: Springer-Verlag 1990, p. 162.
90K1 Kahn, A., Duke, C.B., in: The Structure of Surfaces III, Tong, S.Y., Van Hove, M.A., Takayanagi, K., Xie, X.D. (eds.), Berlin: Springer-Verlag 1990, p. 566.
90K2 Kleinle, G., Moritz, W., Ertl, G.: Surf. Sci. **238** (1990) 119.
90K3 Korte, U., Meyer-Ehmsen, G.: Surf. Sci. **232** (1990) 367.
90K4 Korte, U., Meyer-Ehmsen, G.: Vacuum **41** (1990) 343.
90L1 Larsen, P.K., Meyer-Ehmsen, G.: Surf. Sci. **240** (1990) 168.
90L2 Le Bossé, J.C., Lopez, J., Rousseau, J., Zasada, I., Wojtczak, L.: J. Phys. Condens. Matter **2** (1990) 3143.
90L3 Liepold, S., Elbel, N., Michl, N., Nichtl-Petcher, W., Heinz, K., Mueller, K.: Surf. Sci. **240** (1990) 81.
90L4 Liew, Y.-F., Wang G.-C.: Surf. Sci. **227** (1990) 190.
90L5 Lin, R.F., Li, Y.S., Jona, F., Marcus, P.M.: Phys. Rev. B **42** (1990) 1150.
90L6 Lin, R.F., Marcus, P.M., Jona, F., in: The Structure of Surfaces III, Tong, S.Y., Van Hove, M.A., Takayanagi, K., Xie, X.D. (eds.), Berlin: Springer-Verlag 1990, p. 150.
90M1 Martin, P., Blanckenhagen, P. v., Romahn, U., Schommers, W.: Vacuum **41** (1990) 349.
90M2 McRae, E.G., Malic, R.A.: Phys. Rev. B **42** (1990) 1509.
90M3 Moritz, W., Over, H., Kleinle, G., Ertl, G., in: The Structure of Surfaces III, Tong, S.Y., Van Hove, M.A., Takayanagi, K., Xie, X.D. (eds.), Berlin: Springer-Verlag 1990, p. 128.
90N1 Nagano, S.: Phys. Rev. B **42** (1990) 7363.
90N2 Noonan, J.R., Davis, H.L.: J. Vac. Sci. Technol. A **8** (1990) 2671.
90O Ormeci, A., Hall, B.M., Mills, D.L.: Phys. Rev. B **42** (1990) 4524.
90P1 Pavlovska, A., Steffen, H., Bauer, E.: Surf. Sci. **234** (1990) 143.
90P2 Pendry, J.B., Heinz, K.: Surf. Sci. **230** (1990) 137.
90P3 Pendry, J.B., Heinz, K., Oed, W.: Vacuum **41** (1990) 340.
90P4 Peng, L.-M., Whelan, M.J.: Surf. Sci. **238** (1990) L446.
90P5 Pinkawa, P., Crampin, S.: Surf. Sci. **233** (1990) 27.
90Q Quinn, J., Li, Y.S., Tian, D., Li, H., Jona, F., Marcus, P.M.: Phys. Rev. B **42** (1990) 11348.
90R1 Rous, P.J., in: The Structure of Surfaces III, Tong, S.Y., Van Hove, M.A., Takayanagi, K., Xie, X.D. (eds.), Berlin: Springer-Verlag 1990, p. 118.
90R2 Rous, P.J., Van Hove, M.A., Somorjai, G.A.: Surf. Sci. **226** (1990) 15.
90S1 Schneider, R., Duerr, H., Fauster, Th., Dose, V.: J. Vac. Sci. Technol. A **8** (1990) 3363.
90S2 Stock, M., Meyer-Ehmsen, G.: Surf. Sci. **226** (1990) L59.
90S3 Stock, M., Risse, J., Korte, U., Meyer-Ehmsen, G.: Surf. Sci. **233** (1990) L243.
90U1 Uebing, C.: Surf. Sci. **225** (1990) 97.
90U2 Uebing, C., Viefhaus, H.: Surf. Sci. **236** (1990) 29.

90X Xing, Y.R., Zhang, J.P., Wu, J.A., Liu, C.Z., Wang, C.H.: Surf. Sci. **232** (1990) L215.

90Z1 Zhang, X.-G., Rous, P.J., MacLaren, J.M., Gonis, A., Van Hove, M.A., in: The Structure of Surfaces III, Tong, S.Y., Van Hove, M.A., Takayanagi, K., Xie, X.D. (eds.), Berlin: Springer-Verlag 1990, p. 144.

90Z2 Zhang, X.-G., Rous, P.J., MacLaren, J.M., Gonis, A., Van Hove, M.A., Somorjai, G.A.: Surf. Sci. **239** (1990) 103.

90Z3 Zuo, J.-K., He, Y.-L., Wang, G.-C., Felter, T.E.: J. Vac. Sci. Technol. A **8** (1990) 2474.

91B1 Barrett, S.D., Blyth, R.J.R., Begley, A.M., Dhesi, S.S., Jordan, R.G.: Phys. Rev. B **43** (1991) 4573.

91B2 Barrett, S.D., Blyth, R.J.R., Dhesi, S.S., Newstead, K.: J. Phys. Condens. Matter **3** (1991) 1953.

91B3 Baudoing-Savois, R., Gauthier, Y., Moritz, W.: Phys. Rev. B **44** (1991) 12977.

91B4 Blanchard, D.L., Lessor, D.L., La Femina, J.P., Baer, D.R., Ford, W.K., Guo, T.: J. Vac. Sci. Technol. A **9** (1991) 1814.

91B5 Borbonus, M., Koch, R., Haase, O., Rieder, K.H.: Surf. Sci. **249** (1991) L317.

91C1 Chang, J.P., Blakely, J.M.: J. Vac. Sci. Technol. A **10** (1992) 2154.

91C2 Crampin, S., Rous, P.J.: Surf. Sci. **244** (1991) L137.

91G Gautier, M., Durand, J.P., Phan Van, L., Guittet, M.J.: Surf. Sci. **250** (1991) 71

91H1 Haner, A.N., Ross, P.N., Bardi, U.: Surf. Sci. **250** (1991) 15.

91H2 Heinz, K., Starke, U., Bothe, F.: Surf. Sci. **243** (1991) L70.

91H3 Hetterich, W., Korte, U., Meyer-Ehmsen, G., Heiland, W.: Surf. Sci. **254** (1991) L487.

91H4 Hitchen, G.J., Thurgate, S.M., Jennings, P.J.: Phys. Rev. B **44** (1991) 3939.

91K Kahn, A., Duke, C.B., Wang, Y.R.: Phys. Rev. B **44** (1991) 5606.

91L1 Lessor, D.L., Duke, C.B., Lippel, P.H., Brandes, G.R., Canter, K.F., Horsky, T.N.: J. Vac. Sci. Technol. A **9** (1991) 1874.

91L2 Li, H., Quinn, J., Li, Y.S., Tian, D., Jona, F., Marcus, P.M.: Phys. Rev. B **43** (1991) 7305.

91L3 Li, Y.S., Jona, F., Marcus, P.M.: Phys. Rev. B **43** (1991) 6337.

91L4 Li, Y.S., Jona, F., Marcus, P.M.: Phys. Rev. B **44** (1991) 8267.

91M Mendez, M.A., Palomares, F.J., Cuberes, M.T., Gonzales, M.L., Soria, F.: Surf. Sci. **251/252** (1991) 145.

91O Over, H., Kleinle, G., Ertl, G., Moritz, W., Ernst, K.-H., Wohlgemuth, H., Christmann, K., Schwarz, E.: Surf. Sci. **254** (1991) L469.

91P Powers, J.M., Wander, A., Rous, P.J., Van Hove, M.A., Somorjai, G.A.: Phys. Rev. B **44** (1991) 11159.

91Q Quinn, J., Li, Y.S., Jona, F., Fort, D.: Surf. Sci. **257** (1991) L647.

91S Smith, A.E.: Surf. Sci. **251/252** (1991) 253.

91W Wei, J., Williams, E.D., Park, R.L.: Surf. Sci. **250** (1991) L368

91Z Zhang, X.-G., Van Hove, M.A., Somorjai, G.A., Rous, P.J., Tobin, D., Gonis, A., MacLaren, J.M., Heinz, K., Michl, M., Lindner, H., Mueller, K., Ehsasi, M., Block, J.H.: Phys. Rev. Letters **67** (1991) 1298.

92A1 Atrei, A., Bardi, U., Rovida, G., Torrini, M., Zanazzi, E.: Phys. Rev. B **46** (1992) 1649.

92A2 Atrei, A., Pedocchi, L., Bardi, U., Rovida, G., Torrini, M., Zanazzi, E., Van Hove, M.A., Ross, P.N.: Surf. Sci. **261** (1992) 64.

92D1 Davis, H. L., Hannon, J.B., Ray, K.B., Plummer, E.W.: Phys. Rev. Lett. **68** (1992) 2632.

92D2 De Andres, P.L.: Surf. Sci. **269/270** (1992) 1.

92E Ettema, A.R.H.F., Wiegens, G.A., Haas, C.: Surf. Sci. **269/270** (1992) 1161.

92G1 Gao, Y., Park, K.T.: Phys. Rev. B **46** (1992) 1743.

92G2 Gauthier, Y., Baudoing-Savois, R., Bignard, J.M., Bardi, U., Atrei, A.: Surf. Sci. **276** (1992) 1.

92H1 Hamon, J.B., Plummer, E.W., Wentzcovitch, R.M., Lam,. P.K.: Surf. Sci. **269/270** (1992) 7.

92H2 Horsky, T.N., Brandes, G.R., Canter, K.F., Duke, C.B., Paton, A., Lessor, D.L., Kahn, A., Horng, S.F., Stevens, K., Stiles, K., Mills jr., A.P.: Phys. Rev. B **46** (1992) 4830.

92J Joly, Y.: Phys. Rev. Lett. **68** (1992) 950.

92K1 Kahn, A., Ahsan, S., Chen, W., Dumas, M., Duke, C.B., Paton, A.: Phys. Rev. Lett. **68** (1992) 3200.

92K2 Korte, U., Meyer-Ehmsen, G.: Surf. Sci. **271** (1992) 616.

92K3 Korte, U., Meyer-Ehmsen, G.: Surf. Sci. **277** (1992) 109.
92L1 Le Bossé, J.C., Lopez, J., Rousseau, J., Zasada, I.: J. Phys. Condens. Matter **4** (1992) 1671.
92L2 Le Bossé, J.C., Lopez, J., Rousseau, J., Zasada, I.: J. Phys. Condens. Matter **4** (1992) 1685.
92L3 Lessor, D.L., Duke, C.B., Chess, X.M., Brandes, G.R., Canter, K.F., Ford, W.K.: J. Vac. Sci. Technol. A **10** (1992) 2585.
92L4 Li, Y.S., Quinn, J., Jona, F., Marcus, P.M.: Phys. Rev. B **46** (1992) 4830.
92M1 McCoy, J.M., Korte, U., Maksym, P.A., Meyer-Ehmsen, G.: Surf. Sci. **261** (1992) 29.
92M2 Mitura, Z., Daniluk, A.: Surf. Sci. **277** (1992) 229.
92O Oed, W., Rous, P.J., Pendry, J.B.: Surf. Sci. **273** (1992) 261.
92P1 Peng, L.-M., Whelan, M.J.: Surf. Sci. **268** (1992) L325.
92P2 Powers, J.M., Wander, A., Van Hove, M.A., Somorjai, G.A.: Surf. Sci. **260** (1992) L7.
92Q Quinn, J., Li, Y.S., Jona, F., Fort, D.: Phys. Rev. B **46** (1992) 9694.
92R Rundgren, J., Gauthier, Y., Baudoing-Savois, R., Joly, Y., Johansson, L.I.: Phys. Rev. B **45** (1992) 4445.
92S Schildbach, M.A., Hamza, A.V.: Phys. Rev. B **45** (1992) 6197.
92W1 Wander, A., Pendry, J.B., Van Hove, M.A.: Phys. Rev. B **46** (1992) 9897.
92W2 Wei, C.M., Tong, S.Y.: Surf. Sci. **274** (1992) L577.
92Z1 Zschack, P., Cohen, J.B., Chung, Y.W.: Surf. Sci. **262** (1992) 395.
92Z2 Zuo, J.M., Liu, J.: Surf. Sci. **271** (1992) 253.
93A1 Atrei, A., Bardi, U., Galeotti, M., Rovida, G., Torrini, M., Zanazzi, E.: Proc. Int. Conf. Struct. Surf. IV (ICSOS IV), 1993, p. 539.
93A2 Atrei, A., Bardi, U., Torrini, M., Zanazzi, E., Rovida, G., Kasamura, H., Kudo, M.: J. Phys. Condens. Matter **5** (1993) L207.
93B1 Begley, A.M., Kim, S.K., Jona, F., Marcus, P.M.: Phys. Rev. B **48** (1993) 12326.
93B2 Bugnard, J.M., Baudoing-Savois, R., Gauthier, Y., Hill, E.K.: Surf. Sci. **281** (1993) 62.
93C1 Chen, X.M., Canter, K.F., Duke, C.B., Paton, H., Lessor, D.L., Ford, W.K.: Phys. Rev. B **48** (1993) 2400.
93C2 Chen, W., Paul, J.A.K., Barbieri, A., Van Hove, M.A., Cameron, S., Dwyer, D.J.: J. Phys. Condens. Matter **5** (1993) 4585.
93D1 Daley, R.S., Felter, T.E., Hildner, M.L., Estrup, P.J.: Phys. Rev. Lett. **70** (1993) 1295.
93D2 Dorenbos, G., Boerma, D.O., Buck, T.M., Wheatley, G.H.: Proc. Int. Conf. Struct. Surf. IV (ICSOS IV), 1993, p. 531.
93D3 Dassanayake, U.M., Chen, W., Kahn, A.: J. Vac. Sci. Technol. B **11** (1993) 1467.
93D4 Davis, H.L., Noonam, J.R.: Abstracts of the Int. Conf. Struct. Surf. IV (ICSOS IV), 1993, 51
93D5 Dhanak, V.R., Baraldi, A., Comelli,.G., Paolucci., G., Kiskinova, M., Rosei, R.: Surf. Sci. **295** (1993) 387.
93F1 Ford, W.K.: Surf. Sci. **292** (1993) 342.
93F2 Fritzsche, V., Pendry, J.B., Loeffler, U., Wedler, H., Mendez, M.A., Heinz, K.: Surf. Sci. **289** (1993) 389.
93G Gauthier, Y., Baudoing-Savois, R., Rosink, J.J.W.M., Sotto, M.: Surf. Sci. **297** (1993) 193.
93H1 Herman, J.W., Elsayed-Ali, H.E., Murphy, E.A.: Phys. Rev. Lett. **71** (1993) 400.
93H2 Hu, P., King, D.A.: Abstracts of the Int. Conf. Struct. Surf. IV (ICSOS IV), 1993, p. 89.
93H3 Huang, Y., Cowley, J.M.: Surf. Sci. **285** (1993) 42.
93H4 Hikita, T., Hanada, T., Kudo, M., Kawai, M.: J. Vac. Sci. Technol. A **11** (1993) 2649.
93K Kaukasoina, P., Lindroos, M., Diehl, R., Fisher, D., Chandavarkar, S., Collins, I.R.: J. Phys. Condens. Matter **5** (1993) 2875.
93M1 McCoy, J.M., Korte, U., Maksym, P.A., Meyer-Ehmsen, G.: Phys. Rev. B **48** (1993) 4721.
93M2 Mendez, M.A., Glueck, C., Wagner, M., Loeffler, U., Doell, R., Heinz, K.: Surf. Sci. **289** (1993) 389.
93M3 Moritz, W., Over, H.: Proc. Int. Conf. Struct. Surf. IV (ICSOS IV), 1993, p. 103.
93N1 Nagano, S.: Proc. Int. Conf. Struct. Surf. IV (ICSOS IV), 1993, p. 113.
93N2 Nakayama, K., Usami, S.: Surf. Sci. **287/288** (1993) 355.
93O Over, H., Moritz, W., Ertle, G.: Phys. Rev. Lett. **70** (1993) 315.
93P Peacor, S.D., Hibma, T.: Surf. Sci. **287/288** (1993) 403.

93Q Quinn, J., Wang, C.P., Jona, F., Marcus, P.M.: J. Phys. Condens. Matter **5** (1993) 541.
93R1 Rakova, E.V.: Surf. Sci. **287/288** (1993) 400.
93R2 Rous, P.J.: Surf. Sci. 296 (1993) 358.
93S Sprunger, P.T., Pohl, K., Davis, H.L., Plummer, E.W.: Surf. Sci. **297** (1993) L48.
93T Tong, S.Y., Over, H., Duan, Z.G., Huang, H., Guo, X.Q.: Abstracts of the Int. Conf. Struct. Surf. IV (ICSOS IV), 1993, p. 45.
93V Venus, D.: Surf. Sci. **291** (1993) 418.
93W1 Warren, D.L., Thiel, P.A.: Phys. Rev. B **47** (1993) 10848.
93W2 Wei, C.M., Tong, S.Y.: Abstracts of the Int. Conf. Struct. Surf. IV (ICSOS IV) 1993, p. 88.
93Z1 Zhao, R.G., Jia, J.F., Zhang, Y., Yang, W.S.: Abstracts of the Int. Conf. Struct. Surf. IV (ICSOS IV), 1993, p. 47.
93Z2 Zhao, T.C., Tong, S.Y.: Phys. Rev. B **47** (1993) 3923.
93Z3 Zhou, Y., Campuzano, J.C., Koukasoina, P., Lindroos, M., Jennings, G.: Phys. Rev. B **47** (1993) 10395.

6.2 Inelastic scattering of electrons

6.2.1 Introduction

6.2.1.1 Preliminary remarks

In high resolution electron energy loss spectroscopy (acronyms HR-EELS or EELS) electrons with well defined impact energy and direction of incidence impinge on a crystal. The energy spectrum and often the angular distribution of the scattered electrons are measured and related to the frequency and to the momentum of the excited modes. The technique was applied in the sixties and in the seventies to the investigation of bulk properties using high energy electrons in transmission through thin foils [80R]. Surface effects were observed, too, but the interest in their study did not grow before UHV techniques allowed to prepare reproducible and well defined solid surfaces [88R]. Primary energies of less than 1 keV are particularly advantageous for surface studies as the mean free path of the electrons has in most materials a minimum of about 1nm at 100 eV and increases slowly below and above this energy. When electrons in this energy range are backscattered from a solid the interaction is therefore limited to the outermost atomic layers of the crystal. After some pioneering work on adsorbate vibrations [67P] the potentiality of EELS for surface studies was definitely recognized in the seventies [77I2, 82I]. EELS gave an outstanding contribution to the development of surface chemistry and of the science of catalysis. Precious results were obtained for clean surfaces.

The acronym EELS is used also for studies of interband transitions without analysis of the momentum transfer in which the inelastic signal is collected over a large portion of the solid angle, i.e. integrating over the momentum transfer. This review does not include these studies being limited to measurements performed with high energy resolution and well defined momentum transfers.

Progress in the field of EELS over the past decade has been surveyed, in addition to the above cited reviews of Raether [80R,88R] and Ibach and Mills [82I], by Lucas and Šunjić [71L] and Mills [77M] for the theoretical aspects, by Lüth [83L1, 83L3, 88L] for semiconductor surfaces, by Ho [85H] for time resolved EELS and by Rocca et al. [86R2] for the measurement of surface phonon dispersion.

6.2.1.2 The EELS experiment

Conventionally the electron interaction with the surface is divided into small angle dipole scattering, caused by far reaching electric fields outside of the crystal, and large angle impact scattering, which arises from the short range interaction with the ionic nuclei of the surface. At low impact energy, E_i the dipole or Coulomb interaction is generally dominant over impact scattering and gives rise to the so called dipole lobe around the specular direction. On the other hand impact scattering is dominant at high impact energies and large transferred momenta. A pictorial view of impact scattering intensity and of the dipole lobe is given in Fig. 1.

The most appropriate E_i to be used in the experiment depends therefore on the surface property under investigation, i.e. whether a dipole moment is connected to the excitation or whether momentum transfer is implied in the measurement.

Dipole active vibrations and electronic excitations of the clean surface can most efficiently be studied with E_i below 20 eV. A link exists between some of the results which will be presented here and those derived from the optical methods discussed in chapter 8.1, subvolume d.

Impact scattering is dominant above $E_i \approx 30$ eV and out-of-specular. In this regime EELS was applied to study the surface phonon spectra [86R2] which are reviewed in chapter 4.1, subvolume b. The measurement of inelastic cross section of surface-phonons allowed to derive information on the structure of the surface. Moreover in dipole and impact scattering information could be inferred by making use of the selection rules relevant for the excitation of surface vibrational modes.

An outlook of the kinematics of the electron scattering experiment is given in Fig. 2 with the definition of the symbols used below. The primary electrons with energy E_i and wavevector $\boldsymbol{k}_i$ impinge on the crystal with an angle θ_i and are reflected with an energy E_s and a wavevector $\boldsymbol{k}_s$ at the angle θ_s after suffering an energy loss E_{loss} and having experienced a momentum transfer $\boldsymbol{q}_{\|}$. Conservation of the component of the momentum of the primary electrons parallel to the surface

$$q_{\|} = k_i \sin\theta_i - k_s \sin\theta_s \tag{1}$$

and of the energy

$$E_{loss} = E_i - E_s \tag{2}$$

defines the kinematic of the experiment as can be best visualized by use of the Ewald construction (Fig. 3). For $E_{loss} << E_i$, as is the case for surface phonon excitations, the small electron mass implies nearly vertical cuts in momentum space.

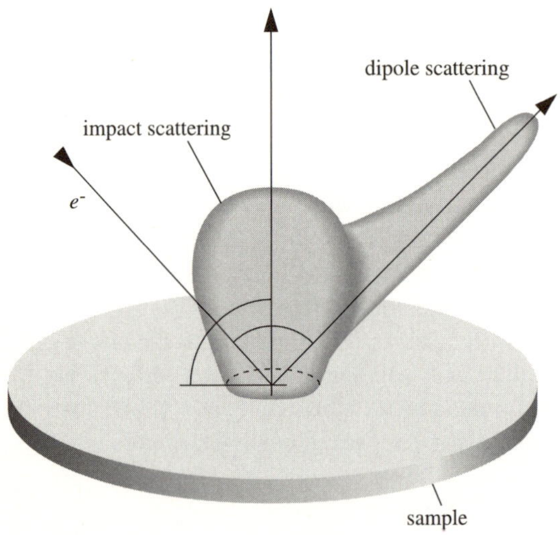

Fig. 1. Impact and dipole scattering intensity for low energy electrons at low E_i. In the forward specular direction dipole scattering is more effective than impact scattering and gives rise to the dipole lobe. Off-specular impact scattering is dominant.

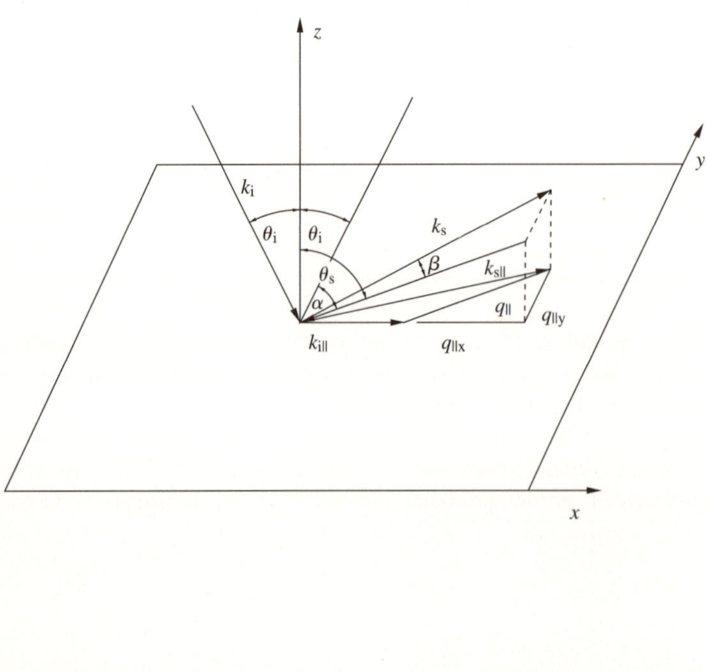

Fig. 2. Kinematical conditions for electron scattering off the surface of a crystal. α and β denote the in-plane and the off-plane angular deviation from the specular direction [91I].

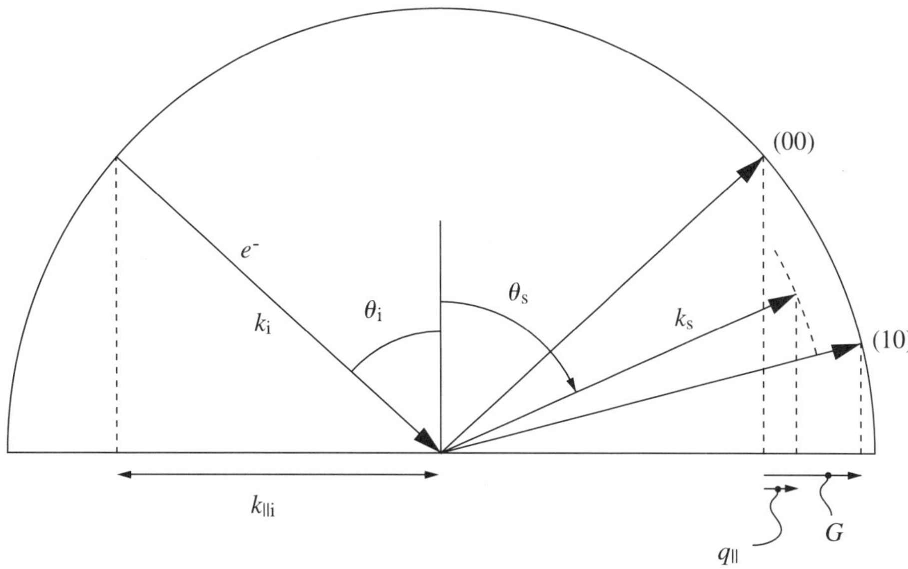

Fig. 3. Energy and momentum conservation illustrated by the Ewald construction for in-plane scattering. The energy loss $E_s - E_i$ and the momentum transfer $\boldsymbol{q}_\|$ define the trajectory of the scattered electron. The elastic diffraction event into the (10) beam with exchange of the reciprocal lattice vector $\boldsymbol{G}$ is also shown.

6.2.1.3 EEL-spectrometer designs

To produce monoenergetic beams of electrons electrostatic deflecting devices are generally used. Various designs of the monochromator and of the analyzer were proposed over the years based on the focusing properties of cylindrical mirror analyzers (CMA 120°) [79A,82A], spherical deflector analyzers (SDA 180°) [85H, 86T, 87P1, 88G, 89J] or cylindrical deflector analyzers (CDA 127°) [75F1, 77R, 79S, 82N, 82I, 83K, 84F1, 86S, 91R2] as illustrated in Fig. 4. Two EEL-spectrometers are shown in Figs. 5 and 6. The monochromator selects the electrons inside of a given energy range from those emitted by a thermoionic source (normally either a tungsten hairpin cathode or a LaB_6 single crystal). A double stage of monochromatization is usually employed to remove the background in the EEL-spectra. In the EELS experiment the monochromatic electron beam is scattered off the sample and the scattered electrons are analyzed in energy by a second electrostatic system. To record the energy loss spectrum the potential of the analyzer is usually shifted rigidly with respect to the monochromator. High levels of energy resolution are reached when the monochromatization and the analysis are performed at low pass energy. The electrons are then accelerated between the exit slit of the monochromator and the entrance slit of the field free scattering chamber, where the champion sits, to reach the desired impact energy, and decelerated thereafter to analyze them. Low energy electrons may easily be deviated from their trajectory by spurious magnetic fields. All components of an EEL spectrometer are therefore constructed of non-magnetic materials and the earth magnetic field has to be suppressed up to some mgauss. This is usually done including the EELS in a mu-metal shield. The obtainable resolution may be limited by the patchy surface potentials of the electrodes which constitute the spectrometer, too. All electrodes are therefore covered by graphite to achieve a uniform work function.

As in any other spectroscopy one has also in EELS a trade-off between energy resolution and available signal. In spectrometer design the matter of particular concern is improving the available current at a given resolution. As generally recognized the current obtainable with an EELS is limited by the space

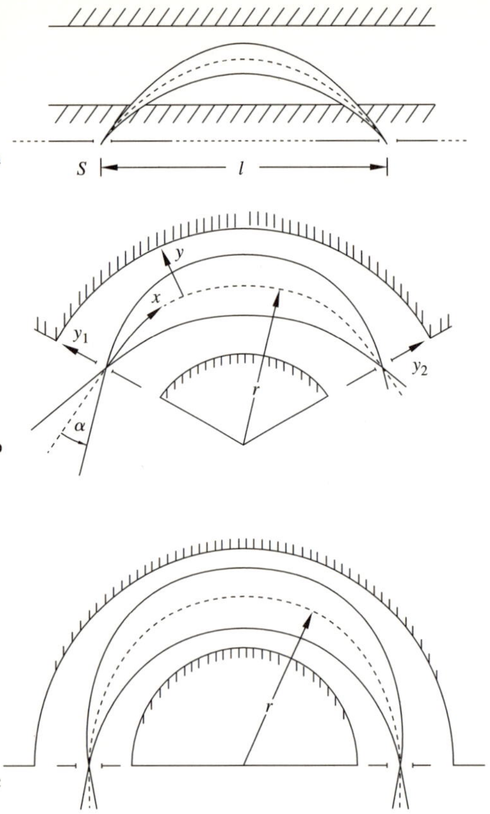

Fig. 4a-c. Electrostatic energy dispersive devices used in low energy EELS: **a** cylindrical mirror analyzer; **b** 127° cylindrical deflector; **c** 180° spherical deflector. Electrons which enter the capacitors normally to the slit follow a trajectory with radius r. The focusing property is defined by the crossing point of the trajectories which enter the device with angular deviation α and the central trajectory. The entrance slit y_1 is imaged on the exit slit y_2 [82I].

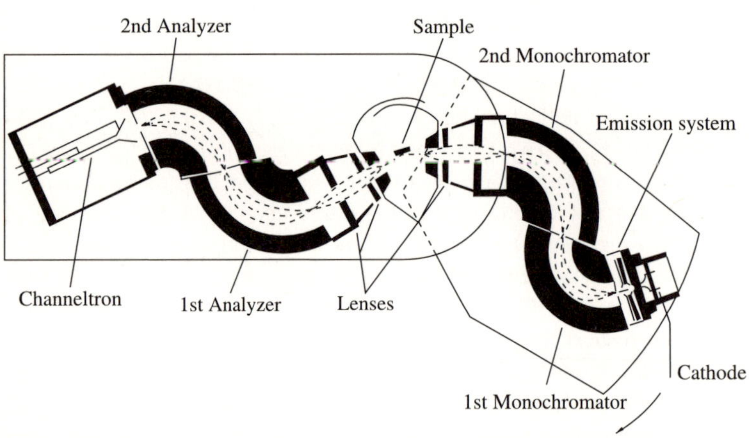

Fig. 5. Double pass CDA EEL-spectrometer developed by H. Ibach and collaborators. The trajectories of the electrons in the spectrometer are shown by dashed lines [91I].

charge in the monochromator. The CDA geometry is in this respect superior to SDA or CMA as the energy resolution depends principally on the slit width and only weakly on its height. Using rectangular slits the charge of the electron beam is therefore distributed over a larger area than would be the case for a circular aperture with the same performance. Most commercial spectrometers were therefore built following the CDA design. The latest development of CDA-EELS is shown in Fig. 5 [91I]. The deflec-

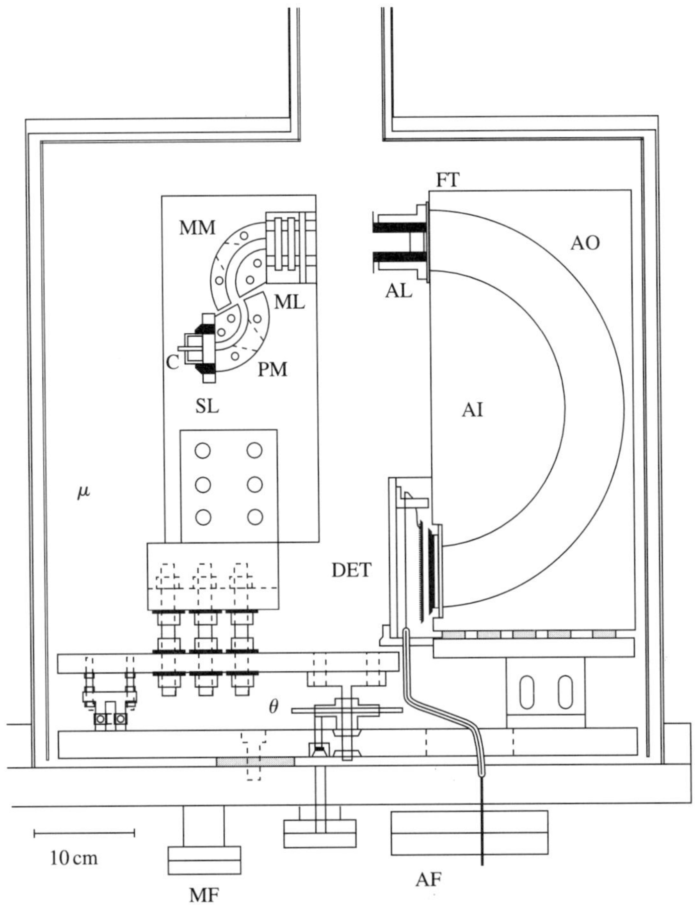

Fig. 6. EEL-spectrometer with parallel detection developed by W. Ho and collaborators [88G]. The double pass monochromator is based on the CDA(127°): cylindrical deflector analyzer design. The single pass analyzer is an SDA(180°): spherical deflector analyzer: μ: mu metal shielding, MF: monochromator feedthrough, C: indirectly heated cathode, SL: slit lenses for electron extraction, PM: premonochromator, MM: main monochromator, ML: monochromator lens elements, AL: analyzer lens elements, FT: field termination wires, AO: and AI: analyzer outer and inner hemisphere, DET: anode detector, AF: analyzer feedthrough. The monochromator is mounted on a rotating table

tion angle of the energy selective devices was optimized following three dimensional trajectory calculations which include effects due to space charge [91I, 92V]. The electron transmission was thus improved compared to previous designs based on two-dimensional trajectory simulations [70R, 83O, 86O] or on analytical corrections [35H]. With this spectrometer resolutions of 1 meV were achieved in dipole scattering, while out-of-specular the best resolution, reported so far in EELS measurements is 2.5 meV. This last value corresponds to a current of 10^{-11} A in the direct beam which is necessary to have a sufficient count rate for surface-phonon measurements.

Electrostatic lenses are inserted between the electron source and the entrance slit of the monochromator to extract and focus the beam as well as between the exit slit of the monochromator and the sample. Computer simulations allowed to optimize the focusing properties of the lens system at low [86S] as well as at high impact energies [84S1, 90S1, 91I].

The detection efficiency of the electrons may be improved on the other hand by parallel detection, i.e. by means of a channelplate [83O, 85F, 86R1, 88G, 89J, 89R]. The SDA geometry seems to be the most apt for this development because of its second order focusing property. A spectrometer based on this design (see Fig. 6) allowed to perform time resolved measurements on the msec time scale [86R1]. The SDA design looks promising also for the production of electron beams with high spacial resolution. A spot diameter of 100 μm was applied to study the spacially non homogeneous surfaces of high T_c superconductors [90P]. This spot diameter has to be compared with an area of 0.3×3 mm^2 illuminated by the electron beam generated by a CDA spectrometer as the one of Fig. 5.

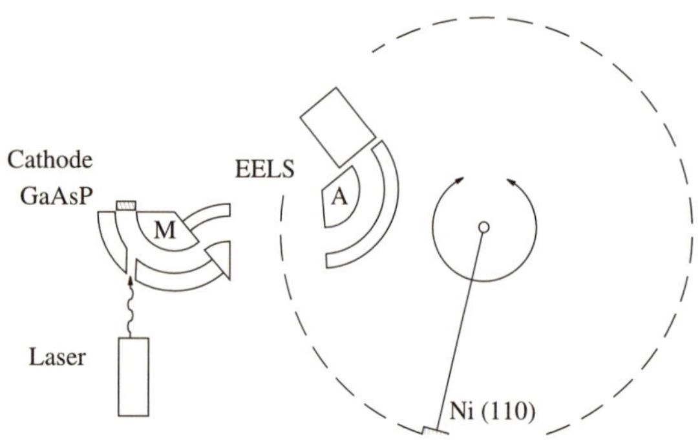

Fig. 7. Schematic design of a spin polarized EELS. The beam is produced by laser irradiation of a GaAsP crystal. The cylindrical condenser placed after the monochromator M corrects the electron trajectories to generate a transverse polarization of the beam. The scattered electrons are analyzed in energy by the analyzer A. The crystal can be moved in front of a CMA for Auger analysis and in front of a LEED (not shown) [84R1].

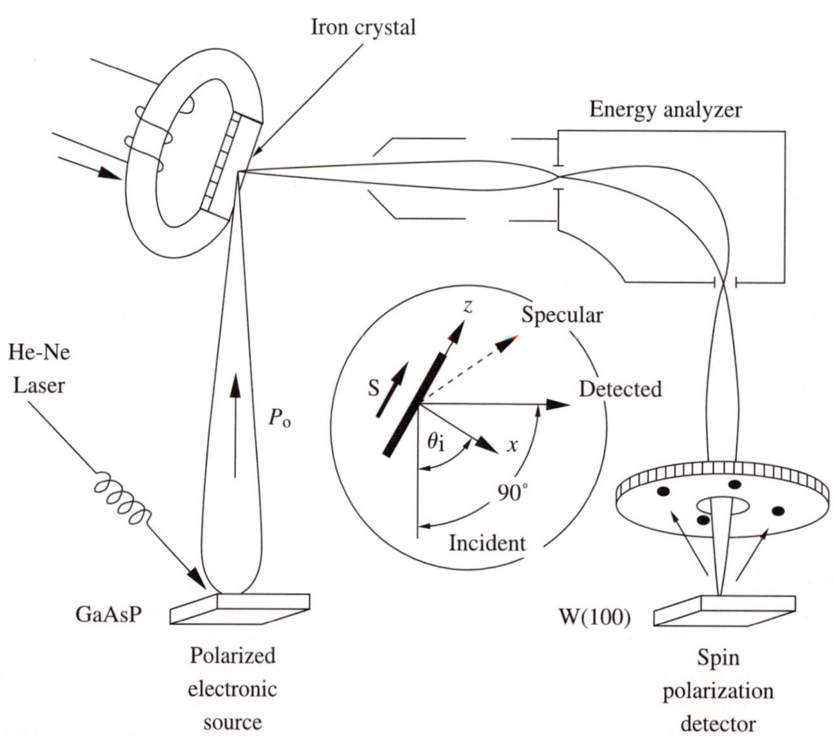

Fig. 8. Schematic picture of the experiment for spin polarized EEL-studies [88V]. Spin polarized electrons are extracted from a GaAsP cathode and scattered from an iron crystal with well defined magnetization. Electrons scattered by 90° are collected by an energy and spin analyzer. The illustrated electron trajectory is distorted for clarity; in reality the electrons are directed into the page as they emerge from the energy analyzer. The scattering geometry and the polarization of the sample S are shown in the inset.

The energy loss spectra are integrated over the angular acceptance of the EELS. A large acceptance is welcome for out of specular measurements in impact scattering because of the low signal to noise ratio. A large angular acceptance can on the other hand come at nuisance for the investigation of surface electronic excitations which may disperse significantly with momentum already over some 10^{-2}Å^{-1}. Zoom lenses may allow in the future to vary the angular acceptance of the EELS.

Further technical improvements include the possibility to rotate the analyzer [89J] or the sample out of plane [92I] to circumvent the selection rules for the excitation of surface phonons. To study insulators the crystal surface was neutralized by flooding the crystal with an intense electron beam at high energy where the secondary yield is unitary using a secondary electron gun [86T].

Magnetic properties were studied with a spin polarized beam produced by substituting the thermoionic cathode with a GaAsP crystal as shown in Fig. 7 [84R1]. In another experiment the scattered electrons were analyzed in spin, too, (Fig. 8) [86K]. Because of the limited count rate, resolutions of 30 meV and of 400 meV were used in the two experiments, respectively.

Alternatively EEL-spectrometers may be constructed according to the principle of dispersion compensation [84K1]. This design is aimed to provide high currents for time resolved studies. This system seems however not to be competitive with the last development of the CDA design because of its sensitivity to small aberrations of the image caused by the lens system which limits severely the obtainable resolution [91I].

Some of the measurements which will be presented were recorded with fast electrons $E_i > 30$ keV. The requirements for such experiments are quite different than at low energy. For details the reader is referred to chapter 11 of [80R].

6.2.1.4 Inelastic cross section

The description of the inelastic scattering process requires to address the nature of the interaction between the incoming electron and the surface. This is, in general, a complex problem since the electron interacts strongly with the substrate. It proves useful to distinguish between dipole and impact scattering, though no sharp and precisely defined boundaries may be traced between them. In dipole scattering approximation the EEL-spectra are quantitatively described in terms of classical dielectric theory [71L, 75M, 85P1]. Full multiple scattering calculations, analogous to those performed for low energy electron diffraction (LEED), are on the other hand necessary to reproduce the scattering cross section in the impact scattering regime [80T].

6.2.1.4.1 Dipole scattering

In dipole scattering the inelastic scattering cross section is related to the dielectric response of the system where the dielectric constant $\epsilon(q_\parallel, \omega)$ characterizes each solid at the frequency ω. The Coulomb field, accompanying the impinging electron interacts with the electron gas at the surface creating a space and time dependent polarization field. This polarization field is damped proportionally to the imaginary part ϵ_2 of the dielectric constant, while the field is screened like $\frac{1}{\epsilon}$ inside the dielectric medium and like $\frac{1}{\epsilon+1}$ by polarization outside of the dielectric medium. The inelastic cross section is proportional to: $\epsilon_2/\epsilon^2 = -\text{Im}(\frac{1}{\epsilon})$ for bulk losses and to $\epsilon_2/(\epsilon+1)^2 = -\text{Im}(\frac{1}{\epsilon+1})$ for surface losses.

The information obtainable with EELS in dipole scattering is thus equivalent to that of optical absorption [77I1, 77F] (see chapter 8.1 of subvolume d). As an example the comparison is shown in Fig. 9 for the optical constant of the surface-state layer of Si(111)2 × 1 as derived by EELS [75F2] with the result of ellipsometry [74M1] and multiple-internal reflection [71C]. When comparing results of the two techniques two aspects must be kept in mind:

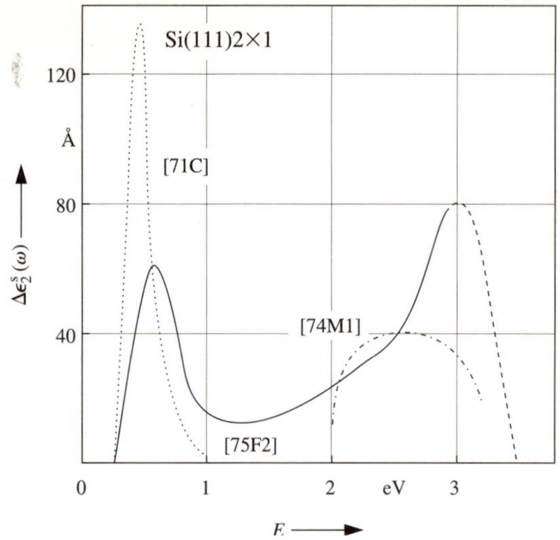

Fig. 9. Si. Comparison between the optical constant of the surface-state layer $\Delta\epsilon_2^s = \epsilon_2^s - \epsilon_2^b$ for Si(111)2 × 1 as determined by EELS (solid line) [75F2] and optical methods: multiple internal reflection (dotted line) [71C] and ellipsometry (dot-dashed line) [74M1]. For energies up to 3 eV $\Delta\epsilon_2^b$ may be neglected. For higher energies the difference between $\Delta\epsilon_2^s$ and $\Delta\epsilon_2^b$ becomes small as indicated by the dashed line.

a) the loss function of the electrons is modulated by $\frac{1}{(\epsilon+1)^2}$ while optical absorption is proportional to ϵ_2, the loss peaks in EELS may therefore be displaced with respect to the corresponding optical data;

b) in optical absorption ϵ, the transverse component of the dielectric function is relevant while in EELS it is the longitudinal component. EELS and optical measurements will coincide only when the longitudinal and the transverse components of the dielectric function of the system under investigation are identical.

The first quantitative analysis of EEL spectra was presented by Lucas and Šunjić [71L, 72L]. A full quantum mechanical treatment of dipole scattering was developed later on by Mills [72E, 73E, 75M, 77M, 82I] and Persson [80P, 83P, 85P1]. Following these authors the differential scattering cross section in the solid angle $\mathrm{d}\Omega$ and in the frequency range $\mathrm{d}\omega$, $\mathrm{d}^2S/\mathrm{d}\omega\,\mathrm{d}\Omega$ is given by:

$$\frac{\mathrm{d}^2 S}{\mathrm{d}\omega\,\mathrm{d}\Omega} = \frac{m^2 e^2 v_\perp^2}{2\pi^2\hbar^5 \cos\theta_i}\frac{k_s}{k_i}\frac{P(q_\parallel,\omega)}{q_\parallel^2} \times \frac{|\, v_\parallel q_\parallel (R_s + R_i) + i(R_i - R_s)(\omega - \boldsymbol{v}_\parallel \cdot \boldsymbol{q}_\parallel)\,|^2}{[v_\perp^2 q_\parallel^2 + (\omega - \boldsymbol{v}_\parallel \cdot \boldsymbol{q}_\parallel)^2]^2} \tag{3}$$

where m and e are the mass and electric charge of the electron. R_i and R_s are the complex reflectivity coefficients before and after suffering the energy loss, $v_\parallel$ and $v_\perp$ are the parallel and normal components of the velocity of the impinging electron with respect to the surface and $P(q_\parallel,\omega)$ is the loss function. The divergence for $\theta_i \to 90°$ is removed for more realistic models, e.g. taking the image potential into account [80P]. The denominator in Eq. (3) has a minimum for $\omega = v_\parallel q_\parallel$. This condition, called surfing condition, corresponds to the interaction of the electron with partial waves of a phase velocity $\omega/q_\parallel$ equal to $v_\parallel$. If we define by θ the deviation of the trajectory of the inelastic electron from the specular direction, then for $\omega << 1$ and $\theta << 1$ the denominator in Eq. (3) can be arranged to read:

$$v_\perp^2 q_\parallel^2 + (\omega - \boldsymbol{v}_\parallel \cdot \boldsymbol{q}_\parallel)^2 = 4E_i^2(\theta^2 + \theta_E^2)\cos^2\theta_i \tag{4}$$

with $\theta_E = \hbar\omega/2E_i$. Eq. (4) highlights the angular dependence of dipole scattering and its concentration in a lobe with halfwidth θ_E. The four amplitudes, which add up in the matrix element of Eq. (3), are represented in Fig. 10. In process (b) the energy loss takes place before the diffraction event while in process (c) the opposite occurs and the electron is first diffracted and then inelastically scattered. Processes (a) and (d) are important only at large angles from the specular direction. If the elastic reflectivity R is constant over the energy range of the experiment one obtains that the loss cross section is proportional to the specular intensity [72I]. This proportionality simplifies the data analysis substancially and contributed to the popularity of EELS.

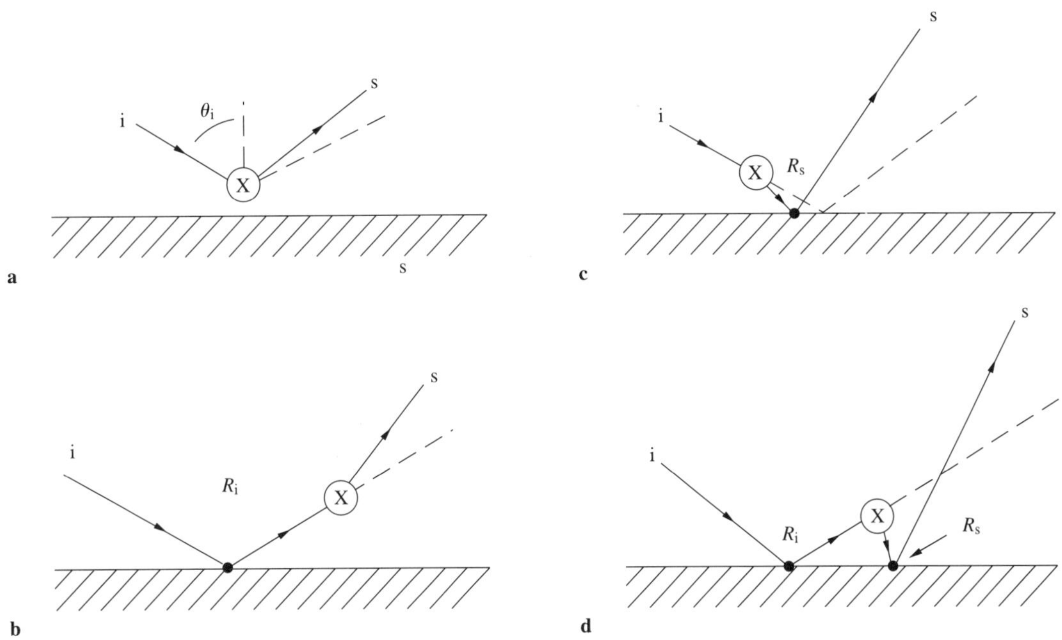

Fig. 10a-d. The four scattering processes which take an electron from the initial state k_i to the final state k_s near the specular direction. The encircled crosses denote the point where the electron scatters from the fluctuating electric field in the vacuum above the crystal. R_i and R_s are the reflection amplitudes at E_i and E_s respectively (after Mills [75M, 82I]).

A useful expression for the loss function was derived for a three layer model description of the crystal (vacuum, surface layer, bulk) and reads [75M, 82I]:

$$P(q_\parallel, \omega) = \frac{2\hbar q_\parallel}{\pi}[1 + n(\omega)]\mathrm{Im}(\frac{-1}{\tilde{\epsilon}(q_\parallel, \omega) + 1}) \tag{5}$$

where $\tilde{\epsilon}$ is the effective dielectric function of the system defined as:

$$\tilde{\epsilon}(q_\parallel, \omega) = \epsilon_s(\omega)[\frac{1 + \Delta(\omega)e^{-2q_\parallel W}}{1 - \Delta(\omega)e^{-2q_\parallel W}}] \tag{6}$$

and

$$\Delta = \frac{\epsilon_b(\omega) - \epsilon_s(\omega)}{\epsilon_b(\omega) + \epsilon_s(\omega)} \tag{7}$$

$n(\omega) = 1/(e^{\hbar\omega/k_B T} - 1)$ is the Bose-Einstein factor and ϵ_s is the dielectric function in the surface layer of thickness W and ϵ_b is the dielectric function in the bulk.

This model can be applied directly to the case of semiconductors. Their surfaces exhibit generally a large density of states in the bulk band gap which cause the pinning of the Fermi level. The mobile carriers will then rearrange themselves in the surface region of the crystal setting up an electrostatic potential which bends the valence and conduction bands making the Fermi level equal everywhere (see Fig. 11). The effect of the bending may be described by a so-called depletion (or accumulation) layer of thickness W. The spatial extension of the space charge layers can reach values as low as 10 Å thus

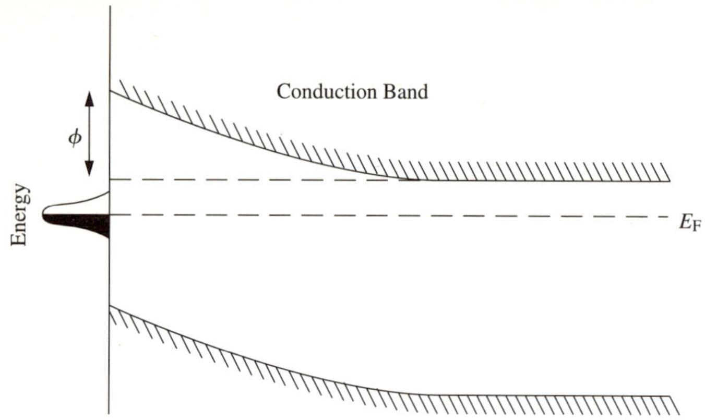

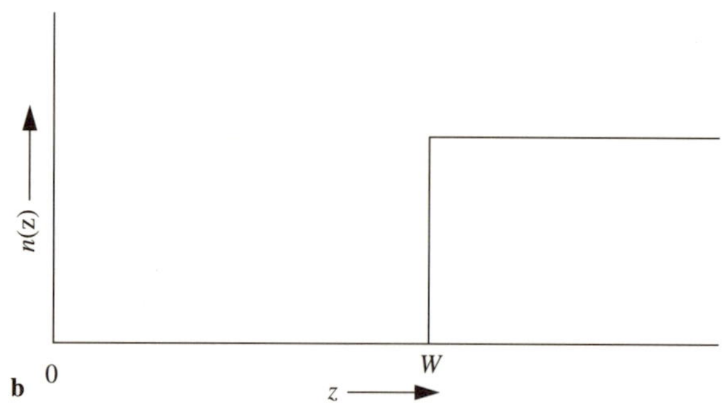

Fig. 11a, b. a band bending at a semiconductor surface; ϕ is the barrier height. **b** free carrier concentration n in the depletion layer approximation. W is the width of the depletion layer [87D3].

realizing a nearly perfect two dimensional electron gas. The surface potential energy barrier is then connected to W by:

$$W = (\epsilon_\circ \phi / 2\pi n e^2)^{\frac{1}{2}} \tag{8}$$

where n is the concentration of free carriers, $\epsilon_\circ$ the static dielectric constant and ϕ the barrier height. Within this model the dielectric properties of the semiconductor crystal are represented by those of a uniform slab with dielectric constant $\epsilon_\circ$ in contact with a semiinfine substrate. On poorly doped samples (low plasma frequency) a quasielastic peak develops in electron scattering due to multiple excitation of the surface-plasmon. Applying the three layer model and assuming a gaussian profile of the scattered electron distribution it was shown that the FWHM of the quasielastic peak reads [87D3, 84P]:

$$\Gamma^2(\mathrm{meV}^2) = 128.6 \frac{k_B T \hbar \omega_{sp}}{(\epsilon_\circ + 1)\cos\theta_i \sqrt{E_i}} F(\theta_i, \bar{\omega}, \xi) \tag{9}$$

where $\bar{\omega} = \omega_p / \sqrt{2\epsilon_\circ}$ and $\xi = (\frac{2\phi m}{E_i m^*})^{\frac{1}{2}}$. E_i is measured in eV. k_B is the Boltzmann constant and m^* the effective mass of the free carriers. F reads:

$$F = \frac{4\sqrt{2}}{\pi^3} \frac{\sqrt{\epsilon_s(\epsilon_s+1)}}{\epsilon_s - 1} \frac{1}{\cos\theta_i} \int_0^\infty \frac{1}{x} f(x, \theta_i) \mathrm{d}x \cdot \int_0^\infty \frac{\frac{\epsilon_s+1}{\epsilon_s-1} \frac{y}{\bar{\omega}T} e^{-xy\xi}}{[\frac{\epsilon_s+1}{\epsilon_s-1}(1-y^2) + e^{-xy\xi}]^2 + [\frac{\epsilon_s+1}{\epsilon_s-1}\frac{y}{\bar{\omega}T}]^2} \mathrm{d}y \tag{10}$$

where f takes into account the finite angular acceptance of the EELS. For a circular acceptance cone:

$$f(x, \theta_i) = \frac{1}{x} \int_0^{\infty} [1 + (\frac{1}{x \cos \theta_i} - \tan \theta_i \cos \Psi)^2]^{-2} d\Psi. \tag{11}$$

The above reported formula was applied to interpret the EELS data for doped GaAs(001) [87D3], Si(111)7 × 7 [87D3] and Ge(111)2 × 1 [89P1]. A more realistic description of the depletion layer implies a non uniform free carrier density profile and a non local description of the response of the electrons. The formulation of the problem was addressed only recently by theorists [87E1, 88E].

A similar analysis was successfully applied also to the quasielastic broadening of the specular peak observed on thin metallic films deposited on semiconductors allowing to determine their resistivity [83B, 85D2, 85P2].

For electronic excitations on a clean surface the loss function reads in the limit of $q_{\|} << 1$ [83P]:

$$\frac{-1}{\tilde{\epsilon}(q_{\|}, \omega) + 1} = \frac{\epsilon_b - 1 - A(q_{\|}, \omega)}{\epsilon_b + 1 + A(q_{\|}, \omega)} \tag{12}$$

where A is:

$$A(q_{\|}, \omega) \simeq [\epsilon_b(\omega) - 1] d_{\perp}(\omega) q_{\|}, \tag{13}$$

$d_{\perp}(\omega)$ is the centroid of the induced charge density [82F].
To compare theory with experiment Eq. (3) has to be integrated over the angular acceptance of the EEL-spectrometer which is generally different for the in-plane and the out-of-plane direction. The integral has to be evaluated for each spectrometer and was tabulated for some of the above discussed designs [82I, 91I]. The dependence of the loss intensity with $\cos \theta_i$ and crystal temperature T (contained in the Bose-Einstein factor) was measured for the surface losses of ZnO [72I, 74F], MgO [84T] and Si(111) [77F] and confirmed the dependence of Eq. (3).

Inserting either Eqs. (5)-(7) or Eqs. (12) and (13) into Eq. (3) the inelastic cross section vanishes at $q_{\|} \to 0$ and the dipole lobe looks like an empty cone. This effect was indeed observed for surface plasmon excitations on metals as then the $q_{\|} = 0$ condition is displaced off-specular because of the large energy transfer. The minimum was found at small but finite $q_{\|}$ [90R1]. Two explanations were proposed for the effect based on the large energy difference between E_i and E_s which implies either $R_i \neq R_s$ [91H] or a different self-energy of the impinging and of the scattered electron [91M1].

For dipole scattering only totally symmetric modes are dipole active for metal surfaces where $\epsilon_b(\omega)$ is large. On semiconductors on the other hand both vertical and longitudinal modes may be dipole active.

6.2.1.4.2 Impact scattering

The differential probability to lose energy in the solid angle $d\Omega$ in impact scattering reads [72R, 80T]:

$$\frac{dS}{d\Omega} = \frac{m}{2\pi^2 \hbar^2} A \frac{\cos^2 \theta_s E_i}{\cos \theta_i} |< \{n_{q_{\|}t}\}_s \,|\, f(\boldsymbol{k}_s, \boldsymbol{k}_i; \{\boldsymbol{R}\}) \,|\, \{n_{q_{\|}t}\}_i >|^2 \tag{14}$$

where $\{n_{q_{\|}t}\}$ is the ensemble of the phonon occupation numbers of the modes t with wavevector $\boldsymbol{q}_{\|}$ before (i) and after (s) the inelastic event. A is the area of the crystal over which coherent scattering takes place.

The scattering amplitude $f(\boldsymbol{k}_s, \boldsymbol{k}_i; \{\boldsymbol{R}\})$ describes the amplitude of the wave scattered off an ensemble of atoms located at $\{\boldsymbol{R}\}$ and contains the multiple scattering effects. It is inversely proportional to $\sqrt{A}$ so that Eq. (11) is, as obvious, independent of A. If thermal disorder is present one can write $\{\boldsymbol{R}_j\} = \{\boldsymbol{R}_j^{(0)}\} + \{\boldsymbol{u}_j\}$ where $\boldsymbol{R}_j^{(0)}$ corresponds to the positions of the atoms in the equilibrium state and

$\boldsymbol{u}_j$ is the displacement caused by the thermal vibration. If the amplitude of the vibrations is small one can expand $f(\boldsymbol{k}_s, \boldsymbol{k}_i; \{\boldsymbol{R}\})$ and keep the leading term:

$$f(\boldsymbol{k}_s, \boldsymbol{k}_i; \{\boldsymbol{R}\}) = f(\boldsymbol{k}_s, \boldsymbol{k}_i; \{\boldsymbol{R}^0\}) + \sum_\alpha \sum_j \left(\frac{\partial f}{\partial R_{j\alpha}}\right)_0 u_{j\alpha} \dots \tag{15}$$

where $\alpha \equiv \{x, y, z\}$. Expressing u in form of the annihilation and the creation operators of vibrational quanta one can see that the second term in Eq. (15) corresponds to inelastic scattering involving single vibrational quanta. As the wavevector of each mode is different one can tune the EELS on the different phonons by choosing the most appropriate values of E_i, θ_i and θ_s. As an example we show in Fig. 12 EELS spectra recorded at $\overline{\mathrm{X}}$ on Ni(001) which display intensity due to the excitation of the S_4 and of the S_6 mode. The variation of the inelastic intensity with E_i was reproduced theoretically [85X]. Experiment and theory are compared in Fig. 13 over a wide energy range to give the reader a feeling of the actual state of the art in the theoretical description of impact scattering. In the calculation the scattering amplitude of the single atom is taken from LEED theory and neglects the deformation of the electron distribution around the atom connected to its displacement due to the vibrational motion. This approach

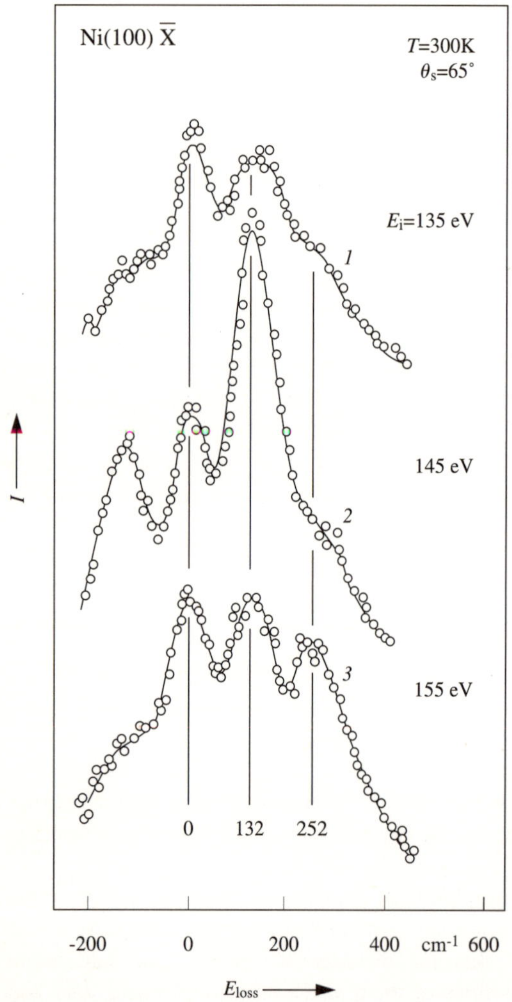

Fig. 12. Ni. EEL-spectra for different primary energies E_i on Ni(001)$\overline{\mathrm{X}}$. Two phonons appear in the spectra, at 16.4 meV (S_4) and 31.2 meV (S_6), whose cross section depends on E_i. The frequency of these modes can be measured with the highest accuracy under the kinematical conditions corresponding to the largest cross section [86R2].

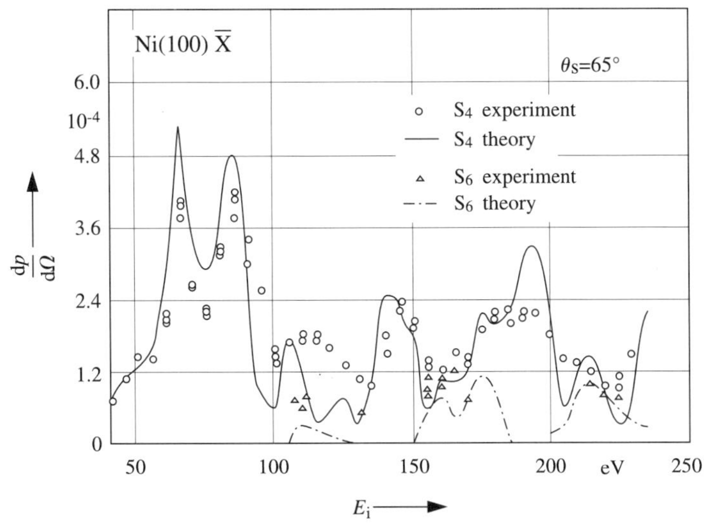

Fig. 13. Ni. Comparison between measured and calculated cross section (phonon loss differential probability $dp/d\Omega$) for S_4 and S_6 modes on Ni(001)$\overline{X}$.

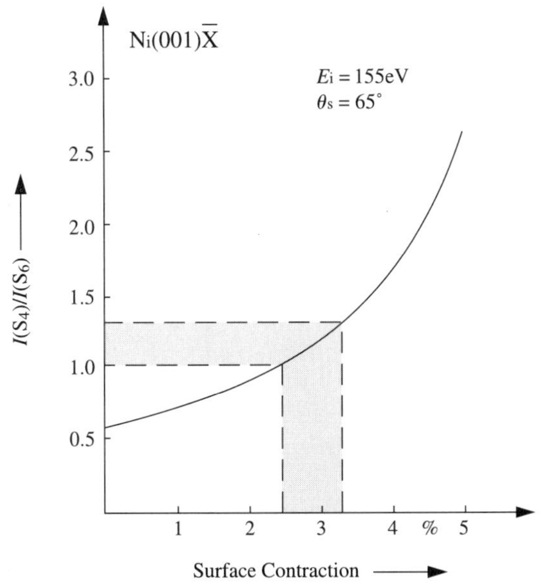

Fig. 14. Ni. Calculated ratio of the cross section of $S_4(\overline{X})$ and $S_6(\overline{X})$ on Ni(001) at given E_i and θ_s as a function of the contraction of the first interlayer spacing [85X, 85R1]. The ratio, determined from spectrum *3* of Fig. 12, indicates a contraction of 3% (see shaded area in the figure). This value is in accord with the 20% stiffering of the force constant between the first and the second layer needed to fit the dispersion curve of the Rayleigh wave with a nearest neighbor lattice dynamical model [83L2].

has the limitation that no electric dipole moment may be generated so that no direct contact with dipole scattering is possible.

An interesting application of the measurement of the inelastic cross sections is shown in Fig. 14. The relative intensity of the $S_4(\overline{X})$ and of the $S_6(\overline{X})$ modes allowed to estimate the contraction of the outermost interlayer distance at the surface in case of Ni(001), which resulted of 2.5% in good agreement with estimations obtained with other techniques [85X, 85R1]. The measured EELS intensities were analyzed for this surface also by Hall and Mills [86H]. It is nowadays established that a reliable description of the surface phonon spectrum is given only after a cross section analysis of the data has confirmed the assignment of the losses to the vibrational modes.

Care must be taken in the experiment as favorable conditions for the excitation of a particular mode may not be met over the whole Brillouin zone and can „switch“ from one mode to another by changing θ_i.

To perform reliable measurements it is moreover necessary to optimize the ratio between one-phonon and multiphonon processes. From the surface Debye temperature one can demonstrate that the most favorable condition is met between 40 eV and 100 eV for metal surfaces [86R2].

To derive the selection rules in impact scattering it is useful to write the scattering amplitude in first Born approximation:

$$f = f_0 \sum_j e^{i(\boldsymbol{k}_s - \boldsymbol{k}_i)_\parallel \cdot \boldsymbol{R}_j} \tag{16}$$

where f_0 is the scattering amplitude of a single atom. Substituting into Eq. (14) and Eq. (15) one can see that the matrix element for phonon excitation contains the scalar product of $(\boldsymbol{k}_s - \boldsymbol{k}_i)_\parallel$ with the eigenvector of the phonon. It follows that in impact scattering the excitation of purely shear horizontal modes is not allowed. In-specular the difference $k_{i\parallel} - k_{s\parallel}$ vanishes for small energy losses and no longitudinally polarized modes can be excited. This pseudo selection rule breaks down if selective adsorption of the primary electrons into the bound states of the image potential takes place [83H].

6.2.2 Data

The data are ordered according to the electric properties of the materials: metals and semimetals, section 6.2.2.1, semiconductors, section 6.2.2.2, and insulators and oxides, section 6.2.2.3. Figures and tables are inserted at the end of the appropriate subsections.

The most significative results are discussed in detail in the text..

6.2.2.1 Metals and semimetals

6.2.2.1.1 Surface phonons

The results for surface phonon dispersion are shown in chapter 4.1 of subvolume 24b and will be omitted here. Data presentation will be limited to the EELS cross section of the surface-phonons and to the evaluation of the effective charges in case of dipole active modes.

An example of the importance of the cross section of the different modes in the EELS measurements of the phonon spectrum was already discussed in section 6.2.1 for Ni(001). The importance of the analysis of the inelastic cross section to understand the surface dynamics will be discussed for the significant case of Ag(001). For this surface a displacement of loss and gain peaks in the EELS spectra was observed as a function of E_i which is caused by the relative contribution of the Rayleigh wave and of unresolved modes to the inelastic signal (Fig. 15). Two modes at nearly the same frequency could have determined the effect: S_2, a vertically polarized vibration localized in the even layers of the crystal (2nd, $4^{th}, \ldots$) which is decoupled by symmetry from S_1 along $\overline{\Gamma} - \overline{M}$, and L_1, a longitudinal mode localized principally in the first layer. The comparison between the measured and the computed shift of the loss peak maximum vs. E_i allowed to establish that L_1 and not S_2 contributes to the inelastic intensity and to discriminate between different lattice dynamical models [91C] (Fig. 16). In other cases and for other surfaces, however, the cross section of S_2 could contribute significantly, too, or even be more important than L_1.

The excitation and annihilation continuum of bulk phonons was shown to be responsible for the tail of the specular peak observed in the case of Cu(001) and Ni(001) [84A] (Figs. 17 and 18). The dipole moment of the bulk phonons was correlated to the oscillatory response of the electrons at the metal surface. An excellent agreement was found with an evaluation performed within the jellium model [84A]. Dipole activity allowed also the study of the structure-induced surface resonance mode present at $\bar{\Gamma}$ for the open fcc (110) and bcc (111) surfaces [84S2, 85S2, 86P2] (Fig. 19). The effective charge connected with this mode for the different crystals is summarized in Table 1. In case of Cu(110) the frequency shift vs T due to anharmonicity was studied [91B] and resulted 4 ... 5 times larger than in the bulk (Fig. 20).

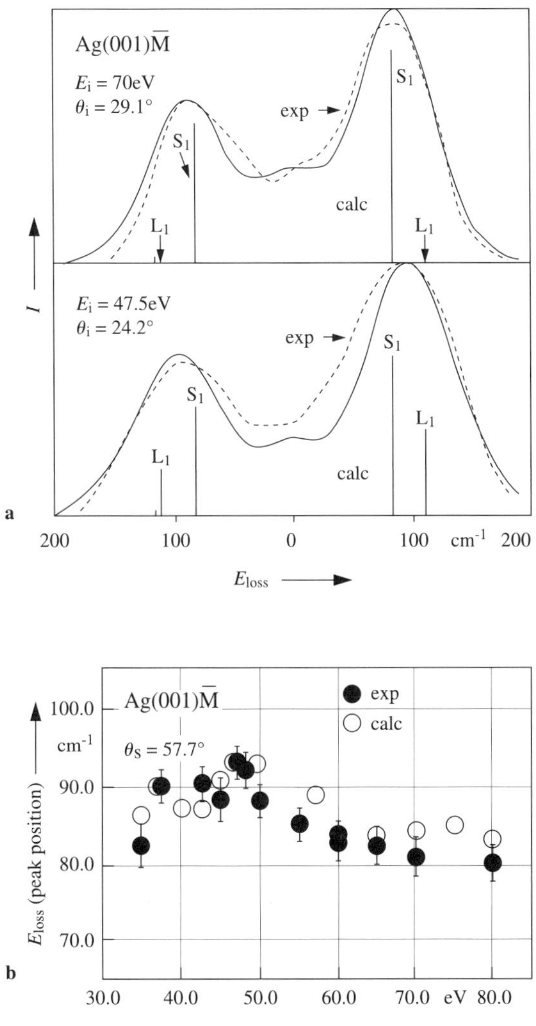

Fig. 15a, b. Ag. **a** two EELS measurements recorded at different E_i for $\theta_s = 57.7°$ are compared with the computed contributions of S_1 and L_1 to the inelastic intensity. **b** calculated and measured loss peak positions for Ag(001) $\overline{M}$ vs E_i [91C]. The modes can't be resolved experimentally because of limited resolution but cause a shift of loss and gain peak position which depends on E_i. The mode responsible for the effect can thus be determined from the dependence of the loss energy with E_i.

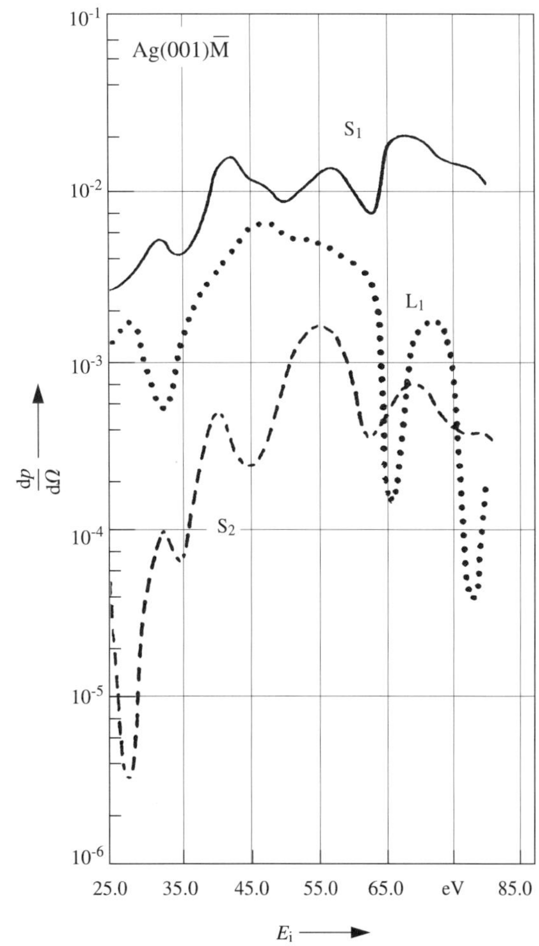

Fig. 16. Ag. Calculated EELS cross section $(dp/d\Omega)$ for individual modes, S_1, L_1 and S_2 for Ag(001)$\overline{M}$ as a function of E_i at $\theta_s = 57.7°$. The mode S_2 does not contribute significantly to the cross section although it might for different scattering conditions or for other systems [91C].

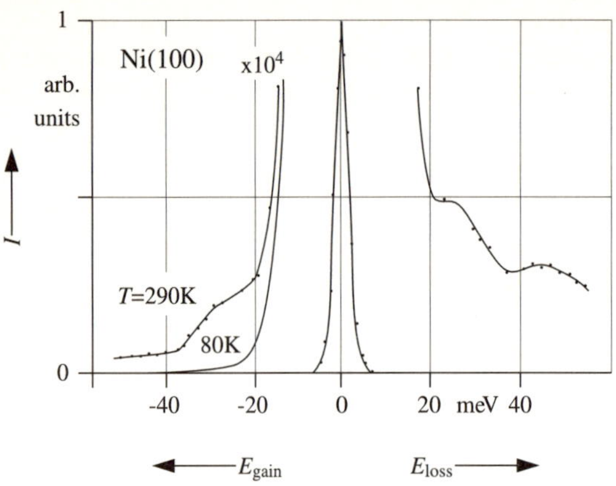

Fig. 17. Ni. Electron gain and loss spectra for the Ni(100) surface recorded in the specular direction for $E_i = 2.3$ eV and $\theta_i = 55°$ at different temperatures. The difference in intensity between the low and the room temperature measurements is due to the contribution of the annihilation of surface phonons to the scattering intensity [84A].

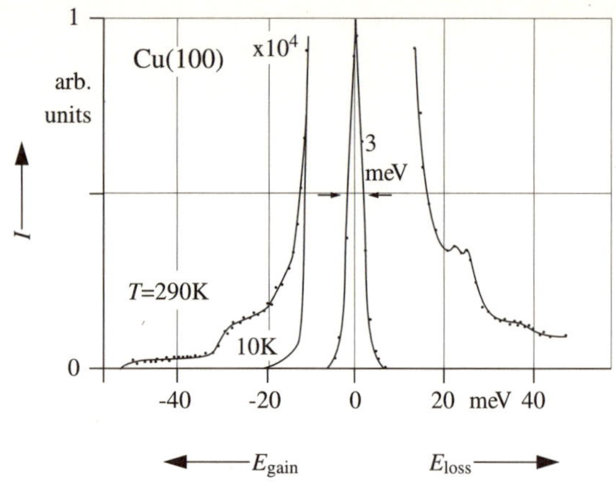

Fig. 18. Cu. Electron gain an loss spectra for the Cu(100) surface recorded in the specular direction for $E_i = 2.3$ eV and $\theta_i = 55°$ at different temperatures. The difference in intensity between the low and the room temperature measurements is due to the contribution of the annihilation of surface phonons to the scattering intensity [84A].

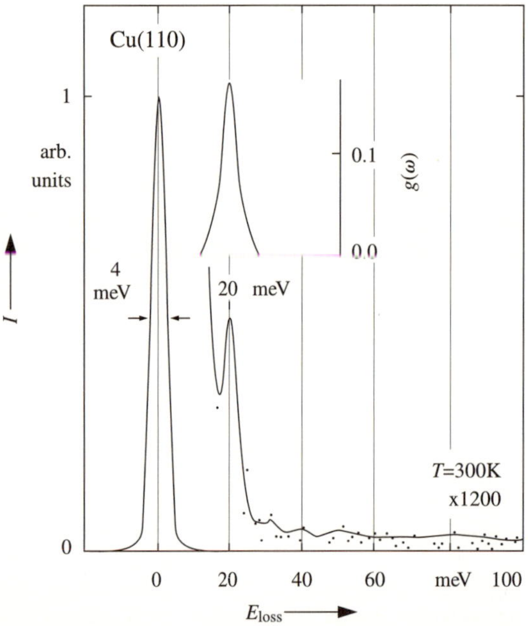

Fig. 19. Cu. EEL-spectrum of Cu(110) at $T = 300$ K in the specular direction showing the intrinsic dipole active phonon mode [86P2]. The inset shows the dipole projected surface density of states $g(\omega)$ convoluted with 4 meV FWHM to account for instrumental resolution. Intrinsic dipole active phonons are expected for all fcc(110) and bcc(111) surfaces.

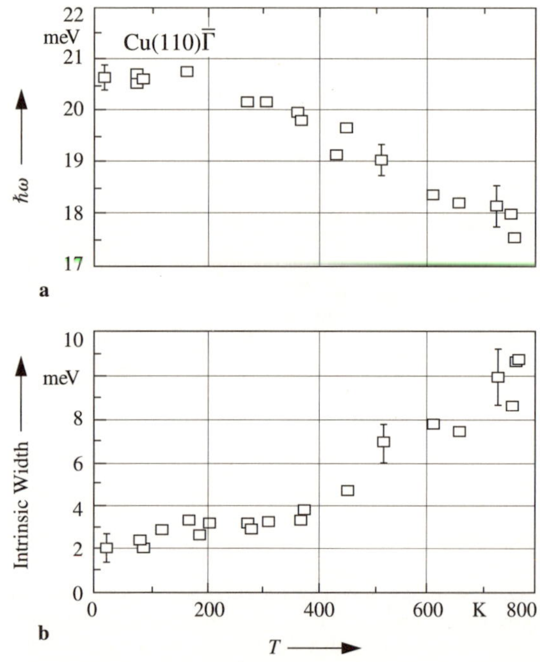

Fig. 20a,b. Cu. Temperature dependence of the dipole active mode for Cu(110): **a** frequency shift, **b** intrinsic width [91B]. The effect is due to anharmonicity and is at the surface 4 . . . 5 times larger than in the bulk.

Table 1. Effective charge connected with the dipole active mode at ω_{ph} present on f.c.c. (110) and b.c.c. (111) metal surfaces at $\bar{\Gamma}$.

metal	e^* [a.u.]	ω_{ph} [meV]
Cu(110)	0.024 (3)	20
Ni(110)	0.028 (3)	24
Fe(111)	0.06	21

6.2.2.1.2 Surface plasmons

Surface plasmons were predicted by Ritchie [57R] and first observed with EELS in transmission experiments through thin foils [66K, 67D, 70Z, 73K, 76K]. They are self-sustained plasma oscillations localized at the surface which decay exponentially both towards the bulk and the vacuum. In the long wavelength limit their frequency is determined by the bulk dielectric function. The surface-plasmon frequency is then given by the condition $\epsilon = -1$. Surface plasmons appear with few exceptions as strongly damped features because of the efficient decay mechanism into electron hole pairs (Landau damping). Surface-plasmon dispersion was measured in transmission for Al, In and Mg [73K, 76K]. More recently surface-plasmon frequency and dispersion were investigated at low impact energy for Al(111) [91T], for thick alkali films [89T, 90T, 91T], for Mg(0001) [92S] and for the single crystal faces of Ag [87C, 89S1, 90R1, 90R2, 91R1, 92R3]. Within the jellium model it was demonstrated that the dispersion of the surface-plasmon with $q_\parallel$ is determined by the position of the centroid of the induced charge density $d_\perp(\omega)$ with respect to the jellium edge [82F, 89F, 87L1, 87L2]:

$$\omega_{sp} = \omega_{sp}(0) - 0.5 d_\perp(\omega) q_\parallel . \tag{17}$$

Calculations were performed using the random phase approximation (RPA) and using the local density approximation (LDA) and predicted a positive $d_\perp(\omega)$ for jellium, i.e. the screening charge is located outside of the jellium edge at the surface-plasmon frequency [82F, 87L1] (Fig. 21).

The width of the surface plasmon is determined by the efficiency of the decay channel into interband transitions. In addition an increase of the loss width is expected with exchanged momentum, due to the broadening of the phase-space at disposition for decay into electron-hole pairs, and with temperature because of the coupling to lattice vibrations (Drude mechanism).

Surface plasmons have monopole character perpendicular to the surface. Additional electronic surface modes with multipole character (Fig. 22) are predicted at $\omega = 0.84\,\omega_p$ (ω_p is the volume plasmon frequency) and were indeed observed. The situation is more complicated for the noble metals as the surface-plasmon frequency is heavily displaced to lower values by strong interband transitions and no theoretical description is possible at the actual state of knowledge.

The major difficulty for the measurement of surface plasmons is the $q_\parallel$ integration caused by the angular acceptance of the EELS. Low energies and grazing incident and grazing scattering angle must be used to achieve reliable results [92R3].

The data for plasmon dispersion show for the alkali metals an initially negative slope while the data for Al are less conclusive. Multipole modes were observed for all the simple metals except for Al, Li, and Mg. The data are summarized in Table 2.

A linear and positive initial dispersion is observed for Ag. While the surface plasmon frequency at vanishing $q_\parallel$ is independent of crystallographic face as required by classical theory [91R1], the dispersion is anisotropic with respect to crystal face [90R1, 90R2, 92R4, 95R] and, for Ag(110), also to crystal azimuth [92R4, 93L]. The available data are summarized in Table 3.

Surface-plasmon damping increases linearly with $q_\parallel$ both for Ag(001) [90R2] and for Ag(110) [92R1, 93R] although with different slopes. Surface-plasmon damping is found to have a quadratic dependence with $q_{\|}$ for K, Cs, Mg and Na. The dependence of surface plasmon energy and damping vs. T was studied for Ag(001). No variation was found for the dispersion coefficient; ω_{sp} is observed to decrease, while the width of the surface plasmon $\Delta\omega_{sp}$ increases with temperature. The dependence of ω_{sp} was related to thermal expansion of the crystal while $\Delta\omega_{sp}$ is suggested to be influenced by the surface-plasmon phonon interaction, too [92R2].

A plasmon associated to the free carriers in the Π-bands was observed on graphite. A strong variation of the plasmon frequency and linewidth was found as a function of temperature and related to the T-dependent concentration of free carriers [91J] (see Figs. 54 and 55 of section 6.2.2.1.3). This behavior is expected to be quite general for semimetals.

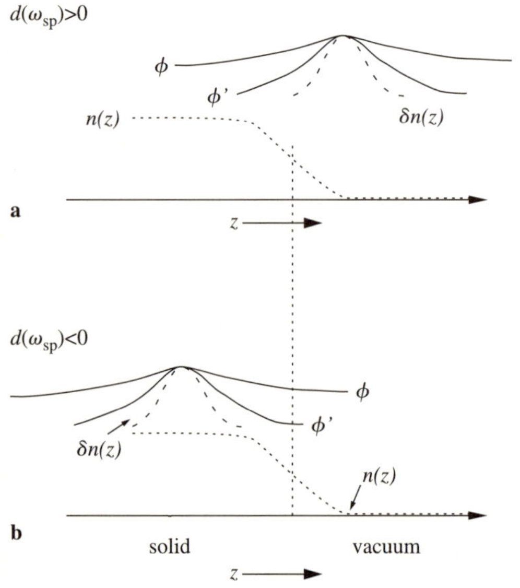

Fig. 21a,b. Schematic illustration of the relation of the position of the induced charge $\delta n(z)$ to the surface plasmon dispersion. **a** $d_{\perp}(\omega_{SP})$ outside of the jellium edge, **b** $d_{\perp}(\omega_{SP})$ inside. ϕ' is the electrostatic potential for a larger value of $q_{\|}$ than for ϕ. The surface plasmon involves in case **a** a larger effective $\delta n(z)$ at low $q_{\|}$ than at large $q_{\|}$ and disperds therefore with a negative slope. For the same reason a positive dispersion is predicted in case **b** [89F].

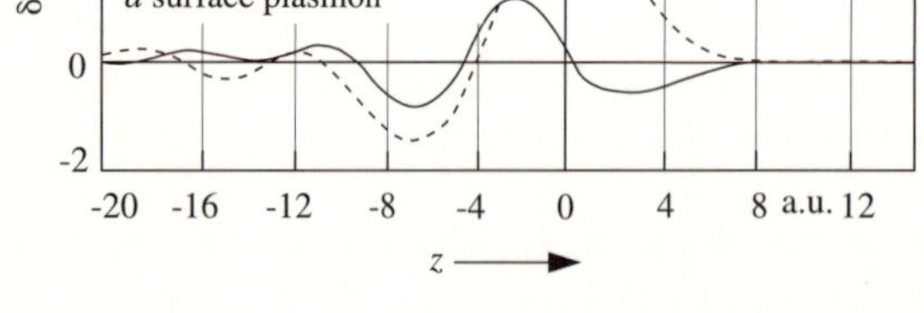

Fig. 22. Calculated induced charge density $\delta n(z,\omega)$ for jellium with an effective density $r_s = 5$ a.u. and at $q_{\|} = 0.15\ k_F$ at the frequencies of the surface plasmon (a) and of the multipole mode (b) as a function of the distance from the surface z [91T]. k_F is the Fermi wavevector at the surface.

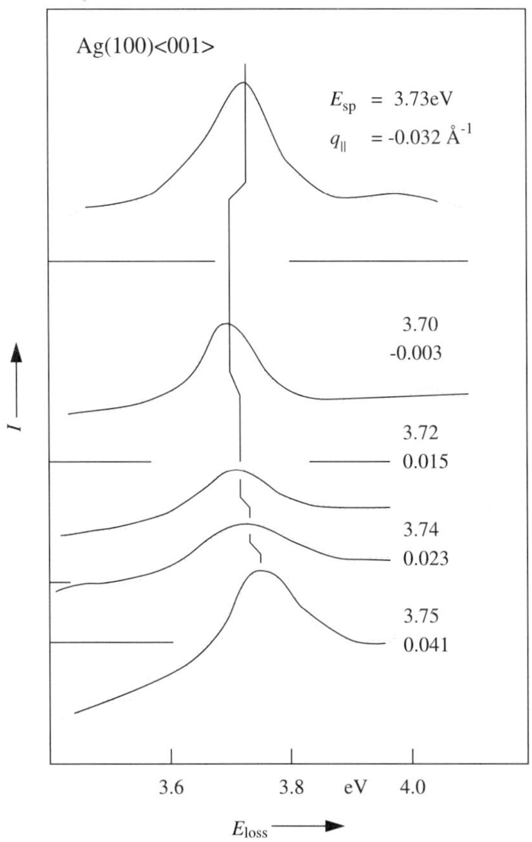

Fig. 23. Ag. Sample EEL-spectra taken for $E_i = 16.0$ eV and $\theta_s = 81.6°$ at different θ_i for Ag(001)<001>. The cross section has a minimum for $q_{\parallel} = 0.023$ Å^{-1}. The broadening of the loss recorded under this condition is due to the finite angular acceptance of the EELS which collects electrons scattered at $q_{\parallel}$ different from the nominal one [90R1].

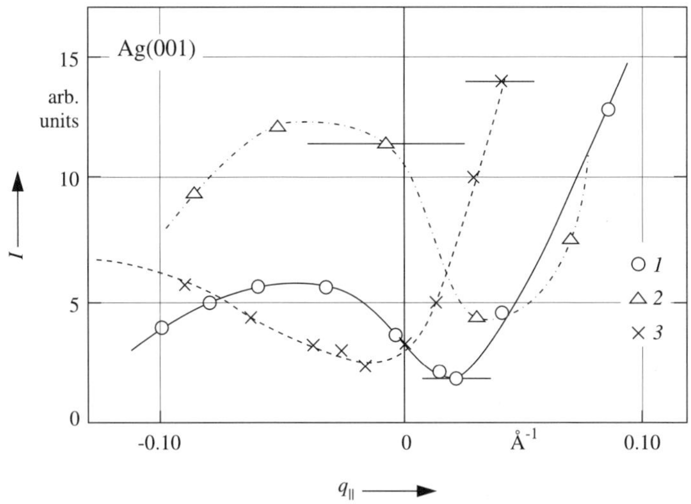

Fig. 24. Ag. Inelastic intensity of the surface-plasmon loss I vs. wavevector transfer $q_{\parallel}$ for electron scattering of Ag(001) (after [90R1]). The bars indicate the window in momentum space over which the experiment is integrated due to the angular acceptance of the EELS. *1*: $\theta_i = 81.6°$, $E_i = 16.0$ eV; *2*: $\theta_i = 60.0°$, $E_i = 16.0$ eV; *3*: $\theta_i = 86.2°$, $E_i = 10.5$ eV.

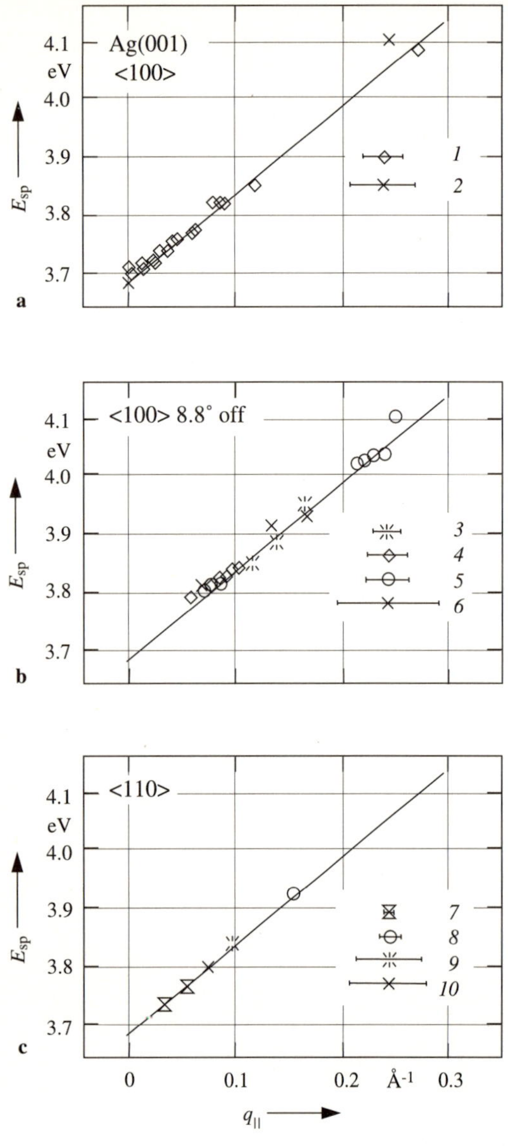

Fig. 25a-c. Ag. Plasmon dispersion for Ag(001) along <110>, <100> and 8.8° off <100> direction.
a <100>, *1*: $E_i = 10.5 \cdots 17$ eV, $\theta_s = 75° \cdots 90°$;
2: $E_i = 10.5 \cdots 17$ eV, $\theta_s = 45° \cdots 60°$;
b 8.8° off <100>,
3: $E_i = 38.5$ or 76 eV, $\theta_s = 75° \cdots 85°$;
4: $E_i = 116 \cdots 127$ eV, $\theta_s = 80° \cdots 85°$;
5: $E_i = 20$ eV, $\theta_s = 75° \cdots 90°$;
6: $E_i = 217$ eV, $\theta_s = 45° \cdots 60°$;
c <110>,
7: $E_i = 217 \cdots 248$ eV, $\theta_s = 41.6° \cdots 69.4°$;
8: $E_i = 36$ eV, $\theta_s = 73.2°$;
9: $E_i = 80$ eV, $\theta_s = 68.4°$;
10: $E_i = 130$ eV, $\theta_s = 69.4°$ The best fit parameters of the curves are reported in Table 3 [90R2].

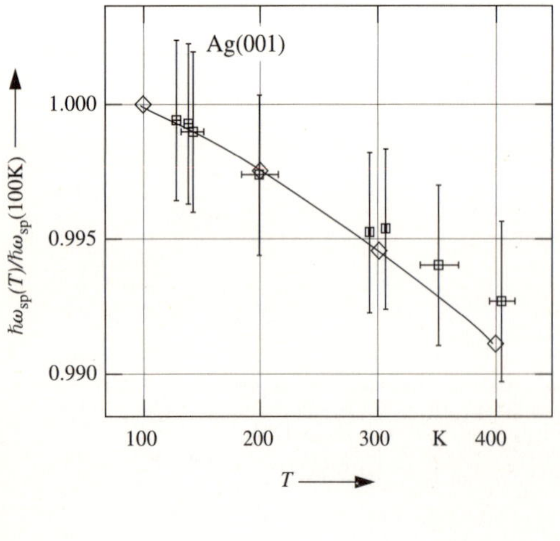

Fig. 26. Ag. Dependence of the surface-plasmon frequency on Ag(001) with crystal temperature. The data □ are compared with the thermal expansion of the crystal ◇ [92R2].

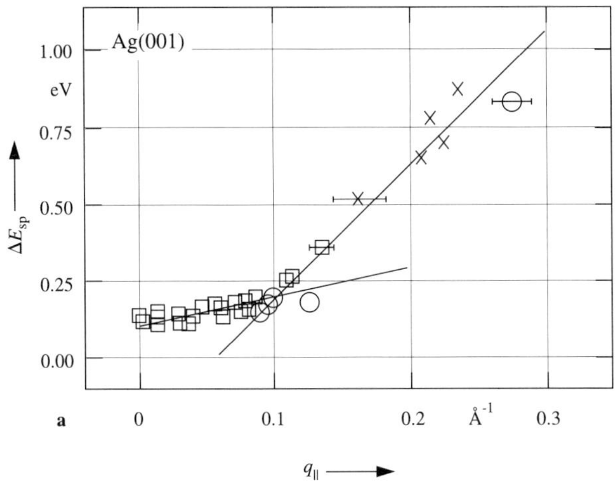

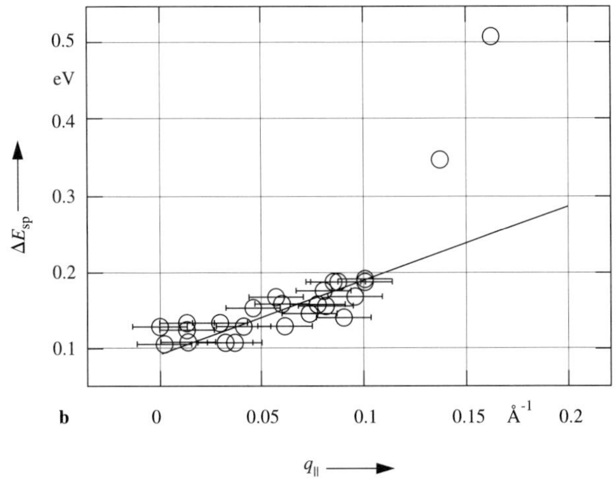

Fig. 27a,b. Ag. **a** collection of plasmon width ΔE_{sp} vs. $q_{\|}$. **b** data below 0.1 Å^{-1}. The best fit parameters in $\Delta E_{sp} = A + Bq_{\|}$ are:
for $q_{\|} < 0.1$ Å^{-1}:
$A = 0.092(5)$ eV, $B = 1.0(1)$ eVÅ;
for $q_{\|} > 0.1$ Å^{-1}:
$A = -0.26(5)$ eV, $B = 4.4(3)$ eVÅ [90R2].
Above $\hbar\omega_{sp} = 3.84$ eV and $q_{\|} = 0.10$Å^{-1} on Ag(001) the decay channel due to interband transitions between surface electronic states opens up causing a strong damping of the plasma wave [92R1].

Table 2. Surface-plasmon ($_{sp}$), volume plasmon ($_{p}$) and multipole mode ($_{MM}$): frequency, loss width and dispersion for simple-metal like materials.

metal	$\hbar\omega_{sp}$[eV]		$\hbar\omega_{p}$[eV]		$\hbar\omega_{MM}$[eV]		$\Delta\hbar\omega_{MM}$[eV]	$d_{\perp}(\omega_{sp})$[Å]
Al	10.3	[91T]	15.0	[89S2]	12.5	[79L]	3.27	
Mg	7.38 (7)	[92S]	10.4	[89V]	8.73 (3)	[92P]	2.3 (25)	0.82
Li	4.3 (1)	[92S]	7.12	[89V]				0.48
Na	3.98	[90T]	5.72	[89S2]	4.67 (4)	[90T]	1.23 (6)	0.73
K	2.73	[90T]	3.72	[89S2]	3.20 (2) 3.15	[90T] [74M2]	0.68 (18) 0.65	0.73
Rb			3.41	[89S2]	2.84	[74M2]	0.41	
Cs	1.99	[91T]	2.90	[89S2]	2.40	[91T]	0.64	0.88

Table 3. Ag. Values of parameters obtained by fitting the experimental dispersion curves of the surface plasmon on Ag. The form $\hbar\omega_{sp} = a + bq_{\parallel} + cq_{\parallel}^2$ is used. The references are: Ag(001) [90R2, 92R2], Ag(110) [92R4], Ag(111) [95R].

face	azimuth	T = 100 K a [eV]	b [eV Å]	c [eV Å^2]	T = 300 K a [eV]	b [eV Å]	c [eV Å^2]
(001)	<100>	3.71 (1)	1.4(1)		3.69 (1)	1.4 (1)	0.006 (44)
	8.8° off <100>				3.68 (2)	1.6 (1)	
					3.67 (1)	1.6 (1)	
(110)	<001>	3.699 (5)	1.13 (2)	1.1 (1)			
	<110>	3.705 (5)	0.42 (2)	2.06 (7)			
(111)		3.708 (6)	0.60 (9)	2.2 (4)	3.692 (5)	0.64 (9)	2.0 (3)

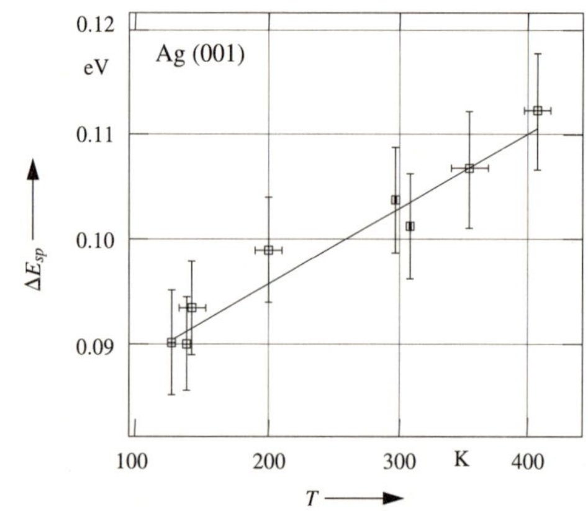

Fig. 28. Ag. Temperature dependence of the width of the surface plasmon on ΔE_{sp} Ag(001) for $q_{\parallel} \approx 0$: the fit yields $\Delta E_{sp} = 81\,(4) + 7\,(2) \cdot 10^{-2}\,T$ (with T in K, ΔE_{sp} in meV) (solid line) [92R2].

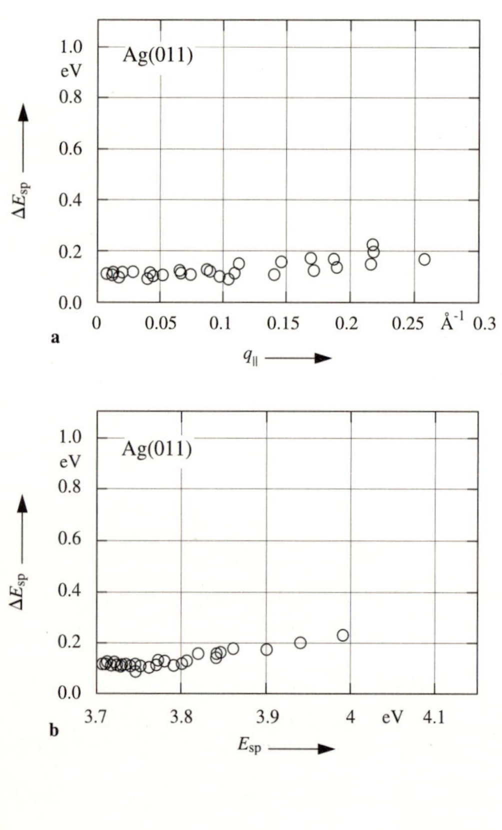

Fig. 29a,b. Ag. **a** dependence of the surface plasmon width ΔE_{sp} vs. $q_{\parallel}$ for Ag(110) [92R1]. **b** Same as **a** vs. plasmon energy E_{sp} [92R1].
According to the theoretical calculations interband transitions involving the d-bands should influence the plasmon damping above 3.72 eV [84E]. No clear evidence for the onset of this mechanism is present although it may be responsible for the larger damping rate observed at higher E_{sp}.

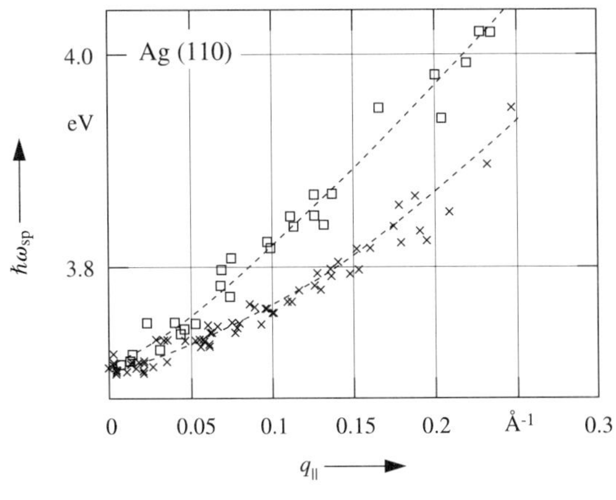

Fig. 30. Ag. Surface-plasmon dispersion on Ag(110) at $T = 100$ K. □ : $q_{\parallel}$ along <001>, ×: $q_{\parallel}$ along <1$\bar{1}$0>. Dashed curves are obtained from the best fit parameters given in Table 3. The dispersion is anisotropic with respect to crystal azimuth [92R2].

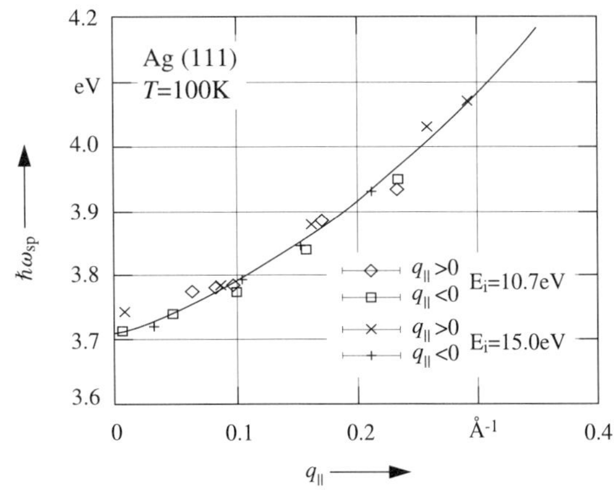

Fig. 31. Ag. Surface plasmon dispersion on Ag(111) [95R].

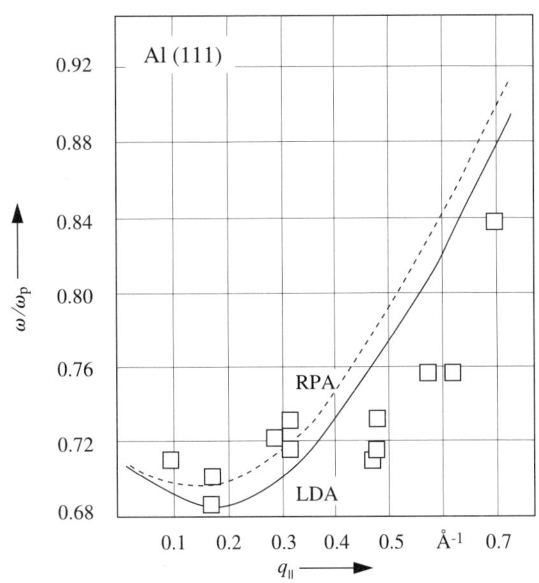

Fig. 32. Al. Dispersion of the surface plasmon on Al(111) measured with low energy EELS compared with the result of RPA and LDA calculations. Contrary to Fig. 29 a negative initial dispersion is observed [91T]. The integration window in $q_{\parallel}$ is comparable to the data for Ag.

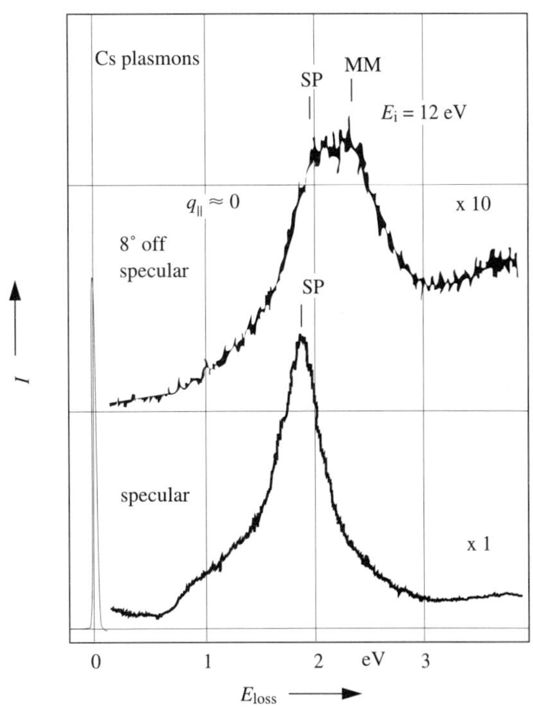

Fig. 33. Cs. Sample spectra for a thick Cs film measured with low energy EELS showing the surface plasmon only and surface plasmon (SP) and a multipole surface-plasmon (MM) [91T].

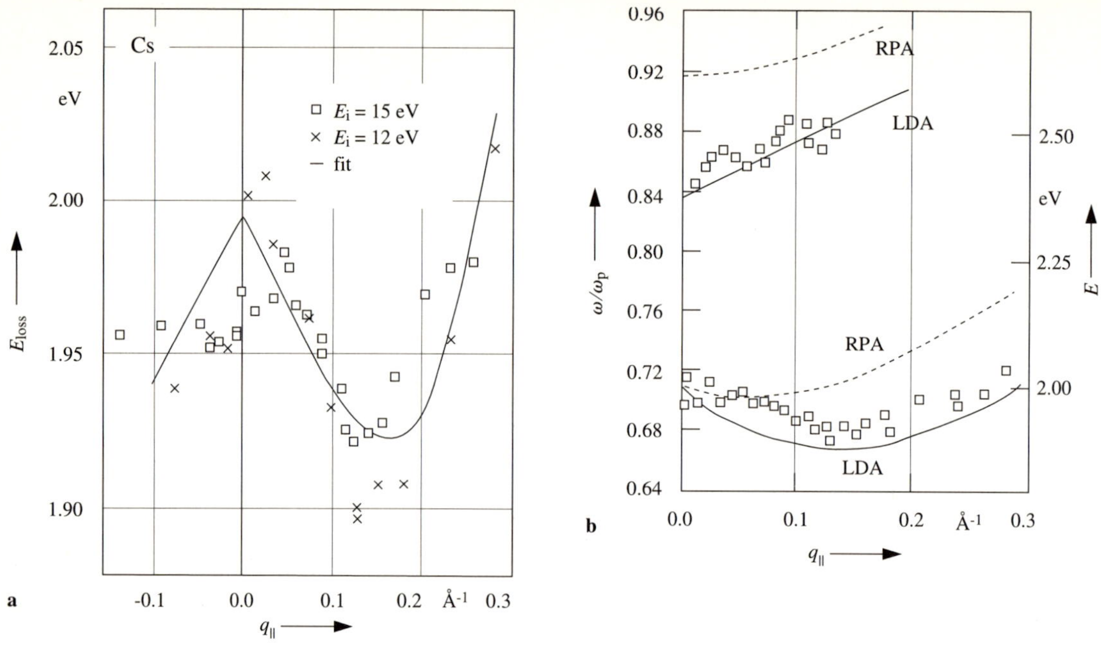

Fig. 34a,b. Cs. Dispersion of the surface plasmon **a** and of the multipole surface plasmon **b** of a thick Cs film measured with low energy EELS. In (**b**) the data are compared to the results of RPA and LDA calculations [91T].

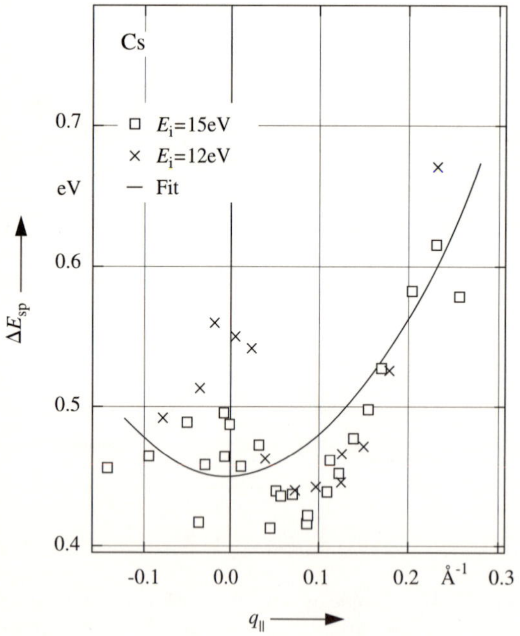

Fig. 35. Cs. Surface plasmon line width for a thick Cs film as measured by low energy EELS [91T].

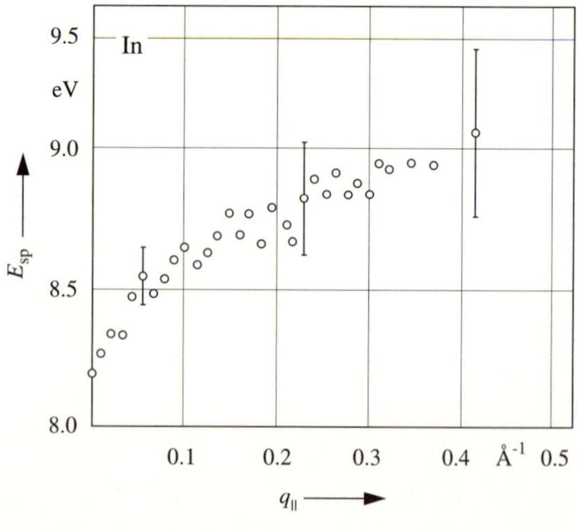

Fig. 36. In. Dispersion of the surface-plasma loss of polycrystalline In measured with 50 keV electrons [76K].

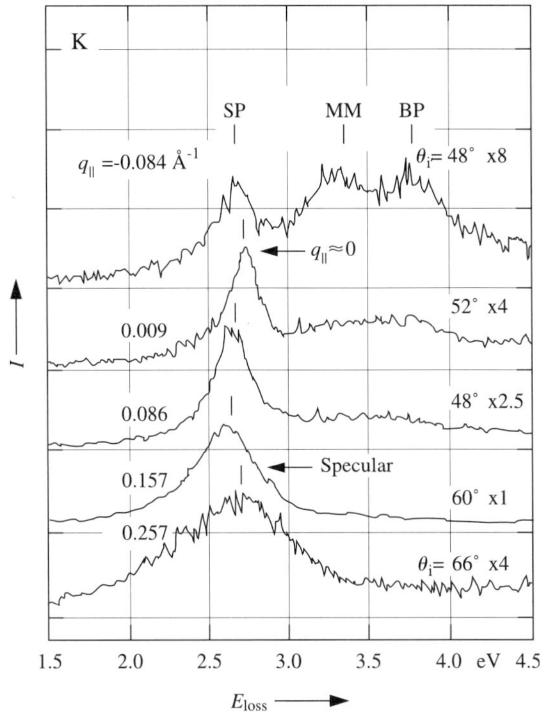

Fig. 37. K. Loss spectra for a thick K film recorded with low energy EELS for $E_i = 15$ eV and $\theta_s = 60°$ at different θ_i (from 48°...66°. The $q_{\|}$ values for the surface plasmon loss are given in the figure together with the multiplication factor of each spectrum. SP, MM and BP indicate the position of the surface plasmon, the multipole mode and the bulk plasmon, respectively [91T].

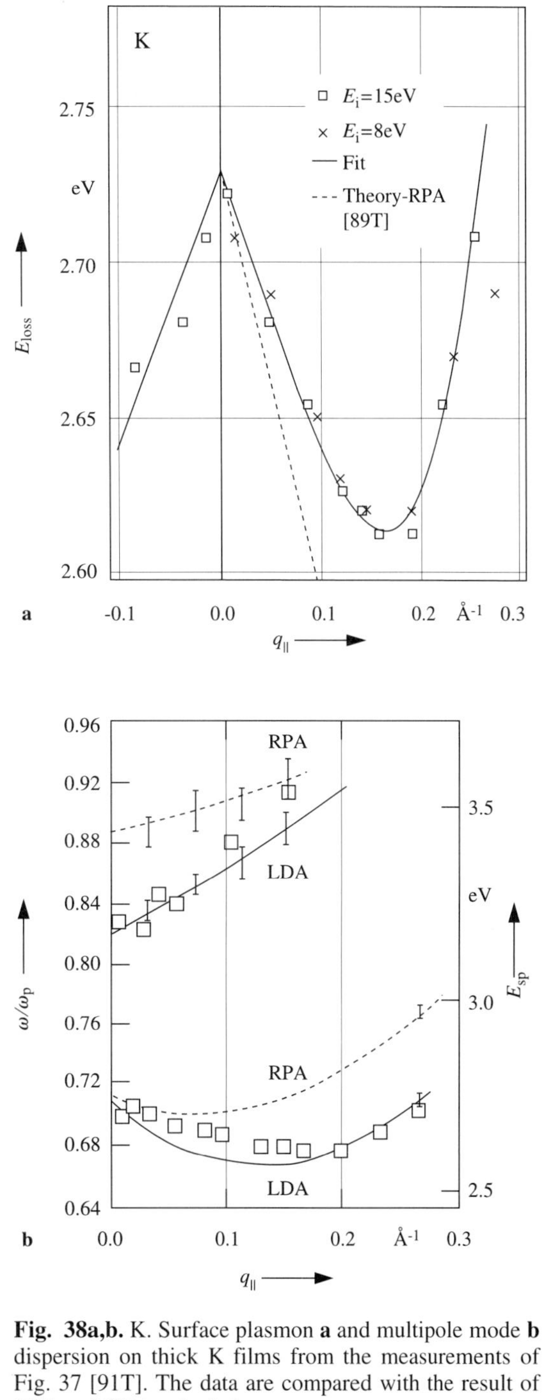

Fig. 38a,b. K. Surface plasmon **a** and multipole mode **b** dispersion on thick K films from the measurements of Fig. 37 [91T]. The data are compared with the result of RPA and LDA calculations.

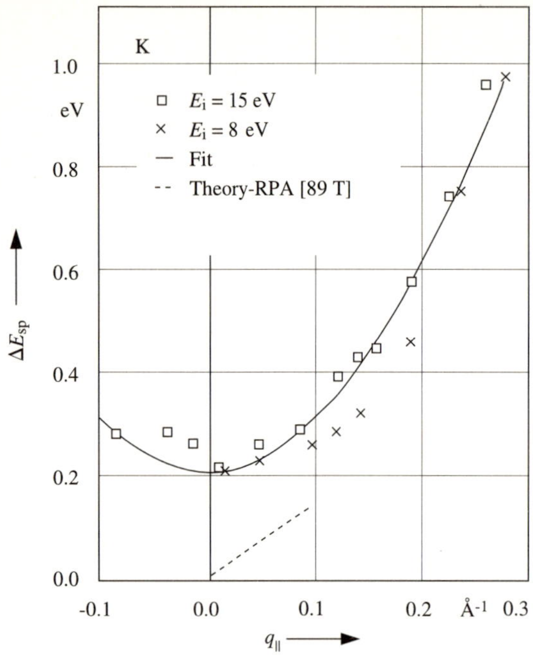

Fig. 39. K. Surface plasmon line width as measured on thick K films [91T] (from the measurements of Fig. 37).

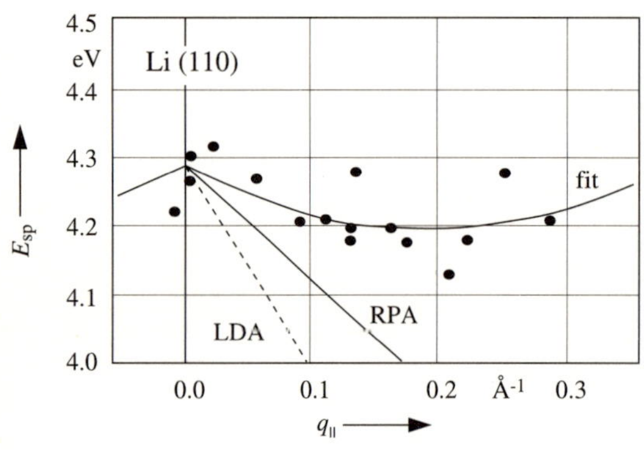

Fig. 40. Li. Surface plasmon dispersion on a (110) oriented thick Li film [92S]. The data are compared with the result of RPA and LDA calculations. The solid line is the best fit to the data points.

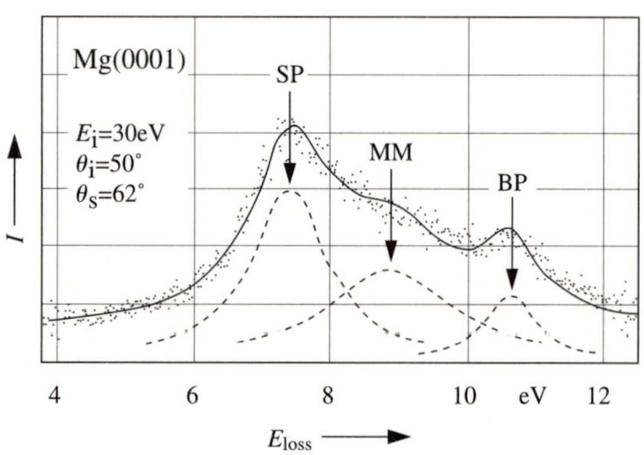

Fig. 41. Mg. EEL-spectrum of the electronic excitations on a Mg(0001) surface [92S]. SP, MM and BP indicate the position of the surface plasmon, the multipole mode and the bulk plasmon, respectively.

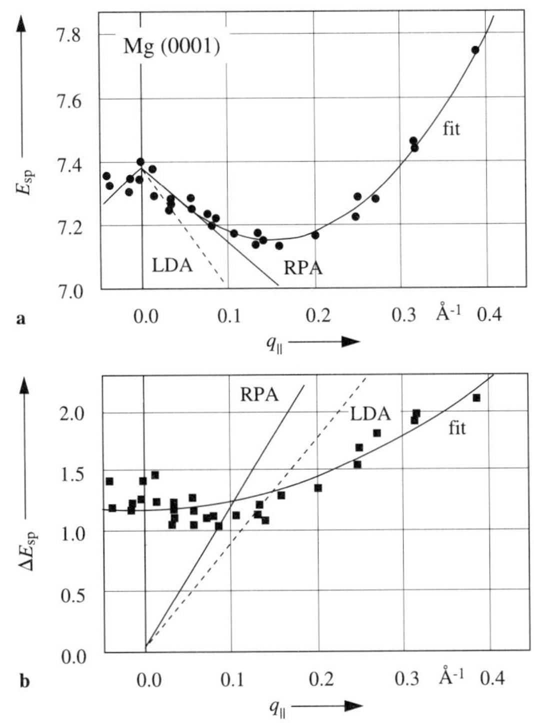

Fig. 42a,b. Mg. Plasmon frequency **a** and linewidth **b** for Mg(0001) compared with the result of LDA and RPA calculations [92S]. The solid line is a best fit curve through the data points.

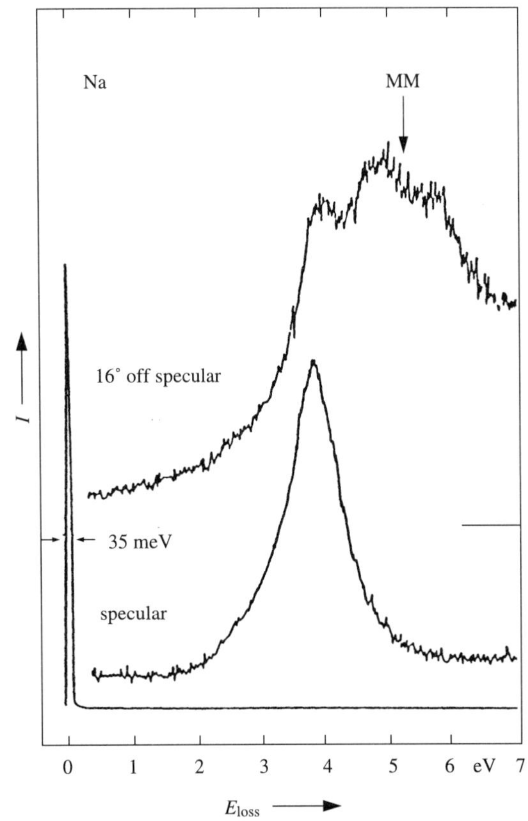

Fig. 43. Na. Sample spectra recorded at $E_i = 15$ eV and $\theta_s = 60°$ for a thick Na film [91T]. MM indicates the multipole mode frequency.

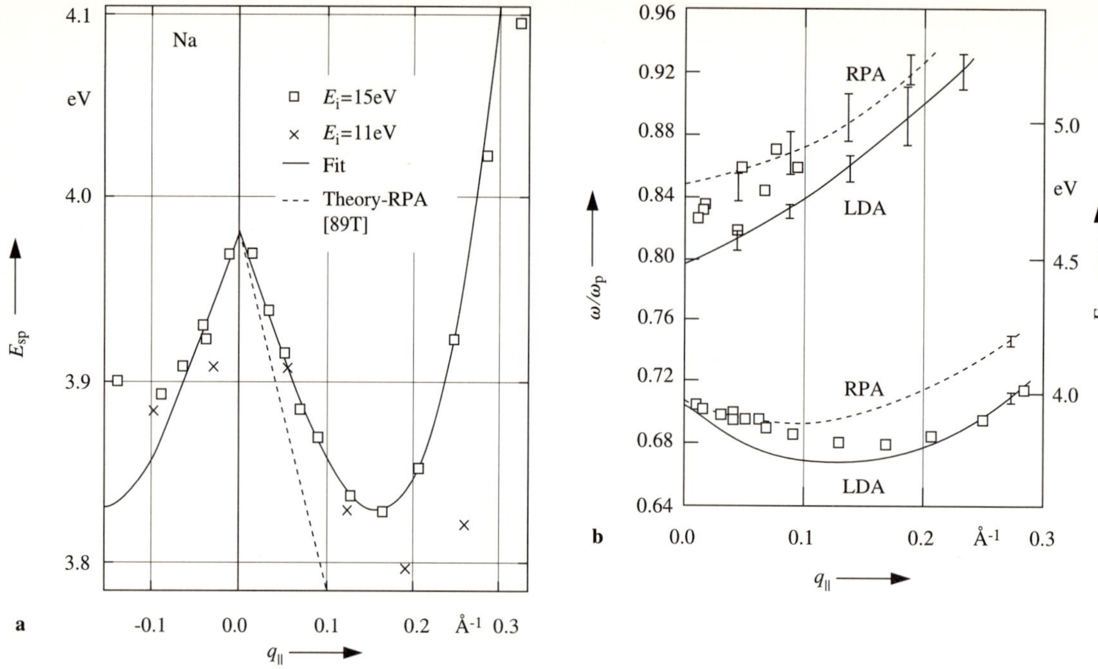

Fig. 44a, b. Na. Dispersion of the surface plasmon **a** and of the multipole mode **b** for the data of Fig. 43 [91T].

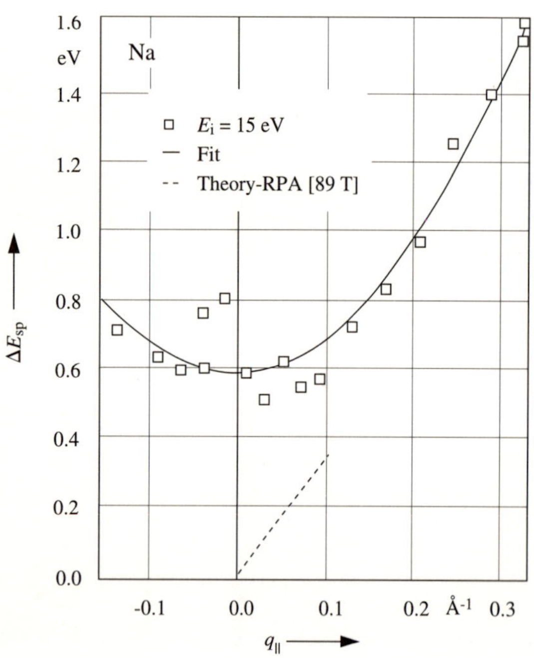

Fig. 45. Na. Surface plasmon line width for the data of Fig. 43 [91T].

6.2.2.1.3 Electron-hole pair and Stoner excitations

Only few measurements of electron-hole pair excitations were reported so far on metal surfaces. The tail of the elastic peak was analyzed in the case of Cu(001) for $E_{\text{loss}} < 0.5$ eV and demonstrated by the model of Eqs. (12) and (13) to be due to electron-hole pair excitations [83A] (see Fig. 46).

An intense and structured background was observed on Ni(110) [85R2] and Ag(001) [90R2]. The modulation is due to the rapid variations of the elastic reflectivity connected with the LEED fine structure (Fig. 47) [85R2] and is therefore a quite general phenomenon. On the other hand an unstructured background was observed for Cu(001) up to $E_{\text{loss}} = 20$ eV [82S].

On Ni(001) a minimum in the inelastic background was reported for energy losses below the onset of the interband transition $X_5 \rightarrow X_2$ [83I] Fig. 48. The structure becomes fainter at high temperatures but is still present above the Curie temperature T_C (Fig. 49). The ratio of the intensity in the minimum and at large E_{loss} remains constant above T_C.

The Stoner excitation spectrum was investigated with spin polarized electron beams for Ni(110) [84K2] and Fe(110) [85K, 86K, 88V]. The energy loss processes, which contribute to the inelastic intensity, are shown in Fig. 51. On Ni the measurements were performed with the EELS shown in Fig. 7 with an energy resolution of 30 meV. The measured asymmetry is defined by

$$A = \frac{1}{|\,P_0\,|} \frac{I^{\uparrow}(E_{\text{loss}}) - I^{\downarrow}(E_{\text{loss}})}{I^{\uparrow}(E_{\text{loss}}) + I^{\downarrow}(E_{\text{loss}})}, \tag{18}$$

where P_0 is the polarization of the incident beam, equal to 35% in the described experiment. The asymmetry is largest at the exchange splitting value (0.3 eV) for Ni. For Fe the spin polarization was measured for both the incident and the scattered electrons with the experimental set up of Fig. 8. The contribution of each of the processes of Fig. 51 was identified by this procedure (Figs. 52, 53). For off-specular scattering the Stoner events comprize up to one third of the total electron-hole pair excitations. Stoner excitations of majority-hole minority-electron character ($F^{\downarrow}$ events) were found to dominate over those of minority-hole majority-electron character ($F^{\uparrow}$ events). At $E_{\text{loss}} \approx 2$ eV a broad peak occurs in the $F^{\downarrow}$ partial intensity which is interpreted as due to the Stoner density of states in Fe. The surface or bulk character of these excitations was not investigated further.

Losses due to electron-hole pair excitation were observed on graphite [87P1, 87P2, 88A2]. The inelastic intensity is caused by the semimetallic band structure of graphite. The modulation of the signal by the kinematical factor causes the apparent peak in off-specular measurements (see Fig. 56).

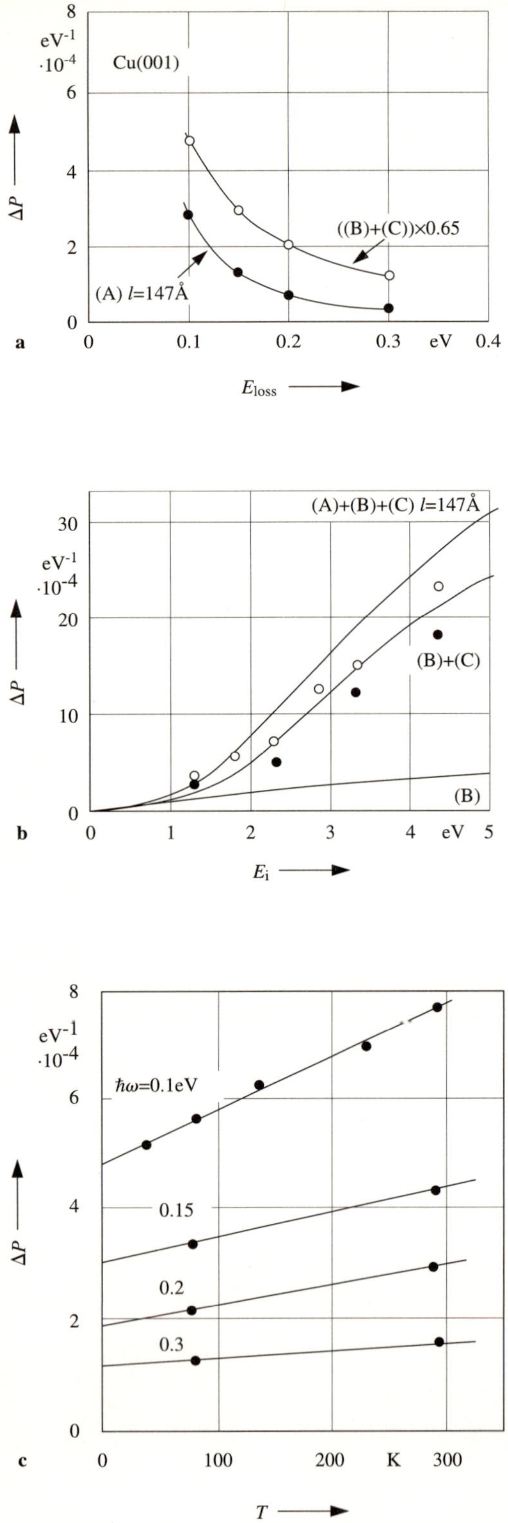

Fig. 46a-c. Cu. Measured and calculated energy loss probability ΔP in the tail of the specular peak for Cu(001) at $\theta_i = 54°$: **a** at $E_i = 2.3$ eV vs. E_{loss}, **b** at $E_{loss} = 100$ meV vs. E_i, and **c** for different losses at $E_i = 2.3$ eV vs. T. In **a** the open and filled circles represent the extrapolation of the inelastic intensity at $T = 0$ K and the difference between room temperature and 0 K, respectively. In **b** the open and filled circles are data recorded at $T = 293$ K and 80 K, respectively. In **c** the dots represent the total intensity for different E_{loss} (in eV). The solid curves in **a** and **b** are predicted for processes involving momentum transfer from interband transitions between bulk states (A), from the surface potential (B), and from the spatial variation of the potential felt by the incident electron (C) (l is the mean free path of the electron in the bulk) [83A].

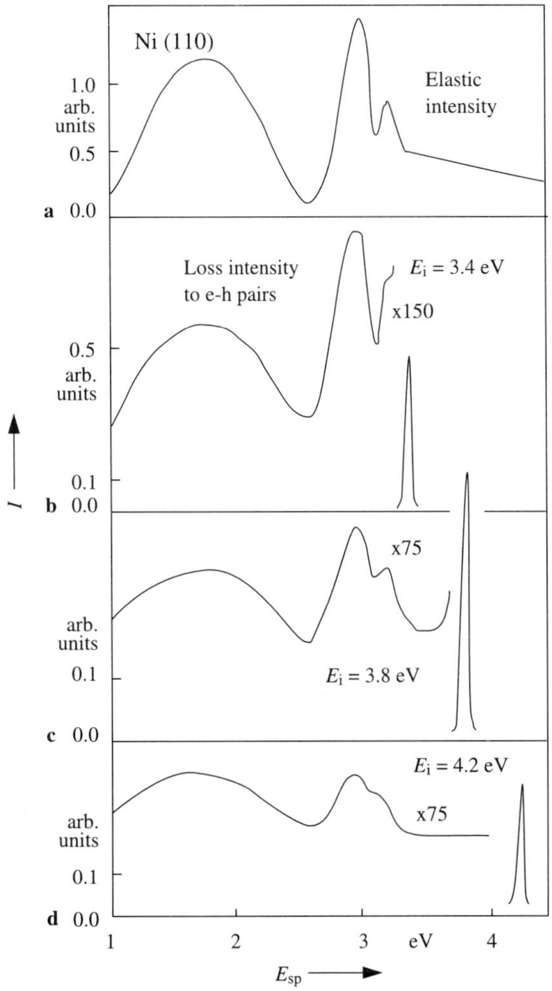

Fig. 47a-d. Ni. Inelastic background due to electron-hole pair excitations on Ni(110) modulated by the elastic reflectivity: **a** elastically diffracted intensity of the (0,0) beam at $\theta_i = 75°$ vs. kinetic energy; **b**–**d** electron energy loss spectra at various primary energies E_i plotted as a function of the kinetic energy of the scattered electrons E_s [85R2].

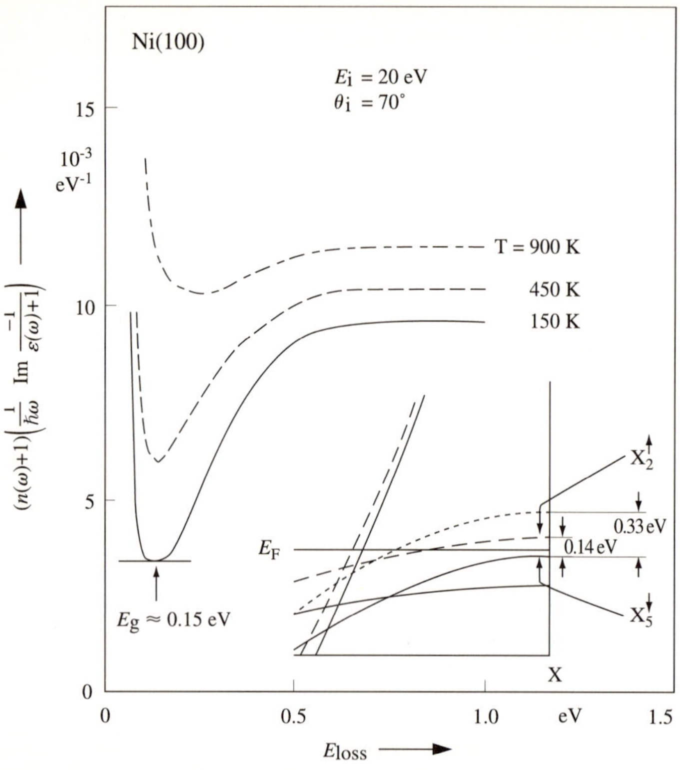

Fig. 48. Ni. Sample loss spectra of Ni(001) recorded at different temperatures. The noise in the spectra is of the order of the linewidth. Data are normalized to yield $\frac{1}{\hbar\omega}\mathrm{Im}(\frac{1}{\epsilon(\omega)+1})$. The threshold at 0.15 eV is interpreted as the onset of the $X_5^\downarrow \rightarrow X_2^\uparrow$ transition. E_g defines the energy gap between these two bands. The insert shows the bulk bandstructure in $\Gamma - X$ direction [78M]. Full lines are majority spin bands. The experiment was performed with a resolution of 30 meV at E_i = 20 eV and 50 eV [83I].

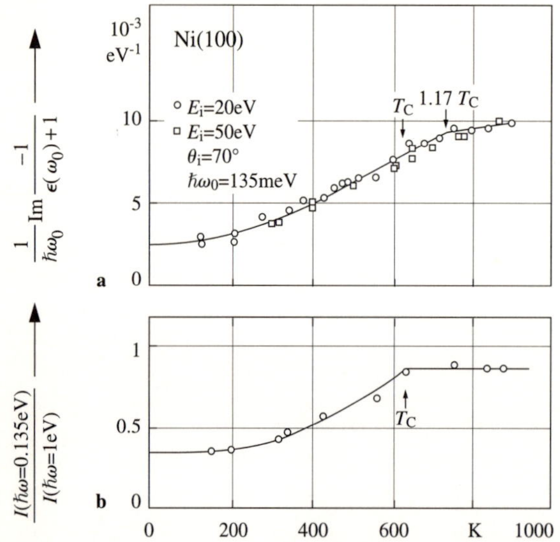

Fig. 49a,b. Ni. **a** loss intensity for Ni(100) at E_i = 20 eV and 50 eV at $\hbar\omega_o$ = 0.135 eV vs. temperature. **b** loss function at $\hbar\omega$ = 0.135 eV divided by the loss function at $\hbar\omega$ = 1 eV. The ratio rises with temperature up to the Curie temperature T_C and remains constant above it [83I].

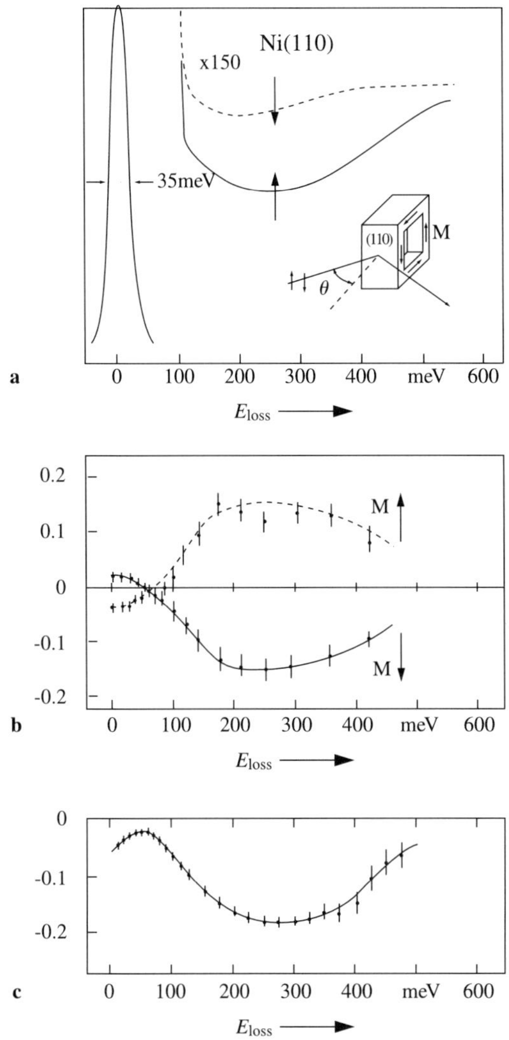

Fig. 50a-c. Ni. Spin polarized EEL-spectra for Ni(110): **a** loss intensity for spin up and spin down orientation corrected for finite polarization of the incident beam; **b** asymmetry A of the scattered electrons as a function of E_{loss} for the two orientations of the magnetization M for $E_i = 9.9$ eV and $\theta_i = 75°$. **c** same as (**b**) but averaged over different E_i between 5 eV and 18 eV. The asymmetry A peaks at $E_{loss} = 0.28$ eV with a FWHM of 0.32 eV. The energy resolution in the experiment is 35 meV [84K2].

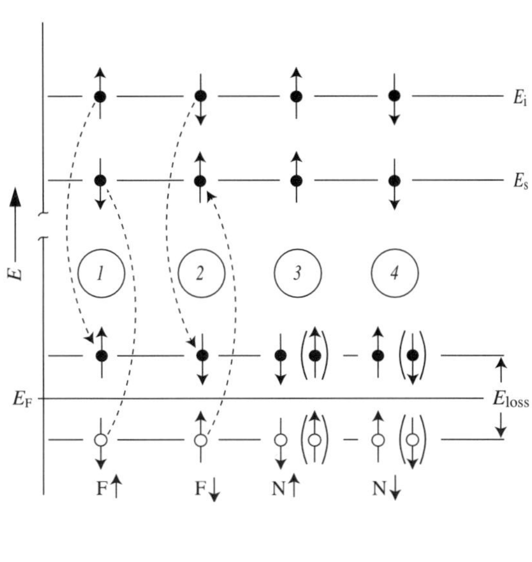

Fig. 51. Schematic representation of two-electron inelastic scattering processes with the spins included. The final states in processes *1* and *2* represent Stoner excitations. Because of the exchange processes (dashed lines) the inelastically scattered electron at E_s appears to have its spin flipped with respect to the primary electron at E_i. The corresponding transitions are termed flip rates F in contrast to the non-flip rates N. E_F is the Fermi energy.

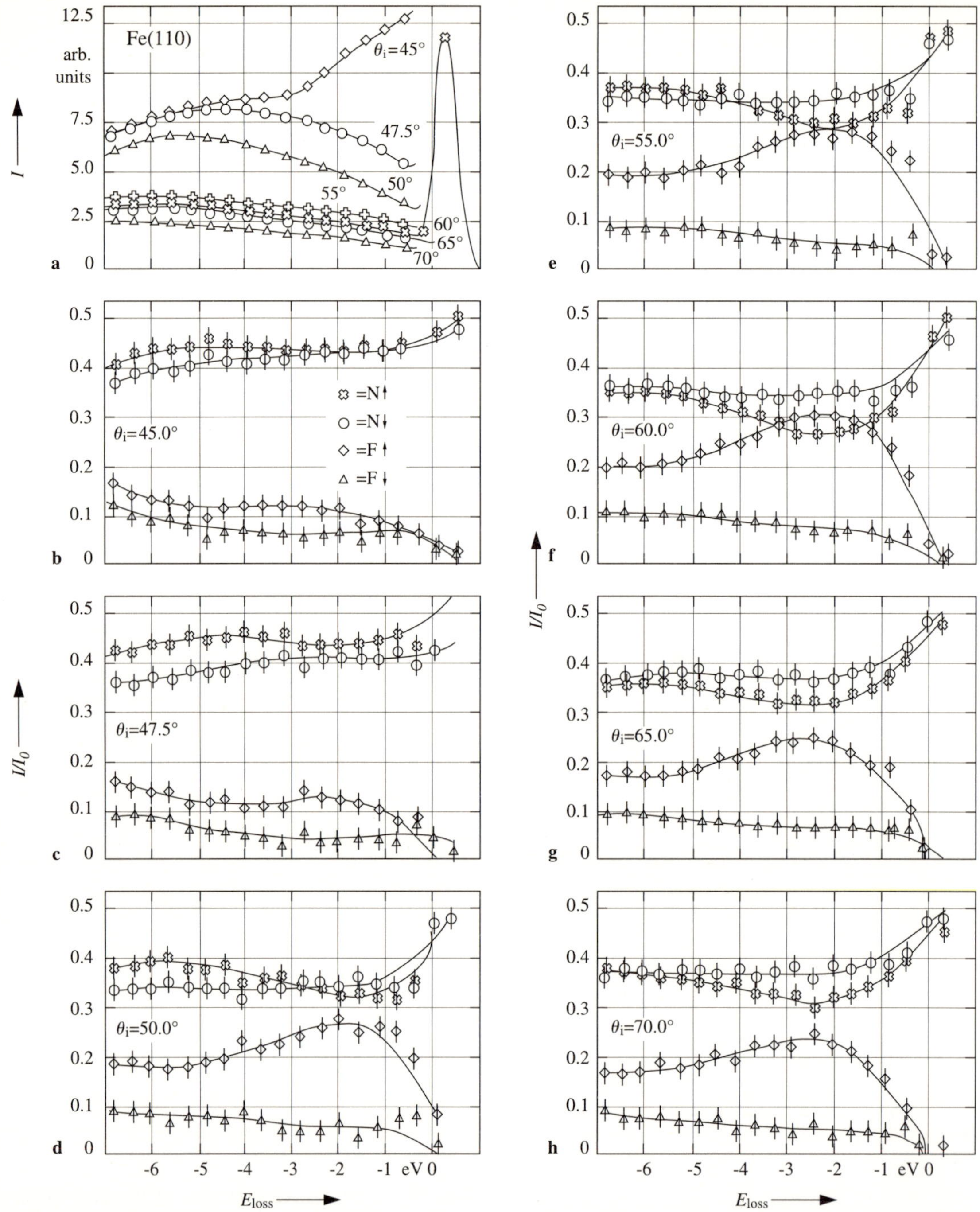

Fig. 52a-h. Fe. **a** Total EEL-intensity for spin polarized electron scattering off Fe(110) for $E_i = 22$ eV for different θ_i. In the experiment the total angle $\theta_s + \theta_i$ was kept fixed. The peak in the EELS intensity for the spectra close to the specular direction is due to the contribution of the efficient dipole scattering mechanism to the loss cross section. **b-h** spin flip ($F^{\uparrow}$ and $F^{\downarrow}$) and non-flip ($N^{\uparrow}$ and $N^{\downarrow}$) contribution to the inelastic intensity as defined in Fig. 51. The $F^{\downarrow}$ partial intensity shows off-specular a maximum at $E_{loss} \approx 2$ eV which is connected to the Stoner excitation spectrum. It shifts to larger loss energy and broadens with $q_{\parallel}$. The experimental energy resolution is 0.4 eV [88V].

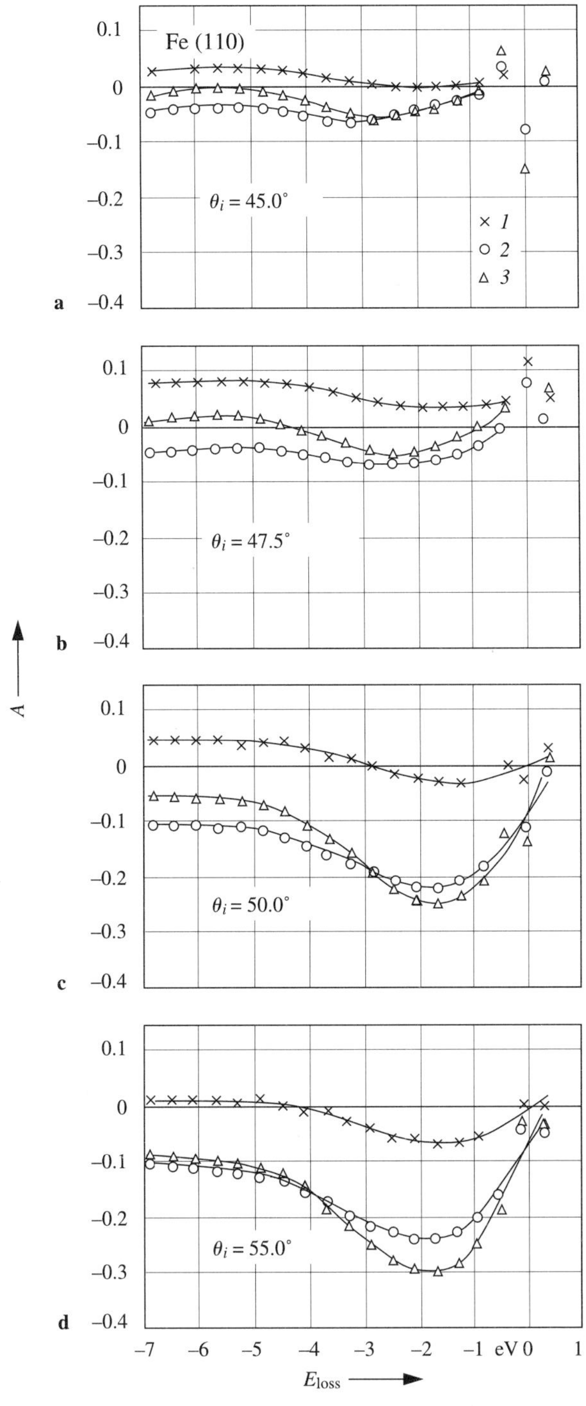

Fig. 53a-d. Fe. Partial and total intensity asymmetries, corrected for the partial degree of spin polarization of the impinging beam, for the indicated values of θ_i. *1* is the non flip intensity asymmetry $(N^{\uparrow} - N^{\downarrow})/J$; *2* is the flip intensity asymmetry $(F^{\uparrow} - F^{\downarrow})/J$; *3* is the total intensity asymmetry $(N^{\uparrow} + F^{\uparrow} - N^{\downarrow} - F^{\downarrow})/J$ where J is the total current. The data close to $E_{loss} = 0$ near to the specular direction ($\theta_i = 45°$) are unreliable because of saturation of the electron counter. Off-specular the peak in the asymmetry near 2 eV is due to the Stoner excitations [88V].

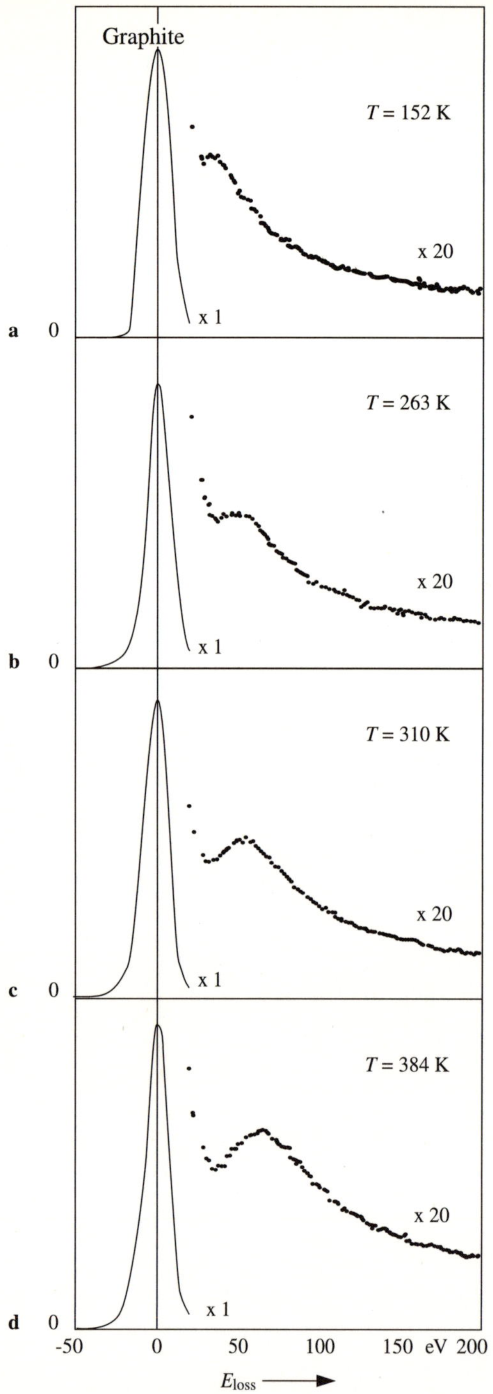

Fig. 54a-d. Graphite. EEL-spectra of graphite measured in-specular at different temperatures of the sample showing a loss due to the excitation of the Π-band plasmon. The shift in frequency is due to thermal excitation of free carriers into the conduction band. In the experiment $E_i = 55$ eV and $\theta_i = 60°$ [91J].

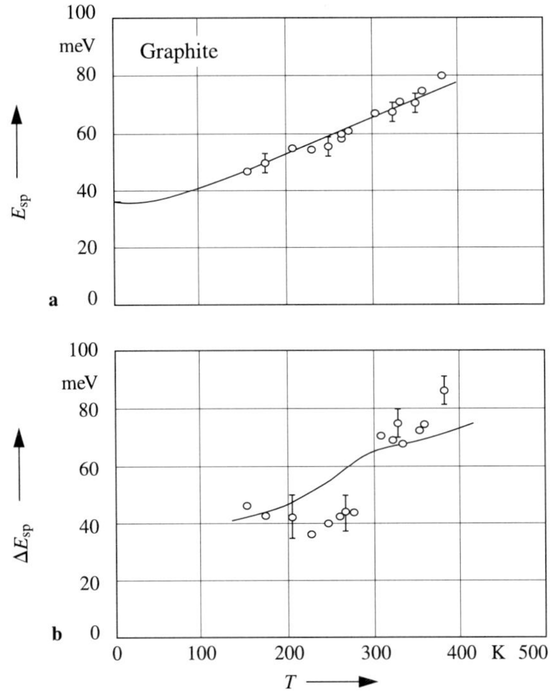

Fig. 55a,b. Graphite. Plasmon frequency **a** and width **b** in graphite vs. crystal temperature T. The solid line shows the plasmon frequency and damping calculated from the graphite band structure [91J].

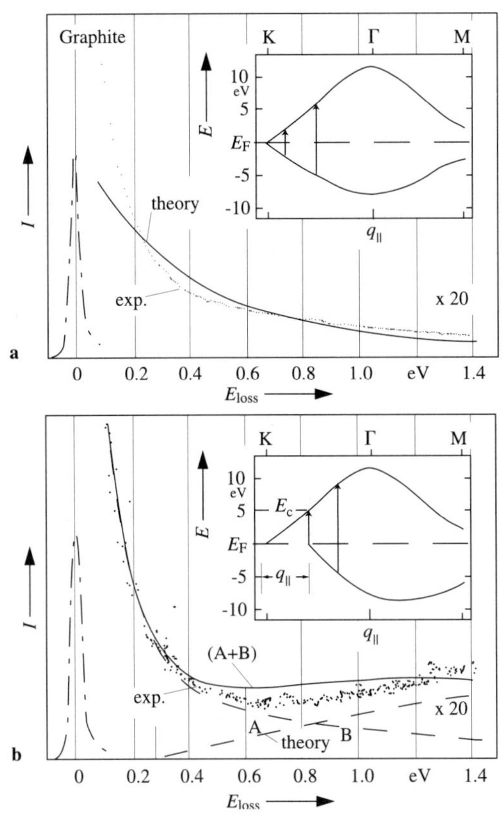

Fig. 56a,b. Graphite. EEL-spectra of graphite recorded at $E_i = 9$ eV and $\theta_i = 60°$ in-specular **a** and 10° off-specular **b**. In the inset the Π bands of graphite showing finite $q_{\|}$ transitions of minimum energy E_c about the Fermi energy E_F. Curve A is the calculated single-scattering loss function and curve B the contribution from diffuse elastic scattering followed by energy loss. Similar backgrounds in EELS are expected for all semimetals [87P1].

6.2.2.2 Semiconductors

Characteristic features of semiconductor surfaces observed with EELS are dipole active surface phonons, interband transitions and surface plasmons or the presence of a quasielastic broadening of the elastically reflected beams.

6.2.2.2.1 III-V semiconductors

Most EELS studies on III-V semiconductor surfaces were performed on the (110) face which may be produced in vacuum by cleavage. These surfaces present a buckled reconstruction with zigzag chains of cations and anions in the $<1\bar{1}0>$ direction which preserves the global symmetry of the ideal (110) plane.

For compound semiconductors the existence of dipole active surface phonons was predicted by theory [65F]. At the surface the dynamical equations for two ionic sublattices together with the Maxwell equations have solutions with both $\text{div}\boldsymbol{P} = 0$ and $\text{curl}\boldsymbol{P} = 0$ where $\boldsymbol{P}$ is the polarization vector. These modes are called Fuchs-Kliever modes and are self-sustained oscillations of the dielectric medium. Like surface plasmons they are accompanied by a long range electric field which decays exponentially into the vacuum and into the crystal with a penetration depth proportional to their wavelength. With $\epsilon_\circ$ the static and ϵ_∞ the high frequency dielectric constant of the crystal the frequency of the mode is given by

$$\omega_{\text{FK}} = \omega_{\text{TO}} \left(\frac{\epsilon_\circ + 1}{\epsilon_\infty + 1} \right)^{\frac{1}{2}}, \tag{19}$$

where ω_{TO} is the frequency of the transverse optical mode in the bulk.

On GaAs (110) two further surface phonons were resolved in $\bar{\Gamma}$ at 17.5 meV and 23.2 meV besides the Fuchs-Kliever mode [89D2].

The surface anisotropy of the reconstructed (110) surfaces is reflected in the anisotropic EEL-cross section of the electronic transition observed on all compound semiconductor surfaces. Measurements were reported for GaAs [75F3, 88D, 89N1], GaP [89N1, 89N2], and InP [91N]. The data for GaAs published by the three groups are not consistent and are therefore omitted.

A study of the space charge layer was performed on poorly doped GaAs(001). The Fermi level pinning position was measured from the temperature dependence of the quasielastic peak (Fig. 58) [87D3]. While below 500 K Eq. (9) describes satisfactorily the temperature dependence of the FWHM of the quasielastic peak, above that temperature a stronger dependence is observed. This effect was related to the shift towards midgap with T of the Fermi level in the bulk (see Fig. 59).

Surface-plasma oscillations give rise to well defined loss peaks if the doping level is large enough [81M]. The dependence of the relative intensities of phonon-like and plasmon-like losses shows the existence of a dead layer for surface plasmons [83E, 84R3, 87E2, 89D2]. The effect is due to the wavelength dependence of the penetration depth of the surface plasmon connected to the selection of momentum transfer with E_i due to the „surfing effect“ in Eq. (3). Thickness and band bending of the depletion layer can thus be determined by EELS [85R3]. The electric field associated with the Fuchs-Kliever phonon interacts with that of the surface plasmon giving rise to modes of mixed character. The coupling is particularly strong at doping levels for which the plasmon frequency is near to ω_{FK} [81M, 89B, 89C1, 89C2].

Hole surface-plasmon were studied for p-doped samples [91M2, 92D].

Some work was performed on (111) and $(\bar{1}\bar{1}\bar{1})$ surfaces, too. Poor attention was, however, devoted so far to the intrinsic properties of these surfaces.

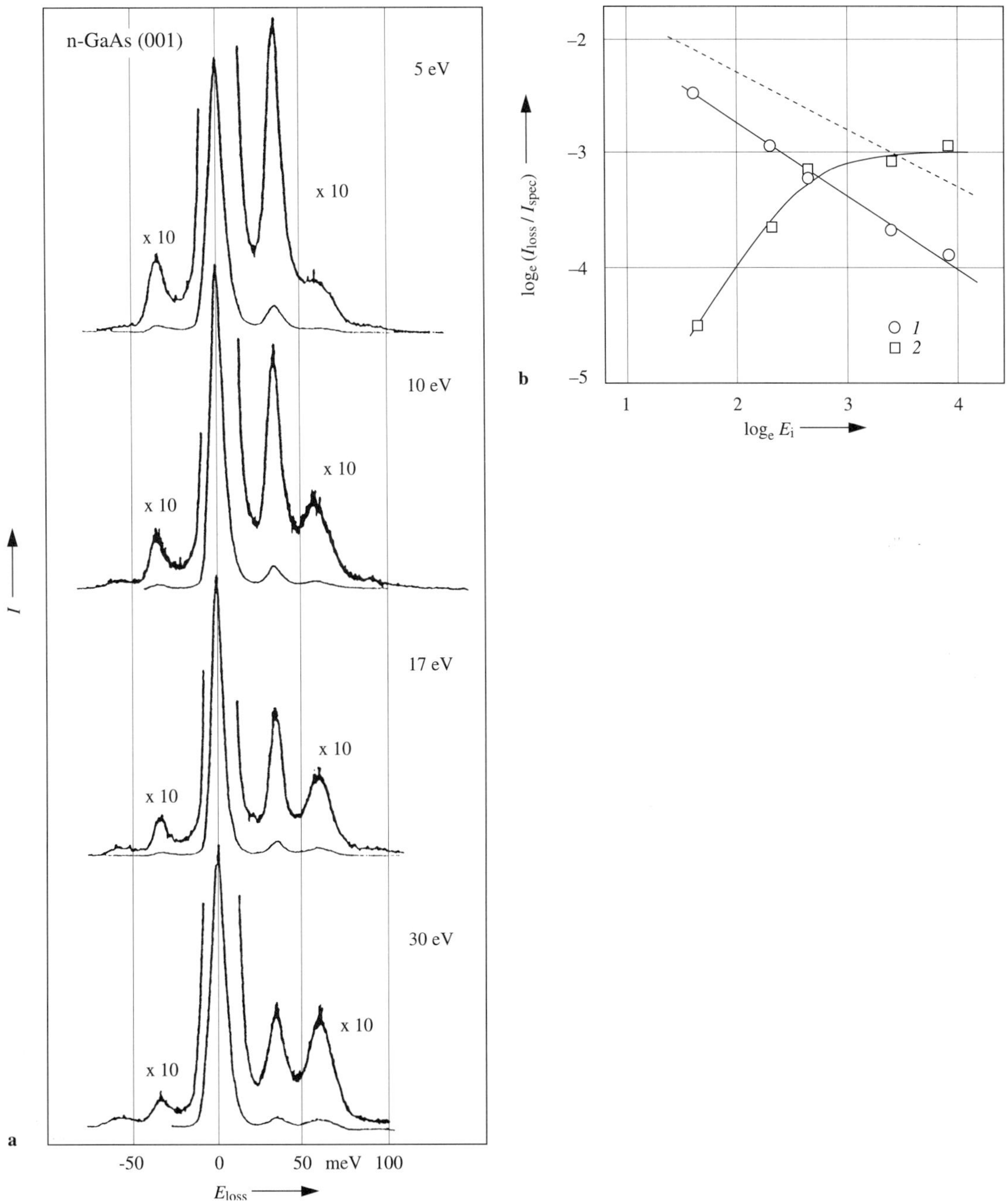

Fig. 57a,b. n-GaAs. EEL-spectra of Si-doped n-GaAs(001) taken at E_i from 5 . . . 30 eV showing the Fuchs-Kliever surface phonon and the plasmon associated with the free carriers ($n = 4 \cdot 10^{18}$ cm^{-3}) **a**. The increasing intensity of the plasmon loss with E_i ($E_{loss} = 60$ meV) is indicative of the existence of a depletion layer at the surface [87G]. **b** variation of the intensity of the phonon loss (*1*) and of the plasmon loss (*2*) with E_i. The reduction of the plasmon loss at low E_i is indicative of a depletion layer at the surface.

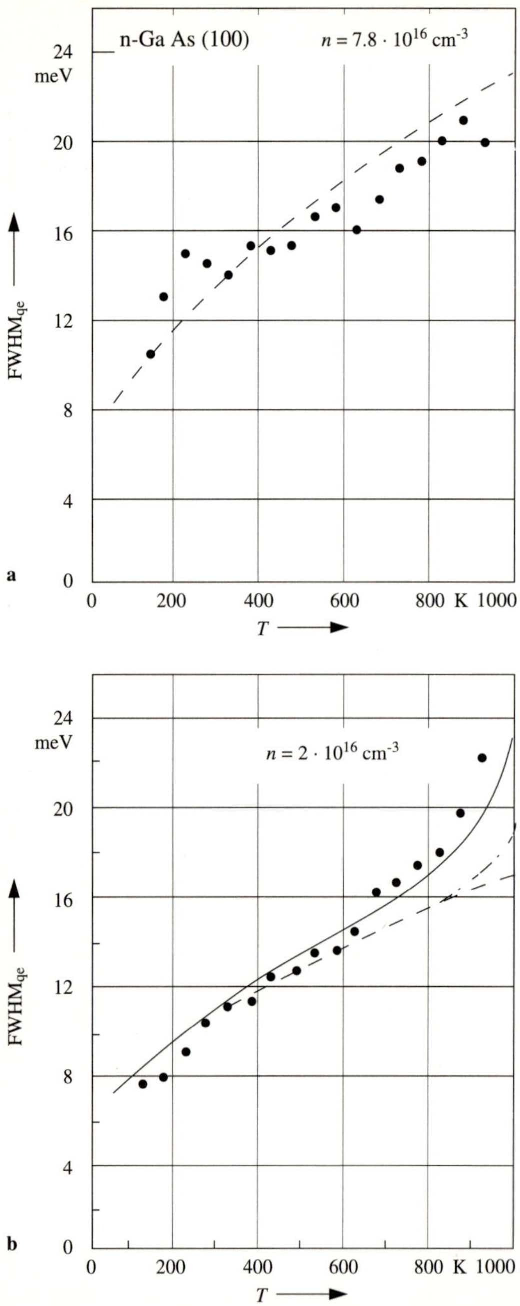

Fig. 58a,b. n-GaAs. Width of the quasielastic peak $FWHM_{qe}$ as a function of crystal temperature for n-GaAs(100) for two different treatments of the surface. **a** the surface was sputtered for less than 30 minutes. **b** same as **a** but after one hour of sputtering and annealing. The dashed curves are calculated following Eq. (9) assuming all terms as temperature independent. With $\epsilon = 12.85$, $m^* = 0.067$ a.u. and $\mu = 4180\ \mathrm{cm^2/V\ sec}$ the barrier height results of 0.81 eV in accord with the value determined by other techniques for GaAs(001). To fit the data in **b** one has to assume that the free carrier concentration is reduced by a factor 4 in the surface region with respect to the value in the bulk. As a result of sputtering and annealing lattice imperfections and impurities have accumulated at the surface and trapped free carriers. The dot-dashed curve considers the effect of thermally excited free carriers. The solid line considers all the above effects plus the temperature induced change in band bending illustrated in Fig. 59 [87D3].

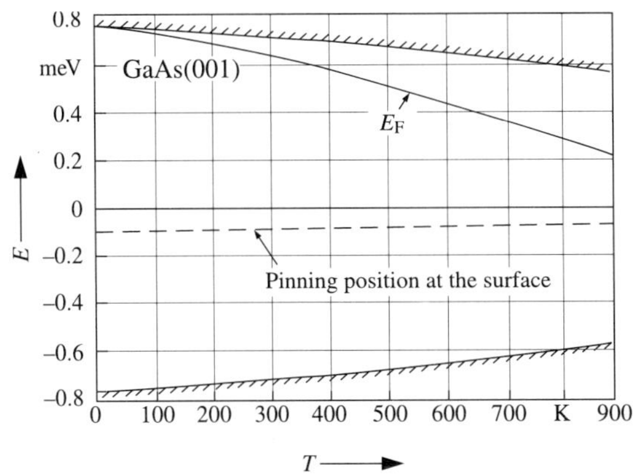

Fig. 59. GaAs. Temperature dependence of the band gap and of the Fermi-level position for $n = 2 \cdot 10^{16}$ doped GaAs(001) crystal according to the analysis of the EELS results of Fig. 58. The dashed line indicates the pinning position of the Fermi level at the surface. E_F is the position of the Fermi level in the bulk with respect to the conduction band. The downwards shift of E_F corresponds to a decrease in band bending and a reduction of the width of the depletion layer [87D1].

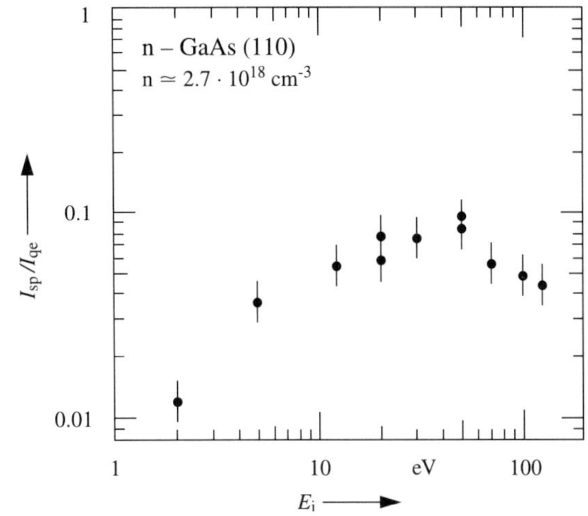

Fig. 60. n-GaAs. Intensity of the plasmon loss I_{sp} (normalized to the area of the quasielastic peak I_{qe}) vs. E_i for n-GaAs(110). The presence of the depletion layer, estimated in some ten Å, is evidenced by the decay of the plasmon loss intensity at low E_i. The maximum of intensity at $E_i \approx 40$ eV corresponds to the condition at which the probing depth of the incoming electron becomes larger than the thickness of the dead layer [89B].

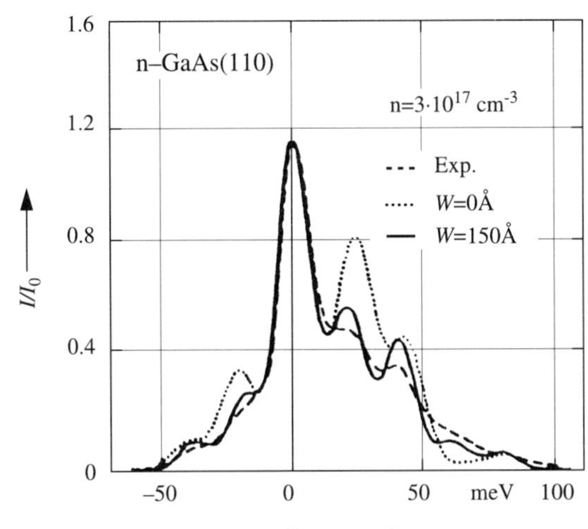

Fig. 61. n-GaAs. EEL-spectrum of n-GaAs(110) together with fits for a layer thickness $W = 0$Å and $W = 150$ Å. The losses at 20 and 40 meV are due to the surface plasmon and to the surface phonon, respectively. The best agreement is for $W = 150$ Å [83E].

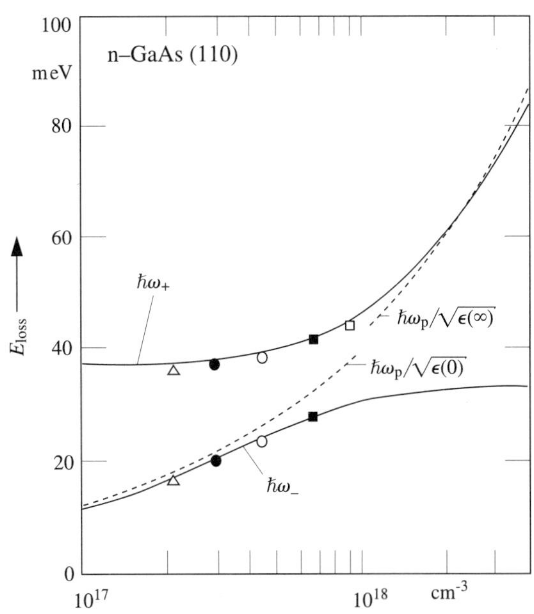

Fig. 62. n-GaAs. Coupling of the surface plasmon and Fuchs-Kliever mode for n-GaAs measured for samples with different doping, i.e. with effective carrier density n. The interaction gives rise to mixed modes ω_+ and ω_- called surface plasmon polaritons [81M].

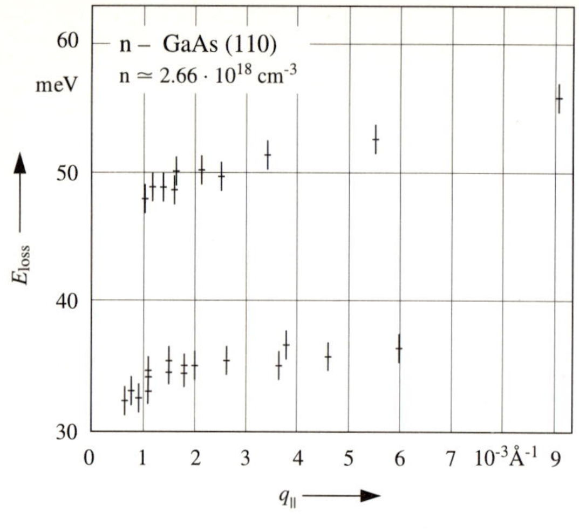

Fig. 63. n-GaAs. Dispersion of the surface plasmon (higher frequency mode) and surface phonon (lower frequency mode) losses vs. $q_{\parallel}$ for heavily n-doped GaAs(110). The upward dispersion of the plasmon is connected to its mixture with the phonon [89B].

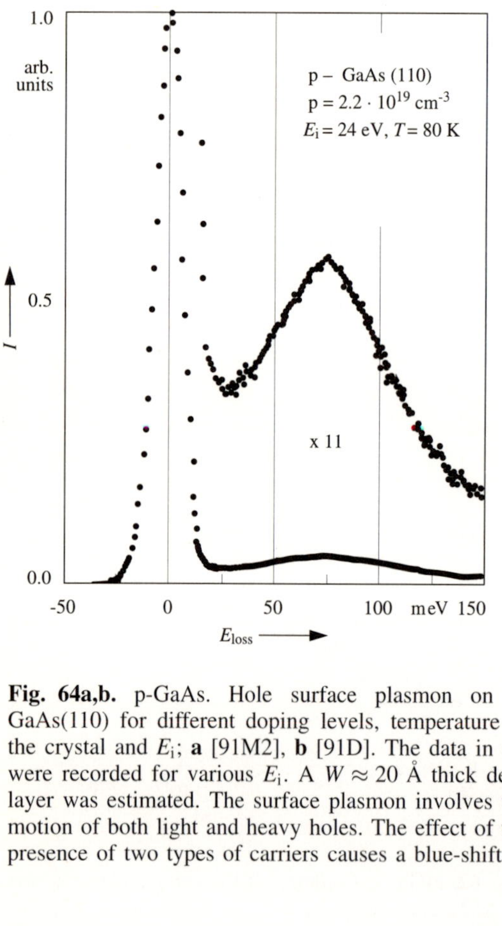

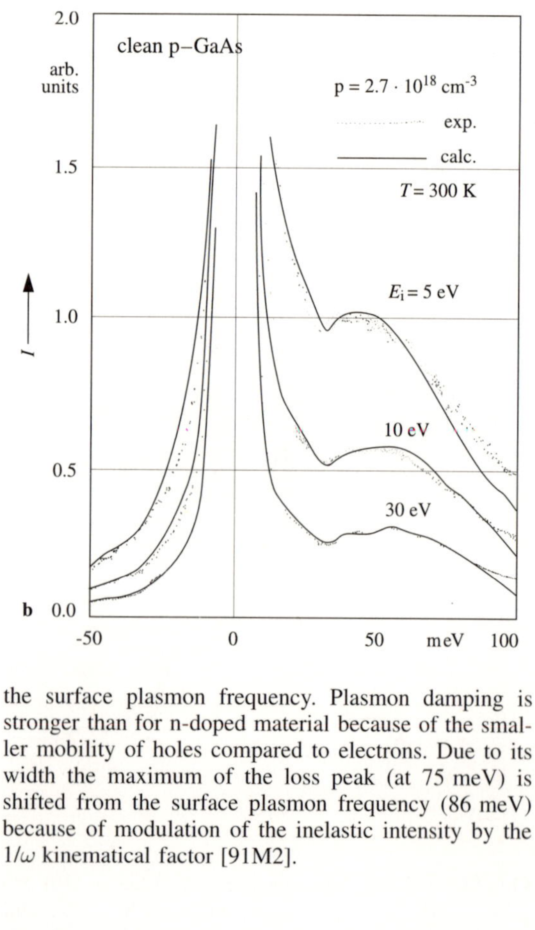

Fig. 64a,b. p-GaAs. Hole surface plasmon on p-GaAs(110) for different doping levels, temperature of the crystal and E_i; **a** [91M2], **b** [91D]. The data in (**b**) were recorded for various E_i. A $W \approx 20$ Å thick dead layer was estimated. The surface plasmon involves the motion of both light and heavy holes. The effect of the presence of two types of carriers causes a blue-shift of the surface plasmon frequency. Plasmon damping is stronger than for n-doped material because of the smaller mobility of holes compared to electrons. Due to its width the maximum of the loss peak (at 75 meV) is shifted from the surface plasmon frequency (86 meV) because of modulation of the inelastic intensity by the $1/\omega$ kinematical factor [91M2].

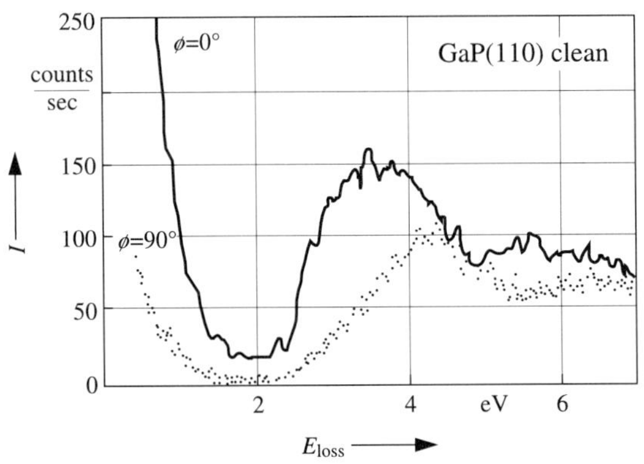

Fig. 65. GaP. Surface interband transition on GaP(110) for scattering along the crystallographic direction <1$\bar{1}$0>($\phi = 0°$) and <001>($\phi = 90°$). The azimuthal dependence of the EEL-spectra is due to the anisotropic surface reconstruction. Two pronounced broad structures are present at 3.5 eV and 5.6 eV ($\phi = 0$) and 4.5 eV ($\phi = 90°$). The surface character of the transition is demonstrated by the sensitivity to H exposure of the surface [89N1, 89N2]. Corresponding optical data are given in [87S2].

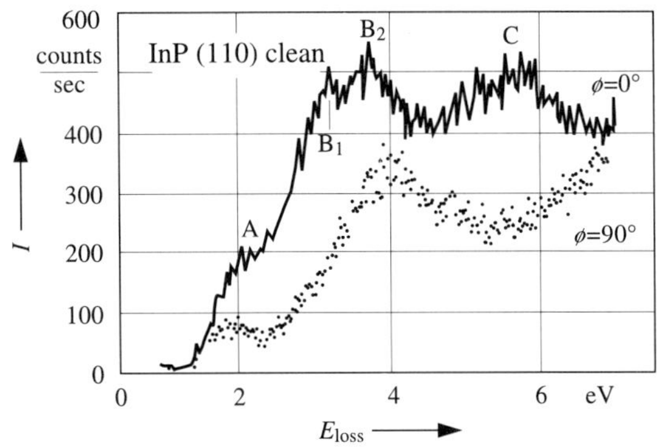

Fig. 66. InP. Same as Fig. 65 for InP(110) [91N]. The transitions are A: 1.9 eV ($\phi = 0°$ and 90°); B_1: 3.2 eV ($\phi = 0°$); B_2: 3.8 eV; ($\phi = 0°$), 4.0 eV ($\phi = 90°$); *C*: 5.6 eV ($\phi = 0°$). Corresponding optical data are reported in [89C3]. The transitions are quenched by exposure to $\approx$ 500 L of hydrogen. ($1\,\mathrm{L} = 10^{-6}$ Torr $\cdot$ s)

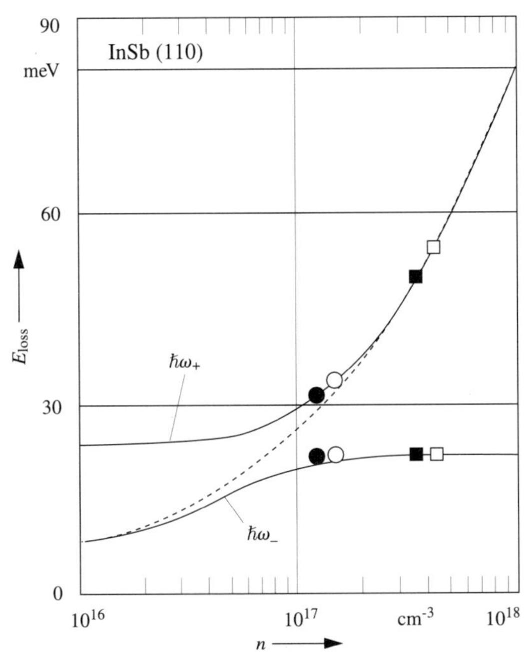

Fig. 67. InSb Dispersion of the mixed surface plasmon and phonon modes (ω_+ and ω_-) vs. effective carrier concentration *n* on InSb(110) [85R3].

6.2.2.2.2 Si and Ge

Only Si(111) and Ge(111) reconstructed surfaces were studied by EELS.

A dipole active surface phonon is present on freshly cleaved (2 × 1) reconstructed Si surfaces [71I, 86D2, 87S1, 89D1] at $\hbar\omega = 56$ meV. The dynamical dipole moment is set up because the reconstruction makes the positions of the surface atoms in the unit mesh non equivalent with respect to symmetry operations.

The frequency of this mode was demonstrated to depend on the direction of the cleave, varying between 48 meV and 56 meV (Fig. 71) [87S]. The effect was explained assuming that the vibration is connected to the π bonded chains (Pandey model [81P, 82P]) at the surface whose tilt angle may be sensitive to the cleave. The azimuthal dependence of the cross section of the phonon mode was used to discriminate between two models of the dynamics of Si(111)2 × 1 which forecasted a longitudinal [85A, 88A1] or a vertical [89M] polarization of the 56 meV mode. The data are shown in Fig. 72 for two independent experiments [86A, 86D2, 89D1] which show that the loss intensity is independent of the azimuth within experimental error for vanishing $q_{\parallel}$ while a dependence is present at $q_{\parallel} \approx 0.2$ Å^{-1}. The EELS experiment supports therefore the theoretical model of [89M].

Surface electronic transition is present on both Si(111)2 × 1 [75F2, 83M, 86D1, 87D2, 85D1] and Ge(111)2 × 1 [75F2, 84R2, 86D3]. Transitions between occupied and empty dangling bond surface bands were observed at 0.25 and 0.4 eV on Si and Ge, respectively [75R], in accord with optical data [71C]. The dispersion of the interband transition with $q_{\parallel}$ was measured by varying the impact energy and moving off-specular (Fig. 76 for Si [83M] and Fig. 70 for Ge [84R2]). In case of Si(111)2 × 1 the dependence of the inelastic cross section was studied with respect to the azimuthal direction [86D1, 87D2] for different $q_{\parallel}$. A strong anisotropy was found in accord with optical data [84C, 84O] and with Pandey's reconstruction model [81P, 82P]. The onset of the electronic transition shifts with temperature by 90 meV and 125 meV for Si [85D1] (Fig. 74) and Ge [86D3] (Fig. 69), respectively, between 20...30 K and room temperature. These temperature dependences of the surface-state gap are larger than those reported for the bulk gap.

The width of the quasielastic peak is considerable in the case of Ge(111)2 × 1 at room temperature [84R2, 86D3]. A quantitatively evaluation of the effect as a function of temperature (Fig. 68) [89P1] indicates that its origin has to be connected to the presence of an accumulation layer due to the pinning of the Fermi level 120...140 meV above the valence band edge at 300 K. The broadening is thus due to the hole plasma occurring in the vicinity of the surface. The pinning position decreases linearly with decreasing temperature and coincides with the valence band edge at zero temperature. This result is at variance with the case of Si(111)2 × 1 where the Fermi level pinning position occurs almost midgap and no broadening is observed.

When Si(111) is heated above a critical temperature the metastable 2 × 1 reconstruction changes to a 7 × 7 superstructure. EELS measurements show a strong broadening of the specular peak [81B1, 83D1, 84P], whose intrinsic nature is demonstrated by its quenching after adsorption of hydrogen or by traces of impurities (C or Ni [87D1, 84F2]).

The quasielastic peak is strongly temperature dependent [83D1, 84P]. At 20 K the EELS spectrum shows five well defined peaks, labeled $T_0 \rightarrow T_4$ in Fig. 77. The first at 63 meV is attributed to a surface phonon. The other peaks are assigned to interband transitions at the surface. The analysis of the data suggests the model depicted in Fig. 78 whose principal feature is the presence of a very localized, narrow (≈ 2 meV) state at E_F which must be half-occupied to provide for free carriers and a 40 meV gap between this occupied state and the more delocalized surface-state band edge whose energy gap is estimated to be ≈ 100 meV by the energy of the transition T_1.

The presence of dipole active surface phonons was confirmed on Si(111)7 × 7 surfaces for which the quasi-elastic background is lifted by traces of impurities (Fig. 79). Modes at 29.8 meV and 70.7 meV [87D1] were observed. The difference in frequency with the result reported in Fig. 77a, b may be due either to the temperature or to the different preparation of the sample. Comparison with ab initio calculations of the vibrational spectrum of a Si_{15} cluster shows that these vibrations are due to silicon adatoms. They have the electronic structure of an acceptor impurity which is more localized at the surface than the dangling bond and constitute therefore a good candidate to explain the extremely narrow metallic band of the clean Si(111)7 × 7 surface [87D1].

Doped samples of Si(111) show an additional broadening of the elastic peak due to multiple excitation of the surface-plasmon associated with the free carriers in the conduction band (Figs. 80 ··· 83). An analysis of the data within a three layer model allows to estimate the width of the depletion layer at the surface which compares well with the depletion layer expected from the surface band bending. The dependence of the broadening with temperature indicates a dependence of the depletion layer width as $\sqrt{n}$ [85S1, 87S3]. The data for the intrinsic and the doped surfaces were also analyzed to extract the temperature dependent surface resistivity (Fig. 84) [86P1].

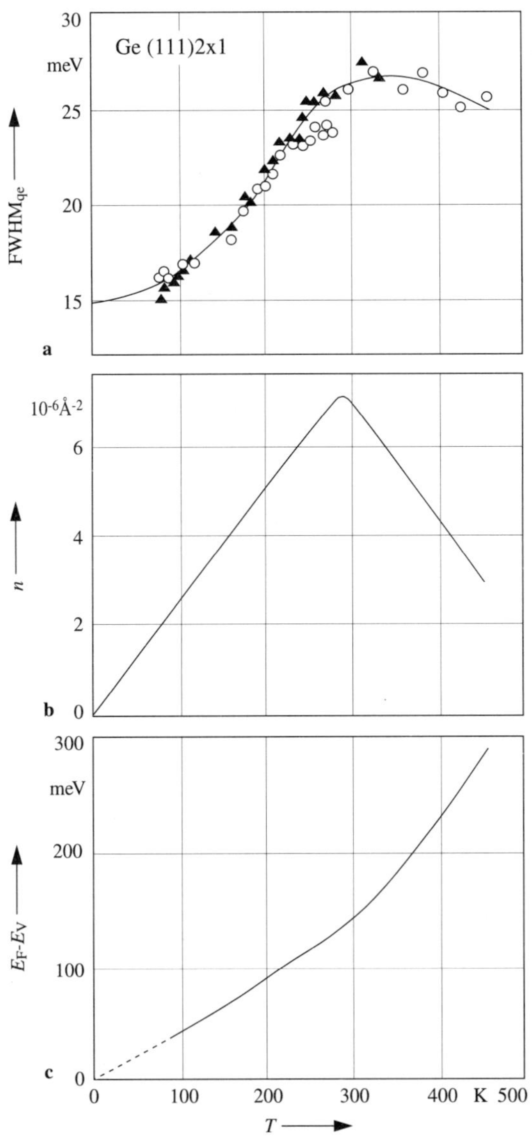

Fig. 68a-c. Ge. **a** width of the quasielastic peak $FWHM_{qe}$ vs. crystal temperature for intrinsic Ge (111)2 × 1. The solid line corresponds to the analysis carried out using Eq. (9) with the carrier density n shown in **b**. In **c** the Fermi level pinning position is reported. E_V is the energy of the valence band edge at the surface [89P1].

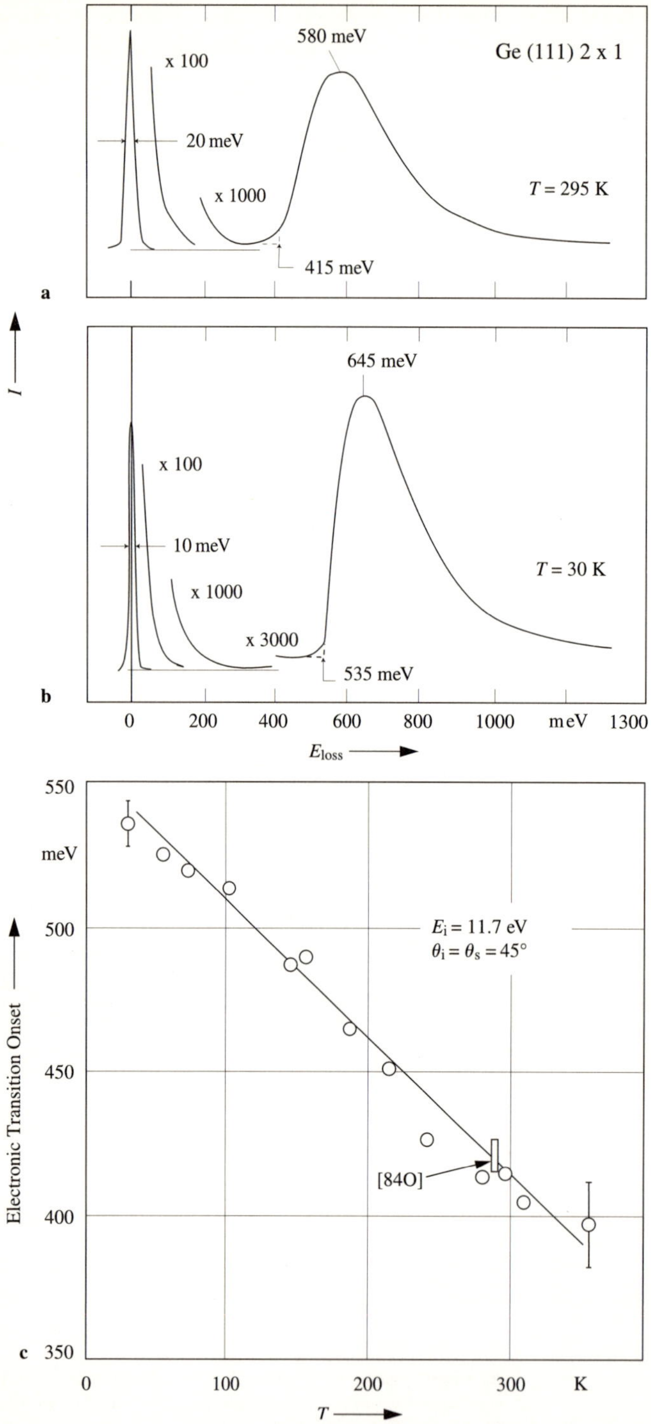

Fig. 69a-c. Ge. EEL-spectra showing the interband transition between dangling-bond surface states on intrinsic Ge(111)2 × 1 at **a** $T = 295$ K and **b** $T = 30$ K. E_i is 11.7 eV, $\theta_i = \theta_s = 45°$ [86D3]. In **c** the dependence of the onset of the interband transition is reported vs. crystal temperature [86D3]. The data are compared with the value of the onset determined by photothermal deflection [84O]. Optical data on this surface are also reported in [87S2].

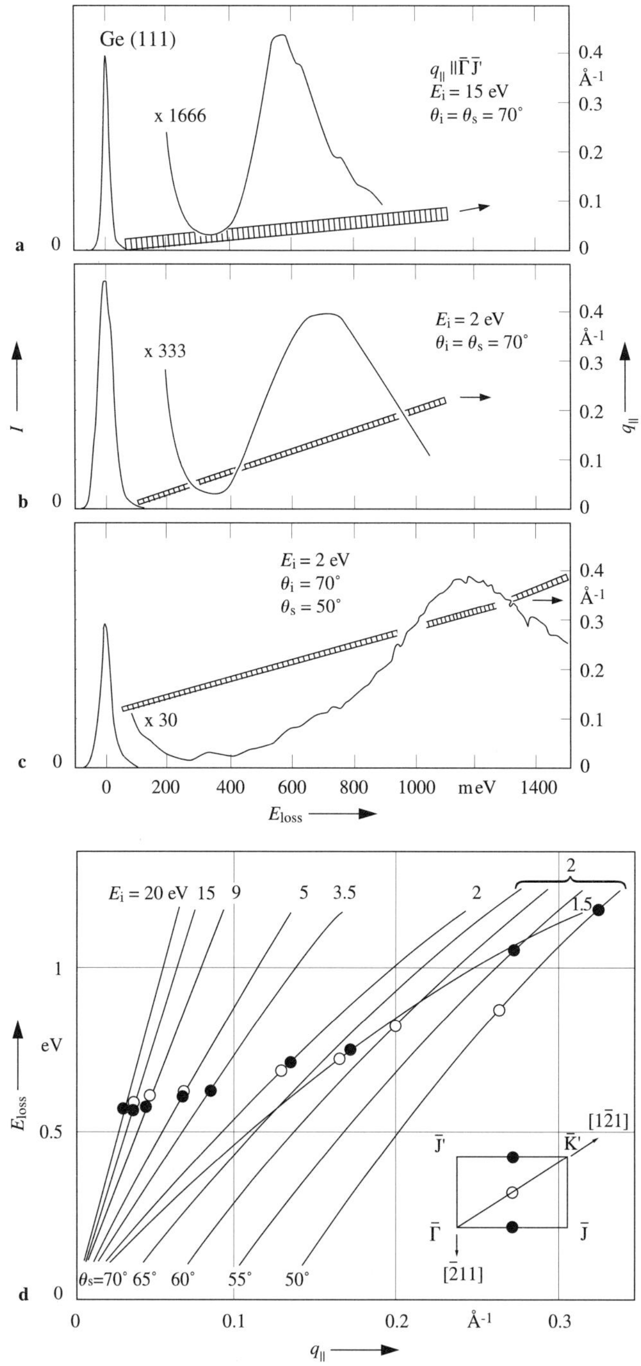

Fig. 70a-d. Ge. **a**–**c** EEL-spectra of the interband transition measured for Ge(111)2 × 1 at different $q_{\|}$ (determined by the choice of θ_s, θ_i, E_i and the conservation equations for energy and momentum). The shaded stripe indicates the uncertainty in reciprocal space due to the angular acceptance of the spectrometer. **d** dispersion of the maximum of intensity associated to the interband transition with $q_{\|}$ with the kinematical conditions of the measurements [84R2]. The open circles in (**d**) were measured with the scattering plane aligned along $\bar{\Gamma}\bar{K}'$ while the full circles were recorded along $\bar{\Gamma}\bar{J}'$ as shown in the inset.

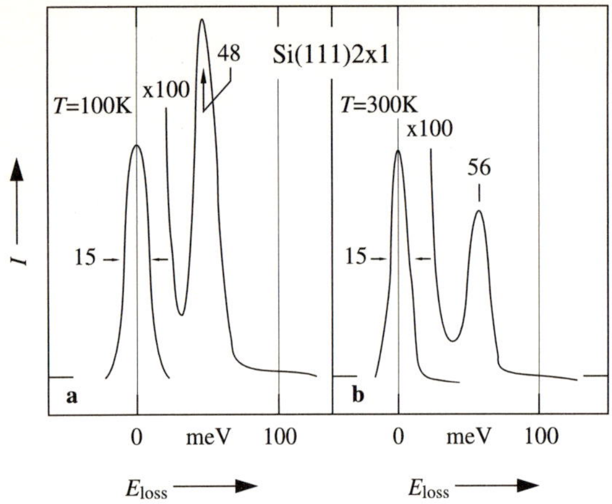

Fig. 71a,b. Si. EEL-spectra of the dipole active surface phonon on Si(111)2 × 1 **a**. cleaved along the [1$\bar{1}$0] direction at T = 100 K, E_i = 6.5 eV, and θ_i = 52.8°. **b** cleaved along the [11$\bar{2}$] direction at T = 300 K, E_i = 5 eV, and θ_i = 45°. The dependence of the phonon frequency on the cleave direction is due to the step density at the surface which influences the tilt angle of the surface dimers [87S1].

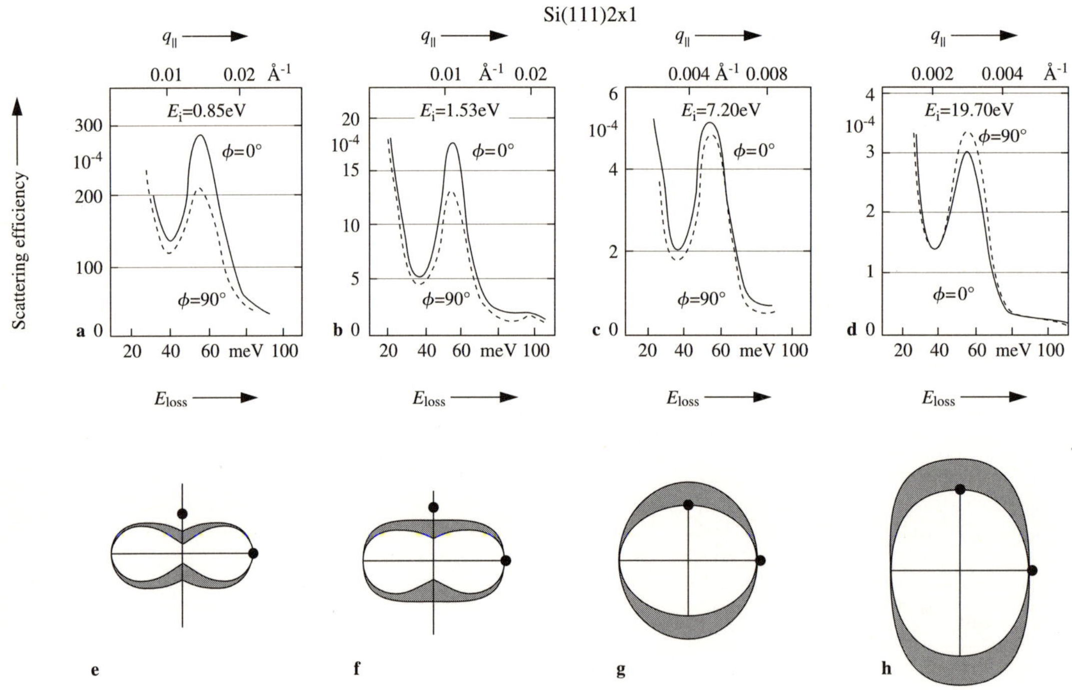

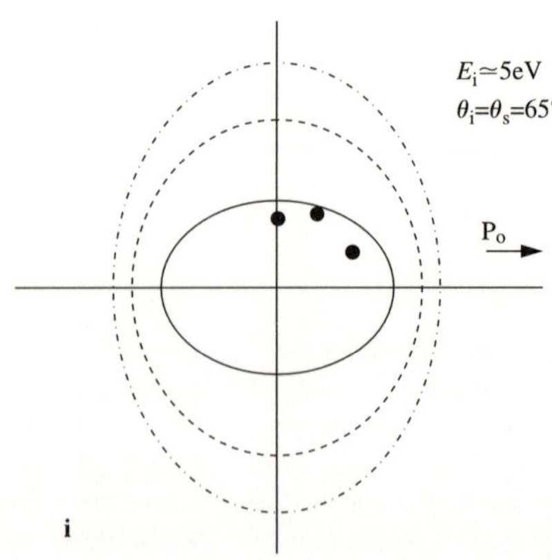

Fig. 72a-i.

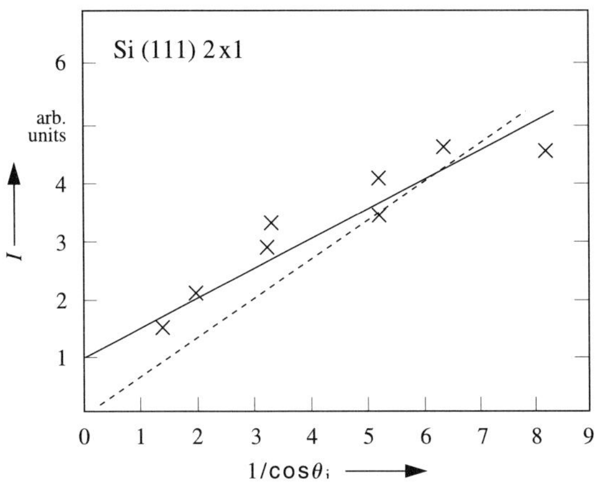

Fig. 73. Si. Loss probability for the 56 meV surface phonon on Si(111)2 $\times$ 1 vs. θ_i. The dashed line indicates the $1/\cos\theta_i$ characteristic. The solid line is a best fit curve [77F].

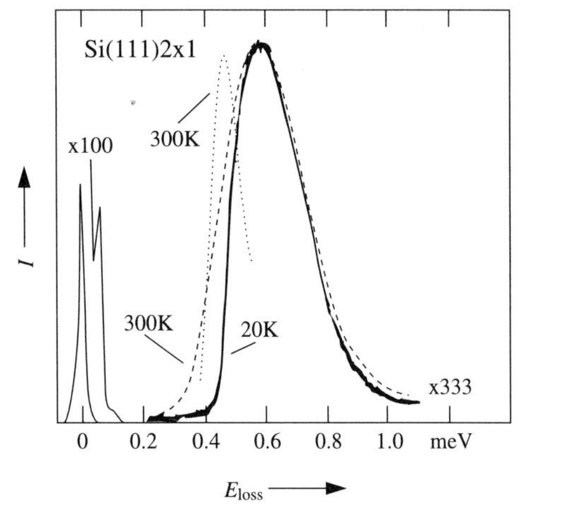

Fig. 74. Si. Temperature dependence of the onset of the interband transition on Si(111)2 $\times$ 1 [85D1]. Surface state peak onset: 0.34 ± 0.01 eV (300 K), 0.41 eV (90 K), 0.43 ± 0.05 eV (20 K); surface state peak maximum 0.58 ± 0.02 eV (independent of temperature). The measurements in the figure are: solid line EEL-spectrum at 20 K; dashed line EEL-spectrum at 300 K; dotted line optical absorption (300 K) [71C, 84O].

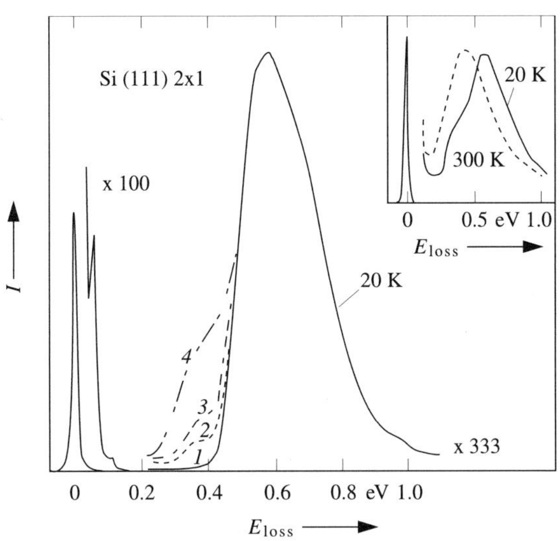

Fig. 75. Si. Interband transition on Si(111)2 $\times$ 1 for 4 cleavages of different quality. The EEL-spectra are represented by the curves *1* $\cdots$ *4* for increasing defect concentration [85D1]. Defect states are present in the gap at a lower energy than the dangling bond states. They can be resolved only at low temperature as demonstrated in the inset by comparison with a room temperature spectrum. In these conditions EELS is an extremely sensitive probe for the quality of the cleavage. The onset of the defect state is at ≈ 0.25 eV and the maximum at ≈ 0.35 eV.

←

Fig. 72a-i. Si. Dependence of the scattering efficiency of the dipole active surface phonon on Si(111)2 $\times$ 1 on the azimuthal direction [89D1]. The measurement was performed at different E_i to discriminate effects connected to the finite $q_\parallel$ involved in the measurements due to energy and momentum conservation (cf. Eqs.(1) and (2)). No anisotropy is present at vanishing $q_\parallel$ (measurement d) leading to the conclusion that at $\bar{\Gamma}$ the mode is polarized vertically [89D1] in accordance with the lattice dynamical model of [89M]. Moving away from $\bar{\Gamma}$ an anisotropy of the cross section sets in, indicating a sagittal polarization of the mode along <1$\bar{1}$0> direction ($\phi = 0°$, momentum exchange between ($\bar{\Gamma}\bar{J}$) in reciprocal space). In the polar diagrams **e**–**h** the intensity of the loss derived from the spectra **a**–**d**, respectively, is reported vs. ϕ. As one can see the strong anisotropy present at low E_i, (large $q_\parallel$) disappears for $q_\parallel \to 0$. In (**i**) similar measurements are reported which were recorded for $E_i = 5$ eV [86D2] showing the same phenomenon.

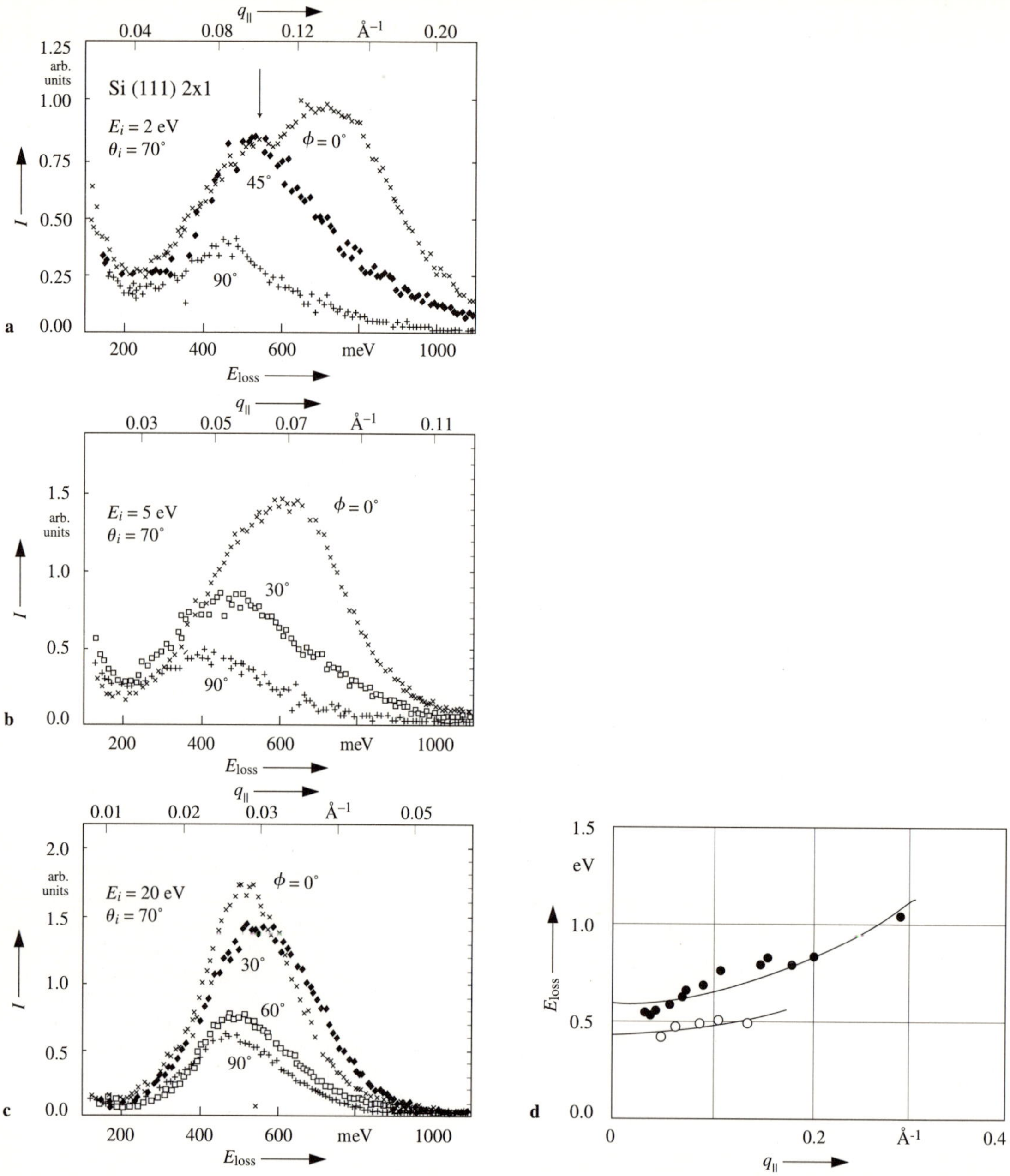

Fig. 76a-d. Si. Dispersion of the interband transition on Si(111)2 × 1 vs. $q_\parallel$: **a–c** EEL-spectra recorded for different azimuthal directions and kinematical conditions (corresponding to differrent $q_\parallel$) along the <1$\bar{1}$0> ($\bar{\Gamma}\bar{J}$) direction ($\phi = 0°$), along $\bar{\Gamma}\bar{K}$ ($\phi = 30°$) and $\bar{\Gamma}\bar{J}'$ ($\phi = 90°$) showing the anisotropy of the transition and a double peak along the <1$\bar{1}$0> direction [86D1]; **d** dispersion of the maximum of the interband transition for $\phi = 0°$ vs. exchanged momentum according to the independent investigation of [83M]. The solid curves in (**d**) are best fit parabolas. The loss peaks are associated to transitions between the dangling-bond states and agree with the prediction of the π-bonded chain model reconstruction [82P]. The transitions take place along $\bar{\Gamma}\bar{J}$ and along $\bar{K}'\bar{J}'$.

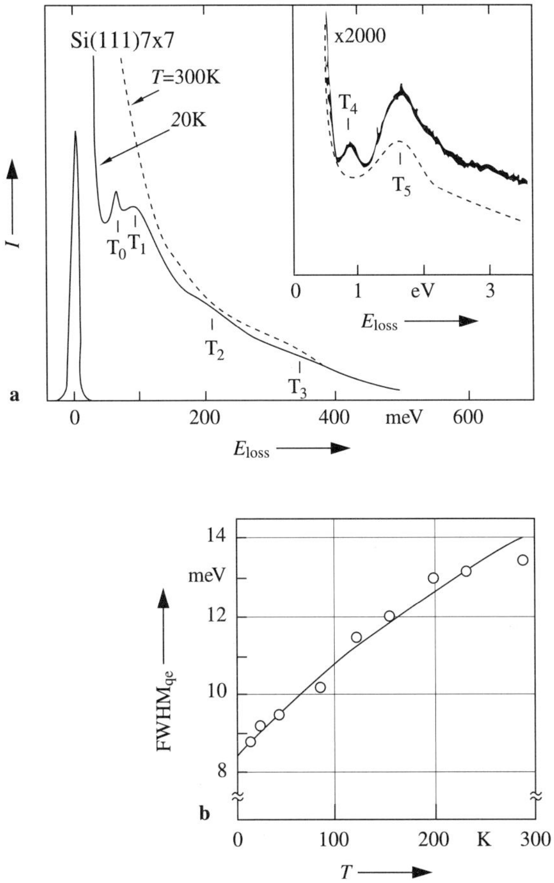

Fig. 77a,b. Si. **a** EEL-spectra recorded on Boron-doped Si(111)7 × 7 ($n = 1.3 \cdot 10^{15}$ cm^{-3}) at low temperature (continuous line) and at $T = 300$ K (dashed line). At 20 K the quasielastic peak is reduced and evidences interband transitions. For the measurement in the inset the resolution was degradated to 20 meV to measure up to $E_{loss} = 4$ eV [83D1]. The loss peak T_0 is attributed to the excitation of a surface phonon. The peaks $T_1 \ldots T_5$ are assigned to interband transitions. T_0 is at 63 meV, T_1 at 95 meV, $T_2 \sim 210$ meV, $T_3 \sim 340$ meV, $T_4 \sim 900$ meV and $T_5 \sim 1.7$ eV. The structures $T_0 \ldots T_4$ are strongly affected by adsorption and have therefore surface character. **b** variation of the FWHM$_{qe}$ of the quasielastic peak vs. crystal temperature. The solid line is the theoretical result for a constant carrier concentration. The analysis of the data lead to the model of Fig. 78 for the electronic band structure in the gap.

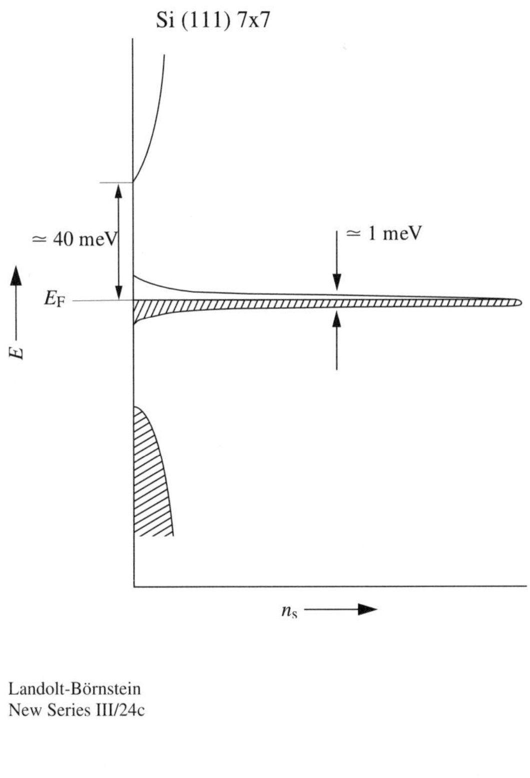

Fig. 78. Si. Model of the states in the gap of Si(111)7 × 7 according to the analysis of the data in Fig. 77 of [84P]. E_F is the Fermi energy.

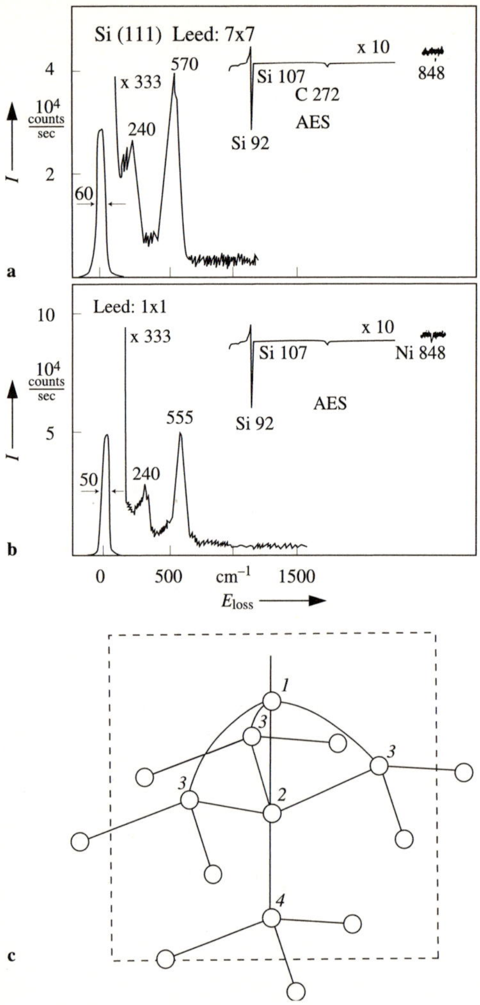

Fig. 79a-c. Si. EEL-spectra of Si(111)7 × 7 surfaces with small contamination of C **a** and 1 × 1 surface stabilized by Ni impurities **b** at $E_i = 4.7$ eV. The impurities quench the quasielastic peak in both cases allowing to observe the loss peaks at 29.8 meV and at 70.7 meV due to Si adatom vibrations. The unusually high frequency of 70.7 meV involves the stretching of two Si-Si bonds and is thus inconsistent with tetrahedral coordination. The data are compared with a model, shown in **c**, for which ab-initio calculations were performed. *1*...*4* indicate non equivalent positions of Si atoms. The adatom *1* is suggested as the site of the extremely localized metallic state shown in Fig. 78 [87D1].

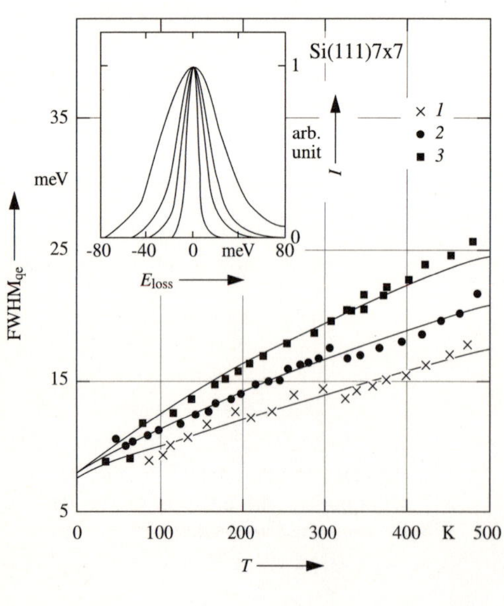

Fig. 80. Si. Temperature dependence of the quasielastic-peak line width FWHM on Si(111)7 × 7 for samples with different doping level. The dashed line is given by the expression for the broadening due to the excitation in a two-dimensional surface state band as it is the case for the intrinsic material. The solid lines are obtained by adding the contribution from the conduction band free carriers with depletion depths of 1050 Å and 1900 Å. The bulk carrier densities are: *1*:$N_d = 6.6 \cdot 10^{12}$ cm^{-3}, *2*: $N_d = 8.9 \cdot 10^{15}$ cm^{-3}, *3*: $N_d = 5.6 \cdot 10^{16}$ cm^{-3}. The inset shows the measured quasielastic peak at 80, 514, 701, 881 K in the case of curve *2* [85S1].

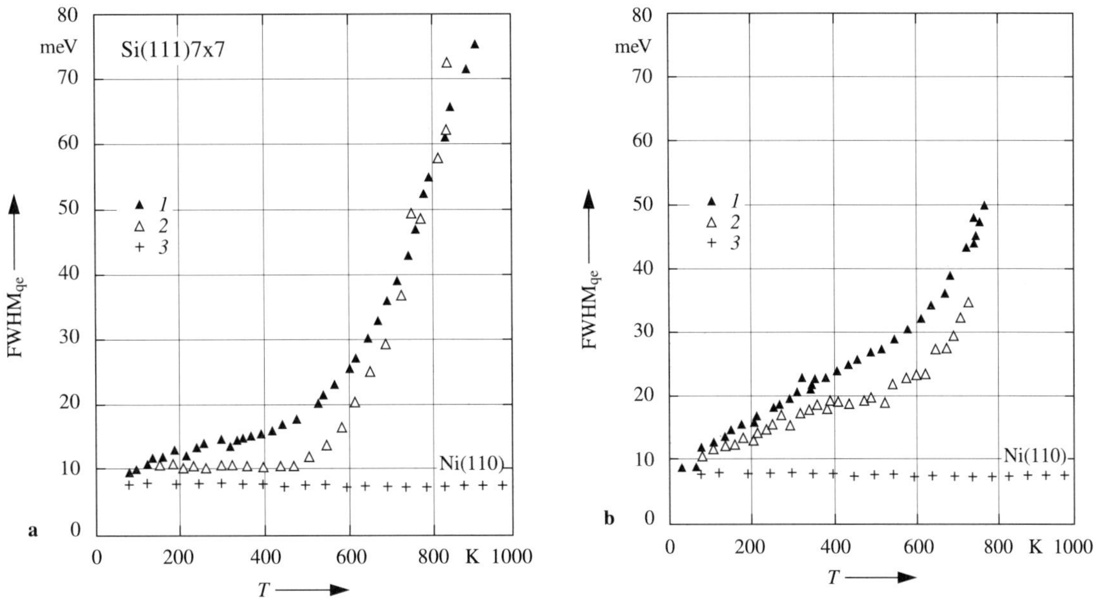

Fig. 81a,b. Si. Temperature dependence of the quasielastic peak width $FWHM_{qe}$ on Si(111)7 × 7 for two bulk impurity concentrations: **a** $N_d = 6.6 \cdot 10^{12}$ cm^{-3} [87S3], **b** $N_d = 5.6 \cdot 10^{16}$ cm^{-3}. *1*: clean surface, *2*: after exposure to 1L of O_2 at 80 K which saturates the surface states and eliminates the intrinsic broadening. The residual broadening is due to the conduction band free carriers. Results from a clean Ni(110) surface *3:* are shown for comparison [87S3].

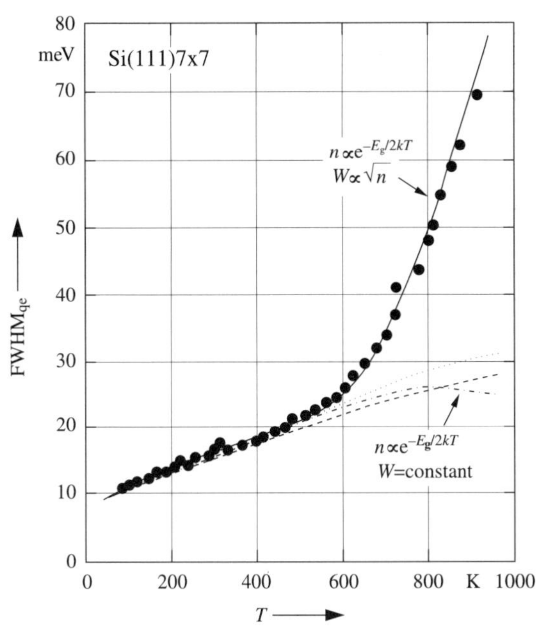

Fig. 82. Si. Same as Fig. 81 for a doping level of $N_d = 8.9 \cdot 10^{15}$ cm^{-3} and theoretical analysis [87S3]. The results of the calculations are for a constant depletion layer and carrier density (dashed line), temperature dependent carrier density and constant depletion layer (dot-dashed line), constant carrier density and temperature dependent depletion layer (dot-dashed line), constant carrier density and temperature dependent depletion layer (dotted line) and temperature dependent carrier density and depletion layer (solid line). The Fermi level pinning position is 0.58 eV below the conduction band edge [87S3]. E_g is the energy gap, W the depletion layer width and n the temperature dependent free carrier concentration.

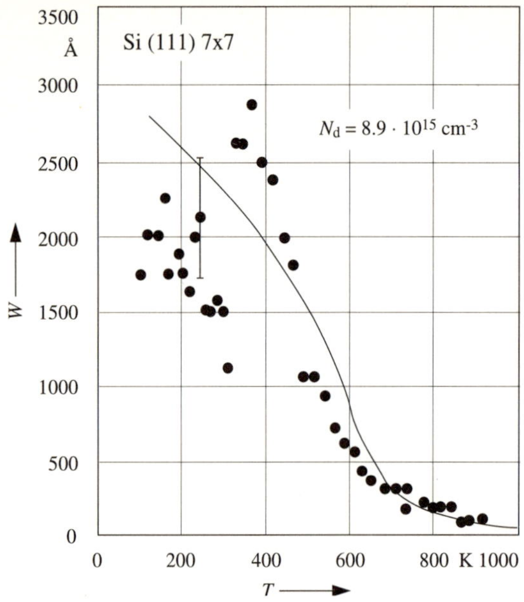

Fig. 83. Si. Temperature dependence of the depletion layer width W for the data of Fig. 82. The Fermi level position at the surface is pinned 0.58 eV below the conduction band edge at 300 K. The error in the depletion layer width corresponds to an uncertainty of ± 1 meV in the quasielastic width [87S3].

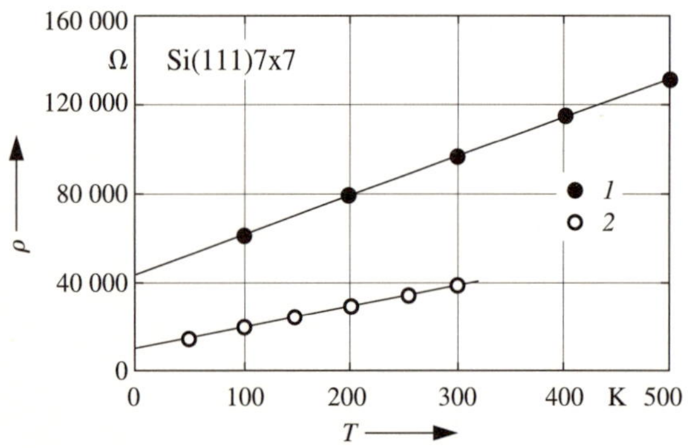

Fig. 84. Si. Temperature dependent resistivity ρ of the Si(111)7 × 7 surface as deduced from the experimental data of [81B1], [84P] (*1*: B-doped crystal, $n = 1.3 \cdot 10^{12}$) and [87S3]. (*2*: nearly undoped crystal , $n = 6.6 \cdot 10^{12}$) from [86P1].

6.2.2.3 Insulators and oxides

ZnO was the first extensively studied surface by EELS [70I, 72I, 74F]. Compound materials are characterized by strong losses due to the excitation of the Fuchs-Kliever mode [65F]. Overtones are usually observed whose intensity decays exponentially according to the Poisson distribution for the probability of multiple excitation of independent oscillators (see Figs. 85 and 86). An electronic transition is present at ≈ 4 eV loss energy on $(1\bar{1}00)$ surface which shows a strong azimuthal anisotropy (Fig. 87).

Insulators may be studied by EELS at primary energies for which the secondary electron yield is unitary [86C] or by neutralizing the sample by flooding it with an auxiliary electron gun [84T, 86T]. The measured and calculated frequencies of the Fuchs-Kliever phonons of various materials are reported in Table 4 [86C]. The Fuchs-Kliever phonons measured on different crystals were discussed also in [86T].

On $NiSi_2$ intrinsic surface-phonons were observed at $\hbar\omega = 33.1$ meV on the (111) and at $\hbar\omega = 28.8$ meV on the (001) surface, respectively, [83D2].

Results for TiS_2 were shown in [90P] while $TiO_2(100)$ was studied by EELS in [81K]. Two modes were observed on clean $TiO_2(100)$ at 54 ± 1 meV and at 95 ± 1 meV. Both modes are suggested to be surface optical phonons. Two modes were observed also for Al_2O_3 at 60.3 meV and 99.5 meV [84T].

For $SrTiO_3$ vibrational modes were observed at 57 meV and 91.8 meV [81B2]. These modes fall just below the LO phonon frequencies found at 58.8 meV and 97.7 meV by infrared reflectivity measurements [66B].

Surface and superconducting properties of superconducting oxides were studied with low energy EELS on $YBa_2Cu_3O_7$ and $Bi_2Sr_2CaCu_2O_8$ single crystals cleaved under vacuum [90D, 90P]. The study is difficult because of the poor homogeneity of the investigated samples which present different cleavage terminations with drastically different conductivities. The surface was scanned looking for superconducting regions thanks to the spacial resolution of the EELS (illuminated area 100μm). In the high conductivity terminations the first conducting layer screens out the electric field of the incoming electrons and the loss spectrum is directly proportional to the resistivity of the top layer. On $YBa_2Cu_3O_7$ a loss feature associated with the energy gap was observed (see Fig. 90) at $E_{\text{loss}} = 60$ meV which corresponds to $7.8 \pm 0.3\ k_B T_c$, where T_c is the transition temperature to the superconducting state. The gap feature decreases continuously with T with little change in the gap position. Above T_c the gap is reduced but does not disappear (Fig. 90). This behavior cannot be explained by BCS theory. The condensation of preexisting pairs is suggested as a hypothesis. On $Bi_2Sr_2CaCu_2O_8$ two modes were observed at 51 meV and 81 meV on low conductivity areas (Fig. 92). They are temptatively assigned to vibrations of the Cu-O planes in the light of available lattice dynamical calculations [89P2].

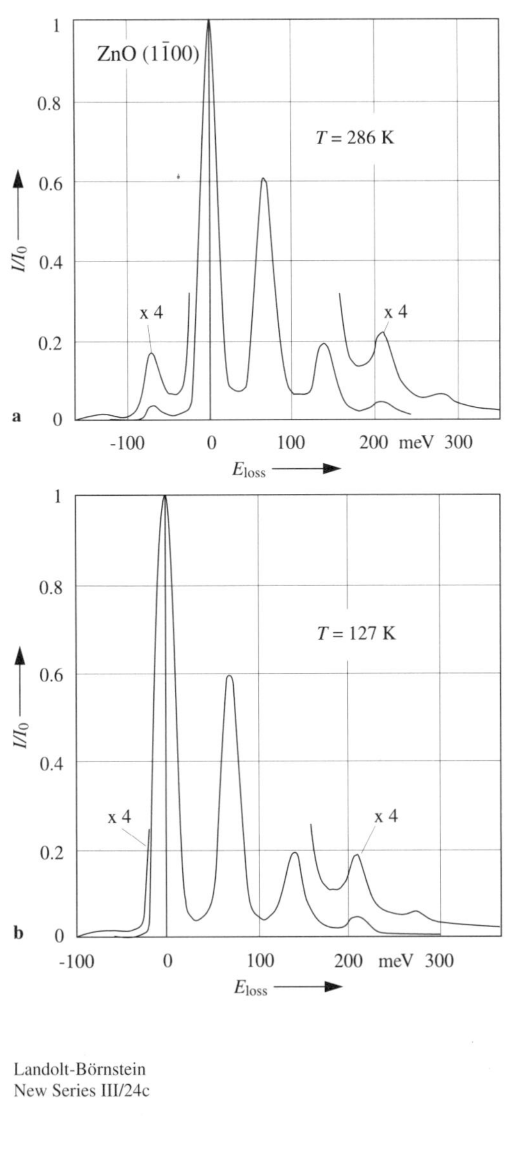

Fig. 85a,b. ZnO. EEL-spectra of ZnO(1$\bar{1}$00) taken at $E_i = 7.5$ eV and $\theta_i = \theta_s = 45°$ at $T = 286$ K **a** and $T = 127$ K **b** showing the multiple excitation of the Fuchs-Kliever surface phonon. At 286 K also gain processes are observed. I_0 is the elastic intensity in the specular direction. The sample is semiconductive so that no neutralization is necessary for the measurements [70I, 72I].

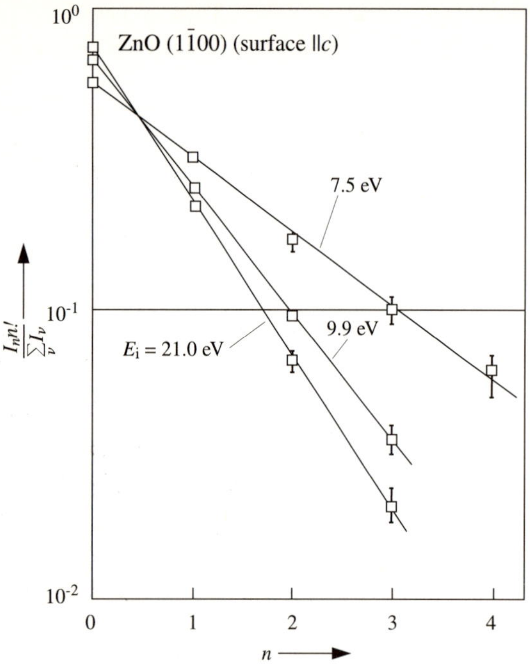

Fig. 86. ZnO. Probability of n-phonon excitation (multiplied by $n!$ in the graph) vs. n for three different E_i. A continuous background has been subtracted. The error is due mainly to the uncertainty of the background subtraction. The intensity of the losses decreases exponentially as expected from the Poisson distribution of independent events [70I].

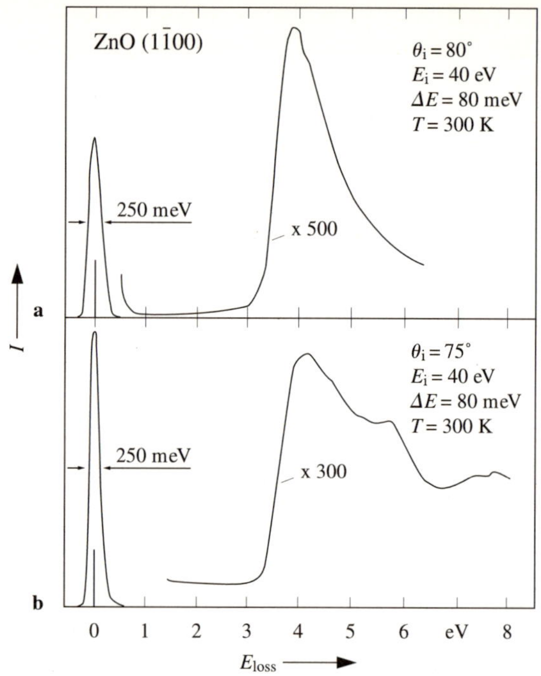

Fig. 87a,b. ZnO. EEL-spectra displaying the interband transition on ZnO (1$\bar{1}$00) with the scattering plane aligned parallel **a** and perpendicular **b** to the c axis of the sample (wurtzite lattice) [74F].

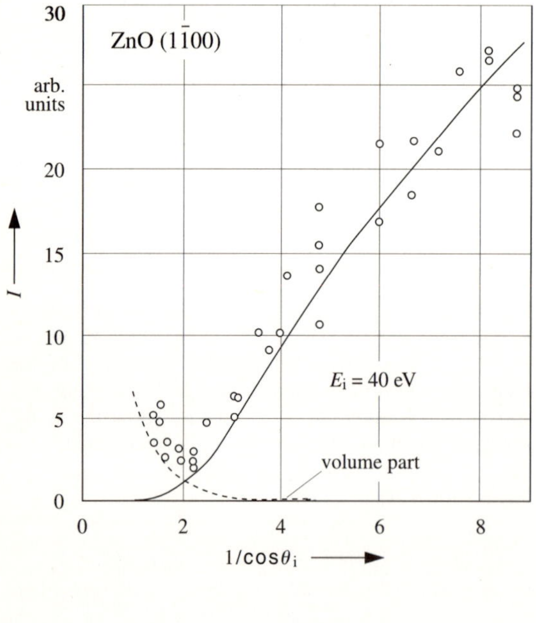

Fig. 88. ZnO. Relative loss intensity of the 4 eV loss due to the interband transition on ZnO vs. $1/\cos\theta_i$. The solid curve has been calculated from the theory for an angular acceptance of 1.4° [71L, 72L]. The broken curve indicates the contribution of volume scattering to the inelastic intensity [74F].

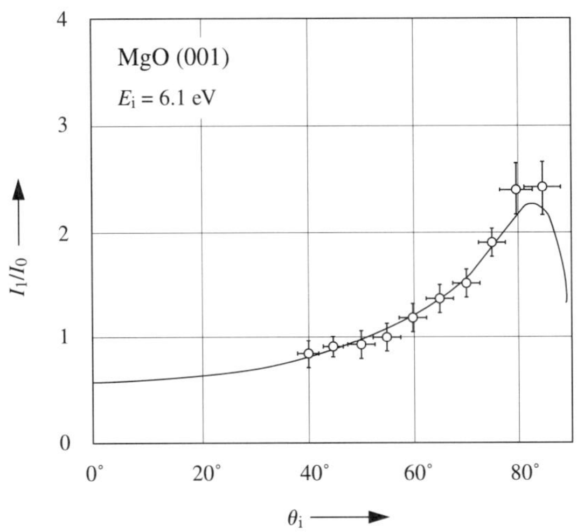

Fig. 89. MgO. Loss probability I_1/I_0 vs. θ_i for the surface phonon on MgO(001). The curve is the dipole scattering intensity calculated with corrections for the image force and spectrometer aperture. I_0 is the specular elastic intensity. An electron gun was used to neutralize the sample during the EELS measurement [84T].

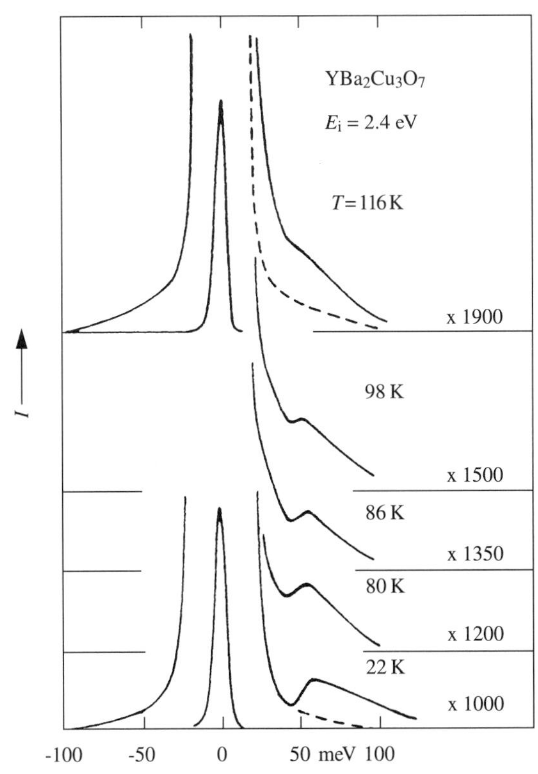

Fig. 90. $YBa_2Cu_3O_7$. Sequence of HREEL-spectra as a function of temperature for the cleaved $YBa_2Cu_3O_7$ surface exhibiting the gap-like loss structure. Experimental conditions: $E_i = 2.6$ eV, $\theta_i = 45°$, angular acceptance 1° [90P]. The multiplication factor is given to the right. The loss structure becomes fainter with rising T but is still present even above the transition temperature to the non superconducting state ($T_c = 75 \ldots 91$ K) in contrast to the behavior expected for a BCS mechanism for superconductivity.

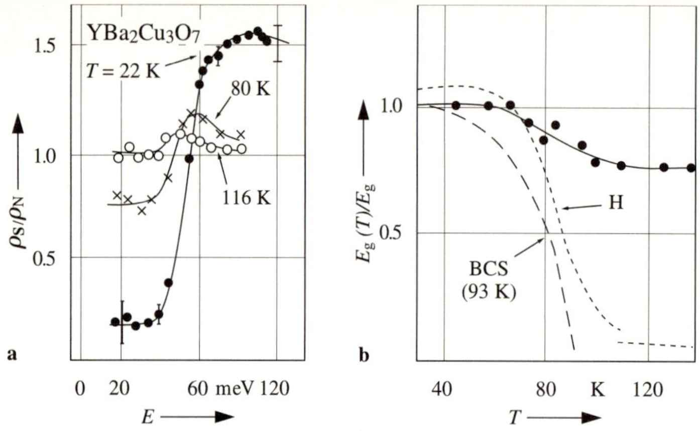

Fig. 91a,b. $YBa_2Cu_3O_7$. **a** ratio of the loss intensity recorded on superconducting $YBa_2Cu_3O_7$ at different temperatures (22K, 80K and 116K) to the normal state EEL-spectrum at 125 K showing the evolution of the gap. The intensity ratio is proportional to the ratio of the resistivity of the surface at T in the superconducting state $\rho_S(T)$ to the resistivity in the normal (non-superconducting) state ρ_N. E_g is the energy necessary to break the Cooper pairs. **b** solid line: evolution of the onset of the transition with temperature. Short dashed line (H): height of the loss feature connected to the superconducting onset in the loss spectra. The discontinuity in H at 116 K arizes from the electron beam damage occurring for consecutive scans above T_c. The resistivity below 22 K depends on the sample and is found to depend on cleavage quality and on the amount of electron beam irradiation received by the crystal above T_c. The prediction for H from BCS theory is given for comparison (dashed line) [90P].

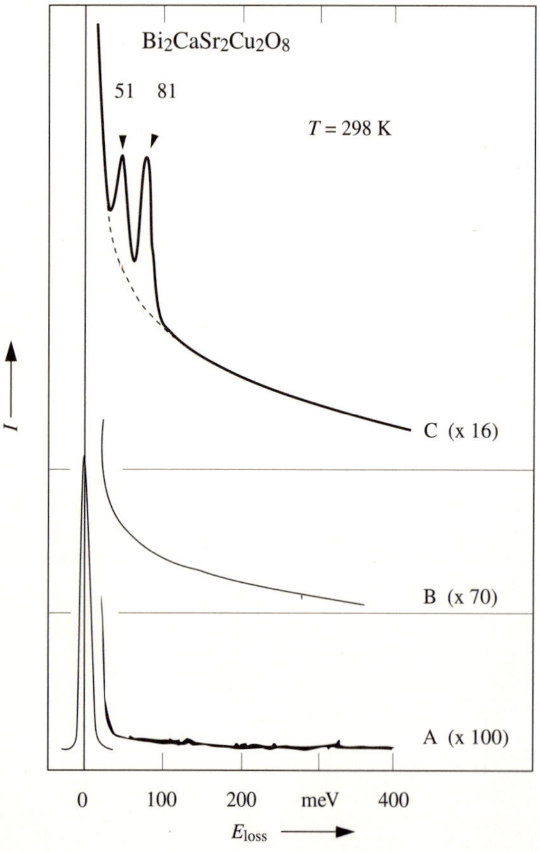

Fig. 92. $Bi_2CaSr_2Cu_2O_8$. HREEL-spectra for $Bi_2Sr_2CaCu_2O_8$ from regions of low (A), medium (B) and high (C) resistivity. The field of the incoming electrons is shielded for the low resistivity regions and no excitation is possible. On the contrary for high resistivity regions dipole active modes connected to the vibration of the Cu-O planes are excited [90P].

Table 4. Surface phonon frequencies of infrared active compounds. For semiconductors the value measured on the intrinsic material is reported.

Compound	Theory [meV]	Experiment [meV]	Ref.
Al_2O_3		62, 101	86T
CaO	63.2	66	86C
GaAs	36.6	36.2	81M
InP		42	90S2
KBr	19.7	20	86C
LiF	69.8	69.6	86C
NaCl	29.2	29.4	86C
MgO	81.0	80.6	86C
NiO	68.3	69.5	86C
TiO_2		54, 95.5	81K
ZnO($1\bar{1}00$)	69.0	68.8 (5)	70I
ZnO(0001),($000\bar{1}$)	67.3	67.5 (5)	70I

6.2.3 References for 6.2

35H Herzog, R.: Z. Phys. **97** (1935) 596.
57R Ritchie, R.H.: Phys. Rev. **106** (1957) 874.
65F Fuchs, R., Kliever, K.L.: Phys. Rev. **140** (1965) A2076.
66B Barker, A.S.: Phys. Rev. **145** (1966) 391.
66K Kunz, C.: Z. Phys. **196** (1966) 311.
67D Daniels, J.: Z. Phys. **203** (1967) 235.
67P Propst, F.M., Piper, T.C.: J. Vac. Sci. Technol. **4** (1967) 53.
70I Ibach, H.: Phys. Rev. Lett. **24** (1970) 1416.
70R Roy, D., Carette, J.D.: Appl. Phys. Lett. **16** (1970) 413.
70Z Zacharias, P.: Z. Phys. **238** (1970) 172.
71C Chiarotti, G., Nannarone, S., Pastore, R., Chiaradia, P.: Phys. Rev. B **4** (1971) 3398.
71I Ibach, H.: Phys. Rev. Lett. **27** (1971) 253.
71L Lucas, A.A., Šunjić, M.: Phys. Rev. Lett. **26** (1971) 229.
72E Evans, E., Mills, D.L.: Phys. Rev. B **5** (1972) 4126.
72I Ibach, H.: J. Vac. Sci. Technol. **9** (1972) 713.
72L Lucas, A.A., Šunjić, M.: Prog. Surf. Sci. **2** (1972) 75.
72R Roundy, V., Mills, D.L.: Phys. Rev. B **5** (1972) 1347.
73E Evans, E., Mills, D.L.: Phys. Rev. B **7** (1973) 853.
73K Kloos, T., Raether, H.: Phys. Lett. A **44** (1973) 157.
74F Froitzheim, H., Ibach, H.: Z. Phys. **269** (1974) 17.
74M1 Meyer, F.: Phys. Rev. B **9** (1974) 3622.
74M2 Monin, J., Boutry, G.A.: Phys. Rev. B **9** (1974) 1309.
75F1 Froitzheim, H., Ibach, H., Lehwald, S.: Rev. Sci. Instrum. **46** (1975) 1325.
75F2 Froitzheim, H., Ibach, H., Mills, D.L.: Phys. Rev. B **11** (1975) 4980.
75F3 Froitzheim, H., Ibach, H.: Surf. Sci. **47** (1975) 713.
75M Mills, D.L.: Surf. Sci. **48** (1975) 59.
75R Rowe, J.E., Froitzheim, H., Ibach, H.: Surf. Sci. **48** (1975) 44.
76K Krane, K.J., Raether, H.: Phys. Rev. Lett. **37** (1976) 1355.

77F Froitzheim, H., in: Electron Spectroscopy for Surface Analysis. Ibach, H. (ed.), Top. Curr. Phys. **4** (1977) 205.
77I1 Ibach, H.: Surf. Sci. **66** (1977) 56.
77I2 Ibach, H.: Electron Spectroscopy for Surface Analysis. Top. Curr. Phys. **4** (1977) 205.
77M Mills, D.L.: Prog. Surf. Sci. **8** (1977) 143.
77R Roy, D., Carette, J.D., in: Electron Spectroscopy for Surface Analysis. Ibach, H. (ed.), Top. Curr. Phys. **4** (1977) 13.
78M Moore, I.D., Pendry, J.B.: J. Phys. C **11** (1978) 4615.
79A Andersson, S.: Surf. Sci. **79** (1979) 385.
79L Levingston, H.J., Plummer, E.W., Feibelman, P.J.: Phys. Rev. Lett. **43** (1979) 5077.
79S Sexton, B.A.: J. Vac. Sci. Technol. **61** (1979) 1033.
80P Persson, B.N.J.: Surf. Sci. **92** (1980) 265.
80R Raether, H: Excitation of Plasmons and Interband Transitions by Electrons. Springer Tracts in Modern Physics **88**, Berlin: Springer Verlag 1980.
80T Tong, S.Y., Li, C.H., Mills, D.L.: Phys. Rev. Lett. **44** (1980) 407.
81B1 Backes, U., Ibach, H.: Solid State Commun. **40** (1981) 575.
81B2 Baden, A.D., Cox, P.A., Egdell, R.G., Orchard, A.F., Willmer, R.J.: J. Phys. C **14** (1981) L1081.
81K Kesmodel, L.L., Gates, J.A., Chung, Y.W.: Phys. Rev. B **23** (1981) 489.
81M Matz, R., Lüth, H.: Phys. Rev. Lett. **46** (1981) 500.
81P Pandey, K.C.: Phys. Rev. Lett. **47** (1981) 1913.
82A Andersson, A., in: Vibrations at Surfaces. Caudano, R., Gilles, J. M., Lucas, A. (eds.). New York: Plenum Press 1982, p. 169.
82F Feibelman, P.J.: Prog. Surf. Sci. **12** (1982) 287.
82I Ibach, H., Mills, D.L.: Electron Energy Loss Spectroscopy and Surface Vibrations. New York: Academic Press 1982.
82N Nishijima, M., Masuda, S., Kobayashi, H., Onchi, M.: Rev. Sci. Instrum. **56** (1982) 790.
82P Pandey, K.C.: Phys. Rev. Lett. **49** (1982) 223.
82S Spitzer, A., Lüth, H.: Surf. Sci. **118** (1982) 121.
83A Andersson, S., Persson, B.N.J.: Phys. Rev. Lett. **50** (1983) 2028.
83B Backes, U., Ibach, H.: Solid State Commun. **48** (1983) 445.
83D1 Demuth, J.E., Persson, B.N.J., Schell-Sorokin, A.J.: Phys. Rev. Lett. **51** (1983) 2214.
83D2 Dubois, L.H., Rowe, J.E.: J. Vac. Sci. Technol. A **1** (1983) 1232.
83E Egri, I., Matz, R., Lüth, H.: Surf. Sci. 128 (1983) 51.
83H Hall, B.M., Tong, S.Y., Mills, D.L.: Phys. Rev. Lett. **50** (1983) 1277.
83I Ibach, H., Lehwald, S.: Solid State Commun. **45** (1983) 633.
83K Kesmodel, L.L.: J. Vac. Sci. Technol. A**1** (1983) 1456.
83L1 Lüth, H.: Physica **117B** & **118B** (1983) 810.
83L2 Lehwald, S., Szeftel, J.M., Ibach, H., Rahman, T.S., Mills, D.L.: Phys. Rev. Lett. **50** (1983) 518.
83L3 Lüth, H.: Surf. Sci. **126** (1983) 126.
83M Matz, R., Lüth, H., Ritz, A.: Solid State Commun. **46** (1983) 343.
83O Oshima, C., Francy, R., Ibach, H.: Rev. Sci. Instrum. **54** (1983) 1042.
83P Persson, B.N.J.: Phys. Rev. Lett. **50** (1983) 1089.
84A Andersson, S., Persson, B.N.J., Persson, M., Lang, N.D.: Phys. Rev. Lett. **52** (1984) 2073.
84C Chiaradia, P., Cricenti, A., Selci, S., Chiarotti, G.: Phys. Rev. Lett. **52** (1984) 1145.
84E Eckhart, H., Fritsche, L., Noffke, J.: J. Phys. F **14** (1984) 97.
84F1 Froitzheim, H.: J. Electron Spectrosc. Relat. Phenom. **34** (1984) 11.
84F2 Froitzheim, H., Köhler, U., Lammering, H.: Phys. Rev. B **30** (1984) 5771.
84K1 Kevan, S.D., Dubois, L.H.: Rev. Sci. Instrum. **55** (1984) 1604.
84K2 Kirschner, K., Rebenstorff, D., Ibach, H.: Phys. Rev. Lett. **53** (1984) 698.
84O Olmstead, M.A., Amer, N.A.: Phys. Rev. Lett. **52** (1984) 1148.

84P Persson, B.N.J., Demuth, J.E.: Phys. Rev. B **30** (1984) 5968.
84R1 Rebenstorff, D.: Kernforschungsanlage Jülich Rep. Jül 1896 (1984).
84R2 Ritz, A., Spitzer, A., Lüth, H.: Surf. Sci. **138** (1984) 390.
84R3 Ritz, A., Lüth, H.: Phys. Rev. Lett. **52** (1984) 1242.
84S1 Strong, R.L., Erskine, J.L.: Rev. Sci. Instrum. **55** (1984) 1304.
84S2 Stroscio, J.A., Persson, M. Bare, S., Ho, W.: J. Vac. Sci. Technol. A**3** (1984) 1627.
84T Thiry, P.A., Liehr, M., Pireaux, J.J., Caudano, R.: Phys. Rev. B **29** (1984) 4824.
85A Alerhand, O.L., Allan, D.C., Mele, E.J.: Phys. Rev. Lett. **55** (1985) 2700.
85D1 DiNardo, N.J., Demuth, J.E., Thompson, W.A., Avouris, P.: Phys. Rev. B **31** (1985) 4077.
85D2 Demuth, J.E., Persson, B.N.J.: Phys. Rev. Lett. **54** (1985) 584.
85F Franchy, R., Ibach, H.: Surf. Sci. **155** (1985) 15.
85H Ho, W.: J. Vac. Sci. Technol. A **3** (1985) 1432.
85K Kirschner, J.: Phys. Rev. Lett. **55** (1985) 973.
85P1 Persson, B.N.J.: Phys. Rev. Lett. **55** (1985) 2957.
85P2 Persson, B.N.J., Demuth, J.E.: Phys. Rev. B **31** (1985) 1856.
85R1 Rocca, M., Lehwald, S., Ibach, H.M., Xu, M.L., Hall, B.M., Tong, S.Y., in: The Structure of Surfaces. Van Hove, M.A., Tong, S.Y. (eds.), Springer Ser. Surf. Sci. **2** (1985) 156.
85R2 Rebenstorff, D., Ibach, H., Kirschner, J.: Solid State Commun. **56** (1985) 885.
85R3 Ritz, A., Lüth, H.: J. Vac. Sci. Technol. B **3** (1985) 1153.
85S1 Stroscio, J.A., Ho, W.: Phys. Rev. Lett. **54** (1985) 1573.
85S2 Stroscio, J.A., Persson, M., Bare, S.R., Ho, W.: Phys. Rev. Lett. **54** (1985) 1428.
85X Xu, M.L., Hall, B.M., Tong, S.Y., Rocca, M., Lehwald, S., Ibach, H., Black, J.: Phys. Rev. Lett. **54** (1985) 1171.
86A Alerhand, O.L., DiNardo, N.J., Mele, E.J.: Surf. Sci. **173** (1986) L659.
86C Cox, P.A., Williams, A.A.: J. Electron Spectrosc. Relat. Phenom. **39** (1986) 45.
86D1 Del Pennino, U., Betti, M.G., Mariani, C., Bertoni, C.M., Nannarone, S., Abbati, I., Braicovich, L., Rizzi, A.: Solid State Commun. **60** (1986) 337.
86D2 DiNardo, N.J., Thompson, W.A., Schell-Sorokin, A.J., Demuth, J.E.: Phys. Rev. B **34** (1986) 3007.
86D3 Demuth, J.E., Imbihl, R., Thompson, W.A.: Phys. Rev. B **34** (1986) 1330.
86H Hall, B.M., Mills, D.L.: Phys. Rev. B **34** (1986) 8318.
86K Kirschner, J., Suga, S.: Surf. Sci. **178** (1986) 327.
86O Oshima, C., Souda, R., Aono, M., Ishizawa, Y.: Rev. Sci. Instrum. **56** (1986) 227.
86P1 Persson, B.N.J.: Phys. Rev. B **34** (1986) 5916.
86P2 Persson, M., Stroscio, J.A., Ho, W.: J. Electron Spectrosc. Relat. Phenom. **38** (1986) 11.
86R1 Richter, L.J., Ho, W.: Rev. Sci. Instrum. **57** (1986) 1469.
86R2 Rocca, M., Lehwald, S., Ibach, H., Rahman, T.S., in: Structure and Dynamics of Surface I. Schommers, W., Von Blankenhagen, P. (eds.), Top. Curr. Phys. **41** (1986) 245.
86S Stroscio, J.A., Ho, W.: Rev. Sci. Instrum. **57** (1986) 1483.
86T Thiry, P.A., Liehr, M., Pireaux, J.J., Caudano, R.: J. Electron Spectrosc. Relat. Phenom. **39** (1986) 69.
87C Contini, R., Layet, J.M.: Solid State Commun. **64** (1987) 1179.
87D1 Daum, W., Ibach, H., Müller, J.E.: Phys. Rev. Lett. **59** (1987) 1593.
87D2 Del Pennino, U., Betti, M.G., Mariani, C., Bertoni, C.M., Nannarone, S., Abbati, I., Braicovich, L., Rizzi, A.: Surf. Sci. **189/190** (1987) 689.
87D3 Dubois, L.H., Zegarski, B.R., Persson, B.N.J.: Phys. Rev. B **35** (1987) 9128.
87E Ehlers, D.H., Mills, D.L.: Phys. Rev. B **36** (1987) 1051.
87E2 Edgell, R.G., Flavell, W.R., Gray-Grychowski, Z.J.T., Joyce, B.A., Neave, J.H.: J. Electron Spectrosc. Relat. Phenom. **45** (1987) 177.

87G Gray-Grychowski, Z.J., Edgell, R.G., Joyce, B.A., Stradling, R.A., Woodbridge, K.: Surf. Sci. **186** (1987) 482.

87L1 Liebsch, A.: Phys. Scr. **35** (1987) 354.

87L2 Liebsch, A.: Phys. Rev. B **36** (1987) 7378.

87P1 Palmer, R.E., Annett, J.F., Willis, R.F.: Phys. Rev. Lett. **58** (1987) 2490.

87P2 Palmer, R.E., Annett, J.F., Willis, R.F.: Surf. Sci. **189/190** (1987) 1009.

87S1 Schaefer, J.A., Anderson, J., Lapeyre, G.J.: J. Electron Spectrosc. Relat. Phenom. **44** (1987) 373.

87S2 Selci, S. Cricenti, A., Ciccacci, F., Felici, A.C., Goletti, C., Zhu, Y., Chiarotti, G.: Surf. Sci. **189/190** (1987) 1023.

87S3 Stroscio, J.A., Ho, W.: Phys. Rev. B **36** (1987) 9736.

88A1 Alerhand, O.L., Mele, E.J.: Phys. Rev. B **37** (1988) 2536.

88A2 Annett, J.F., Palmer, R.E., Willis, R.F.: Phys. Rev. B **37** (1988) 2408.

88D Del Pennino, U., Betti, M.G., Mariani, C., Abbati, I.: Surf. Sci. **207** (1988) 133.

88E Ehlers, D.H.: Phys. Rev. B **38** (1988) 9706.

88G Gurney, B.A., Ho, W., Richter, L.J., Villarubia, J.S.: Rev. Sci. Instrum. **59** (1988) 22.

88L Lüth, H.: Vacuum **38** (1988) 223.

88R Raether, H.: Surface Plasmons. Springer Tracts Modern Physics **111**, Berlin: Springer Verlag 1988.

88V Venus, D., Kirschner, J.: Phys. Rev. B **37** (1988) 2199.

89B Betti, M.G., Del Pennino, U., Mariani, C.: Phys. Rev. B **39** (1989) 5887.

89C1 Chen, Y., Nannarone, S., Schaefer, J., Hermanson, J.C., Lapeyre, G.J.: Phys. Rev. B **39** (1989) 7653.

89C2 Chen, J., Hermanson, J.C., Lapeyre, G.J.: Phys. Rev. B **39** (1989) 12682.

89C3 Cricenti, A., Selci, S., Felici, A.C., Goletti, C., Chiarotti, G.: Surf. Sci. **211/212** (1989) 552.

89D1 Del Pennino, U., Betti, M.G., Mariani, C., Nannarone, S., Bertoni, C.M., Abbati, I., Rizzi, A.: Phys. Rev. B **39** (1989) 10380.

89D2 Del Pennino, U., Betti, M.G., Mariani, C., Abbati, I.: Surf. Sci. **211/212** (1989) 557.

89D3 Dubois, L.H., Schwartz, G.P.: Phys. Rev. B **40** (1989) 8336.

89F Feibelman, P.J.: Phys. Rev. B **40** (1989) 2752.

89J Jeong, E.J., Erskine, E.J.: Rev. Sci. Instrum. **60** (1989) 3139.

89M Miglio, L., Santini, P., Ruggerone, P., Benedek, G.: Phys. Rev. Lett. **62** (1989) 3070.

89N1 Nannarone, S., D'Addato, S., Schaefer, J.A., Chen, Y., Anderson, J., Lapeyre, G.J.: Surf. Sci. **211/212** (1989) 524.

89N2 Nannarone, S., D'Addato, S., Schaefer, J.A., Chen, Y., Lapeyre, G.J.: Phys. Rev. B **39** (1989) 5975.

89P1 Persson, B.N.J., Ping, J.G., Xu, Y.B., Frankel, D., Chen, Y., Lapeyre, G.J.: Phys. Rev. B **40** (1989) 7819.

89P2 Prade, J., Kulkarni, A.D., De Wette, F.W., Schröder, U., Kress, W.: Phys. Rev. B **39** (1989) 2771

89R Richter, L.J., Mieher, W.D., Whitman, L.J., Noonan, W.A., Ho, W.: Rev. Sci. Instrum. **60** (1989) 12.

89S1 Suto, S., Tsuei, K.D., Plummer, E.W., Burstein, E.: Phys. Rev. Lett. **63** (1989) 2590.

89S2 Sprosser-Prou, F., Von Felde, A., Fink, J.: Phys. Rev. B **40** (1989) 5799.

89T Tsuei, K.D., Plummer, E.W., Feibelman, P.J.: Phys. Rev. Lett. **63** (1989) 2256.

89V Von Felde, A., Sprosser-Prou, F., Fink, J.: Phys. Rev. B **40** (1989) 10181.

90D Demuth, J.E., Persson, B.N.J., Holtzberg, F., Chandrasekhar, C.V.: Phys. Rev. B **64** (1990) 603.

90P Persson, B.N.J., Demuth, J.E.: Phys. Rev. B **42** (1990) 8057.

90R1 Rocca, M., Valbusa, U.: Phys. Rev. Lett. **64** (1990) 2398.

90R2 Rocca, M., Biggio, F., Valbusa, U.: Phys. Rev. B **42** (1990) 2835.

90S1 Sellidj, A., Erskine, J.L.: Rev. Sci. Instrum. **61** (1990) 49.

90S2 Schaefer, J.A., Frankel, D.J., Lapeyre, G.J.: Z. Phys. **79** (1990) 259.

90T Tsuei, K.D., Plummer, E.W., Liebsch, A., Kempa, K., Bakshi, P.: Phys. Rev. Lett. **64** (1990) 44.
91B Baddorf, A.P., Plummer, W.: Phys. Rev. Lett. **66** (1991) 2270.
91C Chen, Y., Tong, S.Y., Rocca, M., Moretto, P., Valbusa, U., Bohnen, K.P., Ho, K.M.: Surf. Sci. **250** (1991) L389.
91H Hall, B.M., Mills, D.L.: Phys. Rev. B **44** (1991) 1202.
91I Ibach, H.: High Resolution Electron Energy Loss Spectrometers: The Technology of High Performance. Springer Series Optical Science **63**, Berlin: Springer Verlag 1991.
91J Jensen, E.T., Palmer, R.E., Allison, W., Annett, J.F.: Phys. Rev. Lett. **66** (1991) 492.
91M1 Marvin, A.M., Toigo, F.: Europhys. Lett. **14** (1991) 445.
91M2 Meng Y., Anderson, J.R., Hermanson, J.C., Lapeyre, G.J.: Phys. Rev. B **44** (1991) 4040.
91N Nannarone, S., D'Addato, S., Betti, M.G., Del Pennino, U., Chen, Y., Samonto, P., Lapeyre, G.J.: Phys. Rev. B **43** (1991) 9818.
91R1 Rocca, M., Lazzarino, M., Valbusa, U.: Phys. Rev. Lett. (Comment on reference 89S1) **67** (1991) 3197.
91R2 Rocca, M., Valbusa, U., Gussoni, A., Maloberti, G., Racca, L.: Rev. Sci. Instrum. **62** (1991) 2172.
91T Tsuei, K.D., Plummer, E.W., Liebsch, A., Pehlke, E., Kempa, K., Bakshi, P.: Surf. Sci. **247** (1991) 302.
91V Vattuone, L., Repetto, M., Rocca, M. : COMPEL **11** (1992) 85.
92D Del Pennino, U., Biagi, R., Mariani, C.: Appl. Surf. Sci. **56-58** (1992) 44.
92I Ibach, H., Ballden, M., Bruchmann, D., Lehwald, S.: Surf. Sci. **269/270** (1992) 94.
92R1 Rocca, M., Lazzarino, M., Valbusa, U.: Surf. Sci. **269/270** (1992) 560.
92R2 Rocca, M., Moresco, F., Valbusa, U.: Phys. Rev. B **45** (1992) 1399.
92R3 Rocca, M., Valbusa, U., Moresco, F., in: Surface Science: Lectures on basic concepts and proceedings of the VI Latin American Symposium on Surface Physics. Cardona, M., Ponce, F. (eds.), Springer Proceedings of Physics **62** (1992) 59.
92R4 Rocca, M., Lazzarino, M., Valbusa, U.: Phys. Rev. Lett. **69** (1992) 2122.
92S Sprunger, P., Plummer, E. W., Watson, G.: Surf. Sci. **269/270** (1992) 551.
93L Lee, G., Sprunger, P. T., Plummer, E. W.: Surf. Sci. **286** (1993) L 547.
93R Rocca, M., Valbusa, U.: Surf. Sci. **287/288** (1993) 770
95R Rocca, M, Li, Y., Buatier de Mongeot, F., Valbusa, U.: to be published in Phys. Rev. B, 1995.

6.3 Elastic and inelastic scattering of ions

6.3.1 Introduction

6.3.1.1 Preliminary remarks

Two main experimental techniques for analysis of the geometrical structure of surfaces are low energy electron diffraction (LEED) and ion scattering spectrometry (ISS). The principle of the former technique is based upon the correspondence between the wavelength of the probing particles (30 ... 400 eV electrons) and the interatomic distances in a crystal. In addition, the penetration depth of the electrons is small, making LEED surface sensitive. The principle of the latter technique is based on the negligible small wavelength of the probing particles. Ion scattering from crystals is essentially a classical phenomenon. Although the penetration depth of ions - especially high energy ions - is large, surface sensitivity is achieved by preventing or reducing either subsurface scattering or detection of subsurface scattered projectiles, or both. While LEED is sensitive for long range order in a surface, ion scattering is sensitive for short range order. Hence, the two techniques compete and complement each other.

This chapter contains an overview of studies on elastic and inelastic scattering of ions from single crystal surfaces. Not only studies that provide surface structure data are included, but also studies in which the basics of the ion-surface interaction are investigated. Especially for low energy ion scattering spectrometry (LEIS), the latter type of studies outnumber the former. Section 6.3.1 discusses briefly the principles of ion scattering spectrometry with emphasis, of course, on surface structure determination. Other important subjects are scattering phenomena and charge exchange. The discussion runs roughly parallel to the compilation of the data presented in section 6.3.2. Section 6.3.3 lists the references.

Aspects of ion scattering spectrometry applied to materials or surface analysis are discussed thoroughly in several books. Most notably are the books by Feldman et al. [82F2] (application of ion channeling for materials analysis), by Ziegler et al. [85Z2] (stopping of ions in matter), by Mashkova [85M] (reflection of low and medium energy ions from crystalline surfaces), by Feldman et al. [86F2] (experimental techniques for surface and thin layer analysis), and by Eckstein [91E] (computer simulations of ion-solid interactions). A compilation of surface structure determination by ISS prior to 1988 has been made by Watson [90W1].

6.3.1.2 Low energy ion scattering

Among the many techniques applied in surface science, low energy ion scattering spectrometry (LEIS) has its merits mainly in the field of surface composition and surface structure analysis. The first investigation of a surface using an ion beam was made by Smith in the late sixties [67S]. Initially, the subject of most LEIS studies was either surface composition - often different from the bulk because of segregation or adsorption - or ion scattering phenomena. As a consequence of the complexity of the interaction between low energy ions and crystalline solids, LEIS did not develop rapidly into a routine technique for surface structure analysis. The complexity is a consequence of - strangely enough - the relative large scattering cross sections involved (as compared to the area per atom in a solid) and, in many cases, the neutralization of the scattered projectiles. Large scattering cross sections imply on the one hand that ions do not penetrate deep into a solid, making LEIS a surface sensitive technique. On the other hand, they imply that detected ions have often collided with more than one atom. Consequently, ions reflected into the same direction may have followed very different trajectories inside the crystal or just above the outermost atomic layer. Not surprisingly, Monte Carlo computer simulations are used extensively to study these complex multiple scattering mechanisms. However, the simulations are complex too and are computer-time consuming. Moreover, the underlying physics (viz., interaction potentials and charge exchange mechanisms) is still not known in sufficient detail.

The problem of neutralization is nowadays largely circumvented either by using alkali ion beams [87O2] or by measuring reflected particles with a time-of-flight instead of an electrostatic detector

[78B6]. The problems of multiple scatterings and the ill-known interaction potentials have been tackled by detecting at large (near 180°) scattering angles [81A1].

6.3.1.2.1 Scattering phenomena

In the simplest LEIS experiment ions are reflected in a single (or *binary*) collision from one surface atom. The energy E of the scattered ion is related to its initial or primary energy E_0 by:

$$E = k\,(\theta, M_1/M_2)\,E_0\,, \tag{1}$$

in which θ is the scattering angle and M_1 and M_2 are the masses of the projectile and target atom, respectively; k is called the kinematic factor. It has been shown experimentally that the concept of the binary collision is valid for Ne ions with energies as low as 20 eV [75T2].

The force $F(r)$ between a target atom and a projectile at distance r is usually written as:

$$F(r) = Z_1 Z_2\, e^2/r^2 \cdot \chi(r)\,, \tag{2}$$

in which Z_1 and Z_2 are the atomic numbers of the projectile and target atom, respectively, e is the elementary charge. The screening function $\chi(r)$ describes the reduction of the pure internuclear (Coulomb) force by the electrons of the projectile and the target atom. Several approximation models for the screening function (viz., interaction potential) have been suggested. Unfortunately, results of ion trajectory calculations by computer simulation depend heavily on the model chosen.

Scattering from two surface atoms (viz., double scattering) is the next simplest scattering mechanism. In Fig. 1, (quasi-) single and (quasi-) double scattering are compared. One can prove that in case of double scattering not all reflection angles are possible and that for each possible reflection angle there are two distinct scattering mechanisms: one corresponding to quasi-single and the other to quasi-double scattering. Since the final energy of the ion depends on the scattering mechanism, a double peak is observed in the measured energy spectra (Fig. 2), except near the maximal and minimal reflection angles, where the two mechanisms merge (see also Figs. 19 and 20). The energy of the doubly scattered ions is generally larger than that of the singly scattered ions. More complicated scattering mechanisms are triple, zig-zag and chain scattering, all studied extensively in the literature.

As a consequence of the repulsive interatomic force, there is a region (shadow cone) behind each atom into which ions cannot penetrate. If an atom is located in the shadow cone of an other atom, scattering from the former atom is not possible. Shadowing occurs at all ion energies, see e.g. Fig. 3 which illustrates shadowing for medium or high energy ion scattering. Vice versa, scattered ions cannot be detected if one or more atoms are located between the scattering atom and the analyzer: they block the exit path to the detector. Shadowing and blocking are clearly visible in Fig. 4.

Some surfaces have an open structure, e.g., the (110) surface of fcc crystals. They consist of parallel semichannels. The walls of a semichannel are formed by two neighboring first-layer atomic rows and the bottom by one second-layer row. If an obliquely incident ion beam is aligned parallel to the semichannels, focusing occurs: ions are deflected (viz., focused) by the walls towards the bottom of the semichannel (see Fig. 24). Consequently, second layer scattering is enhanced. Furthermore, the angular distribution of the reflected ions is narrower than for non-aligned directions of incidence (Fig. 23). This focusing phenomenon is called *surface semichanneling*. If the angle of incidence ψ is below a critical angle (of typically 10° for low energy ions), the ions are unable to overcome the potential barrier between the surface semichannel and neighboring, parallel subsurface channels; consequently, they remain in the surface semichannel until they are reflected back into the vacuum. This effect (called surface hyperchanneling) is closely related to axial channeling of medium or high energy ions in the bulk of single crystals.

6.3.1.2.2 Charge exchange

Initially, electrostatic analyzers were used for most LEIS experiments. Since their ionization energy is larger than the work function of most materials, noble gas ions moving in or near a solid are readily neutralized. Consequently, subsurface scattered noble gas ions are almost always neutralized, making LEIS with an electrostatic analyzer and a noble gas ion beam extremely surface sensitive (see Fig. 35). If an alkali ion beam or a time-of-flight detector is used, features in the energy spectra of the reflected projectiles appear that correspond to subsurface scattering; see the low energy tails in Fig. 2a (all particles and ions only) and in Fig 2b (all particles). Because of its consequences for quantitative surface analysis, many authors have investigated the charge state of the reflected projectiles. The charge state distribution can be calculated theoretically by solving the Anderson-Newns Hamiltonian (see, e.g., [90L]). However, this approach is too complicated for all but the most simple scattering experiments to be of any practical use. Instead, charge exchange experiments are analyzed within the semi-empirical framework developed by Hagstrum, about 40 years ago [54H].

There are three distinct mechanisms by which slow positive ions can capture an electron. The first is Auger neutralization whereby an electron from the highest filled band of the solid tunnels to the lowest empty level of the ion while a second electron from the same band is emitted in order to preserve energy. The second mechanism is resonance neutralization: an electron tunnels from the highest band of the solid to an excited empty level of the ion. This process is followed either by Auger deexcitation or by resonance ionization in which the electron tunnels back to the solid. In the former case, the ion is neutralized and in the latter case, it is re-ionized. The third mechanism (quasi-resonant neutralization) occurs if a core level electron of the solid tunnels to the lowest empty level of the ion. This is only possible for those projectile-target combinations where the energy of the two levels involved are approximately equal. The main characteristic of quasi-resonant neutralization is its oscillatory dependence on the ion energy (see Fig. 32).

Hagstrum showed that the ion survival probability P^+ depends on the inverse of the velocity component $\upsilon_\perp$ perpendicular to the surface:

$$P^+ = \exp(-\upsilon_c / \upsilon_\perp)\,, \tag{3}$$

in which υ_c is a constant of the order of 10^5 m/s. The exponential relationship has been confirmed in a number of experiments (see e.g., Fig. 33). Other experiments, however, suggest that neutralization may depend on the ion's trajectory; in particular, on the distances of closest approach to the individual atoms in the solid.

For energies above several keV, other (non-adiabatic or dynamic) neutralization mechanisms become important. Moreover, during a violent collision re-ionization of a neutralized projectile is possible. The final ion fraction is then calculated by considering all possible charge exchange mechanisms along the incoming trajectory, during the violent collision and along the outgoing trajectory [76V2].

6.3.1.2.3 Surface structure determination

In the early seventies it was stated that LEIS could be applied for the study of crystal surface structures. However, because of the complexity of the scattering and charge exchange mechanisms, hardly any structure analysis was published before 1980. For instance, Overbury et al. [81O2] measured angle resolved energy spectra of 600 eV K^+ ions reflected from the reconstructed Au (110)1x2 surface. Only with the aid of extensive computer simulations, Hemme et al. [85H3] were able to show that these data confirmed the (relaxed) missing row reconstruction model. Ellis [78E1] and Aono [81A1] showed that the complexity of the scattering mechanism can be avoided by detecting reflected particles at large scattering angles, while varying the angle, ψ, of incidence. Aono called this variant of LEIS *impact collision ion scattering spectrometry* (ICISS).

For grazing incidence (ψ is small), all atoms in the surface are located in the shadow cone of their neighbors. Since the ions cannot approach the nuclei of the atoms, hard (large-scattering-angle) collisions do not occur and, if the crystal surface is perfect, no backscattered ions are detected. There is a critical angle of incidence, ψ_c, at which all surface atoms are located on the edge of the shadow cone of their neigbors, see Fig. 5. In that case, the yield of backscattered ions is maximal. The interatomic distance at the surface follows directly from the measured critical angle and the shape of the shadow cone. Since these distances depend on the type of reconstruction (if present), ICISS can be used to determine the structure of reconstructed surfaces. By measuring the critical angle for a direction in which the interatomic distance is beyond question, the shape of the shadow cone can be calibrated without the use of a model for the interaction potential. At present, ICISS and its variants (e.g., ALICISS in which alkali ion beams are used, or NICISS in which all scattered particles are detected by the time-of-flight technique) comprize the majority of all surface structure analyses by LEIS. In principle, one can measure small changes in vertical or lateral distances - viz., relaxations - [88N1], but since the shadow cones are relatively wide, ICISS is less suitable than MEIS or HEIS for relaxation determination.

Also surface shadowing and/or blocking (Fig. 4) can be applied for surface structure analysis. Recently, surface shadowing was used for the study of 1xn reconstructed (110) surfaces of some fcc metals. The intensity of reflected ions under grazing incidence conditions was measured as a function of azimuthal angle, as shown in Fig. 6 (see also Fig. 16). The type - i.e., the number n - of the 1xn reconstruction follows directly from the number of minima in the azimuthal scan.

It is noted that surface structure data can be obtained not only from the intensity of the scattered projectiles but also from the intensity of the recoiled atoms. In this respect, TOF-SARS, *time-of-flight scattering and recoiling ion spectrometry*, developed by Rabalais and coworkers [90R1], is probably the most comprehensive variant of LEIS.

6.3.1.3 Medium and high energy ion scattering

Medium and high energy ion scattering spectrometry (MEIS and HEIS) and LEED intensity-voltage (IV) analysis are the two main experimental techniques for surface relaxation determination. While LEED-IV analyses yield a representation of the lattice structure in reciprocal space, ion scattering analyses yield a representation in real space. In particular, LEED reveals long range order and MEIS and HEIS reveal short range order in a surface. Since the penetration depth of the probing ions (usually 0.05 to 3 MeV H^+ or He^+) is between hundreds of nanometers and several micrometers, MEIS and HEIS as such are not surface analysis techniques. HEIS is in use as a bulk materials analysis technique - commonly known as Rutherford backscattering spectrometry (RBS) - since the early sixties. After the discovery of ion channeling in single-crystalline materials, Turkenburg et al. [76T2] and Davies et al. [76D] realized that the surface peak (see Fig. 7) in a RBS spectrum measured under ion channeling conditions, can be used for surface structure analysis; in particular, for surface relaxation determination. Initially, discrepancies between ion scattering and LEED-IV relaxation results were not uncommon. Nowadays, after the development of effective surface preparation procedures and of extensive calculation models for both LEED-IV and for MEIS/HEIS, the results are in most cases in good agreement with each other. More difficult are MEIS/HEIS analyses of reconstructed surfaces. Generally, the atomic positions of these surfaces cannot be determined via ion scattering unless some a priori knowledge is available. Nevertheless, in a number of studies impressive results have been obtained.

The distinction between MEIS and HEIS is mainly instrumental; the underlying physics is the same and, therefore, treated here together. Below about 250 keV, the energy of backscattered ions can be measured using an electrostatic analyzer with a resolution of 1% or better. For higher energies, solid state detectors must be used with a resolution of at best 10 keV, almost independent of energy. Because of its superior energy resolution and because the stopping power of most materials for H^+ and He^+ has its maximum in the 50 ... 200 keV range, MEIS is more surface sensitive than HEIS.

Remarkable for MEIS/HEIS is the simplicity of the underlying physics. Charge exchange, multiple scattering and ill-known interaction potentials do not complicate analyses as they do for LEIS. Not surprisingly, the number of basic papers is relatively low. For medium and high energies, neutralization is not excessively high. Moreover, it is independent of the particle's trajectory [84H2] (see Fig. 48). Of course, if a solid state detector is used, both backscattered ions and neutrals are detected without any discrimination. Furthermore, the collisions of the ions in the crystal can be fully regarded as binary collisions. And, since scattering cross sections are relatively small, almost all detected ions are backscattered in one single hard collision: double and multiple scatterings are rare. The required ion accelerator and beam steering optics are in contrast more complicated and expensive than for LEIS.

6.3.1.3.1 Channeling, shadowing and blocking

At the root of the phenomenon of ion channeling in a crystal lies the repulsive force between the incident ions and the atoms in the lattice. Ions are deflected by this force so that they cannot penetrate into the region - called the shadow cone - behind the atoms. If the ion beam is aligned parallel to a major crystallographic direction, all atoms except the topmost ones are located in the shadow cone of other atoms and the ions travel mainly in the open channels between the atomic rows, see Fig. 3. Consequently, since they cannot approach the nuclei of the lattice atoms, the probability for backscattering is greatly suppressed. This phenomenon is called ion channeling. Due to thermal fluctuations in the actual atomic positions, shadowing is never perfect: the relative backscattering probability decreases gradually from 100% in the first layer to a few percent about five to ten layers deep. It remains at that low level for the greater part of the ion's trajectory in the crystal. A peak (surface peak) is visible in the energy spectrum of the backscattered ions, reflecting the relatively high probability in the surface region and the low probability in deeper layers, i.e., at lower energies, see Fig. 7b. The rate of the decrease depends on experimental conditions (beam type and energy, crystal temperature, crystallographic direction chosen) and on structural parameters of the crystal (atomic number, interatomic distance along the rows, surface and bulk Debye temperature and the presence or absence of surface relaxation or reconstruction). In general, the structure of a surface is derived from the area L of the surface peak. This quantity, expressed in number of visible atoms per row, corresponds to the relative backscattering probability integrated from the first to the tenth (or so) crystal layer (see Fig. 57).

In a double alignment experiment, not only the incident beam but also the scattered beam (viz., the detector) is aligned with a crystallographic direction [85V1]. In that case, blocking occurs [76T2]: the line of sight from subsurface atoms towards the detector is blocked by atoms closer to the surface. Consequently, the number of detected ions backscattered below the surface layer is further reduced, thus improving the surface sensitivity. Most MEIS experiments are performed under double alignment conditions. Since the blocking cone radius and the backscattering cross section decrease with increasing ion energy, low counting statistics make most HEIS double alignment experiments unattractive. Besides, in order to improve depth resolution of HEIS, it is better to detect at a grazing exit direction than at a blocking direction.

One can calculate backscattered ion intensities in single and double alignment experiments analytically with a two- and a three-atom model, respectively, [77F, 80V]. In most cases, however, the contribution from other atoms (viz., deeper layers) cannot be neglected. Since the interaction of energetic H^+ or He^+ ions with atoms in a crystal are adequately described in terms of successive soft (viz., small-scattering-angle) binary collisions, single and double alignment MEIS/HEIS experiments are well suited for numerical (Monte Carlo) simulation [71B, 86F4]. At present, computer simulation is used for almost every MEIS/HEIS surface structure analysis.

Under grazing incidence conditions energetic ions cannot penetrate a crystal beyond the first atomic layer. Therefore, grazing incident MEIS or HEIS is, in principle, extremely surface sensitive. However, since no subsurface signal is measured, this technique is not suitable for surface relaxation studies. It has been used only once in a surface reconstruction study [90P1]. Nevertheless, several investigators have used grazing incidence MEIS or HEIS to study surface step heights and distances, adsorption mechanisms, or the energy loss and charge exchange of ions moving near surfaces.

6.3.1.3.2 Surface structure determination

It has been shown both experimentally and by computer simulations that for bulklike terminated surfaces, the surface peak area L is a simple function of ρ/R, ρ being the two-dimensional vibration amplitude of the atoms in the surface and R the radius of the shadow cone of a first layer atom at the position of a second layer atom, see Figs. 3 and 58. The vibration amplitude of surface atoms is often enhanced as compared to the bulk. Therefore, measured values for L are larger than values calculated for ideal, bulklike surfaces. On the other hand, correlations in the vibration of neighboring atoms reduce L. If the equilibrium positions of the atoms in the outermost layers are shifted laterally or vertically, shadowing may become less perfect, thus increasing the surface peak area. If a surface is reconstructed, the number of visible atoms per row may be unchanged, may differ by one atom per row, or by any other number, depending on the details of the experiment and the type of the reconstruction.

Since the surface peak area L depends on a number of structural parameters, a single measurement using one scattering geometry or energy is seldom sufficient. If a surface is relaxed, the direction with optimal shadowing or blocking in the surface region (viz., with the largest reduction in surface backscattering intensity) deviates from that direction in the bulk, see Fig. 7. The magnitude and sign of the relaxation are established by comparing for a number of incidence or exit directions near a bulk crystallographic direction, the measured surface peak areas with simulated areas. The latter areas are calculated for a set of possible relaxation values, see Fig. 8. In addition, the overall height of the scan of L vs. incidence or exit direction is a good measure for the surface vibration amplitude. Double or multi-layer relaxation is generally measured using two or more different scattering geometries (Fig. 54). Depending on the details of the experiment, the accuracy of single alignment HEIS and double alignment MEIS lies generally between 1% and 2.5% of the interlayer spacing.

Ion scattering is used to measure vertical and/or lateral displacements of atoms in reconstructed surfaces. Displaced atoms become visible when the displacement exceeds the shadow cone radius. However, the shadow cone radius can be varied only slightly (by varying the primary beam energy). Therefore, often only a lower or an upper limit for the displacement can be given. In general, the type of reconstruction cannot be determined by MEIS/HEIS alone. This is so since there is no unambiguous relation between the structure of a reconstructed surface and L, measured as a function of primary energy and/or incidence or exit angle. Instead, ion scattering is used to test proposed structure models. For instance, Fig. 9 shows angular scans of the surface peak area for the Si(100)1x2 surface measured by MEIS. The data are compared with simulation results for various models. The authors [83T3] concluded that the buckled dimer model with subsurface distortions (see Fig. 9, curve C) was best to reproduce the experimental data.

If the temperature of a crystal is varied, transitions between types of reconstructions or between ordered and disordered surface structures can occur. In several studies these transitions have been investigated by ISS. For instance, it is shown by Frenken and van der Veen [85F] that some surfaces melt at temperatures several degrees below the bulk melting temperature.

6.3.2 Data

The tables in this section summarize virtually all studies on elastic and inelastic ion scattering from single crystal surfaces. Although surface structure determination by ion scattering is the central theme, the main topic of many selected papers is often different (in particular, ion-solid charge exchange or ion scattering mechanisms). The related topics are included in this compilation because they are fundamental for the understanding of the technique. The overview is, as far as possible, complete until the end of the third quarter of 1991.

The requirements for studies to be inserted are:

(1) the probing beam is an ion beam with energy above about 50 eV;
(2) the specimen used is a bulk single crystal with a clean and flat surface;
(3) there is no build-up of damage or contamination during the experiment;
(4) scattered particles are detected; and
(5) the study is published in one of the main journals or in a generally available conference proceedings.

Only papers on experimental work or on (computer) simulation of experimental work are included. A few theoretical papers, important for the understanding of the technique, are included as well. Apart from a number of old papers, no studies are listed that deal solely with the surface composition of multi-component crystals. Many charge exchange processes are investigated using amorphous or poly-crystalline materials. Although often highly relevant, these studies do not appear in this overview. Surface structures or ion-surface interactions can be inferred also from secondary particles - such as electrons, photons or sputtered atoms - but also these studies are outside the scope of this overview. There is no upper limit for the energy of the probing beam; however, with only one or two exceptions, the primary energy did never exceed 3 MeV. The interaction between surfaces and very low energy ions (<50 eV) or grazing incident ions with slightly higher energy (<1 keV) is better understood in terms of adsorption than of scattering. Therefore, a lower limit of about 50 eV is set. Although important and fascinating, only a few studies with ion beams below 50 eV are included in the overview. In most cases listed, singly and positively charged ion beams have been used and in a few cases neutral beams. The interaction of highly charged ions with surfaces is the subject of a whole field of research. However, since these studies have not yet resulted in new data on surface structures, they are not included.

There are six tables: two for each ion scattering technique: LEIS, MEIS and HEIS. One table summarizes the surface structure studies; the other summarizes all other, related studies. A study contains only simulation results if the remark starts with "MC" (i.e., Monte Carlo simulation). If it starts with "Theory:", it is purely a theoretical paper. "MC" at the end of a remark indicates that (Monte Carlo) simulation has been used for interpretation of the experimental data. The angles that define the scattering geometry are shown in Fig. 10.

Tables 2, 4, and 6 (with the related studies) contain a number of categories. Although many papers could have been placed in different categories, most papers appear only once. Therefore, in order to find all the literature on one subject (e.g. energy loss of low energy ions), one should not only check the appropriate category (energy loss in Table 2), but also scan the remarks (on energy loss) in the other categories. As an indication of the completeness of review papers, both their length (in number of pages) and the number of given references are mentioned.

Table 1. Low Energy Ion Scattering (0.05 keV < E_0 < 50 keV). Surface structure data.

First column: Surface (in alphabetical order and in order of the Miller index).
Second column: LEED pattern(s) (if mentioned in the publication).
Third column: Primary beam energy E_0.
Fourth column: Primary beam ion.
Fifth column: Main ISS technique used:

- ISS: Ion Scattering Spectrometry,
- ICISS: Impact Collision ISS,
- NICISS: Neutral ICISS,
- TOFSARS: Time-of-Flight Scattering and Recoiling Spectrometry,
- IIAES: Ion-Induced Auger Electron Spectrometry
- sh+bl: shadowing and blocking (also called channeling and blocking)
- oop-sh+bl: out-of-plane shadowing and blocking.
- chan: channeling (also called shadowing),
- transm ch: transmission channeling.

The parameters ψ, θ, ϕ, E_0 and T indicate the main parameters varied (see also Fig. 10):

- ψ : angle between incident beam and surface,
- θ : scattering angle,
- ϕ : azimuthal angle,
- E_0: primary beam energy,
- T: crystal temperature.

Sixth column: Measured change in first to second layer distance, Δd_{12} (in pm = 0.01 Å).
Seventh column: Measured change in second to third layer distance, Δd_{23} (in pm = 0.01 Å).
Eighth column: Measured lateral displacement of first layer atoms, Δy_1 (in pm = 0.01 Å).
Ninth column: Measured ratio between thermal vibration amplitudes in the first layer and in the bulk, ρ_1/ρ_b, or surface Debye temperature, Θ_D.
Tenth column: Additional experimental results. Additional information about the experimental conditions are in italics.
Eleventh column: Figures or tables.
Twelfth column: Reference(s).

Unless otherwise stated, experiments are performed at or just above room temperature (295 ... 325 K).
var. = various.

Table 1 (cont.)

Surface	LEED pattern	E_0 [keV]	Ion	Technique	Δd_{12} [pm]	Δd_{23} [pm]	Δy_1 [pm]	ρ_1/ρ_b or Θ_D	Additional results *or remarks*	Fig./ Table	Ref.
Ag(110)	-	5	He, Li	ICISS; ψ	-11	+6	-	1.5			88D1
	1x1	5	Li	ICISS; ψ	-	-	-	-	10...15% vacancies in 1st layer.		90D2
Au(110)	1x2 / 1x1	0.6	K	ISS; ϕ, θ, ψ, T	-	-	-	-	Indication of relaxed missing rows. Gradual disappearance of reconstruction between 300 and 900 K.	11 [1])	81O2
	1x2 / 1x1	0.6	K	ISS; ϕ, θ, ψ	-25(4)	+10	<10	-	*Data from [81O2].* Support for missing row model.	11	85H3, 87D1
	1x2	2	Ne	NICISS; ψ	-15(10)	-	-	-	Support for missing row model.		86M2
	1x2 / 1x1	2	Ne, Na	NICISS / ICISS; ψ, T	-20(7)	-	-	-	Support for missing row model. Gradual disappearance of reconstruction between 400 and 950 K. Indication of surface steps.		86M3
	1x2 / 1x1	0.3, 0.6	K	ISS; ϕ, T	-	-	-	-	*1x2↔1x1 phase transition at 650 K.* Support for disordered missing row model above 650 K.	12	87D2, 88D2
	1x2 / 1x1	2	Na	ISS; ϕ, T	-	-	-	-	1x2 surface periodicity at 300 K, 1x1 at 700 K.		90D5
	1x2 / 1x1	4.6, 2	Ne, Na	ICISS; ψ, T	-15(5) [2])	-	-	127(4) K [3]) 108(5) K [3])	*1x2↔1x1 phase transition at 650 K.* Support for disordered missing row model above 650 K. Formation of vacancies above 530 K. Formation of steps above 650 K.		90V2

Table 1 (cont.)

Surface	LEED pattern	E_0 [keV]	Ion	Technique	Δd_{12} [pm]	Δd_{23} [pm]	Δy_1 [pm]	ρ_1/ρ_b or Θ_D	Additional results *or remarks*	Fig./ Table	Ref.
CaF_2 (111)	1x1	1	He	ICISS; ψ	+1(8)	-	-	-	Surface is F-terminated.	13	88S3
Cu(100)	1x1	10	Kr	ISS; E_0, T	-	-	-	255 K 4)	*Vibration amplitude deduced from quasi- triple collisions.*		76P3
	1x1	10	Kr	ISS; E_0, T	-	-	-	258(25) K	*Experimental data from [76P3].* If correlations are neglected: Θ_D=303 K.		80M1
	1x1	10	Kr	ISS; E_0, T	-	-	-	250(8) K	*Experimental data from [76P3]; including correlations.*		82W1
	1x1	10	Kr	ISS; E_0, T	-	-	-	279(6) K	*Experimental data from [76P3]; neglecting correlations.*		83P
Cu(110)	1x1	1	Ne, Na	ISS; T	-	-	-	150(30)K	*Between 100 and 600 K.*		85E, 86E
	1x1	5	Li	ICISS; ψ	-13(6)	-	-	1.5			86Y3, 86Y4
	1x1	3	Li	ICISS; ψ	-	-	-	-	Possibility for surface disorder, even in case of a sharp LEED pattern.		86Y7
	-	0.6, 5	Na, Ne	ISS / ICISS; ψ, T	-	-	-	150 K	Surface disorder above 450 K; indication of adatom creation.		87F1
	-	5	Ne	ICISS; ψ, ϕ	-13(4)	+10(8)	-	-		5	88F1
	-	2 ... 5	Ne	ICISS; ψ, ϕ	-4(4)	-	-	-			89V1, 89V2

Table 1 (cont.)

Surface	LEED pattern	E_0 [keV]	Ion	Technique	Δd_{12} [pm]	Δd_{23} [pm]	Δy_1 [pm]	ρ_1/ρ_b or Θ_D	Additional results *or remarks*	Fig./ Table	Ref.
Cu(110) cont.	-	5	Ne	ICISS; ψ	-	-	-	260 K			90V1
Cu(410)	-	21	H	ISS; θ	-18(5) [5]	-	-	-	*At 540 K.*		80A1
	-	8, 11	Ar, Ne	ISS; ψ	-	-	-	-	*At 540 K.* Steps are alternating <130> and <150> rows.		80A2, 80A4
	-	10, 21	He, H	ISS; ψ, θ	-3(4) [6] -18(5) [7]	-	-	-	*At 540 K.* Smoothing of the surface.		82A5
	-	10 ... 20	H	TOF-ISS; θ	-15(3) [5]	-	-	-	*At 540 K.*		86V
Cu_3Au (100)	-	5, 9.5	Ne	TOF-ISS; ϕ, T	-	-	-	-	For $T < 670$ K: 50% Au in 1st layer, < 3% Au in 2nd layer; for $T > 670$K: decreasing Au segregation		83B6
InP (100)	4x2 ?	3	Ne	ISS; ϕ	-	-	-	-	Dimerization of the surface.		89A2
Ir(110)	var.	2	Na	ISS; ϕ / ICISS; ψ	-	-	-	-	Evidence for 1x3 surface periodicity.		89H2, 90D5
	var.	4	Ar	TOFSARS; ψ, θ	-	-	-	-	Primary 1x3 facets (2 missing 1st layer rows, 1 missing 2nd layer row); secondary 1x1 domains.		90B3
	var.	4	Ar	TOFSARS; ψ, θ, ϕ	-11(10)	0(14)	12(10) [8]	-	Primary 1x3 facets and secondary 1x1 domains; Δd_{14} = -12(10) pm.	14, 15	90S2

Table 1 (cont.)

Surface	LEED pattern	E_0 [keV]	Ion	Technique	Δd_{12} [pm]	Δd_{23} [pm]	Δy_1 [pm]	ρ_1/ρ_b or Θ_D	Additional results *or remarks*	Fig./ Table	Ref.
Ir(110) cont.	var.	4	Ar	TOFSARS; ϕ	-	-	-	-	1x3 surface periodicity.		91B1
	var.	2, 3.5	Ne	NICISS; ψ, T	-29 [9])	-	-	140 K [10]) 180 K [10])	*300 ... 1220 K.*		91H2
	var.	1.1...2	Ne	ISS / NICISS; ψ, T	-	-	-	-	*300 ... 1550 K;* 1x3↔1x1 transition at 1055(30) K.		91H6
MgO (100)	1x1	1	He	ICISS; ψ	-31(6)	-	-	-	Negligible 1st layer rumpling.		88N1
Mo (111)	1x1	1	Li	ISS; ψ, θ	-16(2)	+4(4)	-	-			86O3
Ni(110)	-	0.7, 2	Na	ISS; ϕ	-	-	-	-	1x1 surface periodicity.		90D5
	-	4	Ar	TOFSARS; ϕ	-	-	-	-	1x1 surface periodicity.	16	91M1

Table 1 (cont.)

Surface	LEED pattern	E_0 [keV]	Ion	Technique	Δd_{12} [pm]	Δd_{23} [pm]	Δy_1 [pm]	ρ_1/ρ_b or Θ_D	Additional results *or remarks*	Fig./ Table	Ref.
NiAl (100)	1x1	0.5, 1	Li, K	ICISS; ψ / ISS	-	-	-	-	78%(7%) Al in 1st layer, < 5% in 2nd layer; Ni and Al in 1st layer are randomly mixed.		88M4, 88M5
	$c(\sqrt{2}$x $3\sqrt{2})$	0.5, 1	Li, K	ICISS; ψ/ ISS	-	-	-	-	65%(2%) Al in 1st layer, < 5% in 2nd layer; Ni and Al in 1st layer are ordered.	17	88M4, 88M5
NiAl (110)	1x1	0.5	Li	ICISS; ψ	-	-	-	-	1st layer buckling: Al 21(5) pm above Ni.		88M4, 88M5
NiAl (111)	1x1	2	He	NICISS; ψ	-	-	-	-	Surface is Ni-terminated; 2nd layer consists of Al [11]).		90N4
	1x1	0.5...2	Li	ICISS; ψ, θ	-32(4) [12])	-	-	-	*Stepped surface: 1.7° cut from (111) plane.*		90O2
					+7(4) [12])	-	-	-	46(4)% Al in 1st layer, 47(10)% Al in 2nd layer [13]).		
Pt(110)	1x2	2	Na	ICISS; ψ, ϕ	-	-	-	-	Support for missing row model.		84N2
	1x2 / 1x3 [14])	2	Ne	TOFSARS; ϕ	-	-	-	-	Surface periodicity.	6	91M1, 91M3
	1x2	2	Ne	NICISS; ψ	-22(7) [15])	-	-	-	19(13) pm (peak-to-peak) buckling in 3rd layer.	[1])	91M2
PtNi (111)	-	5	Ne	ISS; ϕ / NICISS; ψ	-	-	-	-	*At 350 K;* 88(2)% Pt in 1st layer; no clustering of Ni in 1st layer. Surface buckling (peak-to-peak) < 5 pm.		91V1

Table 1 (cont.)

Surface	LEED pattern	E_0 [keV]	Ion	Technique	Δd_{12} [pm]	Δd_{23} [pm]	Δy_1 [pm]	ρ_1/ρ_b or Θ_D	Additional results *or remarks*	Fig./ Table	Ref.
Pt_3Sn (100)	c(2x2)	1	Ne	ISS	-	-	-	-	> 50% Sn in 1st layer: termination in a compositionally mixed (100) layer. Indication of buckling: Sn displaced outwards.		91H1
Pt_3Sn (111)	2x2	1	Ne	ISS	-	-	-	-	25...50% Sn in 1st layer. Indication of buckling: Sn displaced outwards.		91H1
Si(100)	1x2	1	He	ICISS; ψ, ϕ	-	-	-	-	Support for symmetric and buckled dimer models. Interdimer atomic distance: 2.4(1) Å.		82A6
Si(111)	7x7	0.93	He	ICISS; ψ, ϕ	-	-	-	-	Support for pyramidal cluster model.		83A2, 83A3
	7x7	24	H_2	IIAES; ϕ	-	-	-	-	Support for dimer-adatom-stacking fault model.		87A1
SmB_6 (100)	var.	1	He	ISS; ϕ	-	-	-	-	Surface is mainly B-terminated.		78A2, 79A1
TiC (100)	1x1	1	He	ISS	-	-	-	-	Nearly stoichiometric surface composition.		81O1, 84A1
	1x1	1	He	ICISS; ψ	-	-	-	-	1st layer buckling: C 10(10) pm below Ti.		84A1, 84A2
	1x1	1	He	ICISS; ψ	-	-	-	1.6	Anisotropic vibrations.		84T1

Table 1 (cont.)

Surface	LEED pattern	E_0 [keV]	Ion	Technique	Δd_{12} [pm]	Δd_{23} [pm]	Δy_1 [pm]	ρ_1/ρ_b or Θ_D	Additional results *or remarks*	Fig./ Table	Ref.
TiC (111)	1x1	1	He	ISS	-	-	-	-	Surface is Ti-terminated.		81O1, 84A1
	1x1	1	He	ICISS; ψ, ϕ	-13(10) [16])	-	-	-			81A1, 84A1
	1x1	1	He	ICISS; ψ	-	-	-	2.0			83S2
	1x1	1	He	ICISS; ψ	-	-	-	1.7	Anisotropic vibrations.		84T1
	1x1	1	He	ISS	-	-	-	-	Surface is Ti-terminated.		85Z1
UO_2 (100)	c(2x2)	0.5	He	ICISS; ψ	-	-	-	-	*At 875 K.* Surface is O-terminated; two bridge-bonding positions for O.		80E
	c(2x2)	0.5	He	ICISS; ψ	-	-	-	-	*At 875 K.* Zig-zag O chains at the surface along the [010] or [001] directions. *Additional LEED analysis.*		81T1
UO_2 (110)	var.	0.5	He	ICISS; ψ	-	-	-	-	*At 675 K.* No buckling in 1st layer.		80E
UO_2 (111)	1x1	0.5	He	ICISS; ψ	-	-	-	-	*At 875 K.* Surface is bulklike and O-terminated.	18	78E1, 80E
UO_2 (553)	1x1	0.5	He	ICISS; ψ	-	-	-	-	*At 875K.* Edges are covered by O atoms.		78E1
W(110)	-	6	noble	ISS; ϕ	-	-	< 2% [17])	-			80O1
W(211)	-	4	Ar	TOFSARS; ψ, ϕ, θ	-12(7)	-	10(7) [18])	-			89G, 89R

1) See also Table 10 in chapter 2.2, subvolume III/24a.
2) At 295 K.
3) For the [$\bar{1}$10] and [001] direction, respectively.
4) Originally published value of 147 K has been corrected into 255 K in [83P].
5) For edge atoms only.
6) In the [100] direction; average for the Cu(100) terrace.
7) In the [100] direction; for edge atoms only.
8) For second layer; towards the center of the trough.
9) For the 1x3 phase.
10) For 1x1 and 1x3 phase, respectively.
11) In disagreement with [90O2].
12) For Ni- and Al-terminated domains, respectively.
13) In disagreement with [90N4].
14) Not observed with LEED but with ISS.
15) In disagreement with a HEIS study [82J1].
16) Originally published value of -38 pm has been corrected into -13 pm in [84A1].
17) No deviations (>2%) between expected and observed surface crystallographic directions.
18) In the [1$\bar{1}\bar{1}$] direction.

Table 2. Low Energy Ion Scattering (0.05 keV < E_0 < 50 keV). Related studies, ordered chronologically within each section.

All primary ion beams are of charge state +1 unless otherwise stated.
BCM = binary collision model.
E_0 = primary beam energy.
var. = various.
MC = Monte Carlo simulation.
QS and QD = quasi-single and quasi-double.
T = temperature.
TFM = Thomas-Fermi potential in the Molière approximation.
vs. = versus or 'as a function of '.
Z_1 = ion's atomic number.
ZBL = universal potential of Ziegler, Biersack and Littmark [85Z2].
θ, ψ, ϕ and η: see Fig. 10.
ρ = vibration amplitude.

E_0 [keV]	Ion	Surface	Remarks and main conclusions	Fig.	Ref.
Spatial and energy distribution of scattered ions					
2.2, 3	Ar, Cu	Cu(100)	MC: anisotropy in spatial distribution.		67K1, 68Y2
2.2	noble	Cu(100)	MC: spatial and energy distribution.		67Y, 68Y1
10 ... 30	He	Al, Ge, Cu, Ag, Au	Analytical model for shape of energy spectra.		72M2
10 ... 30	noble	Si(111), Ge(111)	Spatial distribution vs. T. Annealing of radiation damage above 650 K.		72M5, 72M6
0.2	Ar	Cu(110)	Differences in Ar^+ and Ar^0 energy spectra: Ar^+ stems mainly from surface region. MC.		76V1
1.5 ... 20	var.	Ni(110)	Agreement between MC and analytical model.		78H6
-	noble	-	MC: spatial distribution.		78P
6 ... 40	Ar	Cu(100)	Wide range spatial distribution. MC.		80S1
0.5 ... 1.5	H	Si(111)	Spatial and energy distribution. Separate contributions from surface and bulk.		84G1

Table 2 (cont.)

E_0 [keV]	Ion	Surface	Remarks and main conclusions	Fig.	Ref.
0.5	K, Li	Mo(100)	Spatial and energy distributions; MC.		84O
3 ... 20	Li	Cu(100)	Spatial distribution vs. surface structure; MC.		86Y2
0.06 ... 4	Na	Cu(110)	MC: energy spectra. Differences between 2nd layer scattering in $[1\bar{1}0]$ and [001] azimuths.		89M1
0.1 ... 0.4	Na	Cu(110)	MC: energy and spatial distributions; chain model vs. full crystal model; interaction potential; rainbow scattering.	19	91A
Double and multiple scattering					
25 ... 30	noble	Cu(100), (411)	Observation of double scattering.		65M1, 65M2, 66M1, 66M2
30, 28	noble	Ge(111), Si(111)	Double scattering vs. *T*. Disappearance of double scattering below 600 K.		67E
25, 30	noble	Ni(100)	Double scattering increases with Z_1.		67M1
> 30	Ar	Cu(110)	Double scattering vs. E_0.		68V
10 ... 20	Ar	Cu(100)	Double scattering vs. E_0.		69M2
< 10	He	Au(100)	Analytical model compared with MC.		70A3
30	Ar	Cu(110)	Confirmation two-atom model for double scattering.		70M4
5 ... 30	Ar	Cu, Si, Ge	Confirmation two-atom model for double scattering.		72M3
0.3	Ar	Ni(110)	MC: multiple scattering; neutralization.		72V
-	-	-	Theory: two-atom model for double scattering.		73M3
10 ... 35	Ar	Cu(100)	Spatial and energy distributions. Experimental data compared with two-atom model.		77B2
-	-	-	Analytical two-atom model compared with MC.		78S1
-	-	-	Three-atom model for calculation spatial and energy distributions. Comparison with MC.		79P

Table 2 (cont.)

E_0 [keV]	Ion	Surface	Remarks and main conclusions	Fig.	Ref.
2.4	Ne	Ni(100)	Energy spectra. Good agreement with MC.		80B6
-	-	-	Confirmation two-atom model for double scattering. Occurence of zig-zag collisions. Ratio QS to QD scattering intensities vs. interatomic distance.		82A3
4	Ar	Mg(1000)	Observation of zig-zag collisions. MC.		83K1
-	-	-	Theory: two-atom model for 180° scattering.		84Y
-	-	-	MC: ion flux calculation for ICISS.		89D1
-	-	-	Theory: two-atom model for ICISS.		90W2
Chain scattering					
5 ... 30	Ar	Cu(100)	MC: chain scattering model.		67K2
30	Ar	Cu(100)	Test chain scattering model.		69E
0.3 ... 5	Rb, K	Mo(100),(110),(111)	Confirmation chain scattering model.		70A1, 70A2
5 ... 30	Ar	Cu(100)	MC: spatial and energy distributions vs. E_0, T, ψ^1).		70K
0.6	He	Ni(110)	Observation of chain scattering. MC.	20	72T
30	Ar	Cu(100)	MC: inelastic energy loss.		73M1, 73M2
0.1 ... 0.6	Ne	Ni, Ag, W	Test chain scattering model. MC.		76H2
30	Ar	Cu(100)	MC: chain scattering. Role of vacancies.		78N
30	Ar	Cu(100)	MC: chain scattering. Role of vacancies.		79N1
0.6	He	Ni(110)	Limits of chain scattering model. MC.		81J
Shadowing, blocking and focusing					
50	var.	Cu(110)	Reduced reflection for incidence parallel to close packed atomic rows (*shadowing*).	21	63N

Table 2 (cont.)

E_0 [keV]	Ion	Surface	Remarks and main conclusions	Fig.	Ref.
1 ... 10	Cu	Cu	MC: slowing down of channeled particles.		63R1
1.5	Ar	Cu(100)	Reduced reflection for incidence parallel to close packed atomic rows (*shadowing*).		65Y
> 5	H	Cu(100)	MC: reduced reflection for exit directions parallel to close packed atomic rows (*blocking*).		68M2
0.05 ... 0.4	alkalis	W, Mo, KCl	Shadowing.		72A
10 ... 30	Ar	Cu(110)	Spatial distribution scattered ions. Blocking.		75M
30	Ar	Cu(100)	Spatial distribution scattered ions. Blocking.		77B5
20	H	Si(100)	Shadowing; possibility for surface relaxation determination. MC.		78I
4, 7	Ne, H_2O	Cu(110)	Spatial distribution scattered ions. Shadowing and blocking. Neutralization is of minor importance.		81B3, 81B4
0.25 ... 1	K	W(110)	Focusing by 1st layer atoms onto 2nd layer atoms.		81V2, 82V
0.1 ... 2	Li	Mo(100)	MC: energy spectra. Focusing. Characteristics of various scattering mechanisms.		83O3
3 ... 10	Ne	Cu(100)	Shadowing and focusing by 1st layer atoms. MC.		84C
9.5	Ne	Fe(100)	Shadowing and blocking. Possibility for surface structure analysis.		84M2
0.2 ... 2.5	noble	Rh(111)	Shadowing. MC (multiple collision model).		85C1
1	Ne	Ni(110)	Shadowing and blocking for structure analysis.		85H4
Surface semichanneling and focusing					
30	Ar	Cu(100)	'Out-of-plane' scattering [2]). Observation of focusing for incidence parallel to close-packed surface directions.		69M1
30	Ar	Cu(100), (110)	Role of surface semichannels in focusing.		70M1, 71M
10 ... 30	noble	Cu(100), Si(110)	'Out-of-plane' scattering [2]). Spatial and energy distribution of scattered ions.		70M2

Table 2 (cont.)

E_0 [keV]	Ion	Surface	Remarks and main conclusions	Fig.	Ref.
30	Ar	Cu(100)	'Out-of-plane' scattered ion distribution [2]).		70M3
30	Ar	Cu(100), (110)	Intensity scattered ions vs. ϕ. Focusing.		71A
4.5	Ar	Cu(100)	Double peaked distribution for grazing exit directions caused by blocking [3]). MC.	22	71S1, 72Y
30	Ar	Cu(100)	MC: spatial distribution.		72S1
6	H_2	Cu(100)	Channeling and surface semichanneling.		73F
30	Ar	Cu(100), (110)	Difference in spatial distribution of ions scattered from surfaces with and without semichannels.		73S
5 ... 30	Ar	Cu(110)	Focusing vs. E_0.		74M2
3 ... 9	H, He	Cu(100)	Focusing by surface semichannels. MC.		75F2
-	-	-	MC: focusing by surface semichannels.		75S
15 ... 35	Ar	Cu(100)	Focusing effect increases with increasing E_0.	23	76B1, 77B1
30	Ar	Cu(100)	Spatial and energy distribution of scattered ions.		76M
10 ... 35	Ar	Cu(100)	Focusing by surface semichannels vs. E_0. Compared with theory.		79B1
30	Ar	Cu(100)	MC: double peaked distribution for grazing exit directions caused by blocking [4]).		79Y1, 79Y2
-	-	-	Theory: surface semichanneling.		80S2
20 ... 35	Ar	Cu(100)	Focusing vs. E_0.		82M3
-	-	-	MC: focusing vs. E_0 and ϕ.	24	82Y
30	Ar	Cu(100)	Surface semichanneling.		84M3
30	Ar	Cu(100)	Relation between energy spectra and focusing.		84M5

Table 2 (cont.)

E_0 [keV]	Ion	Surface	Remarks and main conclusions	Fig.	Ref.
Rainbow scattering					
0.04	K, Li	W(110)	Spatial and energy distributions of scattered ions. Observation of rainbow scattering.	25	84T2
0.01 ... 0.1	K	W(110)	Spatial and energy distributions of scattered ions. Rainbow scattering.		86T2
0.04	K	W(110)	MC: rainbow scattering.		86T3
-	-	-	Theory: shadowing and blocking; rainbow scattering. Necessity for quantum-mechanical instead of classical cross sections.		90D1
Binary collision and multiple collision simulation models					
> 40	Ar	Cu(100)	Confirmation of binary collision model. Channeling.		64D
0.05 ... 0.5	Ar	Cu(100)	MC: binary vs. multiple collision model. Negligible difference above 100 eV.		71K
0.05 ... 5	Kr	W(100), (110)	MC: multiple collision model. Confirmation of concept of effective mass.		72S2
15	H	Ni(100)	Multiple collision model. MC.		73E
0.2 ... 1.2	noble	Ni(110)	Energy spectra of scattered ions. Binary collision model is not valid for heavy ions below 1 keV. Confirmation of concept of effective mass.		73H2
< 1	He, Ne	Cu	Binary collision model is valid down to 40 eV. No confirmation for concept of effective mass.		75T2
-	-	-	Instrumental effects in measurement of the effective mass.		78H4
0.6	noble	Ni(110)	MC: effect of simultaneous collisions.		79G
0.1 ... 6	Ne, Ar	Rh, Ni	MC: binary collision vs. multiple collision model. Validity BCM depends on interatomic distances along and perpendicular to direction of incidence.	26	88C
0.2 ... 1.1	Cs	W(100)	MC: binary collision vs. multiple collision model.		89E

Table 2 (cont.)

E_0 [keV]	Ion	Surface	Remarks and main conclusions	Fig.	Ref.
Interaction potential					
1	Ne	Cu(110)	Interaction potential derived from multiple scattering. MC.		75D3
0.2 ... 2	Ne	Ni(110), (111)	Screening length for TFM potential derived from chain scattering. MC.		75H2
-	-	-	Theory: correction factor for screening length for TFM interaction potentials.		77O
-	-	-	Screening length for TFM potential derived from width angular distribution, ratio QS to QD intensities and spectrum peak width. MC.		77P2
1	He	TiC(100), (111)	Shadow cone radii of various interaction potentials compared with experimental data. Agreement for TFM with reduced screening length.		82A7
4	Ar	Mg(1000)	Interaction potential derived from ab initio self-consistent field calculations. MC.		82C
1	He	TiC(100), (111)	Screening length for TFM potential.		82K
-	-	-	Theory: universal expression for shadow cone.		83O1
0.25 ... 2.5	alkalis	Mo(100)	Screening length for TFM potential derived from critical angle of incidence.		85O
1 ... 3	var.	var.	Theory: shadow cone radii for TFM and ZBL potentials.		86C1, 87C1
0.5 ... 5	He, Ne	Ni(100)	Screening length for TFM potential derived from critical angle of incidence.	27	86F1
0.6	K	var. metals	Screening length for TFM potential derived from ratio single to double scattering intensities. MC (fast calculation procedure).		87G
-	-	-	Theory: shadow cone radii for TFM and ZBL potentials; inclusion of thermal vibrations.		88F2
0.3 ... 6	K	Ir(110)	Test of TFM and ZBL interaction potentials via critical angle of incidence. Agreement for ZBL.		88H1, 88H2

Table 2 (cont.)

E_0 [keV]	Ion	Surface	Remarks and main conclusions	Fig.	Ref.
1	noble	TaC(100)	Slight disagreement for ZBL in case of He^+ ions. Effect of image potential. MC.		88K4
1, 2	Li, Na	Ag, Cu	Theory: test of models for interaction potential.		89C
0.15 ... 0.25	Na	W(100)	Interaction potential derived from QS and QD scattering energies. Possibility for structure analysis to within 1 pm.		91G1
-	-	-	Theory: interatomic potentials.		91N1
Various ion scattering studies					
6	Ar, Kr	Cu(100)	Possibilities for surface composition and structure analysis.		72B2
1	Ar	Cu(100)	MC: reflection coefficient vs. ψ.		76H3
0.1 ... 5	H	Ni(110)	MC: specular reflection depends on surface structure and E_0. Focusing by semichannels.		78H7, 80H
0.1 ... 7.5	var.	Cu(100)	MC: critical angle ψ for total reflection.		78H8
9	Kr	Cu(410)	Ambiguity in elastic energy loss.		80A3
< 1.5	He	W(100), (110)	MC: reflection. Fair agreement with experimental data.		81R
-	-	-	Theory: ambiguity in elastic energy loss.		82B2
-	Ne	Ni	MC: principles. Compared with experimental data. BCM is as good as multiple collision model.		82H4
3	Li	Cu(100)	MC: possibility for surface structure analysis from angular distribution of scattered ions.		83W1
2.4	Ne	Ni(100)	MC: principles. Compared with experimental data. Binary model gives satisfactory results.		83Y2
5, 9.5	Ne	Cu_3Au(100)	TOF. Subsurface scattering contribution. MC.		84J2
-	light, heavy	-	Theory: dependence of cross section and final energy on θ for inverse mass ratio collisions.		85D3

Table 2 (cont.)

E_0 [keV]	Ion	Surface	Remarks and main conclusions	Fig.	Ref.
-	-	-	MC: principles. Algorithm for determination of the collision sequence.		85L
5	Ne	TiC(111)	Theory (chain model): relation between final energy and total scattering angle near 180°.		85T2
15 ... 30	Ar, Ne	Cu(100)	Reflection and sputtering coefficients vs. ψ.		87M
2, 4	noble	Ir(110)	Scattering depth vs. E_0 and Z_1. MC.		91B2
Grazing incidence and surface hyperchanneling					
6 ... 15	Ar	Ni(110)	Observation of surface hyperchanneling. MC.		78K
6 ... 15	Ar	Ni(110)	Confirmation of surface hyperchanneling. Comparison with surface semichanneling and subsurface channeling.	28	79E1
6	Ar	Ni(100), (110)	MC: continuous string model for surface hyperchanneling simulation.		79E2, 79E3
6	Ar	Ni(110)	Confirmation of surface hyperchanneling. MC.		79E4
6	Ar	Ni(110)	MC: energy loss during surface hyperchanneling.		79K
-	-	-	MC: continuous string model is as good as (but much faster than) a real crystal model for surface hyperchanneling simulation.		81E
10 ... 30	Ar	Cu(100), (110)	Specular reflection. Surface semichanneling and subsurface channeling.		84M4
2 ... 5	He	Ni(110)	Surface (hyper-)channeling. MC.		89D2
3.5	Ne	Au(110)	Surface (hyper-)channeling.		90N3
1 ... 30	H	Al(111)	H^- formation in case of grazing incidence.		91W3
Skipping motion					
30	H	Ni(100)	Theory+MC: prediction of skipping motion.		79O
30	H	Ni(110)	MC: skipping motion; energy loss.		86O1
30	H	Ni(100)	Skipping motion; energy loss.		88K3
0.2 ... 2	Si	Cu(111)	Observation of skipping motion.	29	88S2, 89S2

Table 2 (cont.)

E_0 [keV]	Ion	Surface	Remarks and main conclusions	Fig.	Ref.
0.05 ... 2	Na, K, Si	Cu(111)	Skipping motion. Transient adsorption by elastic and inelastic energy losses.		90S5
3 ... 3.5	var.	Cu(111)	No difference between transient adsorption for reactive and noble gas incident beams.		91M4
Molecular ions					
0.2 ... 2	H_2	Ni(111), W(100)	Dissociation proceeds not via electron loss but via molecular excitation. Probability depends on surface structure.		79H
4	H_2O	Cu(110)	Dissociation. Strong dependence of O^- yield on ϕ.		80B5
30	N_2	Cu(100)	Reduced dissociation of molecular ions incident parallel to surface semichannels.		82B1
0.4	N_2	Ni(110)	Dissociation probabilities.		82H3
30	N_2	Cu(100)	Dissociation probability vs. ϕ: reduced dissociation of molecular ions incident parallel to surface semichannels. Possible link with focusing.		83B1, 83E
30	N_2	Cu(100)	Dissociation probability vs. ϕ: reduced dissociation explained by two-atom model.	30	83B2, 83E
30	N_2	Cu(110)	Reflection of N_2^+ vs. ϕ.		85B
30	N_2	Cu(100)	MC (full multiple interaction model): motion of the two N atoms cannot be treated separately.		85J
0.12 ... 1	N_2, N	Cu(100)	Grazing incidence. Spatial distributions of dissociated and of reflected ions. Dependence on E_0.		86A1
30	N_2	Cu(100), (110)	Reflection of N_2 vs. ϕ. Role of surface semichannels.		86B
1 ... 30	N_2	Cu(100), (110)	MC (continuous string model): relation between molecular reflection and surface semichanneling.		86S1, 86S2
0.2 ... 4.4	N_2	Ni(111)	Main mechanism of dissociation at low energies is charge capture, at higher energies molecular excitation becomes important.		86W

Table 2 (cont.)

E_0 [keV]	Ion	Surface	Remarks and main conclusions	Fig.	Ref.
30	N_2	Cu(110)	Reflected N_2^+ fractions vs. ϕ and θ.		89B1
30	N_2	Cu(100)	MC: relation between focusing by surface semi-channels and reduced dissociation.		90S4
1.5, 3	H, H_2	Ni(110)	Spatial distribution scattered particles. Relation between molecular orientation and dissociation.		91T
Thermal vibrations					
30	Ar	Ni(100)	Energy spectra vs. T: intensity double scattering peak decreases with increasing T.		67M2
8 ... 20	Ar	Cu(100)	Ratio QS to QD scattering intensity vs. T, E_0 and ψ.		69C
2, 30	Ar	Cu, Ni	MC: spatial and energy distribution vs. T.		74K
10	Ar, Kr	Cu(100)	Energy spectra vs. T. Role of neutralization. MC (chain model).		75P
6	Ar	Cu(100)	Energy spectra vs. T. MC (chain model). Multiple scattering can be used for studying thermal vibrations.		75V1
6	Ar	Cu(100)	MC (chain model): strong dependence of spatial and energy distributions on T.		76P1
6	Ar	Cu(100)	Energy spectra vs. T. ρ deduced from quasi-triple (QT) collisions.		76P2
10	Kr	Cu(100)	QT collision vs. T. Θ_D deduced from QT collisions, is independent of neutralization and interaction potential [5]). Comparison with experimental data.		76P3
6 ... 10	noble	Cu(100)	Energy spectra vs. T. MC (chain model). Reasonable agreement between experimental data and simulations. Role of neutralization.	31	77P1
-	-	Cu	Theory: effects of correlated thermal vibrations.		80M1
-	-	-	MC (chain model): correlations in thermal vibrations. Value for Θ_D may depend on interaction potential assumed [6]).		82W1

Table 2 (cont.)

E_0 [keV]	Ion	Surface	Remarks and main conclusions	Fig.	Ref.
10	Kr	Cu(100)	Theory (three-atom model): quasi-triple collisions vs. *T*. General agreement with [82W1] but Θ_D can be measured without exact interaction potential.		83P
1	Ne, Na	Cu(110)	Energy spectra vs. ψ, θ, ϕ and *T*. Drastic temperature effects for special scattering geometries.		84E
1	He	TiC(100), (111)	MC: ICISS. Thermal vibrations.		84T1
50	noble	Cu(110)	Theory: effect of lattice vibrations at 0 K.		85D2
5	Ne	TiC(111)	Theory (chain model): energy spectra vs. θ. Quasi-triple scattering.		86T1
5	Ne	Cu(110)	Vibration amplitude. Theory (three-atom model).		90V1
Charge exchange					
0.1 ... 1	noble	W, Mo, Si(100)	Reflection of ions and neutralized projectiles. Hagstrum's model: charge exchange processes explained via Auger and resonant transitions.		61H
-	-	-	Charge exchange processes: resonance ionization/neutralization, Auger neutralization/de-excitation.		70H
9	Ar	Cu(100)	Comparison spectra scattered Ar^+ and Ar^0. Ions and neutrals are produced by different processes.		72C
< 0.1	He	Ni(100), (110)	Charge exchange processes for $He^+(2s)$ and He^{2+} beams. Role of intermediate He^{0**} state.		73H1
1	Ne	Cu(110)	Test of binary collision model [7]) for charge exchange. Reasonable agreement with experimentsl data. MC.		75D3
0.2 ... 2	noble	Ge, Pb	Oscillatory scattered ion yields: quasi-resonant neutralization.	32	75E
2 ... 10	He	Cu(100)	Scattered ion yields vs. ψ and E_0. Hagstrum's model [8]) is valid for low E_0; at higher E_0 binary collision model [7]) is also valid.		75V2
0.2 ...3	He, Ne	var.	Oscillatory scattered ion yields.		76B3

Table 2 (cont.)

E_0 [keV]	Ion	Surface	Remarks and main conclusions	Fig.	Ref.
> 10	-	-	Theory (binary collision model [7]) : multiple ionization of scattered ions. Relation with surface structure.		76K1
3 ... 10	He^+, He^0	Cu(100)	Scattered ion yields. Confirmation of model in [75V2]. Quantitative results for Auger neutralization and binary collision model [7]).		76V2
0.2 ... 1.7	He	Bi	Oscillatory ion yields vs. θ and E_0. Confirmation of quasi-resonant neutralization.		77D
0.3 ... 2.5	He	var.	Theory of oscillatory neutralization. Qualitative agreement with experimental data.		77T2, 77T3
2 ... 10	He	Cu(100)	Scattered ion fractions. Confirmation of model in [75V2]. Ionization and neutralization in binary collisions via interaction with Cu L-shell.	33	77V
2.4, 5	Ne	Ni(100)	Comparison TOF and ESA detector. Scattering depth dependent neutralization.		79B2
5 ... 10	Ne	Cu(100)	Scattered ion fractions. Re-ionization during binary collision increases with E_0.		79L
5 ... 10	Ne^+, Ne^0	Cu(100)	Difference in charge fractions scattered Ne^+ and Ne^0 beams only at low E_0. Confirmation of [75V2].	34	80L2
5 ... 10	Ar^+, Ar^0	Cu(100)	Ion fractions. Confirmation of model in [76K1].		80L3, 80L5
3 ... 15	Ar, K	Cu(410)	Scattered ion fractions.		80A4
0.5 ... 4	He	Te, CdTe, HgCdTe	Oscillatory ion yields vs. chemical binding of Te.		80B2, 81B2
1 ... 10	H_2	Au(110)	Ions are neutralized mainly in the bulk.		80M2
0.5 ... 5	He	AlF_3, MgF_2, NaCl	Neutralization vs. ionic size of Al, Mg, Na.		80M3
0.1 ... 1.2	He, Li	Ni(110)	All He^+ ions are neutralized in multiple collisions, Li^+ ions may survive.	35	80T
1	He	Ni(100)	Observation of trajectory dependent (Auger) neutralization.		80W
1 ... 7	Ne, H_2O	Cu(110)	Charge exchange and shadowing and blocking. Possibilities for surface structure analysis.		81B3, 81B4

Table 2 (cont.)

E_0 [keV]	Ion	Surface	Remarks and main conclusions	Fig.	Ref.
0.6 ... 2.5	K, Ar	Au(110)	Comparison Ar - K; vs. E_0, ϕ and θ. Differences in ion yields are due to neutralization. MC.		81O3
1	He	Ni(100)	Theory: trajectory dependent Auger neutralization.		81P
2 ... 10	alkalis	Cu(100)	Scattered ion fractions. Alkali ion fractions are independent of the trajectories of the ions.		82A2
0.6	He, Li	Ni(110)	Trajectory dependent neutralization.		82E
3	Ne	Ni(110)	Theory: trajectory dependent neutralization in case of surface semichanneling.		82N3
0.4 ... 2	^3He, ^4He	MoS_2	Neutralization is largest for 2nd layer scattering.		83D1
1 ... 6	He, Ne, H_2O	Ag(110)	Trajectory dependent Auger neutralization		83D2
?	Na	W(110)	Simplified quantum-mechnical model for charge exchange.		83N1
-	He	var.	Auger neutralization dominates over resonance neutralization.		83W2
0.3 ... 1.2	Ne	GaAs(110)	Trajectory dependent neutralization.		84R1
0.5 ... 15	He	Cu(110), Ni(110)	Optical spectrometry. Resonant neutralization.		84T4
6	Ne, H_2O	Cu(100)	Charge exchange and shadowing and blocking.		84V1
0.2 ... 0.6	H_2, N_2	Ni(111)	Neutralization of molecules.	36	84W1
1	He	Ge(111)	Trajectory dependent neutralization. Implications for ICISS.		85S1
0.3 ... 2	He^+, He^0	TaC(100)	Re-ionization of He^0 is possible. Excited He* ions are not re-ionized.		85S5
2 ... 5	Ne	Cu(100)	No difference in ion fractions for 1st and 2nd layer scattering for perpendicular ejection.		85V2
0.2 ... 1.1	H_3	Ni(111)	Neutralization/dissociation via resonant charge exchange.		85W1
0.2 ... 2	$He^{+/0/*}$	var.	Neutralization and (re-)ionization mechanisms.	37	86S4

Table 2 (cont.)

E_0 [keV]	Ion	Surface	Remarks and main conclusions	Fig.	Ref.
1	He	var.	Spatial distribution of valence electrons derived from trajectory dependent neutralization.		86S5
1	noble	TaC(100)	He^0 and Ar^0 are re-ionized during double scattering. Ne^0 remains neutral.		86S6
3.5, 5	He, Li	Cu(110)	Effects of neutralization on ICISS analyses.		86Y6
0.6 ... 1	Ne, Na	Cu(110)	Energy spectra vs. ϕ at 100 ... 600 K. Na scattering reflects surface structure. Ne^+ yield is determined by trajectory dependent neutralization.		87E
2	Ne, Na	Pt(111)	MC (hitting/detection probabilities). Good agreement with ICISS data. Auger neutralization half-life of Ne is 1.3(1) fs.		88E
1	He	NaCl(100)	MC: trajectory dependent neutralization.		89B2
1 ... 5	He	Ni(110)	Scattered ion fraction vs. ϕ. Surface channeling effects. Trajectory dependent Auger neutralization and resonant ionization. MC.		89M4, 89N1
> 4	H	Al(111)	Grazing incidence. Neutralization by Auger and resonant processes is total at low E_0 and low at high E_0. Formation of H^-.		89N3
3 ... 30	H_x ,Na, K	Al(111), Ni(111)	Grazing incidence. Resonant neutralization to excited states.		89V3
> 3	Na, K	Al(111), Ni(111)	Grazing incidence. Resonant neutralization dominates over Auger neutralization. Good agreement with theory.	38	89Z
2 ... 5	He	Ni(110)	Relation between charge exchange and energy loss. Good agreement with theory.		90N1
0.3 ... 1.3	He	InP(110)	Relation between energy loss of scattered ions and neutralization followed by re-ionization.		90S3
< 1	He	var.	Re-ionization and energy loss. Role of molecular orbitals.		90S6

Table 2 (cont.)

E_0 [keV]	Ion	Surface	Remarks and main conclusions	Fig.	Ref.
-	-	-	Theory: trajectory dependent neutralization. Model for multiple neutralization interactions.		90V3
> 2.4	H, Mg	Al(111), Ni(111)	Grazing incidence. Resonant neutralization of Mg and formation of H^-. Good agreement with theory.		90Z
0.1 ... 2	Ar,N, N_2	NaCl(100)	Resonant charge exchange in case of ion energy levels above, within and at the bottom of Cl-valence band. Good agreement with theory.		91G2
4, 8	Ne	Ni(100)	Ion fractions vs. ψ and ϕ. Consequences for ICISS. Layer specific information can be obtained.		91H7
0.3 ... 6	He^+, He^{2+}	Pb(110), (111)	Quantitative results for Auger and resonant neutralization and two-state charge exchange.		91S1
0.2	D^+, D^0	Mo(111)	Neutralization and re-ionization.		91S2
0.04	Na	Ag(111)	Trajectory dependent charge exchange. Good agreement with theory.		91V3
Energy loss					
10 ... 50	Ar	Mo(100)	MC: binary collision model for inelastic energy loss vs. ψ.		74Z
4 ... 10	He^+, He^0	Cu(100)	Inelastic energy loss by ionization of He^0 and by electronic stopping.		76V3
0.5 ... 1.6	He	Si(111)	Inelastic energy loss vs. ψ, θ, and E_0.		80B3
2 ... 6	noble	W(110)	Surface semichanneling. Inelastic energy loss in binary collisions and by electronic stopping.		80O1, 80O2
-	-	-	Theory: inelastic energy loss for chain scattering.		81K1
0.5 ... 1.5	He	Si(111)	Inelastic energy loss is localized.		83B4
-	-	-	Relation between energy loss and neutralization.		83O2
30, 15	N_2, N	Cu(100)	Surface semichanneling. Inelastic energy loss vs. ϕ. Width of energy distribution of reflected ions varies strongly near surface channel directions.		84D

Table 2 (cont.)

E_0 [keV]	Ion	Surface	Remarks and main conclusions	Fig.	Ref.
0.2 ... 1.5	He	Si(111)	Two inelastic energy loss peaks in spectra, explained by ionization via electron promotion.		88S1
2 ... 5	He	Ni(110)	Inelastic energy loss of ions and neutrals. Qualitative agreement with dielectric response model.		89M3
0.2 ... 2	Si	Cu(111)	Inelastic energy loss for skipping motion. Information about charge state during interaction.		89S3
0.26 ... 14	He	Ni(110)	Inelastic energy loss for grazing incidence. Good agreement with dielectric response model.		90N2
2 ... 5	He	Ni(110)	Relation between energy loss and charge exchange.		91N2
3	Ar	Cu(111)	Skipping motion. Discrete energy losses. Evidence for transient adsorption.		91R2

[1]) Spatial distribution in disagreement with [71S1, 72Y].
[2]) Out of the plane defined by the incident beam direction and the surface normal.
[3]) Spatial distribution in disagreement with [70K].
[4]) Resolved discrepancy between [70K] and [71S1, 72Y].
[5]) Partially in disagreement with [82W1].
[6]) Partially in disagreement with [76P3].
[7]) Model for neutralization and (re-)ionization in binary ion-atom collisions, see [69V].
[8]) See [54H, 61H].

General LEIS studies	**Ref.**
Theory: Auger and resonant charge exchange.	54H
MC: channeling.	63R2
Channeling and shadowing.	65L
Theory of ion reflection.	65P
MC: ion reflection and penetration.	70P
Theory of double scattering.	66P

Table 2 (cont.)

General LEIS studies (cont.)	**Ref.**
Scattering and sputtering from metal surfaces.	67S
Ion beam induced damage and annealing at elevated temperatures.	68M1
Surface composition of CdS.	68S, 71E
Surface composition determination.	71S2, 72B1, 73B2, 73B3
Quantitative aspects of surface composition determination.	75N, 83B5, 90Y3
Surface composition and structure determination.	74B, 78B5
Surface structure determination by multiple scattering.	82A4
Detection at large scattering angles.	76B2
Comparison TOF vs. electrostatic analyzer.	78B6, 78L
Comparison alkali vs. noble gas ion scattering.	82N2
ICISS.	81A1
Noble vs. alkali ICISS or NICISS.	83N2, 84N1, 85A2, 85N2, 86N2, 86N3, 86N4, 87N
LEIS with multiply charged ions.	85D1
Method for subtraction of background in energy spectra.	86N1
Inelastic energy loss processes.	87A4
LEIS with neutral primary beams (NBISS).	87S3
LEIS for isotopically labelled surface structures.	90C3
Low energy neutral ionization spectrometry (LENIS).	91V2
Instrumentation	**Ref.**
LEIS instruments.	71S3, 72B3, 75T1, 77T1, 79S1
High-energy resolution LEIS spectrometer.	89S1
Time-of-flight spectrometers (TOF).	77C, 78A1, 84W2

Table 2 (cont.)

Instrumentation (cont.)	**Ref.**
TOF spectrometer vs. stripping cell.	80L1, 80L4
TOF scattering and recoiling spectrometer (TOF-SARS).	83R, 90G
Masked double-pass CMA.	78E2
Energy and angle resolved ion spectrometer (EARISS).	85H2, 88A
Imaging of scattered ion angular distributions.	86Y5
Spectrometer for coaxial ICISS (CAICISS).	88K2, 90S7, 91K
TOF-ICISS (NICISS).	88S4
Coincidence ion-electron spectrometer.	90W4
Reviews	**Ref. (no. of pages/ no. of references)**
LEIS in general.	72M4 (45/ 74), 73M4 (9/ 23), 74M1 (56/ 91), 75B (28/ 75), 77B3 (22/ 44), 78H3 (11/ 76), 84M1 (17/ 39), 91H4 (68/ 192)
LEIS with 1 ... 1000 eV ions.	90K (16/ 48)
LEIS of metal surfaces.	74S (44/ 155)
Computer simulations.	74Y (50/ 51), 89H3 (6/ 98)
Scattering phenomena.	74M3 (11/ 3), 81M (27/ 136)
Multiple scattering.	77B4 (26/ 51)
Elastic and inelastic scattering.	76H1 (11/ 83), 78H1 (17/ 34), 78H2 (27/ 49), 82H2 (4/ 40), 89K (104/ 214)
LEIS at grazing incidence.	80A5 (12/ 12)
ICISS, NICISS, ALICISS and CAICISS.	83A1 (11/ 32), 84A2 (10/ 59), 88F1 (14/ 138), 88N2 (8/ 25), 89N2 (16/ 20), 89A1 (6/ 38)
LEIS and ICISS.	85A1 (14/ 77)
LEIS with alkali ions.	87O2 (13/ 47)

Table 2 (cont.)

Reviews (cont.)	**Ref. (no. of pages/ no. of references)**
TOF-SARS.	90R1 (7/ 31), 91R (7/ 16)
Scattering of molecular ions.	84B1 (10/ 31), 87H1 (6/ 43), 88D3 (45/ 45)
Charge exchange.	77B6 (25/ 51), 82W3 (9/ 26), 84B3 (9/ 34), 88M1 (5/ 16), 90L (58/ 138)
Oscillatory neutralization.	77R (32/ 41), 77H1 (25/ 14)
Charge exchange and energy loss.	77H3 (20/ 39), 91N2 (10/ 63)
Charge exchange and surface chemistry.	91H3 (5/ 34)
Charge states of reflected alkalis.	84B2 (7/ 28)
Surface analysis by LEIS.	76T1 (15/ 66), 80B4 (44/ 52), 85H1 (20/ 66), 90O1 (15/ 16)
Surface composition and structure determination.	78B4 (15/ 36), 79A2 (20/ 46), 81A2 (12/ 65), 81B1 (8/ 40), 81V1 (12/ 111), 88T1 (6/ 47), 89H1 (5/ 70)
Surface structure determination.	77H2 (12/ 46), 82H1 (10/ 51), 82W2 (5/ 19), 85T1 (10/ 38)
Double scattering and surface structure determination.	78M (8/ 17)
Theory ISS for surface structure determination.	82B3 (32/ 84)

Table 3. Medium Energy Ion Scattering (50 keV < E_0 < 300 keV); Surface structure data. [1])

Surface	LEED pattern	E_0 [keV]	Ion	Technique	Δd_{12} [pm]	Δd_{23} [pm]	Δy_1 [pm]	ρ_1/ρ_b or Θ_D	Additional results *or remarks*	Fig./ Table	Ref.
Ag (110)	1x1	51, 98	H	sh+bl	-14(3)	+9(4)	-	1.65	Δd_{13} = -5.0(15) pm. ρ_2/ρ_b = 1.22.	39	87H2
Au (110)	1x2	65, 180	H	sh+bl	-26	+6	-	-	Support for missing-row model. 20 pm peak-to-peak buckling in 3rd layer.	40,41 [2])	86C2, 87C2
C(111)	1x2	99	H	sh+bl	-	-	-	1.65(15)	Support for π-bonded chain model. Dimerization in the chains.	[3])	86D
Cu (100)	1x1	175	He	sh+bl	-4(4)	-	-	1.5	ρ_2/ρ_b = 1.25 .		86A2
	1x1	100	H	sh+bl	-4.3(15)	+2(2)	-	1.5			91J
Cu (110)	1x1	100	H	sh+bl	-10(2)	+3(2)	-	1.55			86C3
Fe(110)	-	55	H	sh+bl	+2(4)	+1(4)	-	-			91X1
Fe(111)	-	55	H	oop-sh+bl	-24(6)	+6(4)	-	1.18			90X, 91X1
Fe(211)	-	55	H	oop-sh+bl	-14(5)	+2(2)	10(4) [4])	-	Δy_2= -7(3) pm [4]).		91X1

Table 3 (cont.)

Surface	LEED pattern	E_0 [keV]	Ion	Technique	Δd_{12} [pm]	Δd_{23} [pm]	Δy_1 [pm]	ρ_1/ρ_b or Θ_D	Additional results *or remarks*	Fig./ Table	Ref.
Fe_{72} - Cr_{28} (110)	-	55 [5])	H [5])	sh+bl	-7(3) [6]) -5(3) [6])	-2(3) [6]) +4(3) [6])	-	-	30(10)% Fe in 1st layer.		91X2
GaAs (110)	1x1	50, 100	H	sh+bl	-	-	-	-	29(3)° rotation of 1st layer Ga-As bond [7]). No bond-length relaxation [8]).	42	85S2
GaSb (110)	1x1	174	He	sh+bl	-	-	-	1.5(2)	29°+7°/-4° rotation of 1st layer Ga-Sb bond [7]). No bond-length relaxation.		84S1, 84S2
	1x1	174	He	sh+bl	-	-	-	1.65(10)	28.5(26)° rotation of 1st layer Ga-Sb bond [7]). No bond-length relaxation.		86S3
Ge (111)	c(2x8)	99 ... 161	H	sh+bl	-	-	-	-	*Direct comparison with Si(111) 7x7.* No lateral displacements. Support for adatom model [9]).		88M3
	c(2x8) / 1x1	60 ... 175	H	sh+bl	-	-	-	-	*At 350 ... 1200 K.* Decrease of adatom density at 575K; order-disorder transition at 1050K.		91D1
InAs (110)	1x1	174	He	sh+bl	-	-	-	1.65(10)	30.0(24)° rotation of 1st layer In-As bond [7]). No bond-length relaxation.		85S4, 86S3
Ni(100)	-	52, 98	H	sh+bl	-6(1)	-	-	1.46(13)	*At 370 K.*		83F3,

Table 3 (cont.)

Surface	LEED pattern	E_0 [keV]	Ion	Technique	Δd_{12} [pm]	Δd_{23} [pm]	Δy_1 [pm]	ρ_1/ρ_b or Θ_D	Additional results	Fig./ Table	Ref.
											83F4
Ni(110)	-	165 ... 200	H	sh+bl	0(2) [10])	-	-	-			78T2
	-	101	H	sh+bl	-5(1) [10])	-	-	-			79V1, 79V3
	1x1	100	H	sh+bl	-5(1) [10])	-	-	-			84T5
	1x1	110	H	sh+bl	-11(1) [11])	+4(2)	-	395 K		43	86G, 86Y1
	1x1	52	H	oop-sh+bl	-10(2) [11])	+5(2)	-	395(10) K			89X
NiAl (110)	1x1	100	H	sh+bl	-14(4) [12])	+2(4) [12])	-	1.6 [12])	Δd_{13} = -12(1) pm [12]).		87Y
					+10(4) [12])	-2(4) [12])	-	1.6 [12])	Δd_{13} = +8(1) pm [12]).		
Pb (110)	1x1	51, 98	H	sh+bl	-28(4)	-	-	1.5(1)	*Comparison with theory.* Δd_{23} + 3/4 Δd_{34} = +1(4) pm. ρ_2/ρ_b = 1.1(1) .		86F5
	1x1	51, 98	H	sh+bl	-30(1)	+14(3)	-	1.5(1)	*At 29 K.*		87F2
					-9(9)	-	-	-	*At 485 K.*		

Table 3 (cont.)

Surface	LEED pattern	E_0 [keV]	Ion	Technique	Δd_{12} [pm]	Δd_{23} [pm]	Δy_1 [pm]	ρ_1/ρ_b or Θ_D	Additional results	Fig./ Table	Ref.
Pt(110)	1x2	65 ... 180	H	sh+bl	-22(4) [13])	+6(4)	-	1.6	Support for missing row model. 10(4) pm (peak-to-peak) buckling in 3rd layer. $\Delta y_2 < 2$ pm in [001] direction.	[2])	88F3
Pt(111)	1x1	173	H	sh+bl	+3(2)	-	-	1.3	*At 420 K.*		78V2, 79V2
PtNi (111)	1x1	100	H	sh+bl	-4(1)	-4(1)	-	1.5	75(2)% Pt in 1st layer; 27(3)% in 2nd; 53(5)% in 3rd layer.		90D3
Si(100)	1x2	50 ... 150	H	sh+bl	-	-	> 45	-	$\lvert\Delta y_{2,3,4}\rvert < 20$ pm. Support for symmetric and buckled dimer models. Evidence for subsurface strain.	[3])	81T2
	1x2	50, 100	H	sh+bl	-	-	-	-	Support for buckled dimer model. Evidence for subsurface strain.		81T3
	1x2	50 ... 150	H	sh+bl	-	-	> 50	-	*Comparison with theory*. Support for buckled dimer model. Evidence for subsurface strain.	9	83T3
	1x2	75	H	IIAES	-	-	-	-	Support for buckled dimer model.		90P1
Si(111)	1x1	99	H	sh+bl	-	-	< 10	-	*Laser irradiated surface. Direct comparison with Si(111)-7x7*. Support for disordered 7x7 model.		82T
	1x2	99	H	sh+bl	-	-	-	1.45	*Cleaved surface*. $\rho_2/\rho_b = 1.45$. Support for π-bonded chain model.	[2]) [3])	83T4

Table 3 (cont.)

Surface	LEED pattern	E_0 [keV]	Ion	Technique	Δd_{12} [pm]	Δd_{23} [pm]	Δy_1 [pm]	ρ_1/ρ_b or Θ_D	Additional results	Fig./ Table	Ref.
Si(111) (cont.)	1x2	99	H	sh+bl	-	-	-	2.1	*Cleaved surface.* $\rho_{2,3,4}/\rho_b$ = 1.65, 1.4, 1.2, respectively. Support for tilted π-bonded chain model.	[2]) [3])	84T6, 85S3
	7x7	54 ... 151	H	sh+bl	-	-	-	-	Support for stacking-fault models.	44	84T7, 85T3
β-SiC (100)	var.	99	H	sh+bl	-	-	-	-	*During annealing at 1340 K.* c(4x2): 100% Si in 1st layer, c(2x2): 100% C in 1st layer. Indication of Si dimer rows for (3x2) and (5x2). Indication of dimerization of C for c(2x2).		90H
W(111)	-	100	H	sh+bl	-15(10)	-	-	-	*Via analysis of energy losses.*		90Y1

[1]) For explanation of column captions, see Table 1.
[2]) See also Tables in chapter 2.2, subvolume III/24a .
[3]) See also Figures in chapter 2.2, subvolume III/24a .
[4]) In the [111] direction.
[5]) Details not given; values assumed to be the same as in [90X] and [91X1].
[6]) Results for Fe and Cr clusters, respectively.
[7]) Out of the surface plane.
[8]) In disagreement with a HEIS study [83G1,84G2].
[9]) In disagreement with a HEIS study [86C4].
[10]) In disagreement with [86G,86Y1,89X].
[11]) In disagreement with [78T2,79V1,79V3,84T5,83F1].
[12]) Results for Ni and Al planes, respectively.
[13]) In disagreement with a HEIS study [82J1].

Table 4. Medium energy ion scattering (50 keV < E_0 < 300 keV). Related studies, ordered chronologically within each section.
All primary ion beams are of charge state +1 unless otherwise stated.
cyl. = cylindrical.
E_0 = primary beam energy.
MC = Monte Carlo simulation.
T_m = bulk melting temperature.
var. = various.
vs. = versus or 'as a function of'.
θ, ψ, ϕ and η: see Fig. 10.

E_0 [keV]	Ion	Surface	Remarks and main conclusions	Fig.	Ref.
Scattering phenomena					
< 80	Ar	Cu(100)	Spatial and energy distribution scattered ions. Confirmation of binary collision model. Channeling.		64D
< 500	H	Cu(100)	MC: reduced reflection for exit directions parallel to close packed atomic rows *(blocking)*.		68M2
< 90	Ar	Cu(110)	Double scattering vs. E_0.		68V
100	Ar	Au(100)	Observation of double scattering. Shadowing. Influence of thermal vibrations.		69D
60, 90	Ar	Cu(100)	Charge exchange in binary (ion-atom) collisions. Comparison with ion-gas atom scattering data.		69V
-	-	-	Analytical two-atom model for blocking.		78V1
300	H	Si(111)	Relative yield for double alignment is the product of relative shadowing and blocking yields.	45	80K
-	-	-	Analytical expression for blocking.		80O3
100	H	Ni(110)	Analytical three-atom model. Compared with MC.		80V
-	-	-	MC: shadowing-blocking correlations.	46	83T5
-	-	-	MC: principles. Efficient multi-parameter approach.		86F4

Table 4 (cont.)

E_0 [keV]	Ion	Surface	Remarks and main conclusions	Fig.	Ref.
Grazing incidence					
< 300	He	W(110)	Total specular reflection for $\psi < 1°$.		70F
275	He	W(110)	Spatial and energy distributions scattered ions vs. θ and ϕ.		72M1
> 75	H, He, B	var.	Surface peak area vs. E_0.		75D2
150 ... 275	H	W(110)	Energy loss and X-ray production vs. ψ and θ. Possibility for hyper-channeling below the surface.		75F1, 75H1
150	H	Ni(110)	Surface semichanneling and penetration into the bulk. Energy losses. MC and analytical model.		76S
195	He	Ni(110)	Surface semichanneling. Light emission. MC.		78V3, 79V4
-	-	-	Theory: surface semichanneling vs. ϕ. Focusing.		80Y1
< 200	Ar	Cu(100)	MC: surface semichanneling vs. ϕ. Focusing.		80Y2
70 ... 200	He	NiFe(110)	Light emission vs. ϕ. Charge exchange processes. MC.		83G2
150	H	Ni(110)	Surface semichanneling. Ion-induced AES. MC.	47	83S1
200	Ar	Cu(100)	MC: surface semichanneling, focusing and blocking.		83Y1
150, 200	H, He	Ni(110)	MC: surface channeling. Energy losses are well described by Oen-Robinson model.		84H3
70 ... 200	He	NiFe(110)	Surface (hyper-)channeling. Measurement of step distances. MC.		85G
< 200	H	Al(111)	Neutralization by Auger and resonant processes is total at low E_0 and low at high E_0. Formation of H^-.		89N3
< 350	Na, K	Al(111), Ni(111)	Resonant neutralization dominates over Auger neutralization. Good agreement with theory.	38	89Z
75	H	Si(100)	Ion-induced AES for surface structure analysis.		90P1
< 280	H, Mg	Al(111), Ni(111)	Resonant charge exchange. Galilei transformations for Doppler effect.		90Z

Table 4 (cont.)

E_0 [keV]	Ion	Surface	Remarks and main conclusions	Fig.	Ref.
Charge exchange					
60, 90	Ar	Cu(100)	Charge exchange in binary (ion-atom) collisions. Comparison with ion-gas atom scattering data.		69V
< 240	H, H_2, He	Cu, Au, Si	Scattered ion fractions increase with increasing ion velocity.		73B4
< 100	-	-	Theory (binary collision model [1]): multiple ionization scattered ions. Relation with surface structure.		76K1
75 ... 180	He	Si(100)	Trajectory depedent neutralization: ions are neutralized exclusively at the surface.	48	84H1, 84H2
30 ... 350	H, He	var.	Scattered H^-, H^0 and He^0 fractions decrease exponentially with E_0. Scattered He^{2+} fractions increase quadratically with E_0.		86R
50 ... 340	D	var.	Scattered ion and neutral fractions.		90R2
> 50	H	Al(111)	Neutralization.		91W1
Energy loss					
175	He	Cu(100)	Impact parameter dependence of the inelastic energy loss. MC.	49	87A3
100	H	W(111)	High energy-resolution measurement. Inelastic energy loss. Discrepancy with Oen-Robinson model for small impact parameters.		88G1
100	H	W(111)	High energy-resolution measurement. Oen-Robinson model for inelastic energy losses. MC.		90Y1
Surface melting and disordering					
98	H	Pb(110)	Observation of surface melting.	50	85F, 86F3
76	H	Pb cyl.	Crystal face dependence of surface melting. Thickness disordered layer increases logarithmically as T_m is approached.		87P
100	H	Pb(110)	Surface melting. Role of short-range and long-range atomic interactions.		89P
100	H	Al(110), (111)	Surface melting is present on the (110) face but not on the (111) face.	51	90D4

Table 4 (cont.)

E_0 [keV]	Ion	Surface	Remarks and main conclusions	Fig.	Ref.
100	H	Pb(110), (111), cyl.	Most crystal faces, except (100) and (111), become disordered near T_m.		90P2
100	H	Pb(110)	Disorder in $[1\bar{1}0]$ surface direction is more pronounced than in the [001] direction.		91D2
100	H	Al(110), (111)	Theory+MC: crystal face dependent surface melting. Effective-medium theory to calculate crystal structure.		91D3

[1]) Model for neutralization and (re-)ionization in binary ion-atom collisions, see [69V].

General MEIS studies	Fig.	Ref.
Theory: Auger and resonant charge exchange.		54H
Channeling: surface composition and disorder.		72B4
Measurement of relaxation by channeling and blocking.	7	76T2
MEIS with surface barrier detector.		78H5
Comparison surface barrier detector with electrostatic analyser.		86G

Instrumentation	Ref.
Position sensitive plastic particle detector.	76K2
MEIS instrument.	76T3
Toroidal electrostatic analyser.	82S2
High energy-resolution apparatus.	86O2, 88K1

Table 4 (cont.)

Instrumentation (cont.)	**Ref.**
Compact electrostatic analyser.	87O1
TOF spectrometer.	89M2
Reviews	**Ref. (no. of pages/ no. of references)**
Neutralization.	77B6 (25/ 51)
MEIS.	82S1 (8/ 27), 85V1 (89/ 351), 88F4 (7/ 22)
Surface structure determination.	83T2 (8/ 36) , 88G2 (8/ 23), 88Y(9/ 59), 90O1 (15/ 16)
Grazing incidence.	88P (7/ 50), 91W2 (33/ 67)

Table 5. High energy ion scattering ($E_0 > 0.2$ MeV). Surface structure data [1]).

Surface	LEED pattern	E_0 [MeV]	Ion	Technique	Δd_{12} [pm]	Δd_{23} [pm]	Δy_1 [pm]	ρ_1/ρ_b or Θ_D	Additional results *or remarks*	Fig./ Table	Ref.
Ag (110)	1x1	0.2 ... 1	He	chan; ψ, E_0	-11(4)	+6(4)	-	1.44	Δd_{13} = -5(1) pm.	52	84K1
Ag (111)	1x1	1	He	chan	0(10)	-	<10	-			81C2
Al(100)	-	0.6 ... 2	He	chan; ψ, E_0	-5(5)	-	-	1.25(5)			90C1, 90C2
Au (100)	5x20	1	He	chan	-	-	> 20	-			75Z, 76A
Au (110)	1x2	0.2 ... 1.6	He	chan; E_0	> 25 / < -25 [2])	-	-	-	$\|\Delta y_2\|$ = 12(4) pm (100 K), 18(4) pm (295 K).		84K2, 84R2
	1x2	0.2 ... 1	He	chan; ψ, E_0	+62 [3]) -25 [3])	0 [3]) +10 [3])	-	1.55(35)	$\|\Delta y_2\|$ = 12(2) pm; ρ_2/ρ_b = 1.55(35).	53	85W2
Cu (110)	1x1	0.3	He	chan; ψ	-7(2)	+4(2)	-	1.27	*Direct comparison with LEED.*	54	82A1, 83S3
Cu (16,1,1)	1x1-steps	0.2 ... 1	He	sh+bl; ϕ, θ, E_0, T	-5(5)	-	-	-	No additional relaxation of edge atoms; 10 pm disorder in 1st layer at 300 K.		84B4
GaAs (100)	c(4x4)	0.7 ... 3	He	chan; ψ, E_0	-	-	52(17) [4])	-	Evidence for subsurface strain.		85N1

Table 5 (cont.)

Surface	LEED pattern	E_0 [MeV]	Ion	Technique	Δd_{12} [pm]	Δd_{23} [pm]	Δy_1 [pm]	ρ_1/ρ_b or Θ_D	Additional results *or remarks*	Fig./ Table	Ref.
GaAs (110)	1x1	1, 2	He	chan; ψ, E_0	-	-	< 10	-	Support for bond-length relaxation model [5]).		83G1, 84G2
Ge (100)	c(2x4)	0.15 ... 1	He	chan; E_0, T	-	-	> 28 [6])	-	*At 100 K.* $\lvert\Delta y_3\rvert$ > 12 pm. Evidence for subsurface strain. Support for dimer model. *Compared with Ge(100)1x2 at 300 K.*		86C4
Ge (111)	c(2x8)	0.15 ... 1	He	chan; E_0, T	-	-	20 [6])	-	*At 130 K.* Evidence for subsurface strain. Support for stacking-fault model [7]).		86C4
Mo (100)	$c(7\sqrt{2} \times \sqrt{2})$	0.65	He	chan	-11(3)	-	12(2)	-	*At 80 K.*		91H5
Ni(100)	1x1	0.75	He	sh+bl; ψ	0(7)	-	-	1.35(25)	ρ_2/ρ_b = 1.20(25).		87A2
Ni(110)	1x1	0.3	He	chan; ψ	-6(2) [8])	+3.0(15)	-	1.15			83F1
Ni(111)	1x1	0.5 ... 2	He	chan; ψ, E_0	0(2)	-	-	1.20			81N1, 82N1
Pd (111)	1x1	1.8	He	chan; ψ	0(5)	-	-	-			83K2

Table 5 (cont.)

Surface	LEED pattern	E_0 [keV]	Ion	Technique	Δd_{12} [pm]	Δd_{23} [pm]	Δy_1 [pm]	ρ_1 / ρ_b or Θ_D	Additional results *or remarks*	Fig./ Table	Ref.
Pt(100)	1x1	1, 2	He	chan; ψ	+1(1)	-	-	-	*0.1 monolayer of H at surface; at 175 K and 295 K.*		78D2, 79N2, 81D
Pt(100)	5x20	1, 2	He	chan; T	-	-	> 10	-			78D2, 79N2, 81N2
Pt(110)	1x2	0.5 ... 2	He	chan; ψ	-	-	-	-	Lateral reordering in 1st and 2nd layer.		78D2
	1x2	2	He	chan; ψ, T	0(1) 9)	-	< 2	-	Support for missing row model with enhanced 1st layer vibrations or 1st layer buckling model (< 7 pm peak-to-peak).		82J1
Pt(111)	-	1, 2	He	chan; ψ, T	+3(1)	-	-	-	*At 150 ... 600 K. 0.1 monolayer of H at surface.*	8	78D1
	1x1	0.4	He	chan; ψ	+2(5)	-	-	-			78B3
	-	2	He	chan; ψ	+1(1)	-	-	-	*At 185 K.*		80D
Si(100)	1x2	0.1 ... 2	He	chan; E_0	-	-	> 20 6)	-	*At 300 ... 850 K.;* $\Delta y_3 > 15$ pm. Indication of subsurface strain.	55	80F1, 80F2, 81S
	1x2	1	He	transm ch, chan	-		> 96 6), 10)	-	Support for buckled dimer model. Indication of a rotation of the dimers in the surface plane.		86J

Table 5 (cont.)

Surface	LEED pattern	E_0 [MeV]	Ion	Technique	Δd_{12} [pm]	Δd_{23} [pm]	Δy_1 [pm]	ρ_1/ρ_b or Θ_D	Additional results *or remarks*	Fig./ Table	Ref.
Si(111)	1x2	1	He	chan; ψ	-	-	> 10 [6])	300 K	*Cleaved crystal.*		84I
	7x7	0.15 ... 1	He	chan; ψ, E_0	-40(10) [11])	-40(10) [11])	< 15 [6])	-	Support for buckled dimer model.	56	80C, 81C1
	7x7	0.1, 0.2	H, He	chan; E_0	-	-	-	-	*Data from the literature.* Support for stacking-fault models.		83B3
	7x7	0.1 ... 1	He	chan; ψ, E_0	-	-	-	-	*Data from the literature*. Support for DAS model.	56	90Y2
SnTe (100)	1x1	0.7	H, He	chan; ψ, E_0	-	-	-	2.15(15)	*Grazing incidence.*		87K1
	1x1	0.5 ... 2	He	chan; ψ, E_0	0(15)	-	-	1.3(1)			88S5
UO_2 (111)	-	3.05	He	chan; ψ	+19(5) [13])	-	-	1.20(15)			84T3
W(100)	-	0.4	H	chan; ψ	-	-	-	-	Indication of surface disorder or enhanced surface vibrations.		67B
	c(2x2)?	0.6 ... 5	He	chan; E_0	0(9)	-	-	1.5	*< 0.1 monolayer of C and O.*		77F, 78Z
	c(2x2)	0.1 ... 2	He	chan; E_0	-	-	23 [12])	-			79F, 79S2,
	c(2x2) / 1x1	0.5 ... 2	He	chan; E_0, T	-	-	14	-	*At 80 ... 600 K.* Order-disorder transition around 300 K		89S4

Table 5 (cont.)

Surface	LEED pattern	E_0 [MeV]	Ion	Technique	Δd_{12} [pm]	Δd_{23} [pm]	Δy_1 [pm]	ρ_1/ρ_b or Θ_D	Additional results *or remarks*	Fig./ Table	Ref.
W(110)	-	0.6 ... 1.8	He	chan; ψ, E_0	0(4)	-	-	2.6 [14]) 1.0 [14])			87S2

[1]) For explanation of column captions, see Table 1.
[2]) No information about the sign has been obtained.
[3]) Both results are consistent with the experimental data.
[4]) Also for 2nd and 3rd layer.
[5]) In disagreement with a MEIS study [85S2].
[6]) Also for 2nd layer.
[7]) In disagreement with a MEIS study [88M3].
[8]) In disagreement with MEIS studies [86G,86Y1,89X].
[9]) In disagreement with a LEIS [91M2] and a MEIS study [88F3].
[10]) For 50% of the atoms.
[11]) Vertical displacement with respect to the bulklike position.
[12]) For 50 ... 67% of the atoms.
[13]) For U atoms only.
[14]) Component perpendicular and component parallel to the surface, respectively.

Table 6. High Energy Ion Scattering ($E_0 > 0.2$ MeV). Related studies, ordered chronologically within each section.

All primary ion beams are of charge state +1 unless otherwise stated.
E_0 = primary beam energy.
MC = Monte Carlo simulation.
var. = various.
vs. = versus or as 'a function of '.
θ, ψ, ϕ and η: see Fig. 10.

E_0 [MeV]	Ion	Surface	Remarks and main conclusions	Fig.	Ref.
Scattering phenomena					
< 0.5	H	Cu(100)	MC: reduced reflection for exit directions parallel to close packed atomic rows (blocking).		68M2
-	-	-	MC: spatial distributions of scattered ions; chain scattering model.		74E
< 1.2	H, He, B	var.	Surface peak area increases with E_0.		75D2
-	-	-	Theory: analytical expression for shadow cone radii, surface relaxation, minimum channeling yield and surface peak area.		76J
2	He	W(100)	MC: surface peak area vs. E_0, compared with two-atom model.		77F
-	-	-	Model to include correlated vibrations in MC.		77J
-	-	-	MC: principles.		78B1
-	-	-	MC: general concepts of surface peak area.	57, 58	78S2
-	-	-	Theory: analytical model for focusing of grazing incident ions. Possible applications for surface structure analysis.		78T1
-	-	-	MC: effect of correlated vibrations on surface peak area and half-angle for channeling.		80B1
1.5	N	Ge(100), (110)	MC: surface semichanneling. Possible applications for surface structure analysis.		80M4
-	-	-	Analytical two-atom model for blocking.		80O3

Table 6 (cont.)

E_0 [MeV]	Ion	Surface	Remarks and main conclusions	Fig.	Ref.
0.1 ... 2	He	Si(100)	MC: effects of thermal vibrations, displacements and relaxations on surface peak area.		81K2
-	-	-	Theory: analytical model for surface peak area. Role of thermal vibrations.		82J2
-	-	-	Theory: analytical model for surface peak area.		82M1, 82M2
1	He	Si(100)	MC: grazing incidence; simulation of energy spectra.		83T1
0.5 ... 2	He	Pt, Au, Mo, Si	Theory: analytical model for the effect of correlated vibrations on surface peak area.		84J1
0.3	He	Au(110)	MC: ambiguity in relaxation determination: reversal of asymmetry in angular scan of surface peak area for large contractions or expansions.	59	85C2
1	He	Si(100), (111)	MC: flux vs. ψ and ϕ. Validity of method for background subtraction.		85I
0.5 ... 1	He	Au(110)	180° backscattering enhancement. Experimental data compared with two-atom model.		86M4
0.1 ... 2	He	Si, Al	MC: surface peak area. General trends.		90S1
Grazing incidence					
0.3 ... 2	He	W(110)	Total specular reflection for $\psi < 1°$.		70F
10	H	W(100)	MC: total reflection.		84V2
0.7 ... 2	H, H_2, He	SnTe(100)	Surface channeling; charge state; large inelastic energy losses.		86M1
0.7	H, He	SnTe(100)	Spatial distribution scattered ions compared with theory. Role of thermal vibrations.	60	87K1
0.7	H, He	SnTe(100)	Specular reflection. Inelastic energy loss by surface and bulk plasmons and in single collisions. Fair agreement with theory [1]).		87K2, 87K3
0.7	He	SnTe(100)	MC: surface hyperchanneling. Calculation of energy spectra.		88T2
0.7	He	SnTe(100)	Surface channeling. Penetration inside the crystal at steps. Possibility for measurement of step densities. MC.		88F5, 88M2

Table 6 (cont.)

E_0 [MeV]	Ion	Surface	Remarks and main conclusions	Fig.	Ref.
0.7	He	SnTe(100)	Charge state of specularly reflected ions. Charge transfer from tail of valence-electron distribution of the solid.		88K5
0.7	He	SnTe(100)	Charge state of specularly reflected ions. Penetration and subsequent subsurface channeling occur at steps.		88K6
0.8 ... 2.4	H_2, HeH	SnTe(100)	Angular and energy distribution of dissociation fragments. Alignment of the interatomic axis between the fragments with the surface plane.		90S8
0.8	H_2	SnTe(100)	Angular and energy distribution of dissociation fragments. Alignment of the interatomic axis between the fragments with the surface plane.		90S9
0.3 ... 1	H	W(100)	Spatial and energy distributions of specularly reflected ions [2]).		90W3
0.7 ... 1.5	He	SnTe(100)	Charge state of specularly reflected ion vs. ψ. Position dependent charge exchange probabilities. Influence of steps.		91F
< 1.25	H	Al(111)	Neutral fraction of specularly reflected H decreases rapidly with increasing E_0. Disagreement with theoretical models.		91W1
Thermal vibrations					
-	-	-	MC: effect of correlated vibrations on blocking.		71R
2	He	Mo	MC: effect of correlated vibrations on surface peak area.		79J
0.6 ... 2	He	Au(110), Pt(110)	Evidence for correlated vibrations in Pt.	61	83J
0.6 ... 1.8	He	W(110)	Observation of anisotropic vibrations.		87S2

[1]) Results for relation between ψ and energy loss in disagreement with [90W3].
[2]) Results for relation between ψ and energy loss in disagreement with [87K2,87K3].

Table 6 (cont.)

General HEIS studies	Ref.
Channeling.	65B, 77A
Theory: analytical model for blocking.	65O
Channeling and blocking at 80 K and 300 K.	65T
Monte Carlo simulation of channeling.	71B
Surface structure analysis by channeling.	72M7
Surface structure and composition analysis by channeling.	73B1
Surface compostion analysis by channeling.	73M5
Surface relaxation analysis by channeling.	75D1, 76D
Surface analysis by channeling.	78B2, 78B3
Instrumentation	**Ref.**
UHV-system for HEIS.	87S1
Magnetic spectrograph.	90B2
Reviews	**Ref. (no. of pages/ no. of references)**
Surface structure determination.	79N3(1/ 10), 81F1(9/ 29), 81F2 (25/ 29), 82U(26/ 48), 83F2(5/ 10)
HEIS vs. electron beam surfaces probes.	82F1(20/ 43)
Materials analysis by ISS.	90B1(14/ 59)

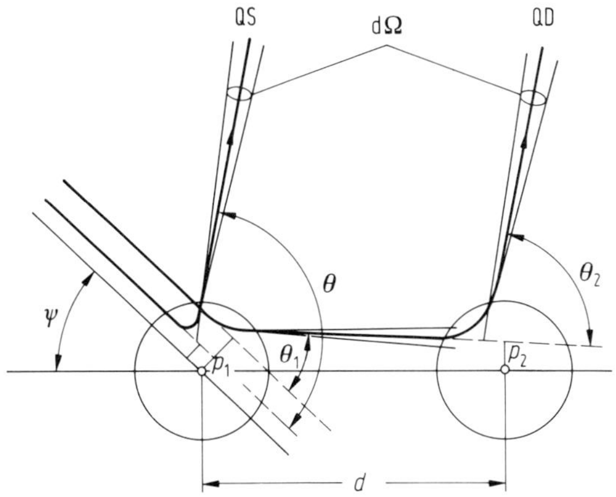

Fig. 1. Quasi-single (QS) and quasi-double (QD) scattering. In both cases, the total reflection angle, θ, is the same. The energy after QS scattering is lower than after QD scattering. All possible reflection angles, θ, can be evaluated by varying the impact parameter, p_1, in the first collision from zero to $d \sin\psi$ [85M, p. 127].

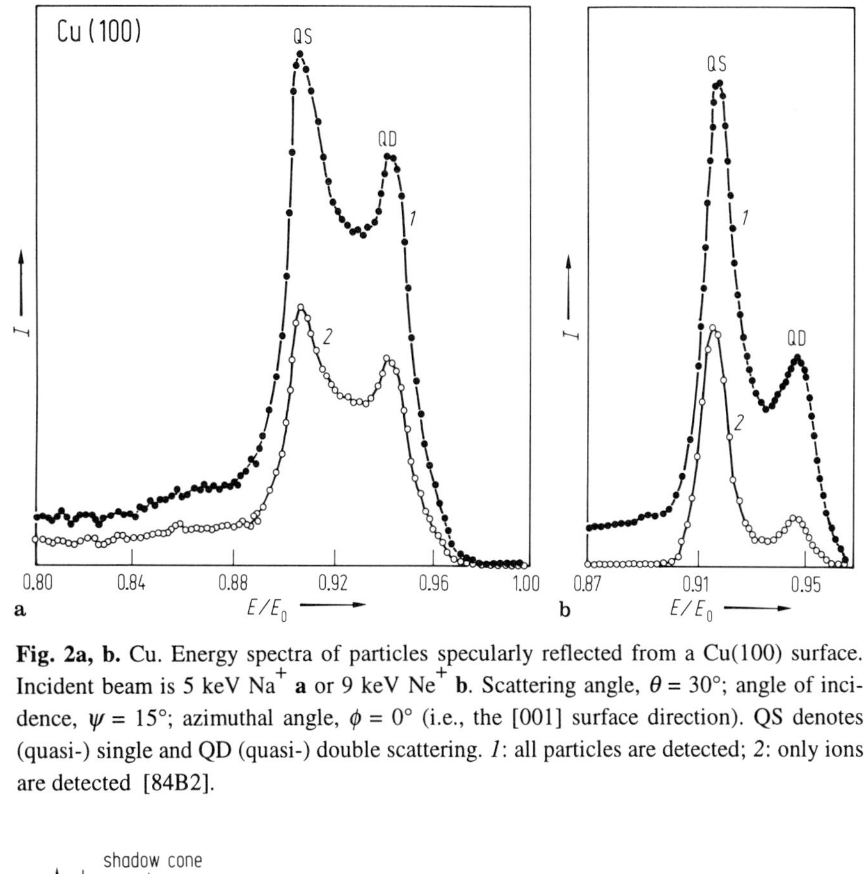

Fig. 2a, b. Cu. Energy spectra of particles specularly reflected from a Cu(100) surface. Incident beam is 5 keV Na^+ **a** or 9 keV Ne^+ **b**. Scattering angle, $\theta = 30°$; angle of incidence, $\psi = 15°$; azimuthal angle, $\phi = 0°$ (i.e., the [001] surface direction). QS denotes (quasi-) single and QD (quasi-) double scattering. *1*: all particles are detected; *2*: only ions are detected [84B2].

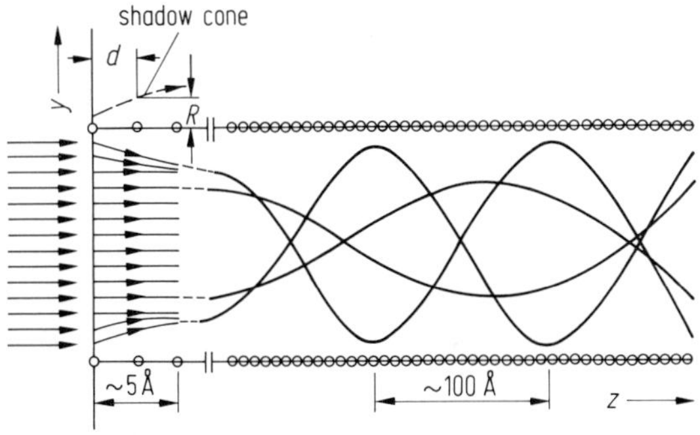

Fig. 3. Ion shadowing. Shadow cones are formed behind surface atoms when a high energy ion beam is incident parallel to the atomic rows. R is the shadow cone radius at the second atom [86F2].

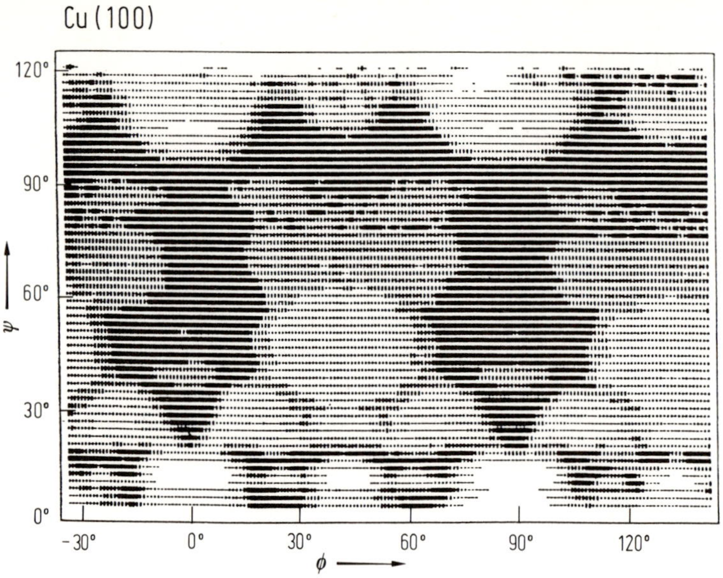

Fig. 4. Cu. Shadowing and blocking. Ion intensities of 6 keV Ne^+ ions reflected from Cu(100), are shown in five grey shades: white indicates low and black high intensity. Scattering angle, $\theta = 120°$. Angle of incidence, ψ, and azimuthal angle, ϕ, are varied; $\phi = 0°$ corresponds to the [011] surface direction. The pattern reflects the structure of the surface: light regions represent atoms [84V1].

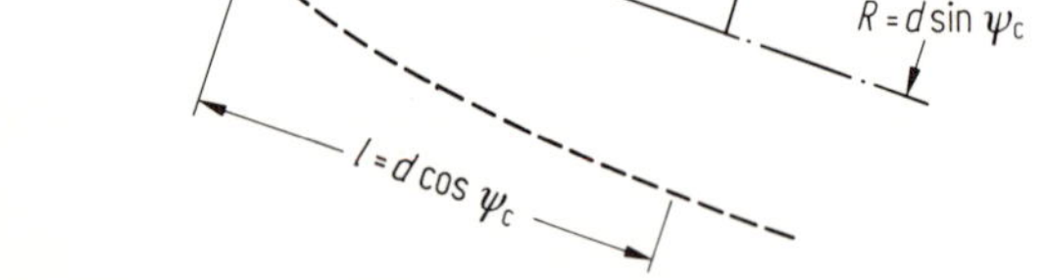

a

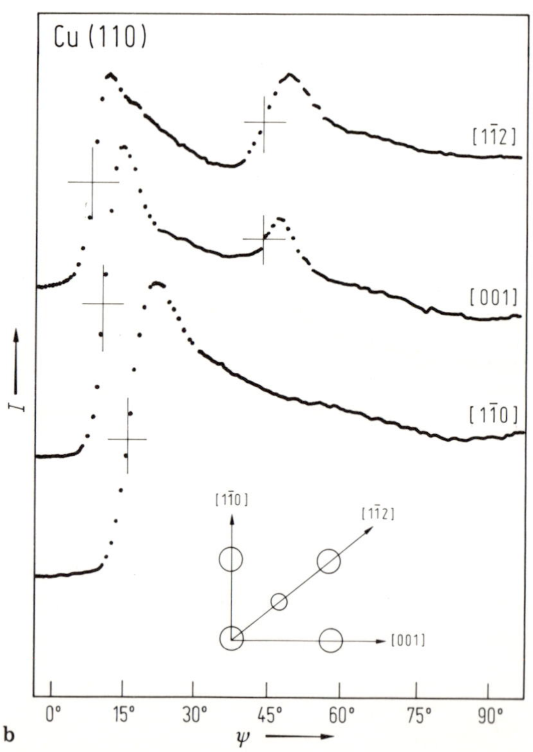

b

Fig. 5a, b. Impact collision ion scattering spectrometry. **a** At the critical angle, ψ_c, the right atom is located on the edge of the shadow cone of the left atom. Simple geometric formulae relate the interatomic distance, d, and the critical angle, ψ_c, to the shadow cone radius, R, at distance l. **b** Example of an ICISS measurement: intensity of 5 keV Ne^+ ions scattered from Cu(110) as a function of ψ for three azimuths. Crosses indicate the critical angles [81A1, 88F1].

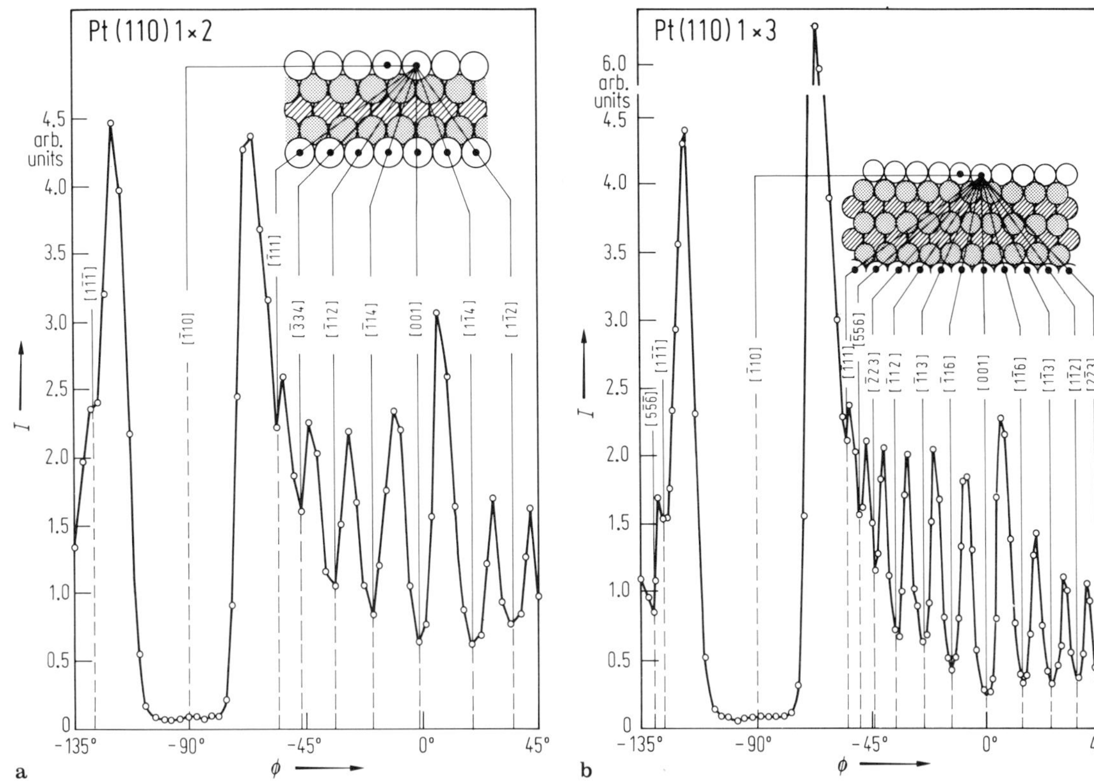

Fig. 6a, b. Pt. Surface periodicity. Scattered particle intensity is measured as a function of azimuthal angle, ϕ, for Pt(110)1x2 **a** and Pt(110)1x3 **b**. The oscillations are caused by surface shadowing. Incident beam is 2 keV Ne^+; scattering angle, θ = 28°; angle of incidence, ψ = 6° **a** and ψ = 4° **b** [91M3].

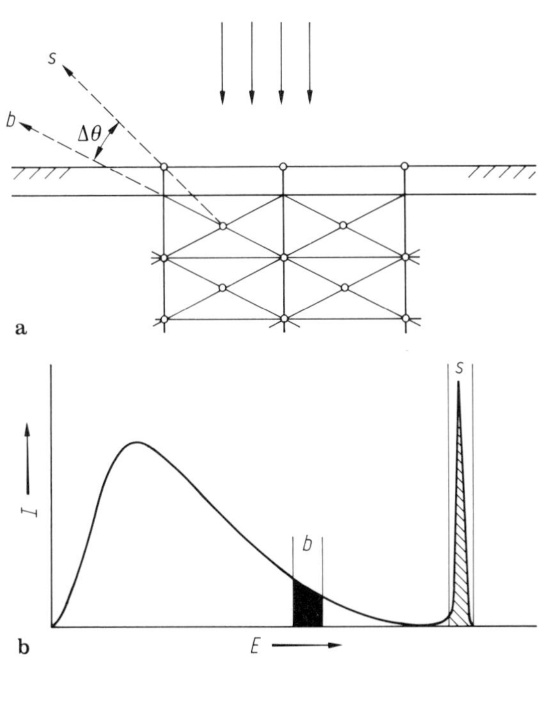

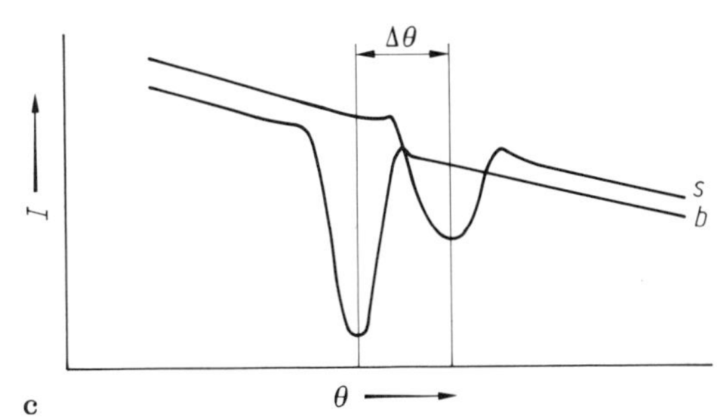

Fig. 7a-c. Surface relaxation determination by shadowing and blocking. **a** The scattering plane. A vertical displacement in the first layer results into different angular positions of the blocking cone for surface (s) and bulk (b) atoms. **b** Idealized energy spectrum of the backscattered ions; the surface peak is located at s. **c** Angular scans of the backscattered intensities for the surface (s) and the bulk (b) [76T2].

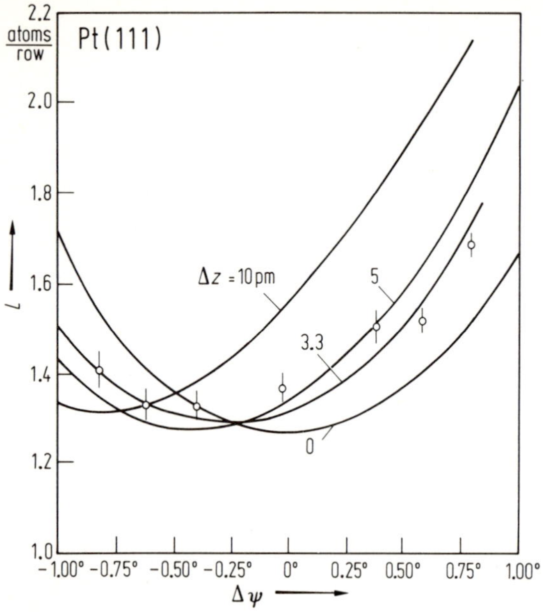

Fig. 8. Pt. Surface relaxation determination by single alignment HEIS. Surface peak area, L, is measured as a function of the direction of incidence, $\Delta\psi$, of 2.0 MeV He^+ ions on Pt(111) at 78 K. Solid curves are calculation results for various values of the first layer relaxation Δz [78D1].

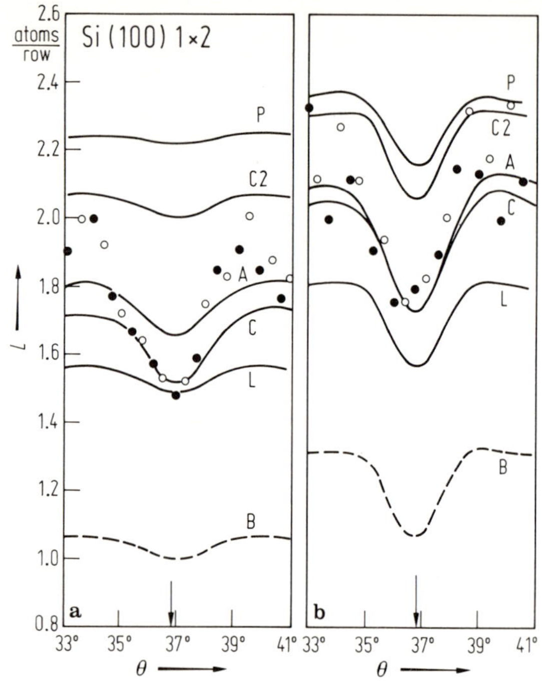

Fig. 9a, b. Si. Surface reconstruction study by shadowing and blocking. Surface blocking profiles for Si(100)1x2 are measured with **a** 50 and **b** 100 keV H^+ ions. Direction of incidence is [$\bar{1}$03], scattering plane is (010) and blocking direction is [103]. The arrows show the bulk crystallographic direction. Curves are computer simulated results: (B) bulklike terminated surface; (L) dimers, no multilayer distortions; (C) buckled dimers with multilayer distortions; (A) dimers with multilayer distortions; (C2) double conjugated chain model; (P) missing row model with multilayer distortions [83T3].

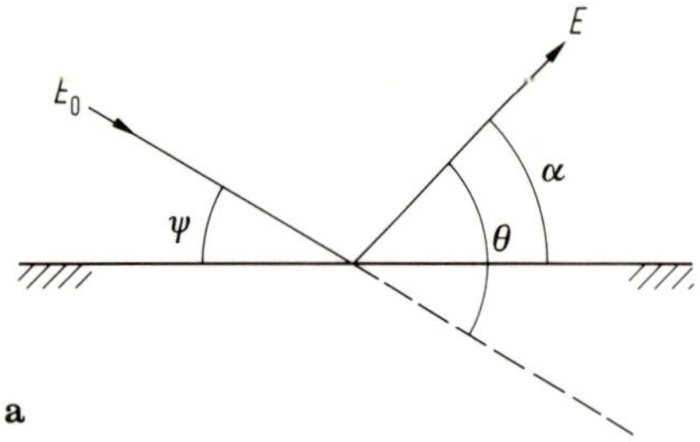

a

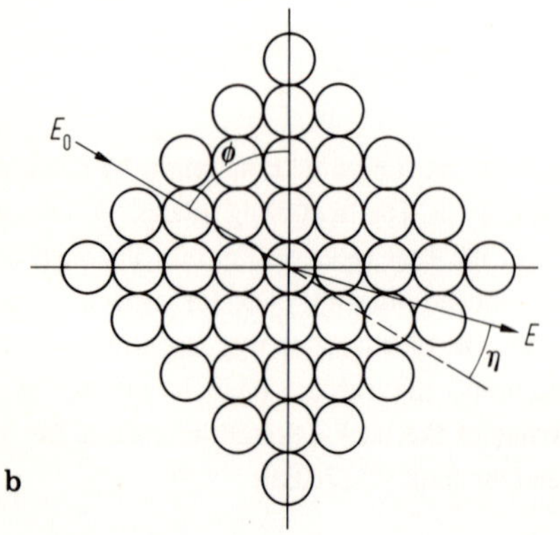

b

Fig. 10a, b. Definition of the angles and energies for ISS. **a** Side view; **b** top view. E_0: primary beam energy; E: detected energy; ψ: angle of incidence; θ: scattering angle; α: exit angle; ϕ: azimuthal angle with respect to a major surface direction; η: azimuthal exit angle. ($\eta = 0°$ in most experiments; only in the out-of-plane detection geometry, $\eta \neq 0°$).

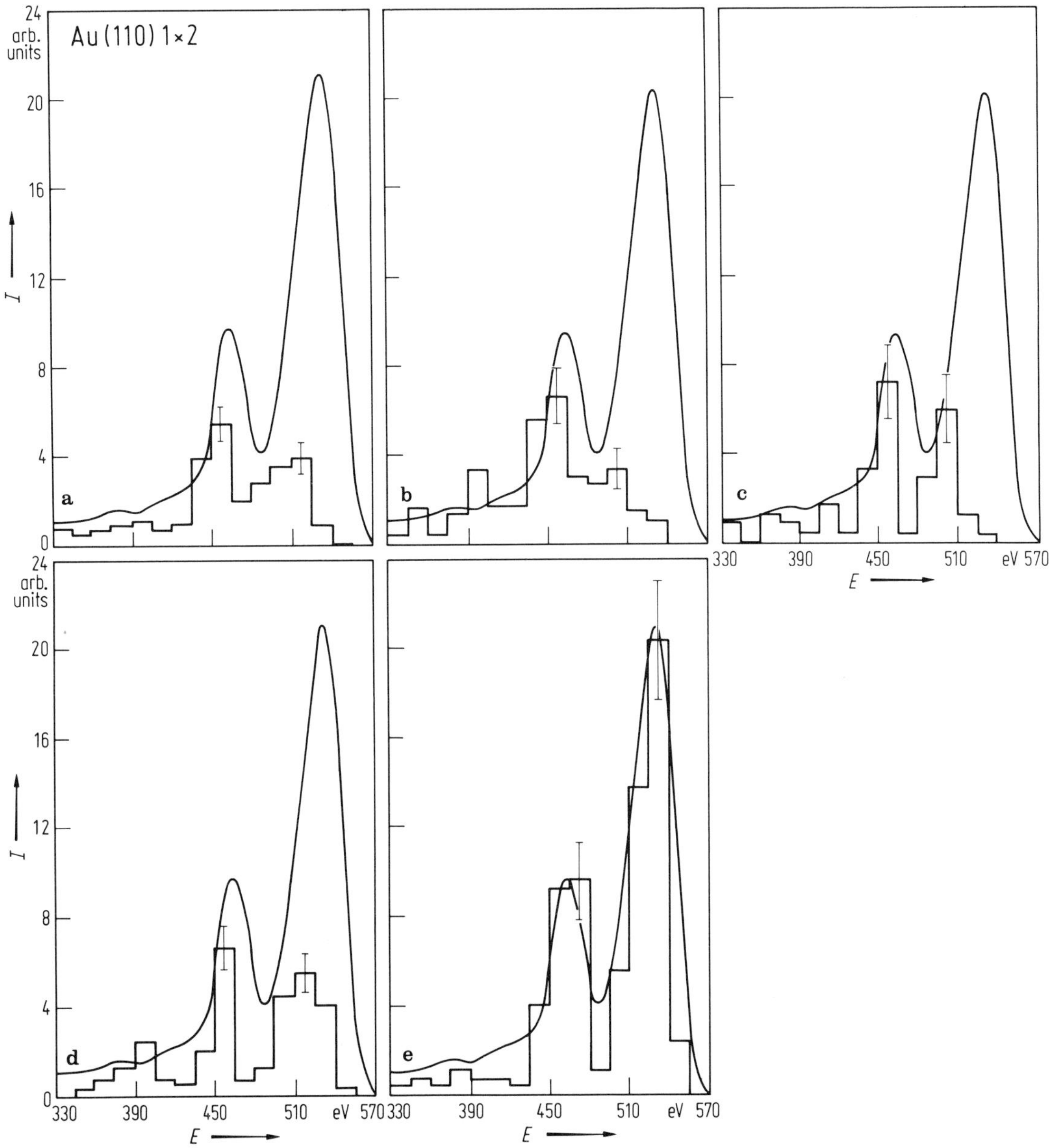

Fig. 11a-e. Au. Reconstruction of Au(110)1x2. Energy spectra of scattered 600 eV K^+ ions ($\psi = 30°$, $\theta = 70°$, $\phi = 27.5°$ with respect to $[1\bar{1}0]$). The full curve is the experimental result [81O2]. Histograms are Monte Carlo results for various surface structure models: **a** saw tooth model; **b** model by Robinson et al. (Δd_{12} = 62 pm); **c** model by Kuk et al. (Δd_{12} = 25 pm); **d** unrelaxed missing row model (Δd_{12} = 0 pm); **e** model by Moritz and Wolf (Δd_{12} = 25 pm). Only the last model is consistent with the experimental data [85H3].

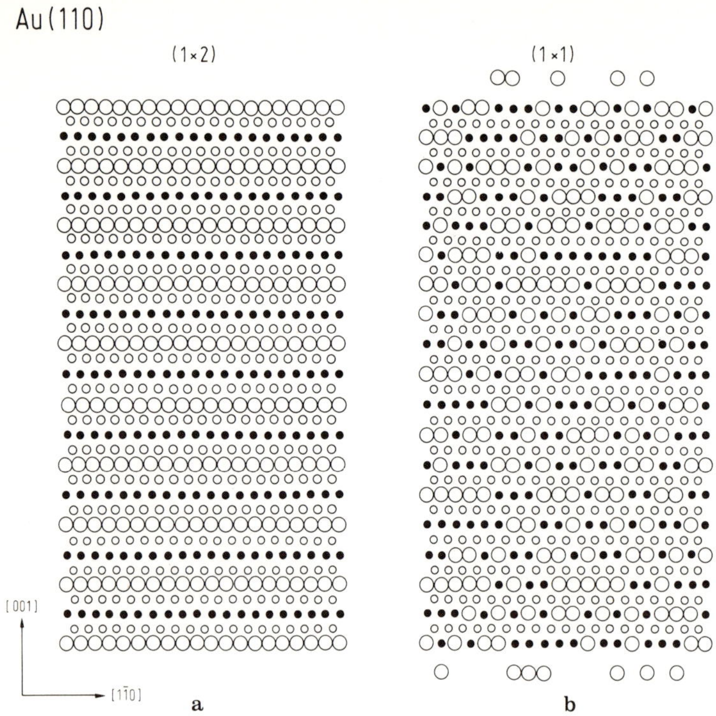

Fig. 12a, b. Au. Reconstruction of Au(110)1x2. **a** Top view of the missing row reconstruction. Large open circles: first layer atoms; small open circles: second layer atoms; filled circles: third layer atoms. **b** Top view of the surface above 650 K. The disordered surface is produced by random jumps of top layer atoms by one lattice unit in the [001] or [00$\bar{1}$] direction [87D2].

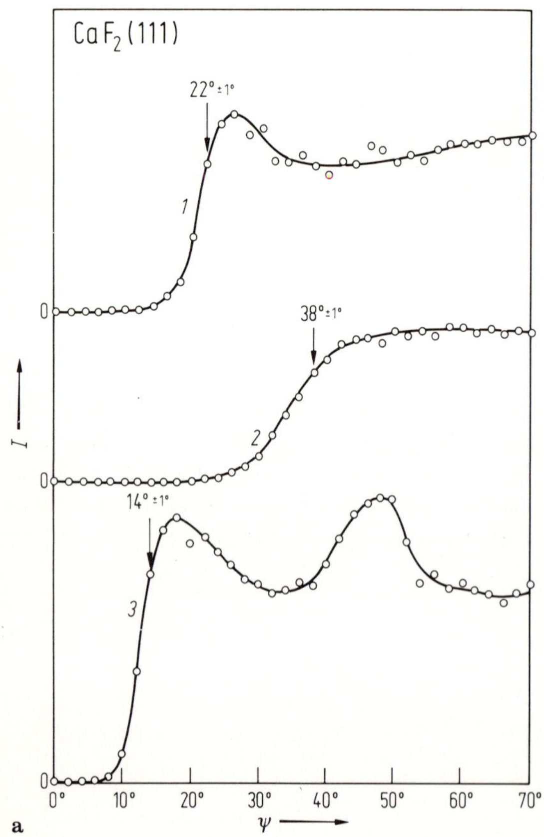

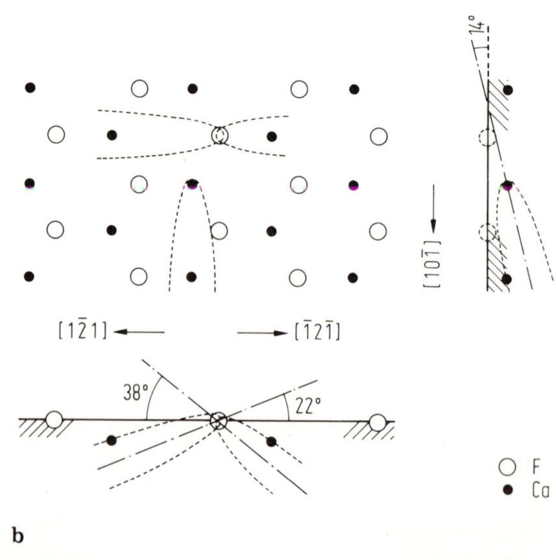

Fig. 13a, b. CaF_2. Surface composition and relaxation of CaF_2(111). **a** ICISS measurement: 1 keV He^+ ions scattered from Ca atoms as a function of ψ. *1*: [1$\bar{2}$1] azimuth; *2*: [$\bar{1}$2$\bar{1}$] azimuth; *3*: [10$\bar{1}$] azimuth. **b** Scattering geometries: top view and side views. The measurements show that the surface is F-terminated and that the relaxation is negligibly small [88S3].

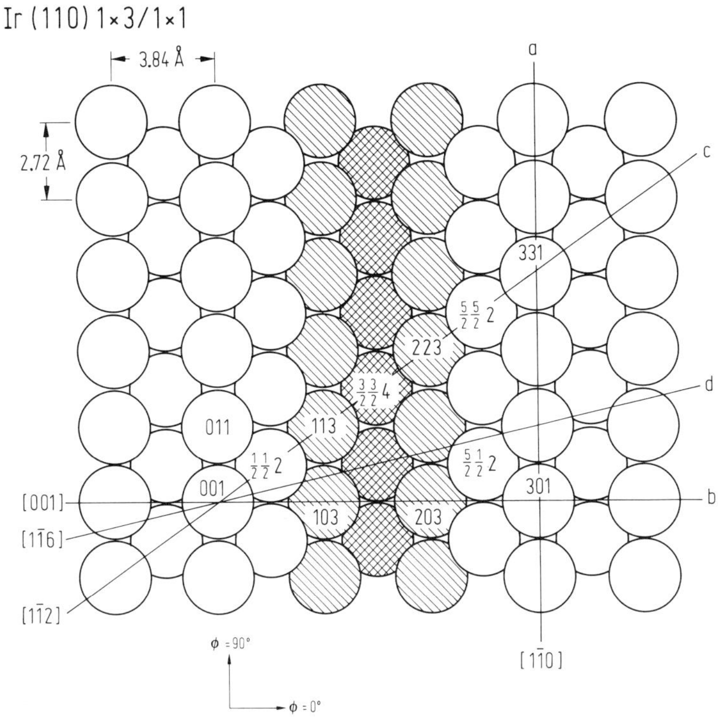

Fig. 14. Ir. Reconstruction of mixed Ir(110)1x3/1x1. Top view of the mixed surface. One atomic row in the third layer and two atomic rows in the second layer are missing. The lines a-d indicate the scattering geometries for the various ICISS measurements of Fig. 15a-d [90S2].

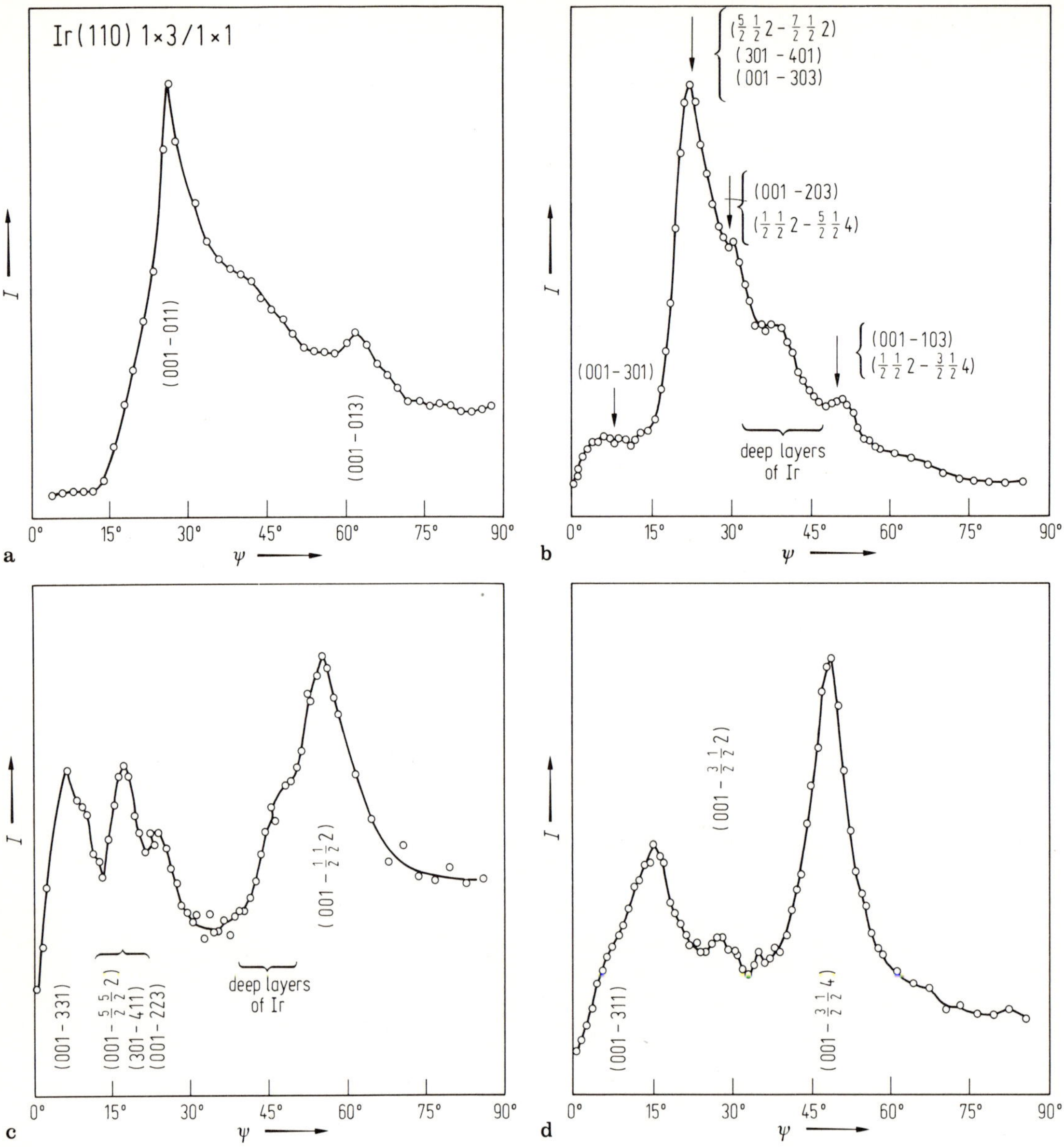

Fig. 15a-d. Ir. Reconstruction of mixed Ir(110)1x3/1x1. ICISS measurements with 4 keV Ar^+; the scattering geometries are shown in Fig. 14. **a** $[1\bar{1}0]$ azimuth; **b** [001] azimuth; **c** $[1\bar{1}2]$ azimuth; **d** $[1\bar{1}6]$ azimuth. The positions for the various critical angles, ψ_c, are labeled with the identities of the two atoms involved in the shadowing and scattering process [90S2].

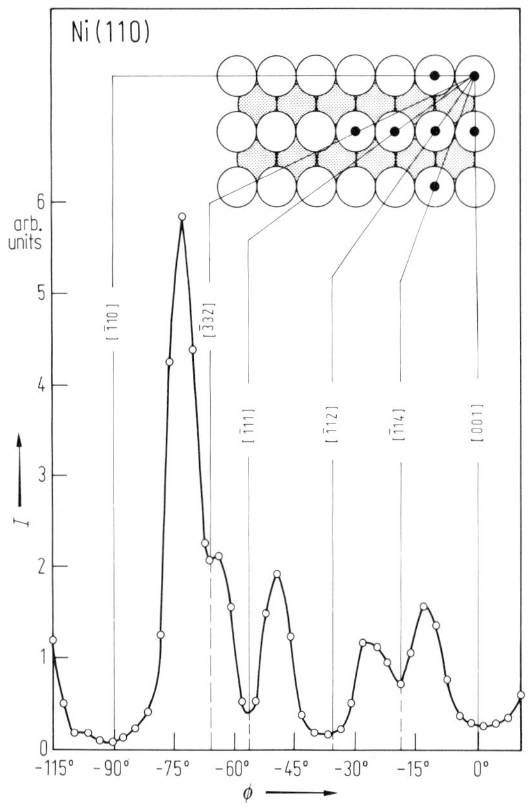

Fig. 16. Ni. Periodicity of Ni(110). Intensity of scattered 4 keV Ar^+ ions as a function of azimuth, ϕ (ψ =7°, θ = 30°). The pattern of the dips and peaks reflects the 1x1 periodicity of the surface [91M1].

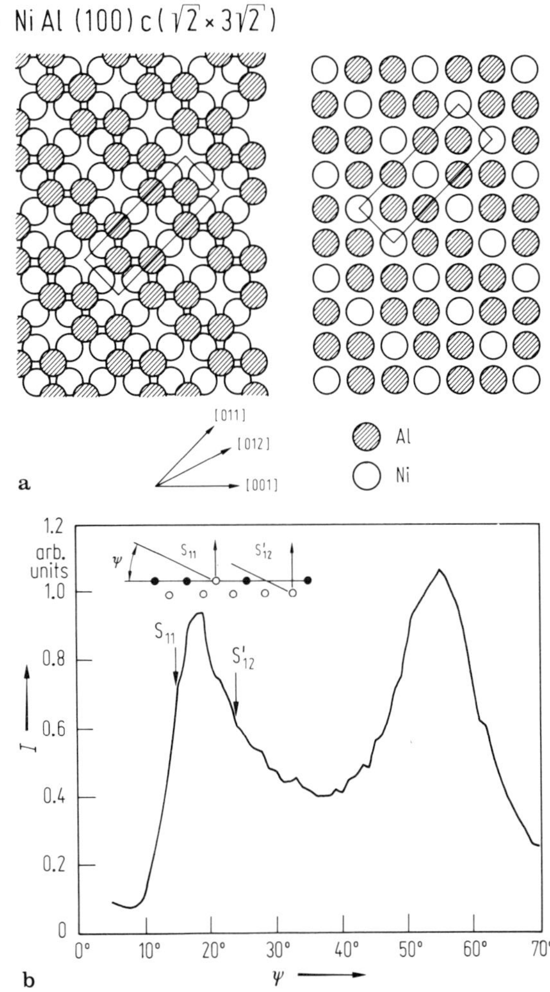

Fig. 17a, b. Reconstruction of NiAl(100)c($\sqrt{2}$x3$\sqrt{2}$). **a** Two possible models: left: the top layer consists of 2/3 Al and 1/3 vacancies; right: the top layer consists of 2/3 Al and 1/3 Ni. Both models have an entirely Ni second layer. **b** ICISS measurement with 0.5 keV Li^+. The inset demonstrates the mechanisms whereby first layer atoms shadow other first layer atoms (S_{11}) and first layer atoms shadow second layer atoms at the edges of vacancies (S'_{12}). The absence of the S'_{12} feature in the scans proves that there are few vacancies in the first layer, thus, confirming the right model in **a** [88M5].

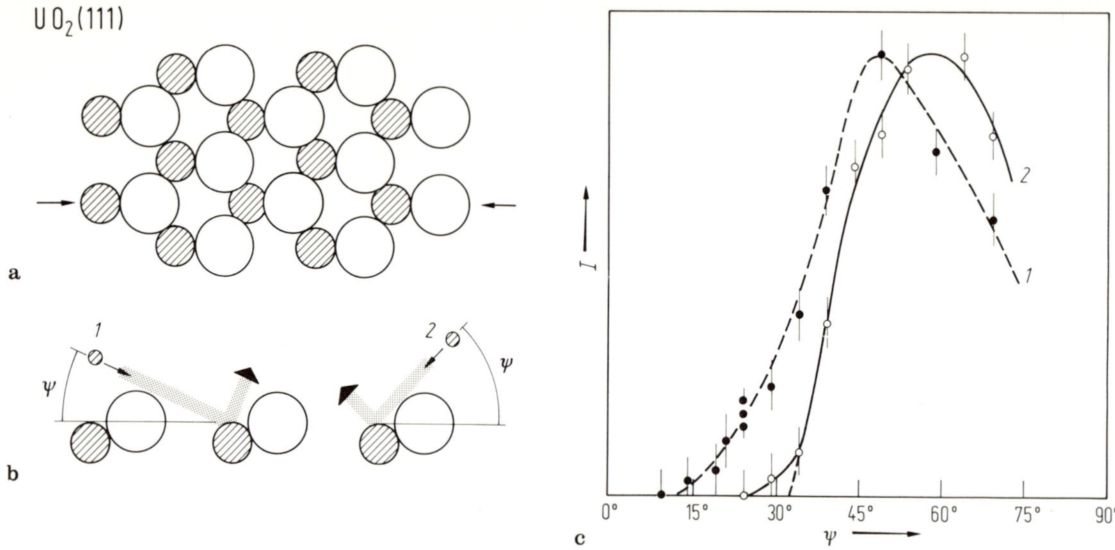

Fig. 18a-c. UO_2. Structure of UO_2 (111). **a** Top view of the surface; **b** two scattering geometries. **c** ICISS measurement with 0.5 keV He^+ backscattered from U atoms ($\theta > 90°$). The cut-off angles near 13° and 33° prove that the surface is bulklike and O-terminated [80E].

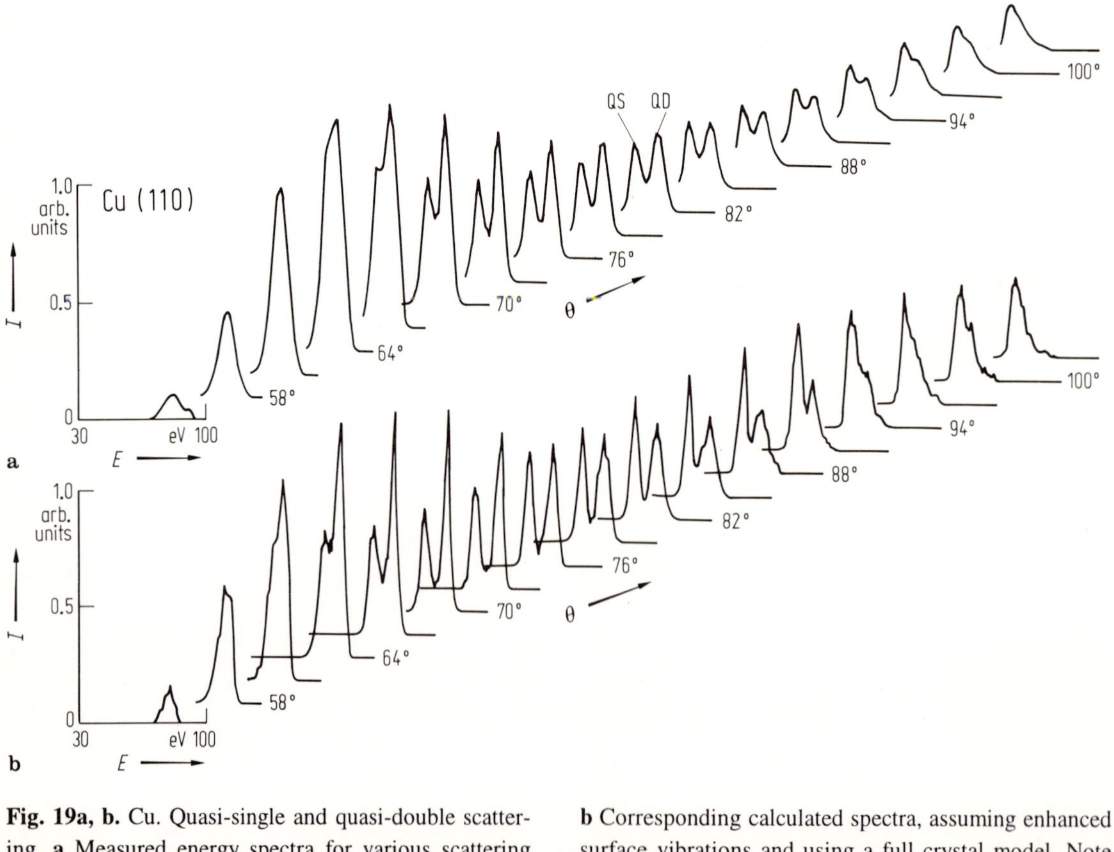

Fig. 19a, b. Cu. Quasi-single and quasi-double scattering. **a** Measured energy spectra for various scattering angles, θ. Incident beam is 100 eV Na^+, surface is Cu(110) ($\psi = 35°$, $\phi = 0°$ with respect to $[1\bar{1}0]$). **b** Corresponding calculated spectra, assuming enhanced surface vibrations and using a full crystal model. Note the development of the quasi-single (QS) and quasi-double (QD) scattering peaks [91A].

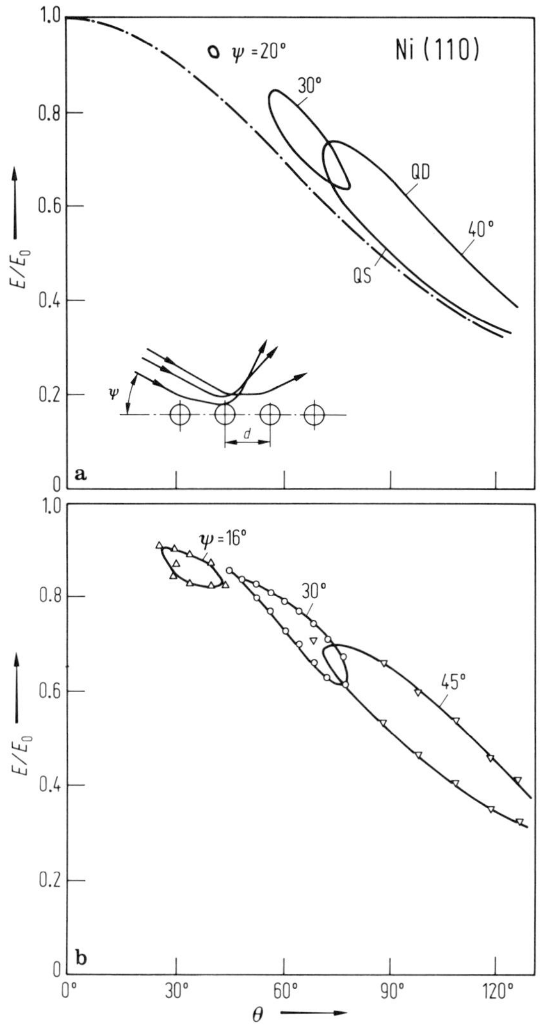

Fig. 20a, b. Ni. Quasi-single and quasi-double scattering of 600 eV He^{+} from the $[1\bar{1}0]$ chains of Ni(110). **a** Model calculations for the relation between total scattering angle, θ, and detected relative energy (E/E_0) for various angles of incidence, ψ. The loops are for the chain model; the dot-dashed curve for the binary collision model. The Born-Mayer potential has been used. QS and QD correspond to the quasi-single and quasi-double scattering, respectively. **b** Experimental results [72T].

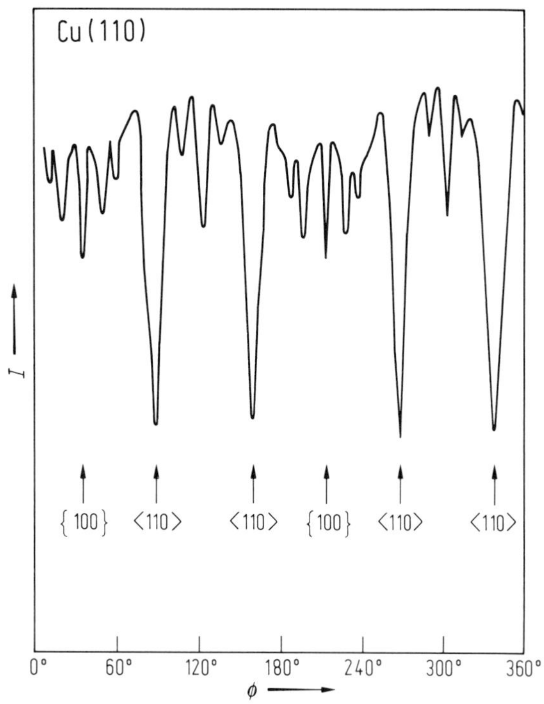

Fig. 21. Cu. Shadowing. Intensity of reflected ions as a function of azimuth, ϕ (50 keV Ne^{+} on Cu(110), ψ = 30°). For incidences parallel to major crystallographic planes or directions, the reflected intensity is reduced by the shadowing effect [63N].

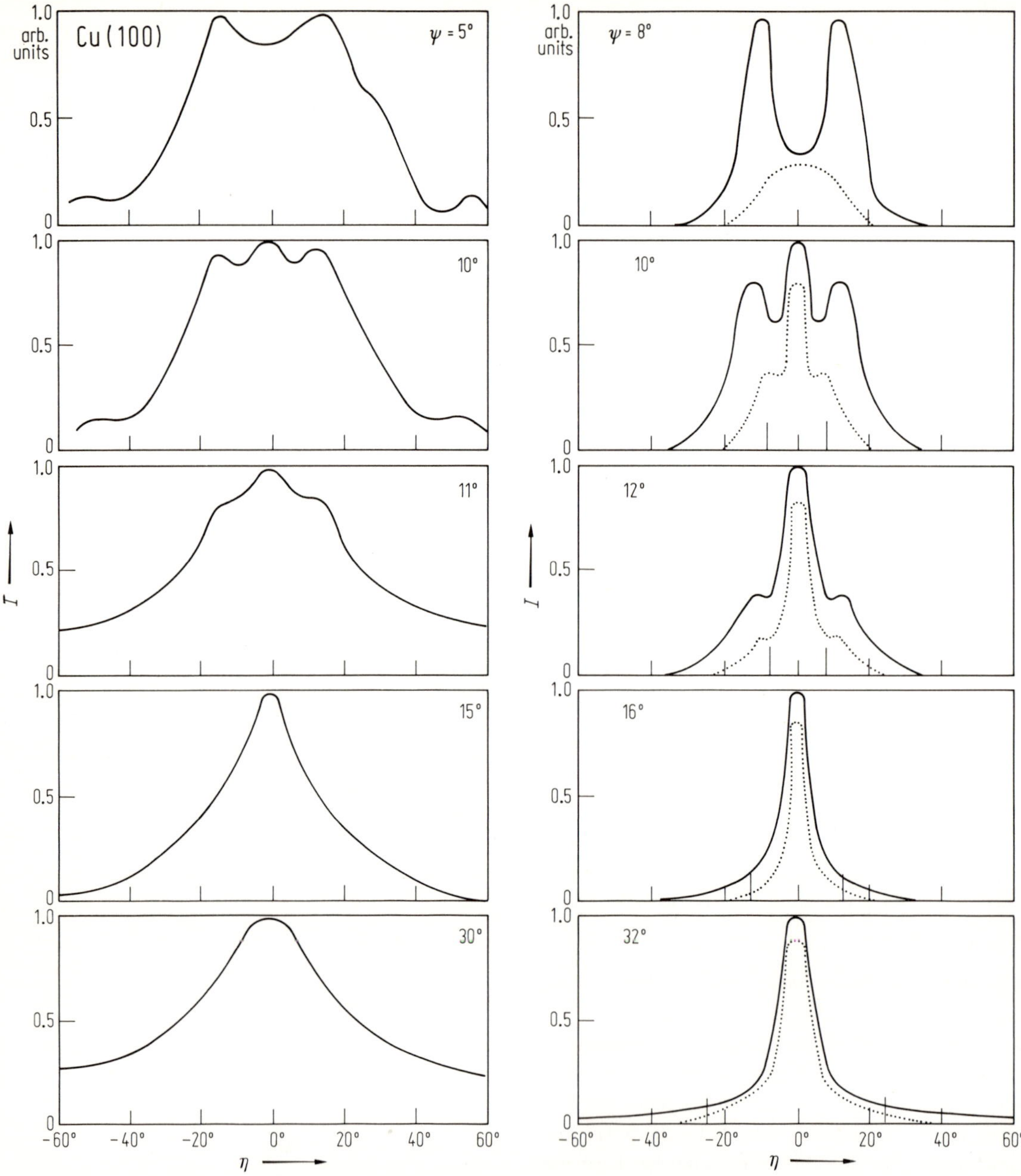

Fig. 22. Cu. Focusing by surface semichannels. Measured (left column) and calculated (right column, Born-Mayer potential, two layer model) spatial distributions of 4.5 keV Ar^+ reflected from Cu(100) ($\phi = 0°$, with respect to [011]). Dotted curves: contribution of reflection from the second layer. At low values of ψ, there is a minimum at $\eta = 0°$ due to blocking by the chains. As ψ increases, ions reflected in the second layer are focused by the first layer chains towards azimuthal exit angles η close to 0° [71S1].

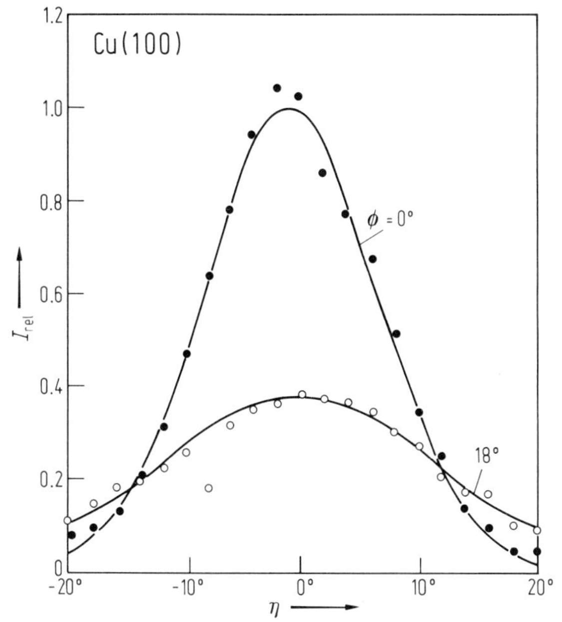

Fig. 23. Cu. Focusing by surface semichannels. Spatial distribution of 15 keV Ar^+ reflected from Cu(100) ($\psi = 12°$, ϕ is related with respect to [011]). For $\phi = 0°$, the reflected ions are focused towards azimuthal exit angles η close to 0° [76B1].

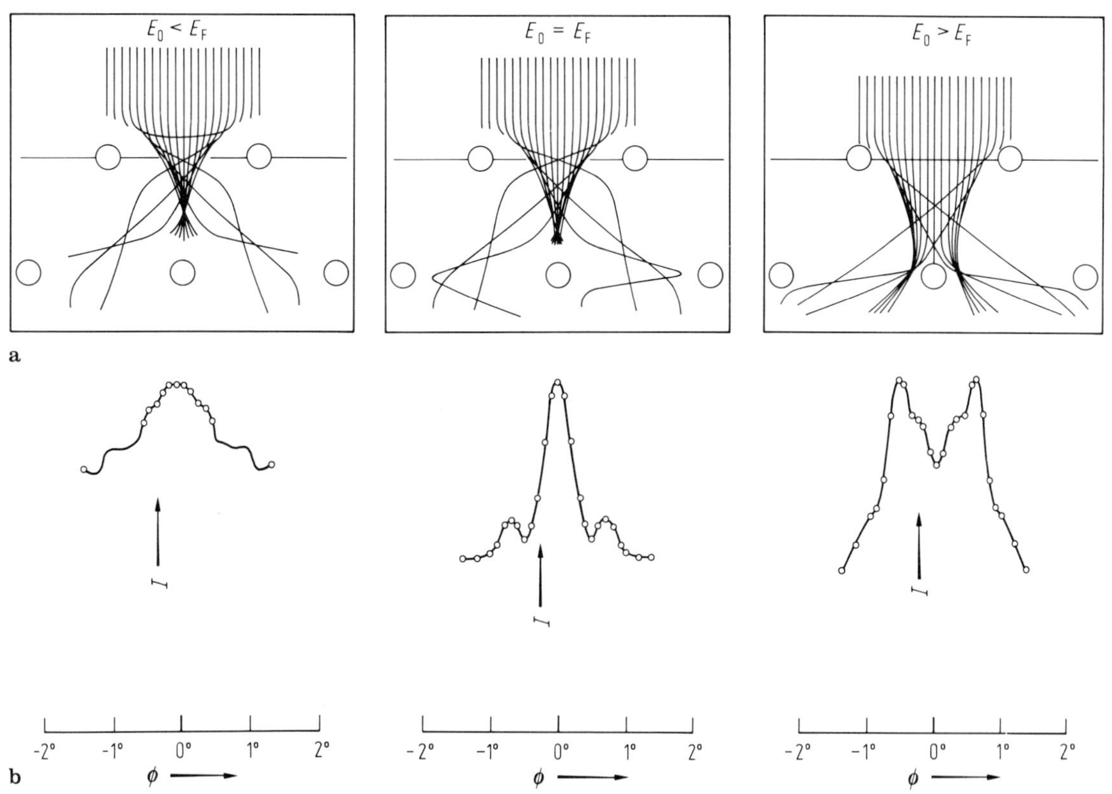

Fig. 24a, b. Focusing by surface semichannels. Ion trajectories in the plane transverse to the [011] chains for different primary energies. At $E_0 = E_F$: **a** focusing by the semichannel walls onto the second-layer chains is maximal; **b** the peak in the calculated spatial distribution of the reflected ions is most pronounced [82Y].

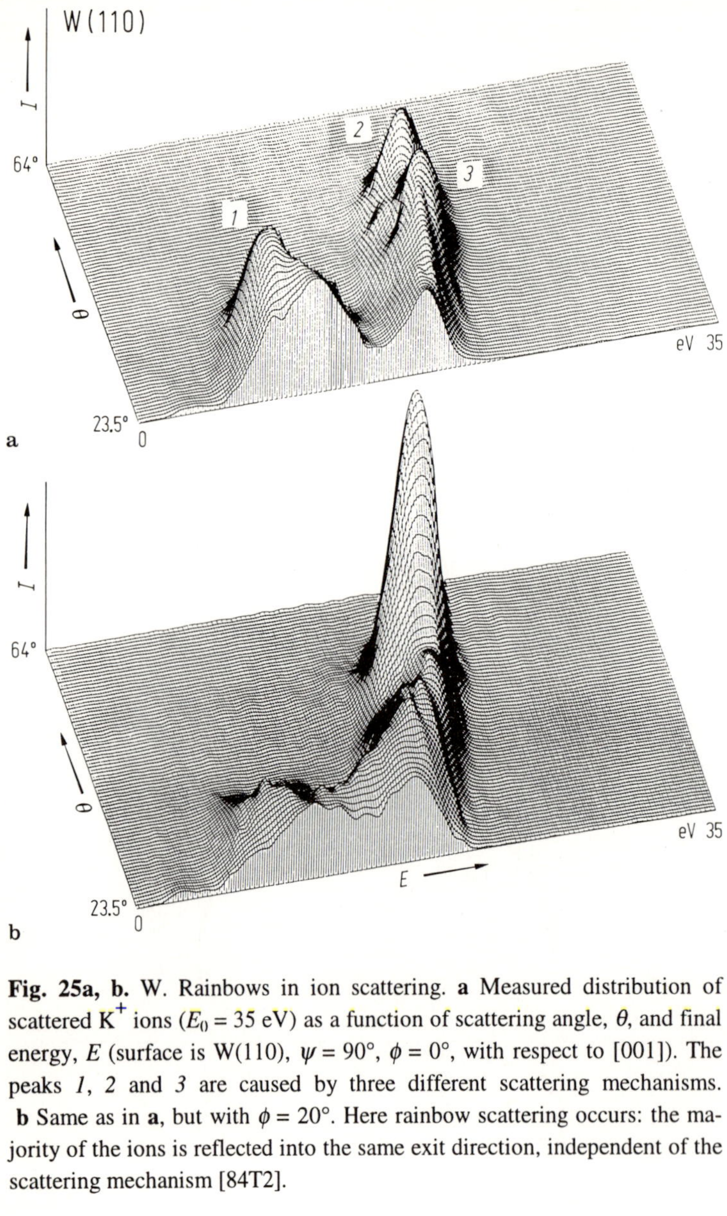

Fig. 25a, b. W. Rainbows in ion scattering. **a** Measured distribution of scattered K^+ ions ($E_0 = 35$ eV) as a function of scattering angle, θ, and final energy, E (surface is W(110), $\psi = 90°$, $\phi = 0°$, with respect to [001]). The peaks *1*, *2* and *3* are caused by three different scattering mechanisms. **b** Same as in **a**, but with $\phi = 20°$. Here rainbow scattering occurs: the majority of the ions is reflected into the same exit direction, independent of the scattering mechanism [84T2].

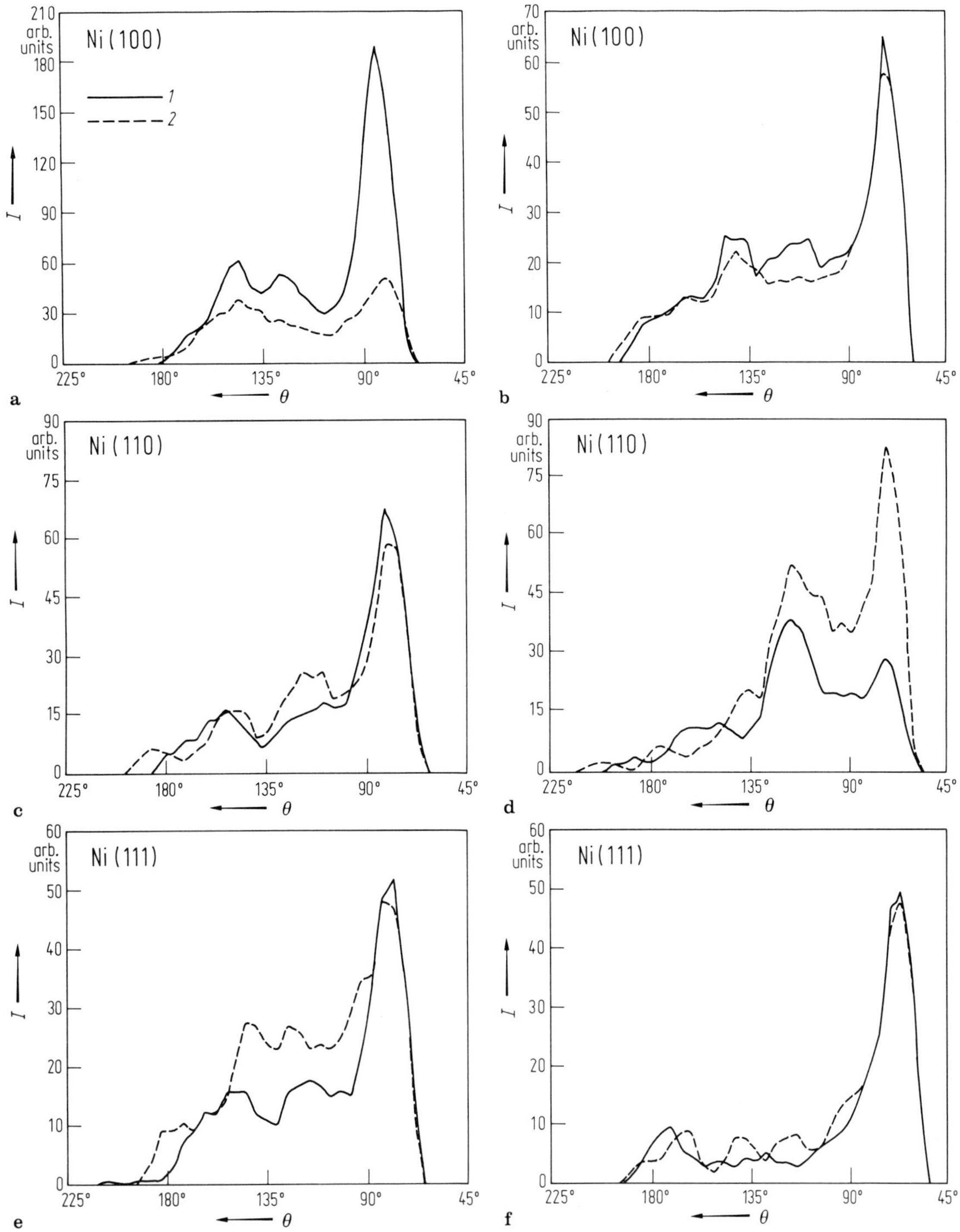

Fig. 26a-f. Ni. Model calculations. Intensities of Ar^+ ions reflected from Ni as a function of scattering angle, θ (E_0 = 2 keV, ψ = 45°). Curve *1*: model with simultaneous multiple collisions. Curve *2*: binary collision model, all collisions are calculated separately. **a** Ni(100) parallel to [011]; **b** Ni(100) parallel to [001]; **c** Ni(110) parallel to $[1\bar{1}0]$; **d** Ni(110) parallel to [001]; **e** Ni(111) parallel to $[1\bar{1}0]$; **f** Ni(111) parallel to $[2\bar{1}\bar{1}]$ [88C].

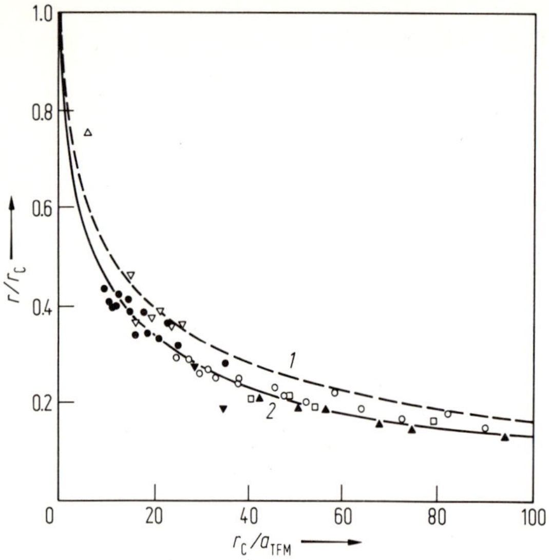

Fig. 27. Scattering potential. Shadow cone radius, r/r_C, as a function of r_C/a_{TFM} for the Thomas-Fermi-Molière potential (curve *1*); r_C is the Coulomb radius and a_{TFM} is the screening length. Curve *2* is the best fit, with a 30% reduction in a_{TFM}, to the experimental data of various authors [86F1].

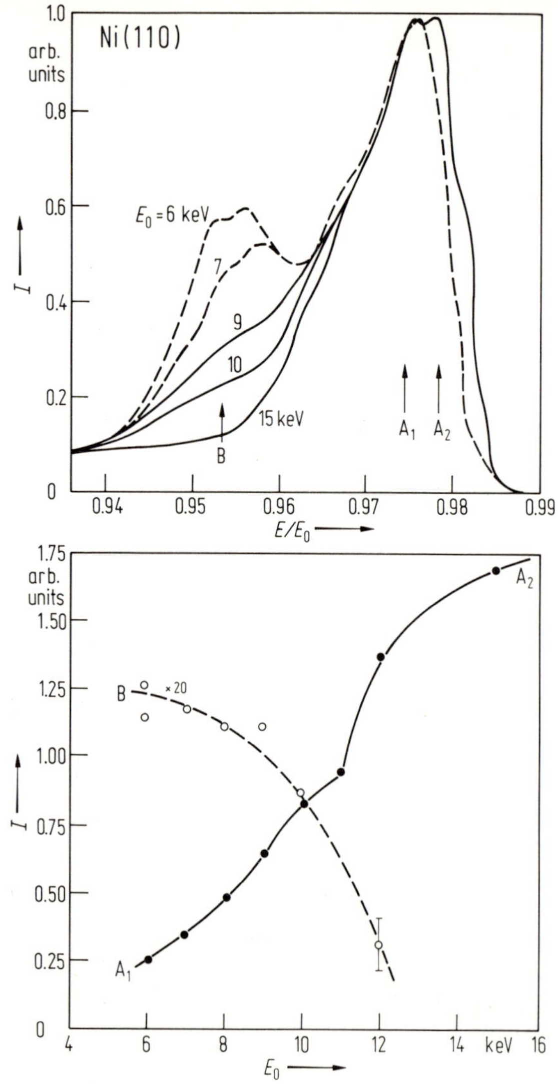

Fig. 28. Ni. Surface hyperchanneling. (Ar^+ reflected from Ni(110), ψ = 6°, θ = 12°, ϕ = 0° with respect to [001].) Peak B is indicative for surface hyperchanneling which, theoretically, occurs if E_0 < 10 keV [79E1].

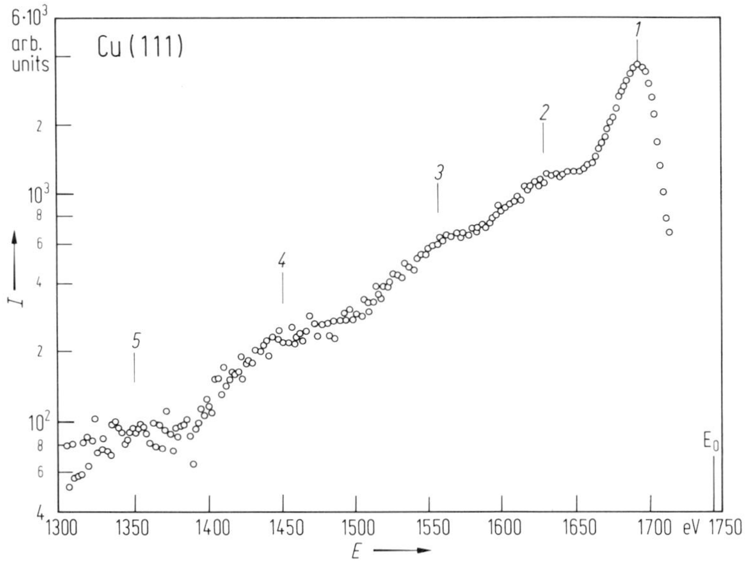

Fig. 29. Cu. Skipping motion or transient adsorption. (Si^+ reflected from Cu(111), E_0 = 1.74 keV, ψ = 4°, θ = 8°, ϕ = 5°, with respect to [$\bar{1}$10]). The peaks in the energy spectrum - see the vertical bars - are due to a number (1 ... 5) of reflections from the surface [88S2].

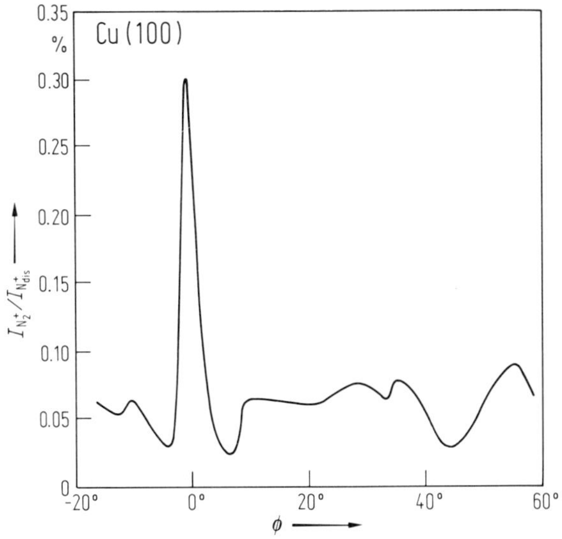

Fig. 30. Cu. Reflection of molecular ions. (Surface is Cu(100), primary beam is 30 keV N_2^+, ψ = 11°,θ = 22,° ϕ is with respect to [1$\bar{1}$0]). Dissociation is strongly reduced when surface semichanneling occurs at ϕ = 0° [83B2].

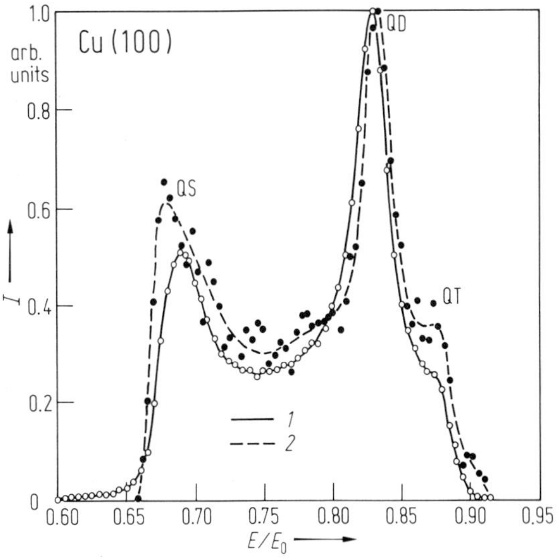

Fig. 31. Cu. Multiple scattering and thermal vibrations. Curve *1*: measured energy spectrum of 10 keV Kr^+ reflected from Cu(100); curve *2*: calculated (chain model) spectrum (ψ = 15°, θ = 30°, ϕ = 0°, with respect to [010], T = 700 K). QS, QD and QT correspond to quasi-single, quase-double and quasi-triple scattering, respectively. The vibration amplitude of the surface atoms can be derived from the intensity of the QT peak [77P1].

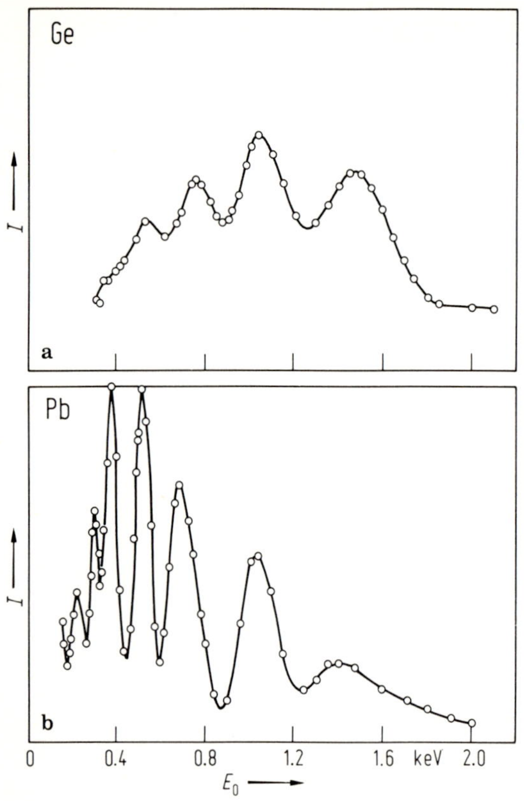

Fig. 32a. b. Ge, Pb. Oscillatory charge exchange. Intensity of He^+ ions reflected from Ge **a** and Pb **b** as a function of primary beam energy, E_0. The oscillations are due to quasi-resonant charge exchange between electronic states of the surface atom and the ion [75E].

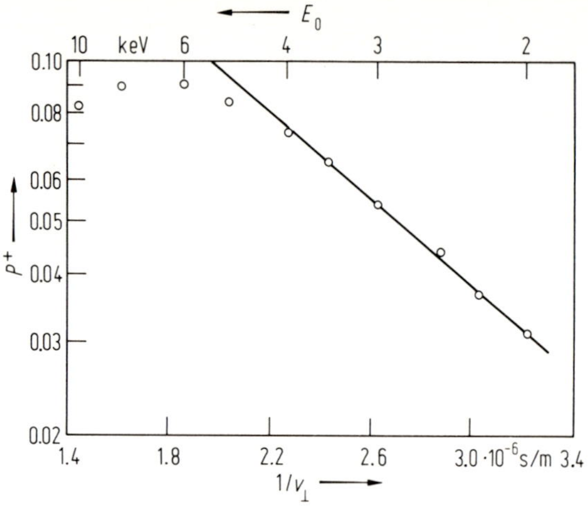

Fig. 33. Resonant and Auger neutralization. Ion survival probability, P^+, as a function of the reciprocal perpendicular velocity, $1/v_\perp$. The linear decrease for high $1/v_\perp$ ($> 2 \cdot 10^{-6}$ s/m) confirms Hagstrum's model for resonant and Auger neutralization. At $1/v_\perp < 2 \cdot 10^{-6}$ s/m, also neutralization during the violent collision can occur [77V].

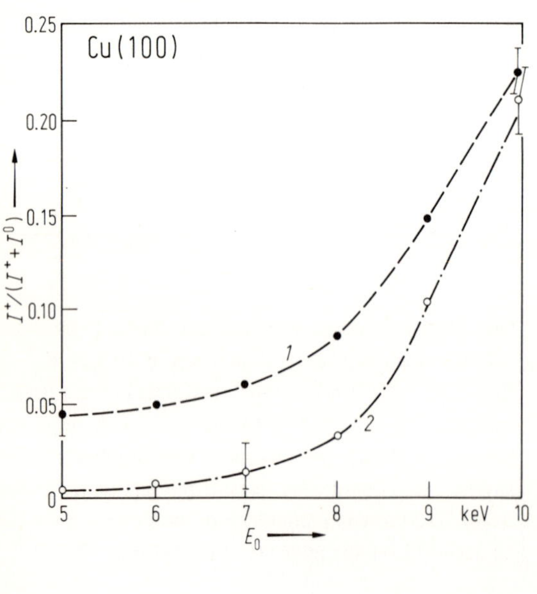

Fig. 34. Cu. Ionization and reionization. Relative intensity of the quasi-singly scattered Ne^+ ions $I^+/(I^+ + I^0)$ as a function of primary energy, E_0 (surface is Cu(100), $\psi = 15°$, $\theta = 30°$, $\phi = 0°$, with respect to [010], $T = 550$ K). *1*: Ne^+ primary beam; *2*: Ne^0 primary beam. The difference at low E_0 is caused by the difference in (re-)ionization probability of neutral and neutralized projectiles: the latter projectiles are in an excited state Ne* [80L2].

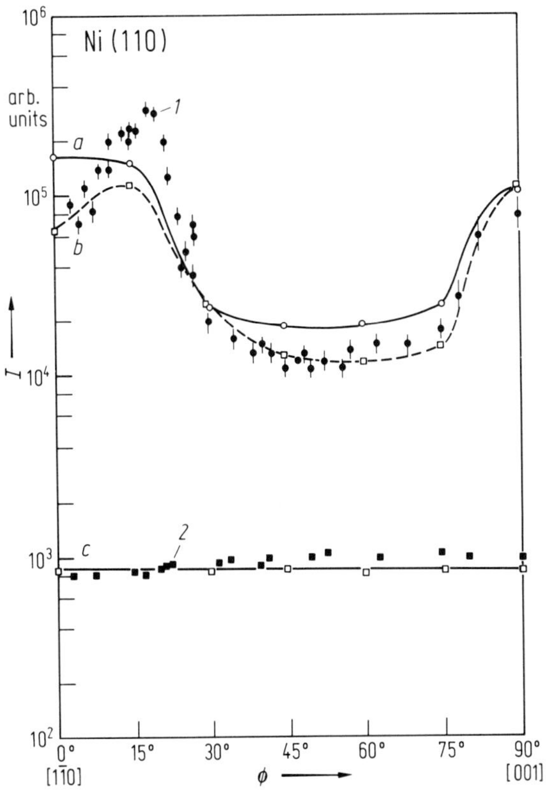

Fig. 35. Ni. Neutralization of noble gas and alkali ions. Measured (filled data points) intensities of Li^+ (*1*) and He^+ (*2*) ions reflected from Ni(110) as a function of azimuth, ϕ ($E_0 = 0.6$ keV, $\psi = 30°$, $\theta = 60°$). Curves and open symbols are intensities calculated with the binary collision model and a two-layer model. The curves *a* and *b* (for Li and He, respectively) are calculated assuming no neutralization; curve *c* is calculated for He assuming a neutralization probability of 89% for (quasi-)single scattering from first layer atoms and a neutralization probability of more than 99.9% for all second layer scatterings and first layer (quasi-)double scattering [80T].

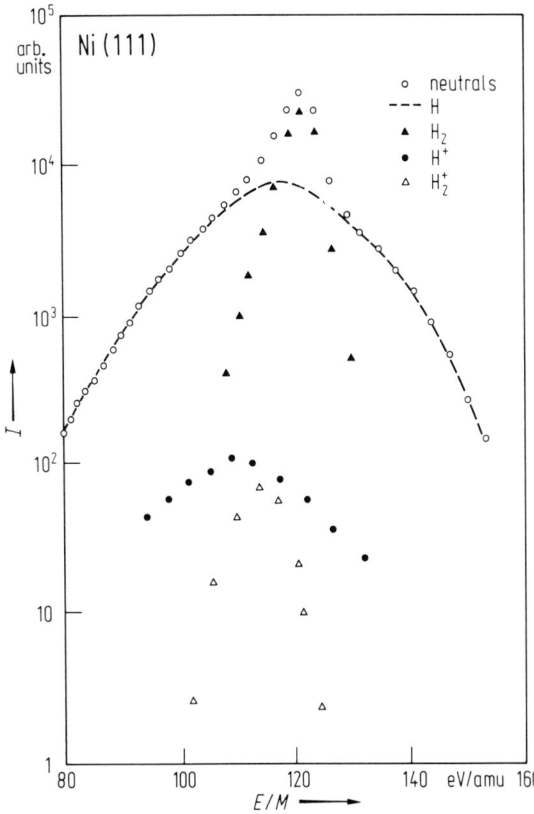

Fig. 36. Ni. Scattering, dissociation and neutralization of H_2^+. (E_0 = 0.26 keV, surface is Ni(111), ψ = 5°, θ = 10°). The energy spectrum of all reflected neutrals, of the reflected H, of the reflected H_2, of the reflected H_2^+, and of the reflected H^+ is given [84W1].

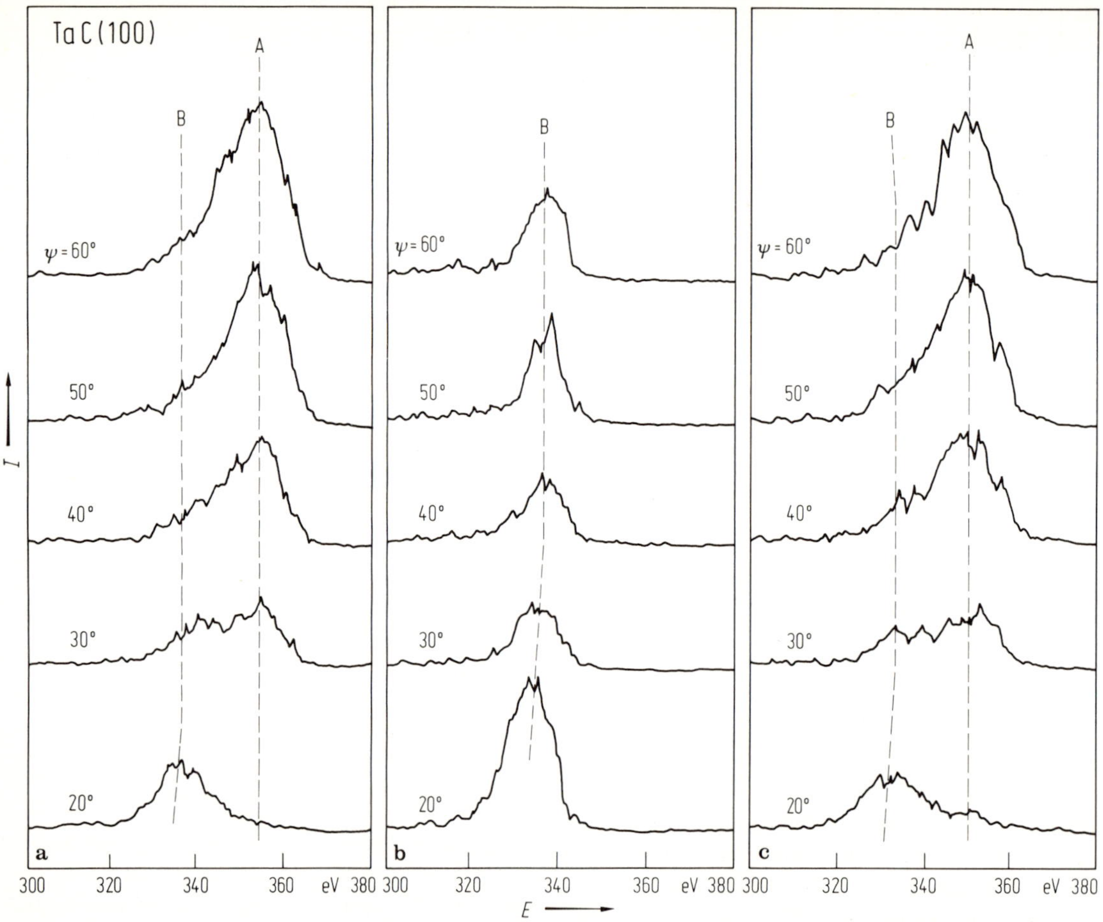

Fig. 37a-c. TaC. Neutralization and (re-)ionization. Energy spectra of ions reflected from Ta atoms in a TaC(100) surface using: **a** a He^+ primary beam; **b** a He^0 beam; **c** a He* beam (E_0 = 390 eV, θ = 120°, ϕ = 0° with respect to [010]). Peak A, corresponding to elastically scattered He^+ ions, does not appear in **b**. Since He* approaching the surface is very easily ionized, the spectra **c** resemble those of **a** [86S4].

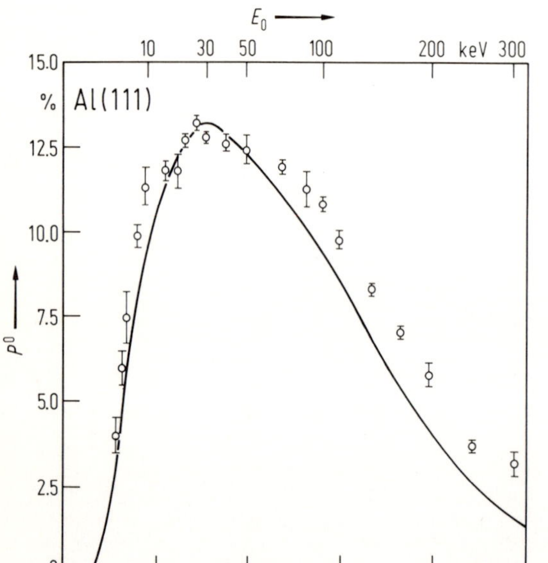

Fig. 38. Al. Neutralization under grazing incidence conditions. Neutralization probability, P^0, of K^+ scattered from Al(111) as a function of the ion's velocity, v_0 (ψ = 0.7°, θ = 1.4°) The solid line represents a theoretical calculation based on resonant tunneling from the Al conduction band to the ionization level of K^+ [89Z].

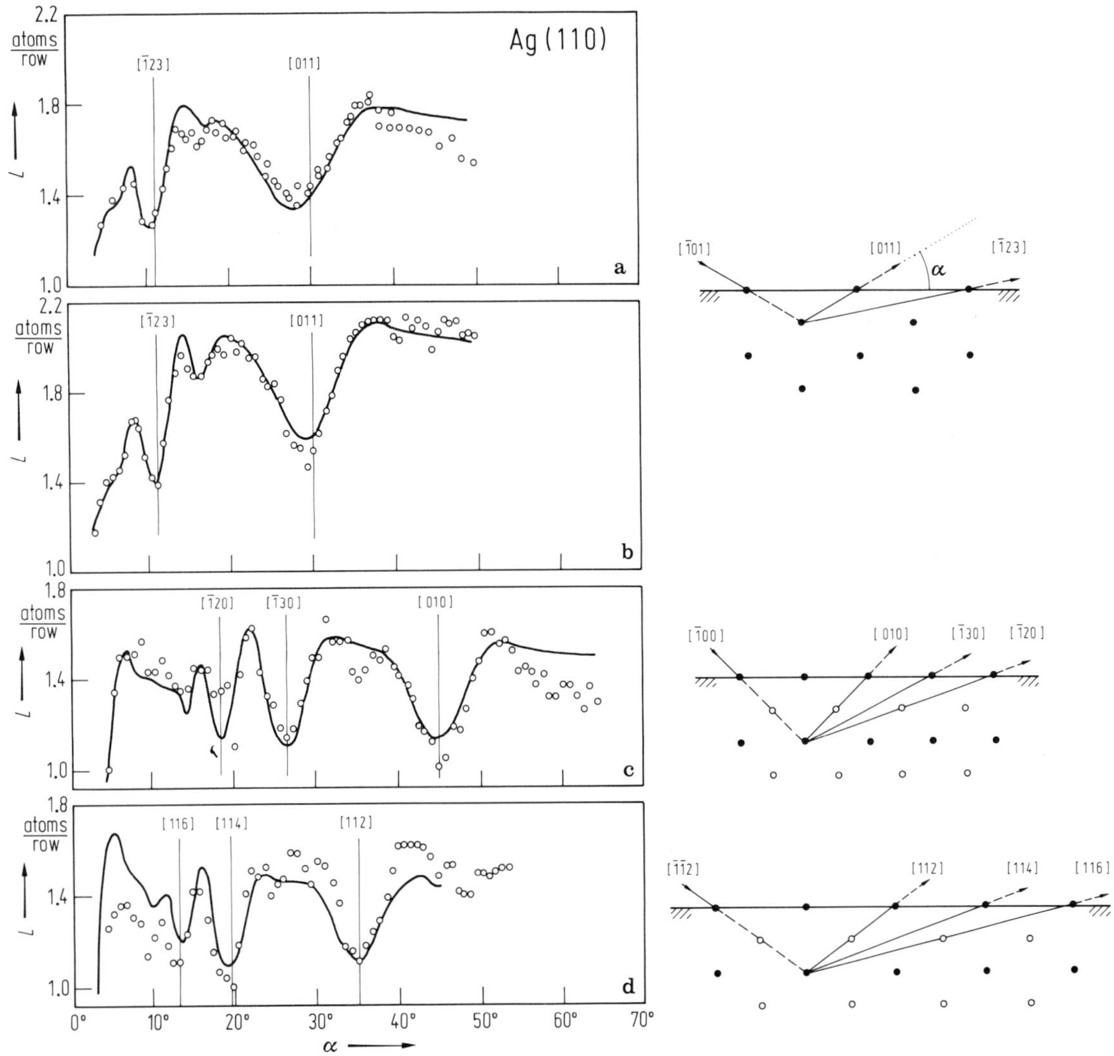

Fig. 39a-d. Ag. Double layer relaxation of Ag(110). Blocking profiles of backscattered H^+ for three scattering geometries (shown at the right hand side. Circles denote atomic positions for an unrelaxed bulklike crystal; open circles are atoms in a parallel, equivalent scattering plane also contributing to the measured blocking profiles.): **a** E_0 = 51 keV, $(1\bar{1}1)$ scattering plane; **b** E_0 = 98 keV, $(1\bar{1}1)$ scattering plane; **c** E_0 = 51 keV, (001) scattering plane; **d** E_0 = 51 keV $(1\bar{1}0)$ scattering plane. The solid curves at the left hand side represent best-fit Monte Carlo simulations for $\Delta d_{12} = -14$ pm and $\Delta d_{23} = +9$ pm [87H2].

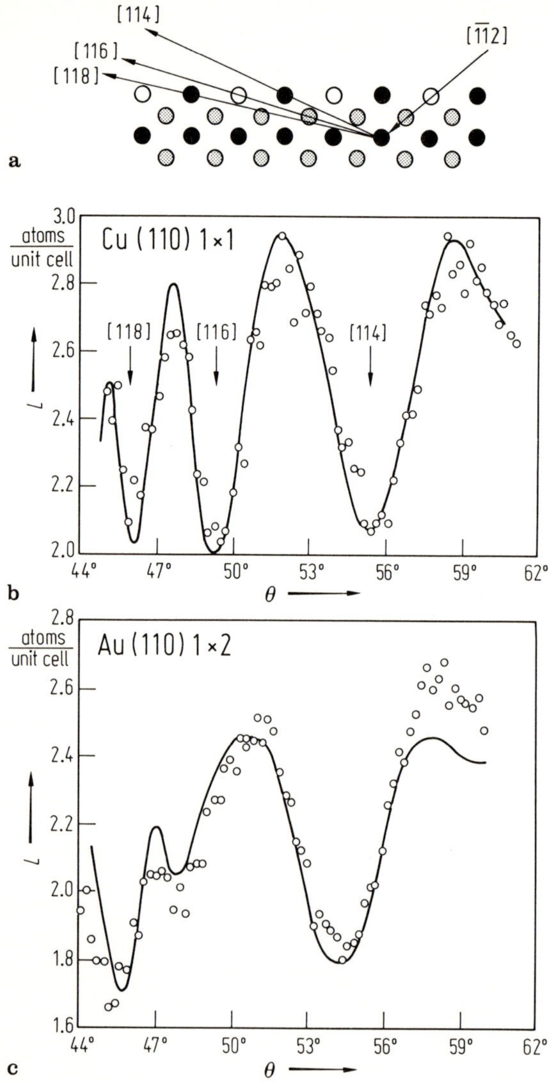

Fig. 40a-c. Cu, Au. Missing row reconstruction. **a** Side view of the ($\bar{1}$10) zone of a fcc-crystal. Two equivalent planes are shown, in black and grey, respectively. Open circles represent possible vacancies. **b** Blocking profile of 100 keV H^+ backscattered from the unreconstructed Cu(110)1x1 surface. Direction of incidence is [$\bar{1}\bar{1}$2]. **c** Same as **b** but for the reconstructed (missing row) Au(110)1x2 surface [86C2].

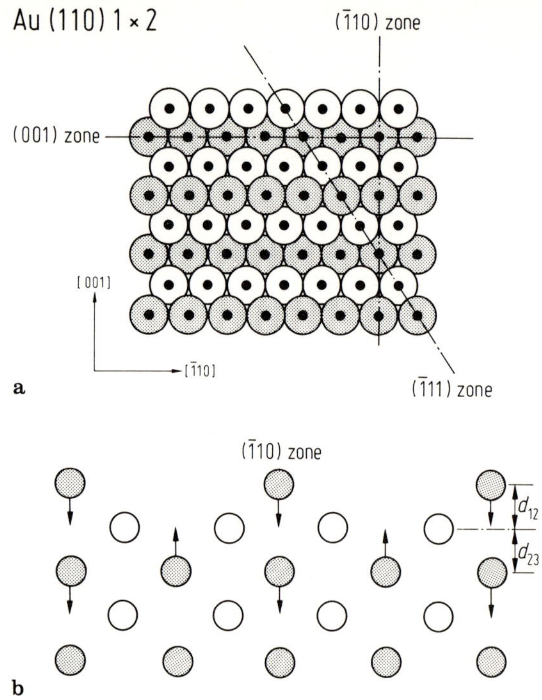

Fig. 41a, b. Au. Missing row reconstruction of Au(110)1x2. **a** Top view; **b** side view. Arrows show the movements of the atoms with respect to their bulk position [86C2].

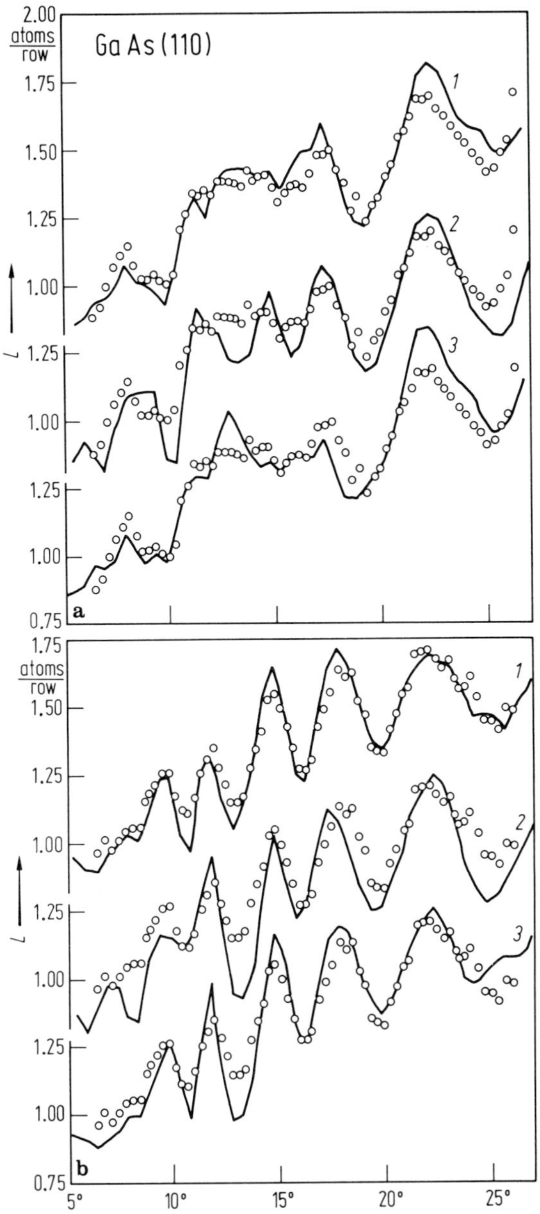

Fig. 42a, b. GaAs. Rotation of Ga-As surface dimers. Blocking profiles of 100 keV H^+ backscattered from GaAs(110). Direction of incidence is **a** [$\bar{1}\bar{1}1$] and **b** [$\bar{1}\bar{1}\bar{1}$]. Full curves are results of Monte Carlo simulations: *1*: bond-length conserving model with rotation angle ω = 30°; *2*: the same but with ω = 10°; *3*: bond-relaxation model. Model *1* fits the experimental data best [85S2].

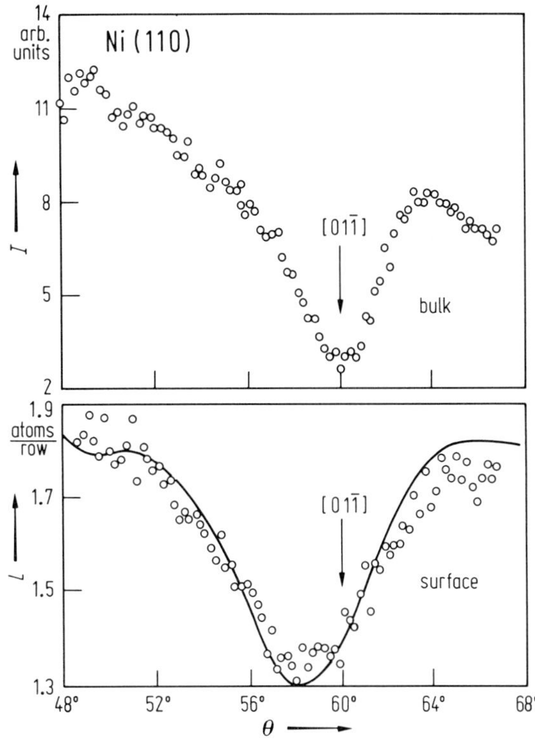

Fig. 43. Ni. Double layer relaxation of Ni(110). Bulk and surface blocking profiles of 100 keV H^+. Direction of incidence is [$\bar{1}0\bar{1}$], scattering plane is ($\bar{1}11$). The full curve is the result of the best-fit Monte Carlo simulation: Δd_{12} = −11 pm and Δd_{23} = +4 pm [86Y1].

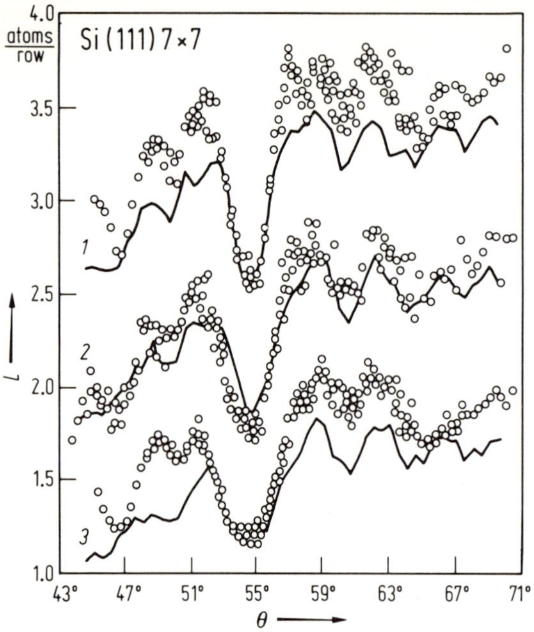

Fig. 44. Si. Si(111)7x7 reconstruction. Blocking profiles for: *1*: 151 keV H^+; *2*: 99 KeV H^+; *3*: 54 keV H^+. (Profiles *1* and *3* have been shifted vertically over +0.5 and −0.5, respectively.) Direction of incidence is $[00\bar{1}]$, scattering plane is $(1\bar{1}0)$. The full curves are calculated for the stacking-fault model by Bennett. This and other stacking-fault models fit the experimental data reasonably well [85T3].

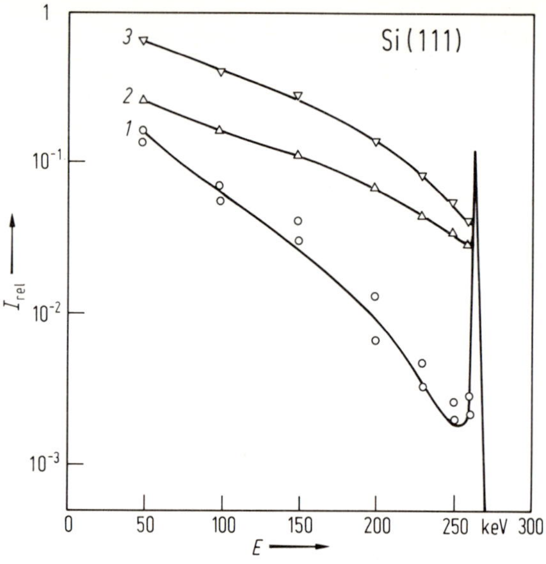

Fig. 45. Si. Single and double alignment. Normalized backscattering spectra of 300 keV H^+ backscattered from Si(111). The normalized spectrum measured for *1*: double alignment is within 10% equal to the product of *2*: the channeling spectrum and *3*: the blocking spectrum [80K].

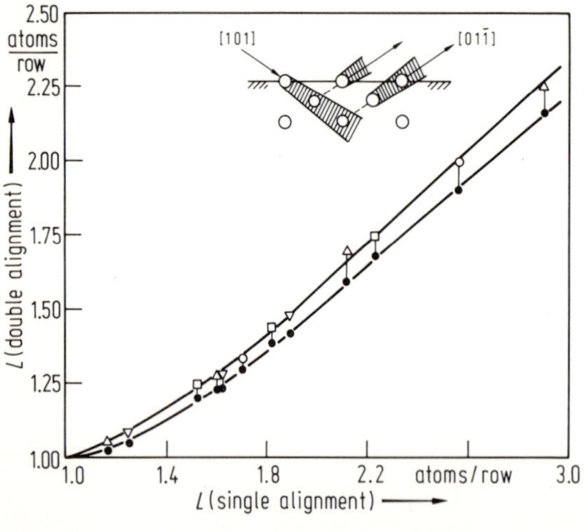

Fig. 46. Monte Carlo simulation of shadowing and blocking. The surface peak area, *L*, is calculated for the double alignment scattering geometry shown at the top. Open symbols are results for various ion-solid combinations with incidence and exit trajectories correlated. The corresponding full circles are results of the conventional Monte Carlo method, viz., without correlations between shadowing and blocking [83T5].

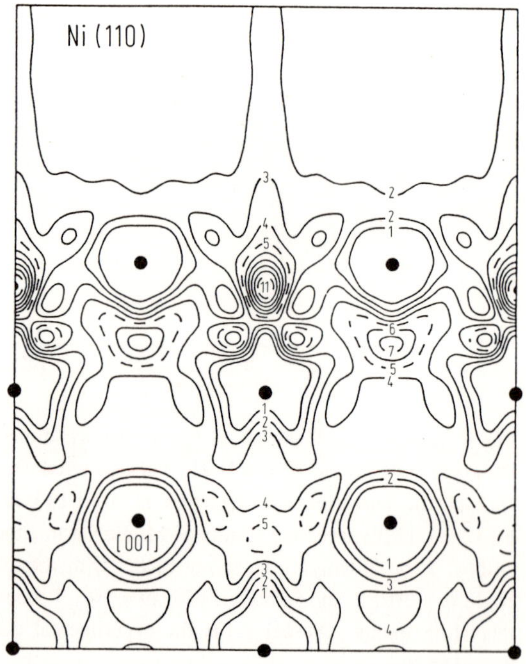

Fig. 47. Ni. Surface semichanneling. Calculated flux density contours in the transverse plane (normalized to the incident flux) of 150 keV H^+ incident on a Ni(110) crystal. (ψ = 1°, ϕ = 0° with respect to [001]). Filled circles indicate atoms [83S1].

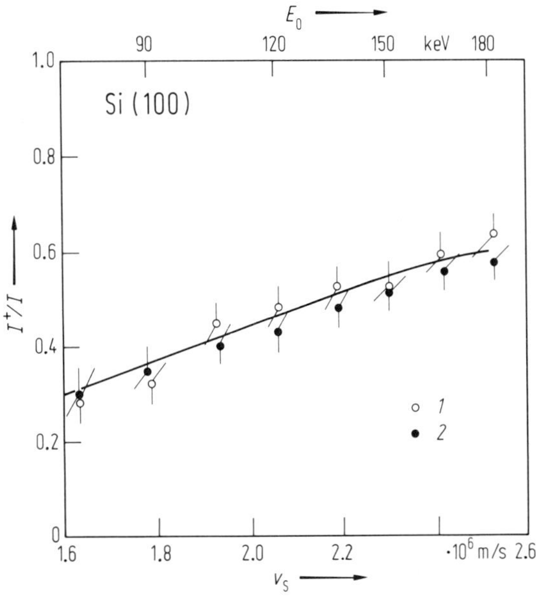

Fig. 48. Si. Neutralization. Ion fraction of He backscattered from Si(100) as a function of scattered velocity v_s. *1*: direction of incidence is [100]; *2*: random direction of incidence [84H2].

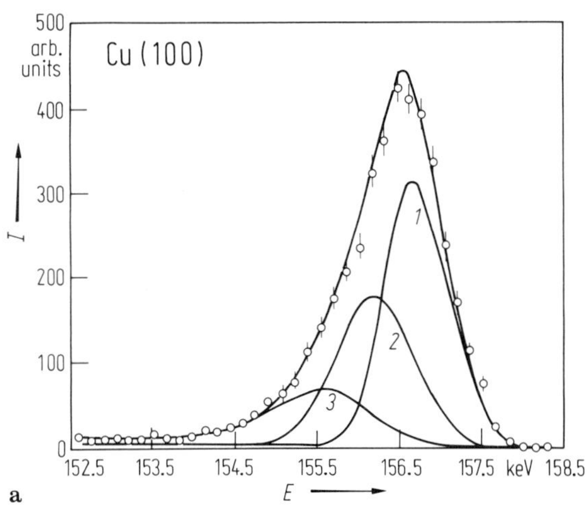

a

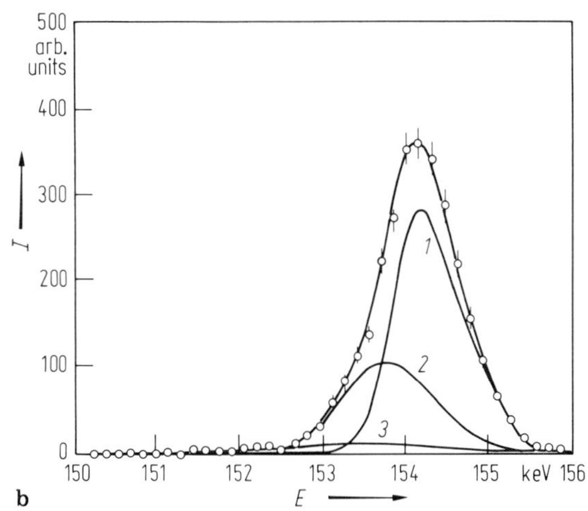

b

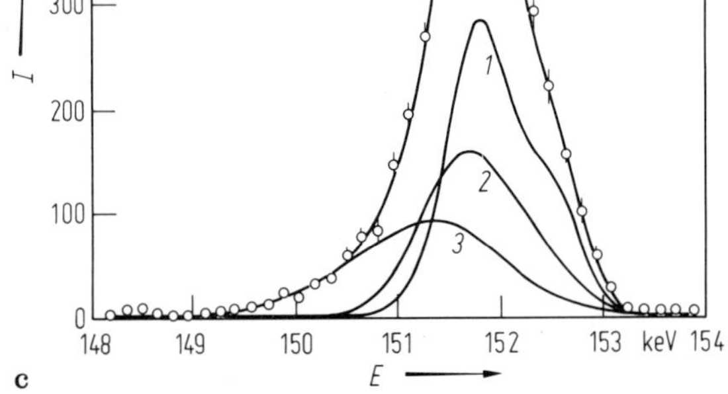

c

Fig. 49a-c. Cu. Elastic and inelastic energy losses. Energy spectra of 175 keV He^+ backscattered from Cu(100). Direction of incidence is $[\bar{1}\,\bar{1}\,0]$, scattering plane is (001); **a** $\theta = 83°$; **b** $\theta = 90°$ (exit direction is $[1\,\bar{1}\,0]$); **c** $\theta = 97°$. Full curves are results of Monte Carlo calculations in which elastic and (impact parameter dependent) inelastic energy losses are included. Contributions of backscattering from various layers *1 ... 3* are indicated as well [87A3].

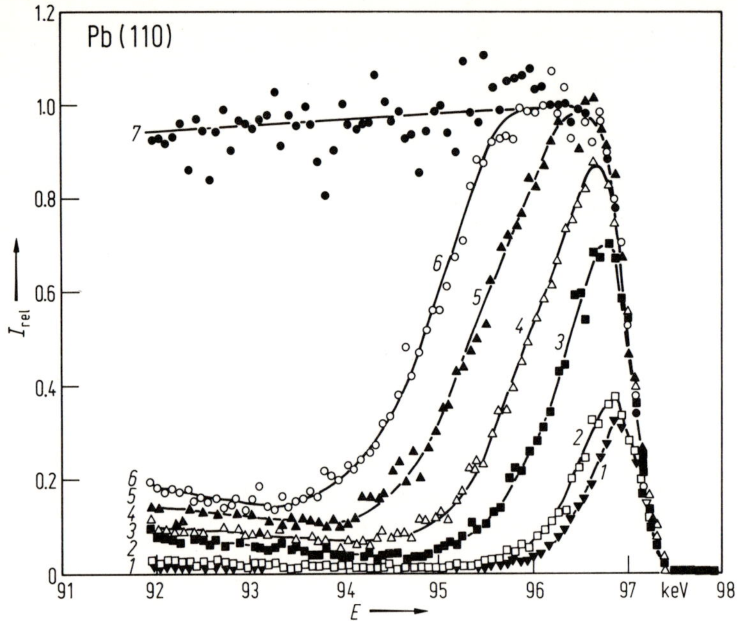

Fig. 50. Pb. Surface melting of Pb(110). Energy spectra of 98 keV H^+ backscattered from Pb(110) at various temperatures. Direction of incidence is [$\bar{1}$01]; exit direction is [011]. *1*: 295 K; *2*: 452 K; *3*: 581 K; *4*: 597 K; *5*: 599.7 K; *6*: 600.5 K; *7*: 600.8 K. Full curves serve to guide the eye. The thickness of the disordered layer increases as the bulk melting temperature of 600.7 K is approached [86F3].

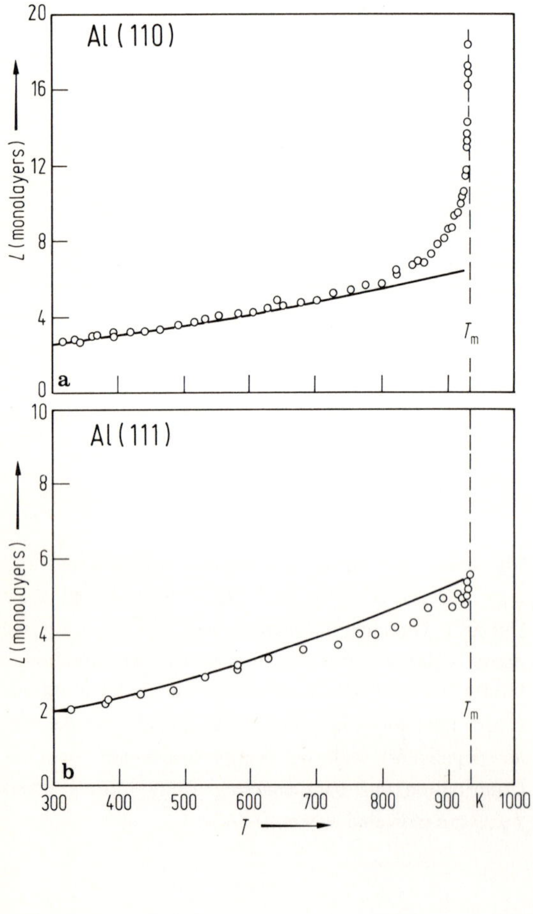

Fig. 51a, b. Al. Surface melting of Al. Surface peak area, *L*, for 100 keV H^+ backscattered from Al(110) **a** and Al(111) **b** as a function of temperature. Directions of incidence are **a** [$\bar{1}$01], and **b** [$\bar{1}\,\bar{1}$0]. Exit directions are **a** [011], and **b** [001]. Full curves are surface peak areas calculated assuming only an increase in vibration amplitude with increasing temperature. The deviation of the experimental data for Al(110) near T_m is caused by surface melting [90D4].

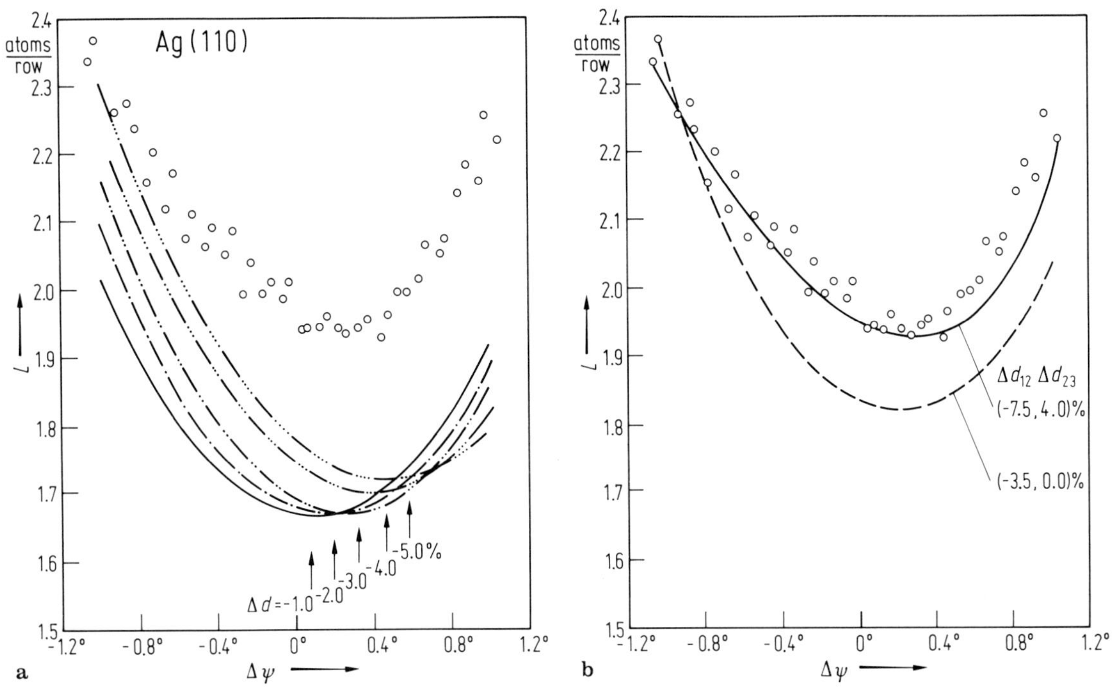

Fig. 52a, b. Ag. Double layer relaxation of Ag(110). Angular scan of the surface peak area, L, measured in the $(1\bar{1}\bar{1})$ plane near the $[\bar{1}0\bar{1}]$ direction for 0.4 MeV He^+. The curves are calculated for **a** single-layer relaxations Δd between −1% and −5%; and **b** double layer relaxations [84K1].

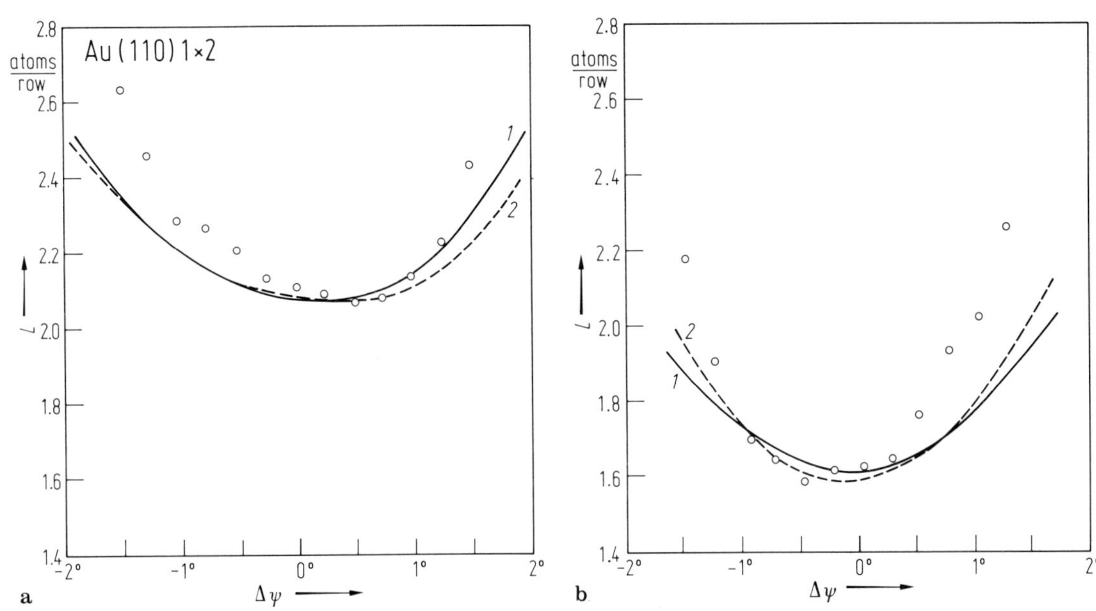

Fig. 53a, b. Au. Relaxation of Au(110)1x2. Angular scan of the surface peak area, L, for 0.3 MeV He^+ measured in **a** the $(\bar{1}11)$ plane near the [101] direction, and **b** the (001) plane near the [100] direction. The curves are calculated for *1*: 62 pm first layer expansion; and *2*: 25 pm first layer contraction (see also Fig. 59) [85W2].

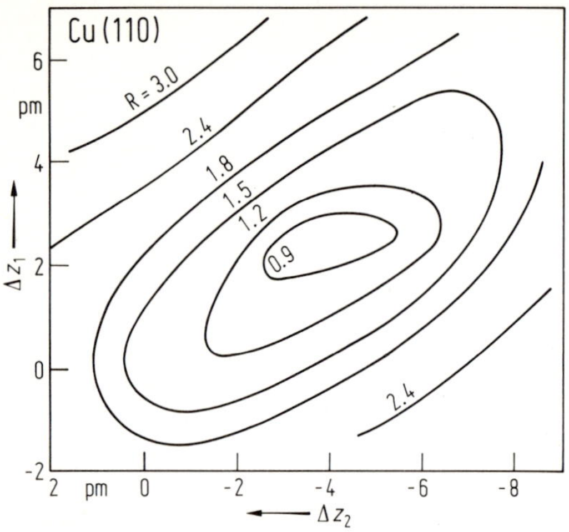

Fig. 54. Cu. Double layer relaxation of Cu(110). Contour plot of goodness-of-fit (R-factor) vs. first, Δz_1, and second layer displacements, Δz_2, from bulk positions. Primary beam is 0.3 MeV He^+. The figure is a combination of results obtained near [101] and near [100] directions of incidence [83S3].

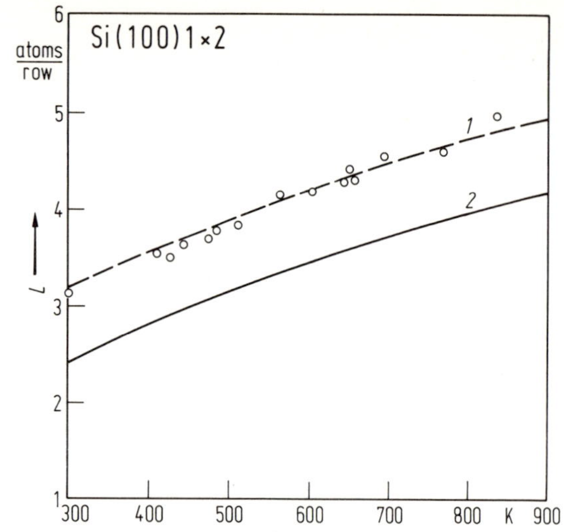

Fig. 55. Si. Lateral displacements in Si(100)1x2. Surface peak area, *L*, as a function of temperature for normal incident 0.6 MeV He^+ ions. Circles and curve *1* are the experimental data. As compared to the bulklike terminated surface (curve *2*) additional monolayers are visible, indicating lateral displacements exceeding 15 pm in the outermost three monolayers [81S].

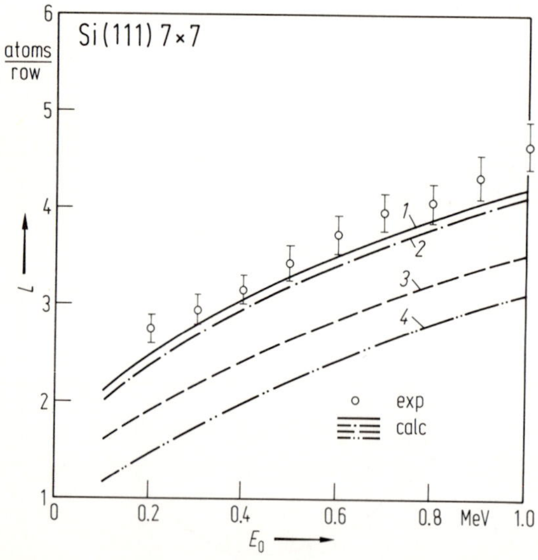

Fig. 56. Si. Si(111)7x7 reconstruction. Surface peak area, *L*, as a function of E_0 for He^+ ions. Direction of incidence is $[00\bar{1}]$. Curves are Monte Carlo results for several variants of the dimer-adatom-stacking fault (DAS) model: *1*: of Tong et al. ; 2: of Qian et al. ; *3*: bulk DAS model; *4*: bulklike terminated surface. Calculations are from [90Y2], experimental data are from [80C].

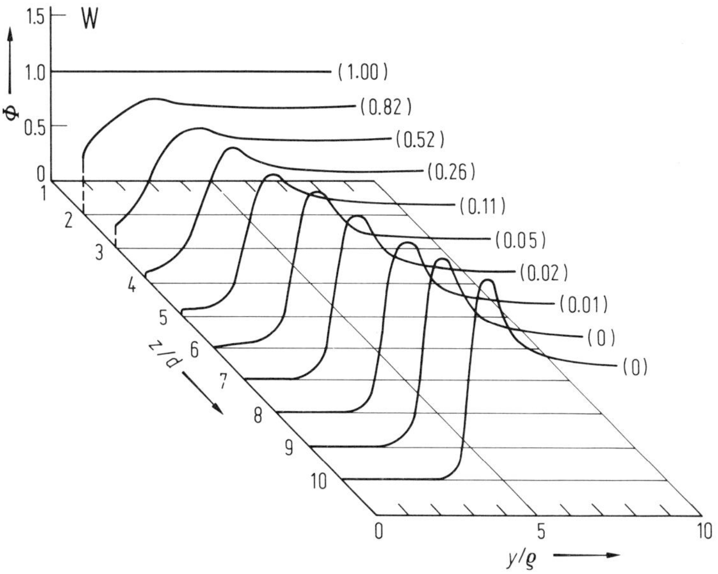

Fig. 57. W. Channeling. Calculated flux distribution near the W[111] direction for 3 MeV H^+ ions. z is the depth, d is the interatomic distance, y is the (lateral) distance from the atoms rows and ρ is the vibration amplitude. The relative backscattering probabilities for each of the ten layers are indicated in parentheses [78S2].

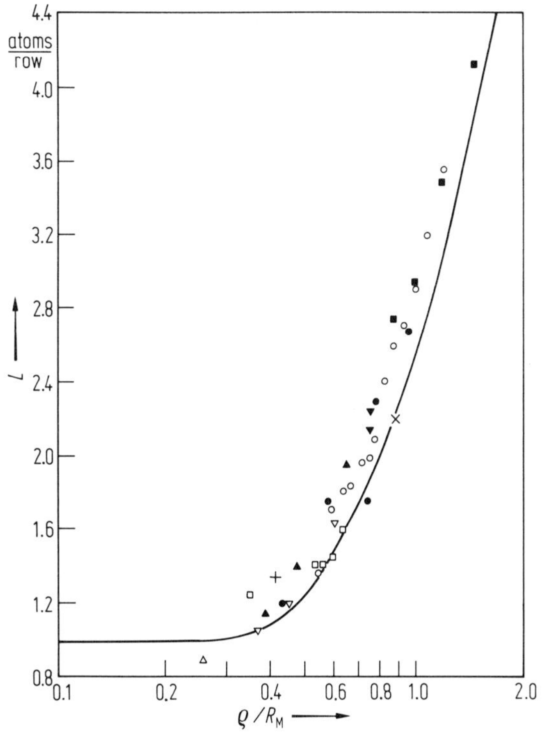

Fig. 58. Universal surface peak area, L. Experimental data (for a number of bulklike terminated surfaces and scattering geometries) and the result of a simulation (full curve) are plotted as a function of scaled vibration amplitude, ρ / R_M R_M is the Molière shadow cone radius [78S2].

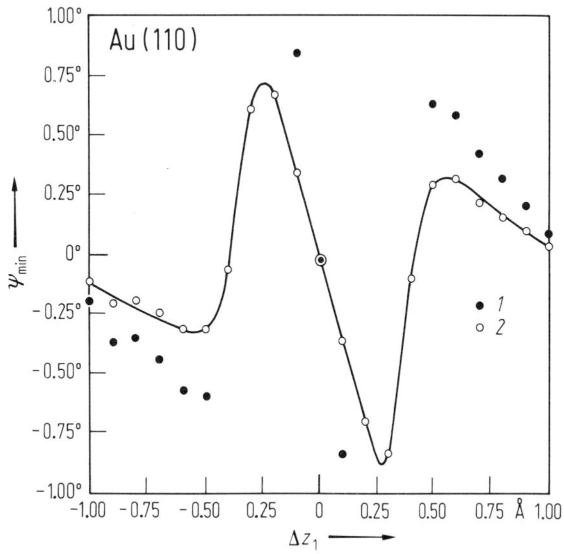

Fig. 59. Au. Ambiguity in the determination of surface relaxation. Dependence of ψ_{min} at which the minimum surface peak area occurs on the expansion Δz_1 of the surface layer. *1*: unreconstructed surface; *2*: reconstructed surface (missing row model). Calculated for 0.3 MeV He^+ incident along the [101] axis of Au(110). The calculation shows that the surface relaxation cannot be derived unambiguously from ψ_{min} (see also Fig. 53) [85C2].

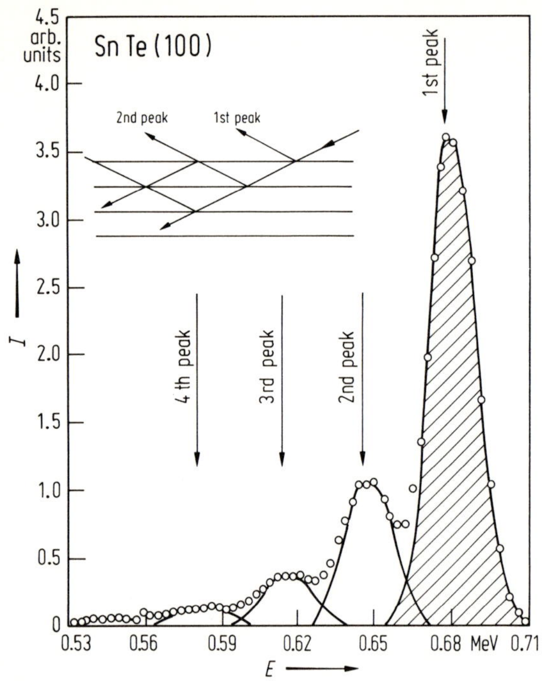

Fig. 60. SnTe. Energy loss of specularly reflected ions. Energy spectrum of 0.7 MeV He^+ reflected from SnTe(100) ($\psi = 0.28°$, $\theta = 0.60°$). The inset shows the reflection of the ions from various atomic planes. Although ψ is smaller than the critical angle for surface (hyper-)channeling, penetration through the surface layer is possible due to thermal vibration of surface atoms, electronic scattering and surface defects [87K1].

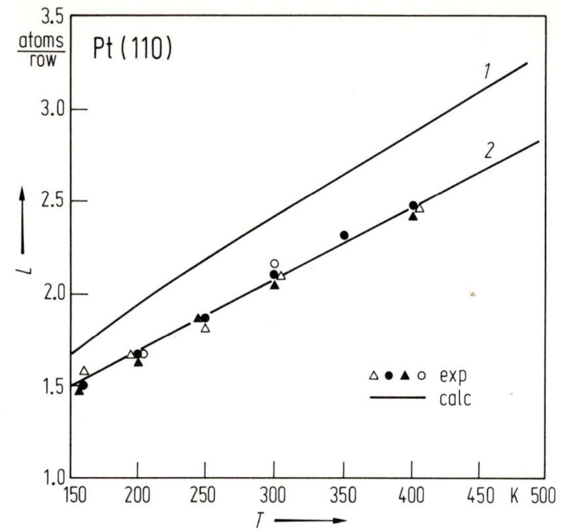

Fig. 61. Pt. Correlated atomic vibrations. Measured surface peak area, *L*, (data points) for 2 MeV He^+ backscattered from Pt(110) as a function of temperature ($\psi = 90°$). *1*: Surface peak area calculated without correlations in the vibration of neighbouring atoms; *2*: surface peak area calculated with correlations [83J].

6.3.3 References for 6.3

54H Hagstrum, H.D.: Phys. Rev. **96** (1954) 336.
61H Hagstrum, H.D.: Phys. Rev. **123** (1961) 758.
63N Nelson, R.S., Thompson, M.W.: Philos. Mag. **8** (1963) 1677.
63R1 Robinson, M.T., Oen, O.S.: Appl. Phys. Lett. **2** (1963) 30.
63R2 Robinson, M.T., Oen, O.S.: Phys. Rev. **132** (1963) 2385.
64D Datz, S., Snoek, C.: Phys. Rev. A **134** (1964) 347.
65B Bøgh, E., Uggerhøj, E.: Nucl. Instrum. Methods **38** (1965) 216.
65L Lindhard, J.: K. Dan. Vidensk. Selsk. Mat. Fys. Medd. **34** (1965) no. 14.
65M1 Mashkova, E.S., Molchanov, V.A., Parilis, E.S., Turaev, N.Y.: Phys. Lett. **18** (1965) 7.
65M2 Mashkova, E.S., Molchanov, V.A., Soshka, V.: Sov. Phys. Dokl. (English Transl.) **10** (1965) 323.
65O Oen, O.S.: Phys. Lett. **19** (1965) 358.
65P Parilis, E.S., Turaev, N.Y.: Sov. Phys. Dokl. (English Transl.) **10** (1965) 212.
65T Tulinov, A.F., Kulikauskas, V.S., Malov, M.M.: Phys. Lett. **18** (1965) 304.
65Y Yurasova, V.E., Brzhezinskii, V.A., Ivanov, G.M.: Sov. Phys. JETP (English Transl.) **20** (1965) 313.
66M1 Mashkova, E.S., Molchanov, V.A.: Sov. Phys. Solid State (English Transl.) **8** (1966) 1206.
66M2 Mashkova, E.S., Molchanov, V.A., Parilis, E.S., Turaev, N.Y.: Sov. Phys. Dokl. (English Transl.) **11** (1966) 52.
66P Parilis, E.S., in: Proc. 7th Intern. Conf. Phenomena Ionized Gases, Belgrade, 1966, p. 129.
67B Bøgh, E.: Phys. Rev. Lett. **19** (1967) 61.
67E Evdokimov, I.N., Mashkova, E.S., Molchanov, V.A.: Phys. Lett. A **25** (1967) 619.
67K1 Karpuzov, D.S., El'tekov, V.A., Yurasova, V.E.: Sov. Phys. Solid State (English Transl.) **8** (1967) 1726.
67K2 Kivilis, V.M., Parilis, E.S., Turaev, N.Y.: Sov. Phys. Dokl. (English Transl.) **12** (1967) 328.
67M1 Mashkova, E.S., Molchanov, V.A.: Sov. Phys. Dokl. (English Transl.) **12** (1967) 133.
67M2 Mashkova, E.S., Molchanov, V.A., Soshka, V.: Phys. Status Solidi **19** (1967) 425.
67S Smith, D.P.: J. Appl. Phys. **38** (1967) 340.
67Y Yurasova, V.E., Shul'ga, V.I., Karpuzov, D.S., in: Proc. 8th Intern. Conf. Phenomena in Ionized Gases, Vienna 1967, p. 50.
68M1 Mashkova, E.S., Molchanov, V.A.: Can. J. Phys. **46** (1968) 713.
68M2 Morgan, D.V., Van Vliet, D.: Can. J. Phys. **46** (1968) 503.
68S Strehlow, W.H., Smith, D.P.: Appl. Phys. Lett. **13** (1968) 34.
68V van der Weg, W.F., Bierman, D.J.: Physica **38** (1968) 406.
68Y1 Yurasova, V.E., Shul'ga, V.I., Karpuzov, D.S.: Can. J. Phys. **46** (1968) 759.
68Y2 Yurasova, V.E., Karpuzov, D.S.: Sov. Phys. Solid State (English Transl.) **9** (1968) 1971.
69C Chicherov, V.M.: Sov. Phys. JETP (English Transl.) **28** (1969) 13.
69D Dahl, P., Sandager, N.: Surf. Sci. **14** (1969) 305.
69E Evdokimov, I.N., Mashkova, E.S., Molchanov, V.A.: Sov. Phys. Dokl. (English Transl.) **14** (1969) 467.
69M1 Mashkova, E.S., Molchanov, V.A., Skripka, Y.G.: Phys. Lett. A **29** (1969) 645.
69M2 Molchanov, V.A., Snisar, V.A., Chicherov, V.M.: Sov. Phys. Dokl. (English Transl.) **14** (1969) 43.
69V van der Weg, W.F., Bierman, D.J.: Physica **44** (1969) 177.
70A1 Arifov, U.A., Aliev, A.A.: Sov. Phys. Dokl. (English Transl.) **14** (1970) 1166.
70A2 Arifov, U.A., Aliev, A.A.: Sov. Phys. JETP (English Transl.) **30** (1970) 1015.
70A3 Armour, D.G., Carter, G., Smith, A.G.: Radiat. Eff. **3** (1970) 175.

70F Farmery, B.W., Marwick, A.D., Thompson, M.W., in: Atomic Collision Phenomena in Solids. Palmer, D.W., Thompson, M.W., Townsend, P.D. (eds.), Amsterdam: North-Holland 1970, p. 589.
70H Hagstrum, H.D., in: The Physics of Ionized Gases. Navinsek, B. (ed.), Ljubljana: Institute J. Stefan 1970, p. 190.
70K Kivilis, V.M., Parilis, E.S., Turaev, N.Y.: Sov. Phys. Dokl. (English Transl.) **15** (1970) 587.
70M1 Mashkova, E.S., Molchanov, V.A., Skripka, Y.G.: Phys. Lett. A **33** (1970) 373.
70M2 Mashkova, E.S., Molchanov, V.A., Skripka, Y.G., Snisar, V.A., Soshka, W., in: Atomic Collision Phenomena in Solids. Palmer, D.W., Thompson, M.W.,Townsend, P.D. (eds.), Amsterdam: North-Holland 1970, p. 606.
70M3 Mashkova, E.S., Molchanov, V.A., Skripka, Y.G.: Sov. Phys. Dokl. (English Transl.) **15** (1970) 26.
70M4 Müller-Jahreis, U.: Sov. Phys. Dokl. (English Transl.) **15** (1970) 353.
70P Pryde, M., Smith, A.G., Carter, G., in: Atomic Collision Phenomena in Solids. Palmer, D.W., Thompson, M.W., Townsend, P.D. (eds.), Amsterdam: North-Holland 1970, p. 573.
71A Aissa, M., Mashkova, E.S., Molchanov, V.A., Skripka, Y.G.: Sov. Phys. Solid State (English Transl.) **12** (1971) 1643.
71B Barrett, J.H.: Phys. Rev. B **3** (1971) 1527.
71E Efremenkova, V.M., Bunin, I.G., Karpuzov, D.S., Pavlychenko, A.A., Yurasova, V.E.: Bull. Acad. Sci. USSR, Phys. Series (English Transl.) **35** (1971) 375.
71K Karpuzov, D.S., Yurasova, V.E.: Phys. Status Solidi (b) **47** (1971) 41.
71M Mashkova, E.S., Molchanov, V.A., Skripka, Y.G.: Sov. Phys. Dokl. (English Transl.) **16** (1971) 440.
71R Ryabov, V.A.: Sov. Phys. Solid State (English Transl.) **12** (1971) 2216.
71S1 Shul’ga, V.I., Bunin, I.G., Yurasova, V.E., Andreev, V.V., Mamaev, B.M.: Phys. Lett. A **37** (1971) 181.
71S2 Smith, D.P.: Surf. Sci. **25** (1971) 171.
71S3 Suurmeijer, E.P.T.M., Boers, A.L.: J. Phys. E **4** (1971) 663.
72A Arifov, U.A., Gruich, D.D., Ermakov, G.E., Khalmirzayev, E.: Radiat. Eff. **12** (1972) 181.
72B1 Ball, D.J., Buck, T.M., MacNair, D., Wheatley, G.H.: Surf. Sci. **30** (1972) 69.
72B2 Begemann, S.H.A., Boers, A.L.: Surf. Sci. **30** (1972) 134.
72B3 Begemann, S.H.A., Boers, A.L.: Surf. Sci. **32** (1972) 607.
72B4 Buck, T.M., Wheatley, G.H.: Surf. Sci. **33** (1972) 35.
72C Chicherov, V.M.: JETP Lett. (English Transl.) **16** (1972) 231.
72M1 Marwick, A.D., Thompson, M.W., Farmery, B.W., Harbinson, G.S.: Radiat. Eff. **15** (1972) 195.
72M2 Mashkova, E.S., Molchanov, V.A.: Radiat. Eff. **13** (1972) 131.
72M3 Mashkova, E.S., Molchanov, V.A.: Radiat. Eff. **13** (1972) 183.
72M4 Mashkova, E.S., Molchanov, V.A.: Radiat. Eff. **16** (1972) 143.
72M5 Molchanov, V.A., Snisar, V.A.: Radiat. Eff. **12** (1972) 105.
72M6 Molchanov, V.A., Snisar, V.A.: Surf. Sci. **33** (1972) 221.
72M7 Morgan, D.V., Bøgh, E.: Surf. Sci. **32** (1972) 278.
72S1 Shul'ga, V.I., Yurasova, V.E.: Sov. Phys. Solid State (English Transl.) **13** (1972) 1867.
72S2 Smith, A.G., Carter, G.: Radiat. Eff. **12** (1972) 63.
72T Taglauer, E., Heiland, W.: Surf. Sci. **33** (1972) 27.
72V van Veen, A., Haak, J.: Phys. Lett. A **40** (1972) 378.
72Y Yurasova, V.E., Bunin, I.G., Shul’ga, V.I., Mamaev, B.M.: Radiat. Eff. **12** (1972) 175.
73B1 Bøgh, E., in: Channeling. Morgan, D.V. (ed.), London: Wiley 1973, p. 435.
73B2 Brongersma, H.H., Mul, P.M.: Surf. Sci. **35** (1973) 393.
73B3 Brongersma, H.H., Mul, P.M.: Chem. Phys. Lett. **19** (1973) 217.
73B4 Buck, T.M., Wheatley, G.H., Feldman, L.C.: Surf. Sci. **35** (1973) 345.
73E Eckstein, W., Schäffler, H.G., Verbeek, H.: Radiat. Eff. **18** (1973) 263.

73F Feijen, H.H.W., Verhey, L.K., Boers, A.L.: Phys. Lett. A **45** (1973) 31.
73H1 Hagstrum, H.D., Becker, G.E.: Phys. Rev. B **8** (1973) 107.
73H2 Heiland, W., Schäffler, H.G., Taglauer, E.: Surf. Sci. **35** (1973) 381.
73M1 Marchenko, S.D., Parilis, E.S., Turaev, N.Y.: Radiat. Eff. **19** (1973) 147.
73M2 Marchenko, S.D., Parilis, E.S., Turaev, N.Y.: Sov. Phys. Dokl. (English Transl.) **17** (1973) 877.
73M3 Martynenko, Y.V.: Radiat. Eff. **20** (1973) 211.
73M4 Mashkova, E.S., Molchanov, V.A.: Radiat. Eff. **19** (1973) 29.
73M5 Morgan, D.V., Wood, D.R.: Proc. R. Soc. London A **335** (1973) 509.
73S Skripka, Y.G.: Radiat. Eff. **19** (1973) 211.
74B Brongersma, H.H.: J. Vac. Sci. Technol. **11** (1974) 231.
74E Edge, R.D., Sizmann, R., Varelas, C.: Radiat. Eff. **22** (1974) 79.
74K Karpuzov, D.S.: Surf. Sci. **45** (1974) 342.
74M1 Mashkova, E.S., Molchanov, V.A.: Radiat. Eff. **23** (1974) 215.
74M2 Mashkova, E.S., Molchanov, V.A., Pavlova, A.D.: Sov. Phys. Tech. Phys. (English Transl.) **19** (1974) 540.
74M3 Molchanov, V.A.: Radiat. Eff. **23** (1974) 197.
74S Suurmeijer, E.P.T.M., Boers, A.L.: Surf. Sci. **43** (1974) 309.
74Y Yurasova, V.E., in: The Physics of Ionized Gases-74. Vujnovic, V. (ed), Zagreb: Institute of Physics 1974, p. 427.
74Z Zyabrev, A.F., Sotnikov, V.M.: Sov. Phys. Solid State (English Transl.) **16** (1974) 289.
75B Buck, T.M., in: Methods of Surface Analysis. Czanderna, A.W. (ed.), Amsterdam: Elsevier 1975, p. 75.
75D1 Davies, J.A., Jackson, D.P., Mitchell, J.B., Norton, P.R., Tapping, R.L.: Phys. Lett. A **54** (1975) 239.
75D2 Della Mea, G., Drigo, A.V., Lo Russo, S., Mazzoldi, P., Yamaguchi, S., Bentini, G.G.: Appl. Phys. Lett. **26** (1975) 147.
75D3 de Wit, A.G.J., van der Schootbrugge, G.A., Fluit, J.M.: Surf. Sci. **47** (1975) 258.
75E Erickson, R.L., Smith, D.P.: Phys. Rev. Lett. **34** (1975) 297.
75F1 Farmery, B.W., Harbinson, G.S., Marwick, A.D., Pabst, H., Thompson, M.W., in: Atomic Collisions in Solids, Vol. 2. Datz, S., Appleton, B.R., Moak, C.D. (eds.), New York: Plenum Press 1975, p. 595.
75F2 Feijen, H.H.W, Verhey, L.K., Suurmeijer, E.P.T.M., Boers, A.L., in: Atomic Collisions in Solids, Vol. 2. Datz, S.D., Appleton, B.R., Moak, C.D. (eds.), New York: Plenum Press 1975, p. 573.
75H1 Harbinson, G.S., Farmery, B.W., Pabst, H.J., Thompson, M.W.: Radiat. Eff. **27** (1975) 97.
75H2 Heiland, W., Schäffler, H.G., Taglauer, E., in: Atomic Collisions in Solids, Vol. 2. Datz, S., Appleton, B.R., Moak, C.D. (eds.), New York: Plenum Press 1975, p. 599.
75M Mashkova, E.S., Molchanov, V.A.: Radiat. Eff. **25** (1975) 33.
75N Niehus, H., Bauer, E.: Surf. Sci. **47** (1975) 222.
75P Poelsema, B., Verhey, L.K., Boers, A.L.: Surf. Sci. **47** (1975) 256.
75S Shul'ga, V.I.: Radiat. Eff. **26** (1975) 61.
75T1 Taglauer, E., Melchior, W., Schuster, F., Heiland, W.: J. Phys. E **8** (1975) 768.
75T2 Tongson, L.L., Cooper, C.B.: Surf. Sci. **52** (1975) 263.
75V1 Verhey, L.K., Boers, A.L., in: Atomic Collisions in Solids, Vol. 2. Datz, S., Appleton, B.R., Moak, C.D. (eds.), New York: Plenum Press 1975, p. 583.
75V2 Verhey, L.K., Poelsema, B., Boers, A.L.: Radiat. Eff. **27** (1975) 47.
75Z Zehner, D.M., Appleton, B.R., Noggle, T.S., Miller, J.W., Barrett, J.H., Jenkins, L.H., Schow, O.E.: J. Vac. Sci. Technol. **12** (1975) 454.

76A Appleton, B.R., Zehner, D.M., Noggle, T.S., Miller, J.W., Schow, O.E., Jenkins, L.H., Barrett, J.H., in: Ion Beam Surface Layer Analysis, Vol. 2. Meyer, O., Linker, G., Käppeler, F. (eds.), New York: Plenum Press 1976, p. 607.
76B1 Balashova, L.L., Mashkova, E.S., Molchanov, V.A.: Bull. Acad. Sci. USSR, Phys. Ser. (English Transl.) **40** (1976) 125.
76B2 Brongersma, H.H., Hazewindus, N., van Nieuwland, J.M., Otten, A.M.M., Smets, A.J.: J. Vac. Sci. Technol. **13** (1976) 670.
76B3 Brongersma, H.H., Buck, T.M.: Nucl. Instrum. Methods **132** (1976) 559.
76D Davies, J.A., Jackson, D.P., Mitchell, J.B., Norton, P.R., Tapping, R.L.: Nucl. Instrum. Methods **132** (1976) 609.
76H1 Heiland, W., Taglauer, E.: Nucl. Instrum. Methods **132** (1976) 535.
76H2 Heiland, W., Taglauer, E., Robinson, M.T.: Nucl. Instrum. Methods **132** (1976) 655.
76H3 Hou, M., Robinson, M.T.: Nucl. Instrum. Methods **132** (1976) 641.
76J Jackson, D.P.: Nucl. Instrum. Methods **132** (1976) 603.
76K1 Kishinevsky, L.M., Parilis, E.S., Verleger, V.K.: Radiat. Eff. **29** (1976) 215.
76K2 Koerner, C., Varelas, C., Sizmann, R.: Radiat. Eff. **30** (1976) 65.
76M Mashkova, E.S., Molchanov, V.A., Serdobol'skaya, T.M.: Radiat. Eff. **28** (1976) 241.
76P1 Poelsema, B., Verhey, L.K., Boers, A.L.: Surf. Sci. **55** (1976) 445.
76P2 Poelsema, B., Verhey, L.K., Boers, A.L.: Surf. Sci. **60** (1976) 485.
76P3 Poelsema, B., Verhey, L.K., Boers, A.L.: Nucl. Instrum. Methods **132** (1976) 623.
76S Sizmann, R., Varelas, C.: Nucl. Instrum. Methods **132** (1976) 633.
76T1 Taglauer, E., Heiland, W.: Appl. Phys. **9** (1976) 261.
76T2 Turkenburg, W.C., Soszka, W., Saris, F.W., Kersten, H.H., Colenbrander, B.G.: Nucl. Instrum. Methods **132** (1976) 587.
76T3 Turkenburg, W.C., Kersten, H.H., Colenbrander, B.G., de Jongh, A.P., Saris, F.W.: Nucl. Instrum. Methods **138** (1976) 271.
76V1 van Veen, A., Fluit, J.M.: Nucl. Instrum. Methods **132** (1976) 573.
76V2 Verhey, L.K., Poelsema, B., Boers, A.L.: Nucl. Instrum. Methods **132** (1976) 565.
76V3 Verhey, L.K., Poelsema, B., Boers, A.L.: Radiat. Eff. **31** (1976) 23.
77A Appleton, B.R., Foti, G., in: Ion Beam Handbook for Material Analysis. Mayer, J.W., Rimini, E. (eds.), New York: Academic Press 1977, p. 67.
77B1 Balashova, L.L., Mashkova, E.S., Molchanov, V.A.: Phys. Lett. A **60** (1977) 67.
77B2 Balashova, L.L., Borisov, A.M., Mashkova, E.S., Molchanov, V.A.: Radiat. Eff. **31** (1977) 249.
77B3 Baun, W.L.: Appl. Surf. Sci. **1** (1977) 81.
77B4 Boers, A.L.: Surf. Sci. **63** (1977) 475.
77B5 Borisov, A.M., Mashkova, E.S., Molchanov, V.A., Serdobol'skaya, T.M.: Radiat. Eff. **32** (1977) 143.
77B6 Buck, T.M., in: Inelastic Ion-Surface Collisions. Tolk, N.H., Tully, J.C., Heiland, W., White, C.W. (eds.), New York: Academic Press 1977, p. 47.
77C Chen, Y.S., Miller, G.L., Robinson, D.A.H., Wheatley, G.H., Buck, T.M.: Surf. Sci. **62** (1977) 133.
77D Delannay, F., Bertrand, P., Streydio, J.-M.: Phys. Rev. B **16** (1977) 3903.
77F Feldman, L.C., Kauffman, R.L., Silverman, P.J., Zuhr, R.A., Barrett, J.H.: Phys. Rev. Lett. **39** (1977) 38.
77H1 Hagstrum, H.D., in: Inelastic Ion-Surface Collisions. Tolk, N.H., Tully, J.C., Heiland, W., White, C.W. (eds.), New York: Academic Press 1977, p. 1.
77H2 Heiland, W., Taglauer, E.: Surf. Sci. **68** (1977) 96.
77H3 Heiland, W., Taglauer, E., in: Inelastic Ion-Surface Collisions. Tolk, N.H., Tully, J.C., Heiland, W., White, C.W. (eds.), New York: Academic Press 1977, p. 27.
77J Jackson, D.P., Barrett, J.H.: Comput. Phys. Commun. **13** (1977) 157.

77O O'Connor, D.J., MacDonald, R.J.: Radiat. Eff. **34** (1977) 247.
77P1 Poelsema, B., Verhey, L.K., Boers, A.L.: Surf. Sci. **64** (1977) 537.
77P2 Poelsema, B., Verhey, L.K., Boers, A.L.: Surf. Sci. **64** (1977) 554.
77R Rusch, T.W., Erickson, R.L., in: Inelastic Ion-Surface Collisions. Tolk, N.H., Tully, J.C., Heiland, W., White, C.W. (eds.), New York: Academic Press 1977, p. 73.
77T1 Terzic, I., Goncic, B., Ciric, D., Vukanic, J.: J. Phys. E **10** (1977) 420.
77T2 Tully, J.C., Tolk, N.H., in: Inelastic Ion-Surface Collisions. Tolk, N.H., Tully, J.C., Heiland, W., White, C.W. (eds.), New York: Academic Press 1977, p.105.
77T3 Tully, J.C.: Phys. Rev. B **16** (1977) 4324.
77V Verheij, L.K., Poelsema, B., Boers, A.L.: Radiat. Eff. **34** (1977) 163.
78A1 Agamy, S.A., Robinson, J.E.: Nucl. Instrum. Methods **149** (1978) 595.
78A2 Aono, M., Nishitani, R., Tanaka, T., Bannai, E., Kawai, S.: Solid State Commun. **28** (1978) 409.
78B1 Barrett, J.H.: Nucl. Instrum. Methods **149** (1978) 341.
78B2 Bøgh, E., Jørgensen, H.W., Nielsen, G.J., Stensgaard, I., in: Proc. 7th Intern. Vacuum Congr. & 3rd Intern. Conf. Solid Surfaces, Vol. 3, Vienna 1978, p. 2707.
78B3 Bøgh, E., Stensgaard, I.: Phys. Lett. A **65** (1978) 357.
78B4 Brongersma, H.H., Beirens, L.C.M., van der Ligt, G.C.J., in: Material Characterization Using Ion Beams. Thomas, J. P., Cachard, A. (eds.), NATO ASI, Series B, Vol. 28, London: Plenum Press 1978, p. 65.
78B5 Brongersma, H.H., Buck, T.M.: Nucl. Instrum. Methods **149** (1978) 569.
78B6 Buck, T.M., Wheatley, G.H., Miller, G.L., Robinson, D.A.H., Chen, Y.-S.: Nucl. Instrum. Methods **149** (1978) 591.
78D1 Davies, J.A., Jackson, D.P., Matsunami, N., Norton, P.R., Andersen, J.U.: Surf. Sci. **78** (1978) 274.
78D2 Davies, J.A., Jackson, D.P., Norton, P.R., in: Proc. 7th Intern. Vacuum Congr. & 3rd Intern. Conf. Solid Surfaces, Vol. 3, Vienna 1978, p. 2527.
78E1 Ellis, W.P., Taylor, T.N.: Surf. Sci. **75** (1978) 279 (errat.: Surf. Sci. **77** (1978) 658).
78E2 Ellis, W.P., Taylor, T.N.: J. Vac. Sci. Technol. **15** (1978) 1769.
78H1 Heiland, W., Taglauer, E., in: The Physics of Ionized Gases-78. Janev, R.K.(ed.), Beograd: Institute of Physics 1978, p. 239.
78H2 Heiland, W., Taglauer, E., in: The Physics of Ionized Gases-78. Janev, R.K. (ed.), Beograd: Institute of Physics 1978, p. 257.
78H3 Heiland, W., Taglauer, E., in: Low-Energy Ion Beams. Stephens, K.G., Wilson, I.H., Moruzzi, J.L. (eds.), Inst. Phys. Conf. Ser. No. 38, 1978, p. 287.
78H4 Helbig, H.F., Adelmann, P.J., Miller, A.C., Czanderna, A.W.: Nucl. Instrum. Methods **149** (1978) 581.
78H5 Hoonhout, D., Kerkdijk, C.B.W., Bhattacharya, R.S., Kersten, H.H., Turkenburg, W.C., Saris, F.W.: Nucl. Instrum. Methods **149** (1978) 355.
78H6 Hou, M., Eckstein, W., Verbeek, H.: Radiat. Eff. **39** (1978) 107.
78H7 Hou, M., Robinson, M.T.: Appl. Phys. **17** (1978) 295.
78H8 Hou, M., Robinson, M.T.: Appl. Phys. **17** (1978) 371.
78I Ichinokawa, T., Kawamura, T., Nakamura, H., Fujimoto, F., Komaki, K.: Nucl. Instrum. Methods **149** (1978) 577.
78K Karpuzov, D.S., Evdokimov, I.N., Armour, D.G., Carter, G.: Phys. Lett. A **68** (1978) 485.
78L Luitjens, S.B., Verbeek, T.R., Algra, A.J., Boers, A.L.: Surf. Sci. **76** (1978) L609.
78M Mashkova, E.S., Molchanov, V.A., in: Low Energy Ion Beams. Stephens, K.G., Wilson, I.H., Moruzzi, J.L. (eds.), Inst. Phys. Conf. Ser. No. 38, 1978, p. 313.
78N Nizhnaya, S.L., Parilis, E.S., Turaev, N.Y., in: Proc. 7th Intern. Vacuum Congr & 3rd Intern. Conf. Solid Surfaces, Vol. 2, Vienna 1978, p. 1397.
78P Preuss, E.: Radiat. Eff. **38** (1978) 151.

78S1 Shul'ga, V.I.: Radiat. Eff. **37** (1978) 1.
78S2 Stensgaard, I., Feldman, L.C., Silverman, P.J.: Surf. Sci. **77** (1978) 513.
78T1 Thompson, M.W., Pabst, H.J.: Radiat. Eff. **37** (1978) 105.
78T2 Turkenburg, W.C., Smeenk, R.G., Saris, F.W.: Surf. Sci. **74** (1978) 181.
78V1 van der Veen, J.F., Sanders, J.B., Saris, F.W.: Surf. Sci. **77** (1978) 337.
78V2 van der Veen, J.F., Smeenk, R.G., Saris, F.W., in: Proc. 7th Intern. Vacuum Congr. & 3rd Intern. Conf. Solid Surfaces, Vol. 3, Vienna 1978, p. 2515.
78V3 Varelas, C., Sizmann, R.: Surf. Sci. **71** (1978) 51.
78Z Zuhr, R.A., Feldman, L.C., Kauffman, R.L., Silverman, P.J.: Nucl. Instrum Methods **149** (1978) 349.
79A1 Aono, M., Nishitani, R., Oshima, C., Tanaka, T., Bannai, E., Kawai, S.: Surf. Sci. **86** (1979) 631.
79A2 Armour, D.G., Van den Berg, J.A., Verheij, L.K.: J. Radioanal. Chem. **48** (1979) 359.
79B1 Balashova, L.L., Evdokimov, I.N., Mashkova, E.S., Molchanov, V.A.: Bull. Acad. Sci. USSR, Phys. Ser. (English Transl.) **43** (1979) 86.
79B2 Buck, T.M., Wheatley, G.H., Verheij, L.K.: Surf. Sci. **90** (1979) 635.
79E1 Evdokimov, I.N., Van den Berg, J.A., Armour, D.G.: Radiat. Eff. **41** (1979) 33.
79E2 Evdokimov, I.N., Webb, R., Armour, D.G., Karpuzov, D.S.: Radiat. Eff. **42** (1979) 83.
79E3 Evdokimov, I.N.: Phys. Lett. A **73** (1979) 417.
79E4 Evdokimov, I.N.: Bull. Acad. Sci. USSR, Phys. Ser. (English Transl.) **43** (1979) 108.
79F Feldman, L.C., Silverman, P.J., Stensgaard, I.: Surf. Sci. **87** (1979) 410.
79G Garrison, B.J.: Surf. Sci. **87** (1979) 683.
79H Heiland, W., Beitat, U., Taglauer, E.: Phys. Rev. B **19** (1979) 1677.
79J Jackson, D.P., Barrett, J.H.: Phys. Lett. A **71** (1979) 359.
79K Karpuzov, D.S., Armour, D.G., Evdokimov, I.N.: Radiat. Eff. **41** (1979) 141.
79L Luitjens, S.B., Algra, A.J., Boers, A.L.: Surf. Sci. **80** (1979) 566.
79N1 Nizhnaya, S.L., Parilis, E.S., Turaev, N.Y.: Radiat. Eff. **40** (1979) 43.
79N2 Norton, P.R., Davies, J.A., Jackson, D.P., Matsunami, N.: Surf. Sci. **85** (1979) 269.
79N3 Norton, P.R.: J. Vac. Sci. Technol. **16** (1979) 469.
79O Ohtsuki, Y.H., Koyama, K., Yamamura, Y.: Phys. Rev. B **20** (1979) 5044.
79P Poelsema, B., Boers, A.L.: Radiat. Eff. **41** (1979) 229.
79S1 Shoji, F., Oura, K., Hanawa, T.: J. Vac. Soc. Jpn. **22** (1979) 179.
79S2 Stensgaard, I., Feldman, L.C., Silverman, P.J.: Phys. Rev. Lett. **42** (1979) 247.
79V1 van der Veen, J.F., Smeenk, R.G., Tromp, R.M., Saris, F.W.: Surf. Sci. **79** (1979) 212.
79V2 van der Veen, J.F., Smeenk, R.G., Tromp, R.M., Saris, F.W.: Surf. Sci. **79** (1979) 219.
79V3 van der Veen, J.F., Tromp, R.M., Smeenk, R.G., Saris, F.W.: Surf. Sci. **82** (1979) 468.
79V4 Varelas, C., Goltz, K., Sizmann, R.: Surf. Sci. **80** (1979) 524.
79Y1 Yamamura, Y., Takeuchi, W.: Phys. Lett. A **70** (1979) 215.
79Y2 Yamamura, Y., Takeuchi, W.: Radiat. Eff. **42** (1979) 55.
80A1 Algra, A.J., Luitjens, S.B., Suurmeijer, E.P.T.M., Boers, A.L.: Phys. Lett. A **75** (1980) 496.
80A2 Algra, A.J., Luitjens, S.B., Suurmeijer, E.P.T.M., Boers, A.L.: Surf. Sci. **100** (1980) 329.
80A3 Algra, A.J., Maaskant, P.P., Luitjens, S.B., Suurmeijer, E.P.T.M., Boers, A.L.: J. Phys. D **13** (1980) 2363.
80A4 Algra, A.J., Luitjens, S.B., Suurmeijer, E.P.T.M., Boers, A.L., in: Proc. 4th Intern. Conf. Solid Surfaces & 3rd Europ. Conf. Surface Science, Vol. 1, Cannes 1980, p. 703.
80A5 Andrä, H.J., Winter, H., Fröhling, R., Kirchner, N., Plöhn, H.J., Wittmann, W., Graser, W., Varelas, C.: Nucl. Instrum. Methods **170** (1980) 527.
80B1 Barrett, J.H., Jackson, D.P.: Nucl. Instrum. Methods **170** (1980) 115.
80B2 Baun, W.L.: Surf. Sci. **100** (1980) L491.
80B3 Bertrand, P.: Nucl. Instrum. Methods **170** (1980) 489.

80B4 Boers, A.L., in: The Physics of Ionized Gases-80. Matic, M. (ed.), Beograd: Boris Kidric Institute 1980, p. 271.
80B5 Bronckers, R.P.N., de Wit, A.G.J., Hupkens, T.M., Fluit, J.M., Niehaus, A., in: Proc. 4th Intern. Conf. Solid Surfaces & 3rd Europ. Conf. Surface Science, Vol. 2, Cannes 1980, p. 1229.
80B6 Buck, T.M., Stensgaard, I., Wheatley, G.H., Marchut, L.: Nucl. Instrum. Methods **170** (1980) 519.
80C Culbertson, R.J., Feldman, L.C., Silverman, P.J.: Phys. Rev. Lett. **45** (1980) 2043.
80D Davies, J.A., Jackson, D.P., Norton, P.R., Posner, D.E., Unertl, W.N.: Solid State Commun. **34** (1980) 41.
80E Ellis, W.P., Taylor, T.N.: Surf. Sci. **91** (1980) 409.
80F1 Feldman, L.C., Silverman, P.J., Stensgaard, I.: Nucl. Instrum. Methods **168** (1980) 589.
80F2 Feldman, L.C., Kauffman, R.L., Silverman, P.J., Stensgaard, I., in: Proc. 7th Intern. Conf. Atomic Collisions in Solids, Vol. 2, Moscow 1980, p. 112.
80H Hou, M., Robinson, M.T., in: Proc. 7th Intern. Conf. Atomic Collisions in Solids, Vol. 2, Moscow 1980, p. 132.
80K Kerkow, H., Pietsch, H., Täubner, F.: Radiat. Eff. Lett. **50** (1980) 169.
80L1 Luitjens, S.B., Algra, A.J., Suurmeijer, E.P.T.M., Boers, A.L.: Appl. Phys. **21** (1980) 205.
80L2 Luitjens, S.B., Algra, A.J., Suurmeijer, E.P.T.M., Boers, A.L.: Surf. Sci. **99** (1980) 631.
80L3 Luitjens, S.B., Algra, A.J., Suurmeijer, E.P.T.M., Boers, A.L.: Surf. Sci. **99** (1980) 652.
80L4 Luitjens, S.B., Algra, A.J., Suurmeijer, E.P.T.M., Boers, A.L.: J. Phys. E **13** (1980) 665.
80L5 Luitjens, S.B., Boers, A.L., in: Proc. 7th Intern. Conf. Atomic Collisions in Solids, Vol. 2, Moscow 1980, p. 103.
80M1 Martin, D.J.: Surf. Sci. **97** (1980) 586.
80M2 Matschke, F.E.P., Eckstein, W., Verbeek, H., in: Proc. 7th Intern. Conf. Atomic Collisions in Solids, Vol. 2, Moscow 1980, p. 98.
80M3 McCune, R.C.: Appl. Surf. Sci. **5** (1980) 275.
80M4 Morita, K.: Radiat. Eff. **52** (1980) 235.
80O1 O'Connor, D.J., MacDonald, R.J.: Radiat. Eff. **45** (1980) 205.
80O2 O'Connor, D.J., MacDonald, R.J.: Nucl. Instrum. Methods **170** (1980) 495.
80O3 Oen, O.S., in: Proc. 7th Intern. Conf. Atomic Collisions in Solids, Vol. 2, Moscow 1980, p. 124.
80S1 Shelyakin, L.B., Schaarschmidt, G., Yurasova, V.E., in: Proc. 7th Intern. Conf. Atomic Collisions in Solids, Vol. 2, Moscow 1980, p. 119.
80S2 Shul'ga, V.I., in: Proc. 7th Intern. Conf. Atomic Collisions in Solids, Vol. 2, Moscow 1980, p. 128.
80T Taglauer, E., Englert, W., Heiland, W., Jackson, D.P.: Phys. Rev. Lett. **45** (1980) 740.
80V van der Veen, J.F., Tromp, R.M., Smeenk, R.G., Saris, F.W.: Nucl. Instrum. Methods **171** (1980) 143.
80W Woodruff, D.P., Godfrey, D.J.: Solid State Commun. **34** (1980) 679.
80Y1 Yamamura, Y., Takeuchi, W.: Radiat. Eff. **49** (1980) 251.
80Y2 Yamamura, Y., Takeuchi, W.: Phys. Lett. A **78** (1980) 105.
81A1 Aono, M., Oshima, C., Zaima, S., Otani, S., Ishizawa, Y.: Jpn. J. Appl. Phys. **20** (1981) L829.
81A2 Armour, D.G.: Vacuum **31** (1981) 417.
81B1 Baun, W.L.: Surf. Interface Anal. **3** (1981) 243.
81B2 Baun, W.L.: J. Vac. Sci. Technol. **18** (1981) 711.
81B3 Bronckers, R.P.N., de Wit, A.G.J.: Surf. Sci. **104** (1981) 384.
81B4 Bronckers, R.P.N., de Wit, A.G.J.: Surf. Sci. **112** (1981) 111.
81C1 Culbertson, R.J., Feldman, L.C., Silverman, P.J.: J. Vac. Sci. Technol. **18** (1981) 871.
81C2 Culbertson, R.J., Feldman, L.C., Silverman, P.J., Boehm, H.: Phys. Rev. Lett. **47** (1981) 657.
81D Davies, J.A., Jackman, T.E., Jackson, D.P., Norton, P.R.: Surf. Sci. **109** (1981) 20.

81E Evdokimov, I.N., Webb, R.P., Armour, D.G.: Radiat. Eff. Lett. **58** (1981) 59.

81F1 Feldman, L.C.: Nucl. Instrum. Methods **191** (1981) 211.

81F2 Feldman, L.C.: CRC Crit. Rev. Solid State Mater. Sci. **10** (1981) 143, and in:Chemistry and Physics of Solid Surfaces, Vol. III. Vanselow, R., England, W.(eds.), Boca Raton: CRC Press 1982, p. 221.

81J Jackson, D.P., Heiland, W., Taglauer, E.: Phys. Rev. B **24** (1981) 4198.

81K1 Karpuzov, D.S.: Radiat. Eff. **54** (1981) 87.

81K2 Kinoshita, K., Narusawa, T., Gibson, W.M.: Surf. Sci. **110** (1981) 369.

81M Mashkova, E.S.: Radiat. Eff. **54** (1981) 1.

81N1 Narusawa, T., Gibson, W.M., Törnqvist, E.: Phys. Rev. Lett. **47** (1981) 417.

81N2 Norton, P.R., Davies, J.A., Creber, D.K., Sitter, C.W., Jackman, T.E.: Surf. Sci. **108** (1981) 205.

81O1 Oshima, C., Aono, M., Zaima, S., Shibata, Y., Kawai, S.: J. Less-Common Met. **82** (1981) 69.

81O2 Overbury, S.H., Heiland, W., Zehner, D.M., Datz, S., Thoe, R.S.: Surf. Sci. **109** (1981) 239.

81O3 Overbury, S.H.: Surf. Sci. **112** (1981) 23.

81P Preuss, E.: Surf. Sci. **110** (1981) 287.

81R Robinson, M.T.: J. Nucl. Mater. **103/104** (1981) 525.

81S Stensgaard, I., Feldman, L.C., Silverman, P.J.: Surf. Sci. **102** (1981) 1.

81T1 Taylor, T.N., Ellis, W.P.: Surf. Sci. **107** (1981) 249.

81T2 Tromp, R.M., Smeenk, R.G., Saris, F.W.: Phys. Rev. Lett. **46** (1981) 939.

81T3 Tromp, R.M., Smeenk, R.G., Saris, F.W.: Solid State Commun. **39** (1981) 755.

81V1 Van den Berg, J.A., Armour, D.G.: Vacuum **31** (1981) 259.

81V2 von dem Hagen, T., Bauer, E.: Phys. Rev. Lett. **47** (1981) 579.

82A1 Adams, D.L., Nielsen, H.B., Andersen, J.N., Stensgaard, I., Feidenhans'l, R., Sørensen, J.E.: Phys. Rev. Lett. **49** (1982) 669.

82A2 Algra, A.J., van Loenen, E., Suurmeijer, E.P.T.M., Boers, A.L.: Radiat. Eff. **60** (1982) 173.

82A3 Algra, A.J., Luitjens, S.B., Borggreve, H., Suurmeijer, E.P.T.M., Boers, A.L.: Radiat. Eff. **62** (1982) 7.

82A4 Algra, A.J., Luitjens, S.B., Suurmeijer, E.P.T.M., Boers, A.L.: Nucl. Instrum. Methods **203** (1982) 515.

82A5 Algra, A.J., Luitjens, S.B., Suurmeijer, E.P.T.M., Boers, A.L.: Appl. Surf. Sci. **10** (1982) 273.

82A6 Aono, M., Hou, Y., Oshima, C., Ishizawa, Y.: Phys. Rev. Lett. **49** (1982) 567.

82A7 Aono, M., Hou, Y., Souda, R., Oshima, C., Otani, S., Ishizawa, Y., Matsuda, K., Shimizu, R.: Jpn. J. Appl. Phys. **21** (1982) L670.

82B1 Balashova, L.L., Garin, S.N., Dodonov, A.I., Mashkova, E.S., Molchanov, V.A.: Surf. Sci. **119** (1982) L378.

82B2 Balashova, L.L., Molchanov, V.A.: Radiat. Eff. **60** (1982) 49.

82B3 Bauer, E., in: Interfacial Aspects of Phase Transformations. Mutaftschiev, B.(ed.), NATO ASI, Series C, Vol. 87, Dordrecht: Reidel, 1982 p. 1.

82C Cruz, S.A., Alonso, E.V., Walker, R.P., Martin, D.J., Armour, D.G.: Nucl. Instrum. Methods **194** (1982) 659.

82E Englert, W., Taglauer, E., Heiland, W.: Nucl. Instrum. Methods **194** (1982) 663.

82F1 Feldman, L.C.: Appl. Surf. Sci. **13** (1982) 211.

82F2 Feldman, L.C., Mayer, J.W., Picraux, S.T.: Materials Analysis by Ion Channeling, New York: Academic Press 1982.

82H1 Heiland, W.: Appl. Surf. Sci. **13** (1982) 282.

82H2 Heiland, W.: Vacuum **32** (1982) 539.

82H3 Heiland, W., Taglauer, E.: Nucl. Instrum. Methods **194** (1982) 667.

82H4 Helbig, H.F., Linder, M.W., Morris, G.A., Steward, S.A.: Surf. Sci. **114** (1982) 251.

82J1 Jackman, T.E., Davies, J.A., Jackson, D.P., Unertl, W.N., Norton, P.R.: Surf. Sci. **120** (1982) 389.

82J2 Jackson, D.P., Davies, J.A., Jackman, T.E., Norton, P.R., Unertl, W.N.: Nucl. Instrum. Methods **194** (1982) 143.

82K Komaki, K., Ootuka, A., Fujimoto, F.: Jpn. J. Appl. Phys. **21** (1982) L521.

82M1 Makoshi, K., Hatada, M.: Surf. Sci. **114** (1982) 673.

82M2 Makoshi, K., Hatada, M., Yoshimori, A.: Surf. Sci. **119** (1982) 305.

82M3 Mashkova, E.S., Molchanov, V.A., Shul'ga, V.I.: Radiat. Eff. **62** (1982) 107.

82N1 Narusawa, T., Gibson, W.M., Törnqvist, E.: Surf. Sci. **114** (1982) 331.

82N2 Niehus, H., Preuss, E.: Surf. Sci. **119** (1982) 349.

82N3 Nizhnaya, S.L., Parilis, E.S., Verleger, V.K.: Radiat. Eff. **62** (1982) 173.

82S1 Saris, F.W.: Nucl. Instrum. Methods **194** (1982) 625.

82S2 Smeenk, R.G., Tromp, R.M., Kersten, H.H., Boerboom, A.J.H., Saris, F.W.: Nucl. Instrum. Methods **195** (1982) 581.

82T Tromp, R.M., van Loenen, E.J., Iwami, M., Saris, F.W.: Solid State Commun. **44** (1982) 971.

82U Unertl, W.N.: Appl. Surf. Sci. **11/12** (1982) 64.

82V von dem Hagen, T., Hou, M., Bauer, E.: Surf. Sci. **117** (1982) 134.

82W1 Walker, R.P., Martin, D.J.: Surf. Sci. **118** (1982) 659.

82W2 Williams, R.S.: J. Vac. Sci. Technol. **20** (1982) 770.

82W3 Woodruff, D.P.: Nucl. Instrum. Methods **194** (1982) 639.

82Y Yamamura, Y., Takeuchi, W.: Radiat. Eff. **61** (1982) 247.

83A1 Aono, M.: J. Vac. Soc. Jpn **26** (1983) 136.

83A2 Aono, M., Souda, R., Oshima, C., Ishizawa, Y.: Phys. Rev. Lett. **51** (1983) 801.

83A3 Aono, M., Souda, R., Oshima, C., Ishizawa, Y.: Nucl. Instrum. Methods Phys. Res. **218** (1983) 241.

83B1 Balashova, L.L., Dodonov, A.I., Garin, S.N., Mashkova, E.S., Molchanov, V.A.: J. Phys. B **16** (1983) 2609.

83B2 Balashova, L.L., Dodonov, A.I., Firsov, O.B., Garin, S.N., Mashkova, E.S., Molchanov, V.A.: Radiat. Eff. **77** (1983) 67.

83B3 Bennett, P.A., Feldman, L.C., Kuk, Y., McRae, E.G., Rowe, J.E.: Phys. Rev. B **28** (1983) 3656.

83B4 Bertrand, P., Ghalim, M.: Phys. Scr. T **6** (1983) 168.

83B5 Buck, T.M., Wheatley, G.H., Jackson, D.P.: Nucl. Instrum. Methods Phys. Res. **218** (1983) 257.

83B6 Buck, T.M., Wheatley, G.H., Marchut, L.: Phys. Rev. Lett. **51** (1983) 43.

83D1 Davis, S.M., Carver, J.C., Wold, A.: Surf. Sci. **124** (1983) L12.

83D2 de Wit, A.G.J., van Zoest, J.M., Fluit, J.M.: Surf. Sci. **126** (1983) 589.

83E Evdokimov, I.N.: Phys. Chem. Mech. Surfaces (English Transl.) **2** (1983) 1342.

83F1 Feidenhans'l, R., Sørensen, J.E., Stensgaard, I.: Surf. Sci. **134** (1983) 329.

83F2 Feldman, L.C.: Phys. Scr. **28** (1983) 303.

83F3 Frenken, J.W.M., van der Veen, J.F., Allan, G.: Phys. Rev. Lett. **51** (1983) 1876.

83F4 Frenken, J.W.M., Smeenk, R.G., van der Veen, J.F.: Surf. Sci. **135** (1983) 147.

83G1 Gossmann, H.-J., Gibson, W.M., Itoh, T., Feldman, L.C.: J. Vac. Sci. Technol. A **1** (1983) 1059.

83G2 Graser, W., Varelas, C.: Phys. Scr. T **6** (1983) 153.

83J Jackson, D.P., Jackman, T.E., Davies, J.A., Unertl, W.N., Norton, P.R.: Surf. Sci. **126** (1983) 226.

83K1 Karpuzov, D.S., Alonso, E.V., Walker, R.P., Armour, D.G., Martin, D.J.: Surf. Sci. **129** (1983) L271.

83K2 Kuk, Y., Feldman, L.C., Silverman, P.J.: Phys. Rev. Lett. **50** (1983) 511.

83N1 Newns, D.M., Makoshi, K., Brako, R., van Wunnik, J.N.M.: Phys. Scr. T **6** (1983) 5.

83N2 Niehus, H.: Nucl. Instrum. Methods Phys. Res. **218** (1983) 230.

83O1 Oen, O.S.: Surf. Sci. **131** (1983) L407.

83O2 Ohtsuki, Y.: J. Vac. Soc. Jpn **26** (1983) 61.
83O3 Overbury, S.H., Stair, P.C., Agron, P.A.: Surf. Sci. **125** (1983) 377.
83P Poelsema, B., Verheij, L.K., Boers, A.L.: Surf. Sci. **133** (1983) 344.
83R Rabalais, J.W., Schultz, J.A., Kumar, R.: Nucl. Instrum. Methods Phys. Res. **218** (1983) 719.
83S1 Schuster, M., Varelas, C.: Surf. Sci. **134** (1983) 195.
83S2 Souda, R., Aono, M., Oshima, C., Otani, S., Ishizawa, Y.: Surf. Sci. **128** (1983) L236.
83S3 Stensgaard, I., Feidenhans'l, R., Sørensen, J.E.: Surf. Sci. **128** (1983) 281.
83T1 Treacy, P.B., Foster, C.: Radiat. Eff. **77** (1983) 167.
83T2 Tromp, R.M.: J. Vac. Sci. Technol. A **1** (1983) 1047.
83T3 Tromp, R.M., Smeenk, R.G., Saris, F.W., Chadi, D.J.: Surf. Sci. **133** (1983) 137.
83T4 Tromp, R.M., Smit, L., van der Veen, J.F.: Phys. Rev. Lett. **51** (1983) 1672.
83T5 Tromp, R.M., van der Veen, J.F.: Surf. Sci. **133** (1983) 159.
83W1 Williams, R.S., Yarmoff, J.A.: Nucl. Instrum. Methods Phys. Res. **218** (1983) 235.
83W2 Woodruff, D.P.: Vacuum **33** (1983) 651.
83Y1 Yamamura, Y., Takeuchi, W.: Phys. Lett. A **94** (1983) 109.
83Y2 Yarmoff, J.A., Williams, R.S.: Surf. Sci. **127** (1983) 461.
84A1 Aono, M.: J. Vac. Sci. Technol. A **2** (1984) 635.
84A2 Aono, M.: Nucl. Instrum. Methods Phys. Res. B **2** (1984) 374.
84B1 Bitensky, I.S., Parilis, E.S.: Nucl. Instrum. Methods Phys. Res. B **2** (1984) 364.
84B2 Boers, A.L.: Nucl. Instrum. Methods Phys. Res. B **2** (1984) 353.
84B3 Boers, A.L.: Nucl. Instrum. Methods Phys. Res. B **4** (1984) 98.
84B4 Boulliard, J.C., Cohen, C., Domange, J.L., Drigo, A.V., L'Hoir, A., Moulin, J., Sotto, M.: Phys. Rev. B **30** (1984) 2470.
84C Coudray, C., Bernheim, M., Slodzian, G.: Nucl. Instrum. Methods Phys. Res. B **2** (1984) 431.
84D Dodonov, A.I., Garin, S.N., Mashkova, E.S., Molchanov, V.A.: Surf. Sci. **140** (1984) L244.
84E Engelmann, G., Taglauer, E.: Nucl. Instrum. Methods Phys. Res. B **2** (1984) 436.
84G1 Ghalim, M., Bertrand, P.: Nucl. Instrum. Methods Phys. Res. B **2** (1984) 427.
84G2 Gossmann, H.-J., Gibson, W.M.: Surf. Sci. **139** (1984) 239.
84H1 Haight, R., Feldman, L.C., Buck, T.M., Gibson, W.M.: Nucl. Instrum. Methods Phys. Res. B **2** (1984) 501.
84H2 Haight, R., Feldman, L.C., Buck, T.M., Gibson, W.M.: Phys. Rev. B **30** (1984) 734.
84H3 Hou, M., Varelas, C.: Appl. Phys. A **33** (1984) 121.
84I Ito, T., Aristov, V.Y., Gibson, W.M.: J. Vac. Sci. Technol. A **2** (1984) 864.
84J1 Jackson, D.P., Barrett, J.H.: Nucl. Instrum. Methods Phys. Res. B **2** (1984) 318.
84J2 Jackson, D.P., Buck, T.M., Wheatley, G.H.: Nucl. Instrum. Methods Phys. Res. B **2** (1984) 440.
84K1 Kuk, Y., Feldman, L.C.: Phys. Rev. B **30** (1984) 5811.
84K2 Kuk, Y., Feldman, L.C., Robinson, I.K.: Surf. Sci. **138** (1984) L168.
84M1 MacDonald, R.J., O'Connor, D.J.: Aust. J. Phys. **37** (1984) 389.
84M2 Marchut, L., Buck, T.M., Wheatley, G.H., McMahon, C.J.: Surf. Sci. **141** (1984) 549.
84M3 Mashkova, E.S., Fleurov, V.B.: Phys. Chem. Mech. Surfaces (English Transl.) **3** (1984) 1018.
84M4 Mashkova, E.S., Fleurov, V.B.: Radiat. Eff. **80** (1984) 227.
84M5 Mashkova, E.S., Fleurov, V.B.: Radiat. Eff. Lett. **86** (1984) 115.
84N1 Niehus, H., Comsa, G.: Surf. Sci. **140** (1984) 18.
84N2 Niehus, H.: Surf. Sci. **145** (1984) 407.
84O Overbury, S.H.: Nucl. Instrum. Methods Phys. Res. B **2** (1984) 448.
84R1 Richard, A., Eschenbacher, H.: Nucl. Instrum. Methods Phys. Res. B **2** (1984) 444.
84R2 Robinson, I.K., Kuk, Y., Feldman, L.C.: Phys. Rev. B **29** (1984) 4762.
84S1 Smit, L., Tromp, R.M., van der Veen, J.F.: Nucl. Instrum. Methods Phys. Res. B **2** (1984) 322.
84S2 Smit, L., Tromp, R.M., van der Veen, J.F.: Phys. Rev. B **29** (1984) 4814.
84T1 Takeuchi, W., Yamamura, Y.: Nucl. Instrum. Methods Phys. Res. B **2** (1984) 336.

84T2 Tenner, A.D., Gillen, K.T., Horn, T.C.M., Los, J., Kleyn, A.W.: Phys. Rev. Lett. **52** (1984) 2183.
84T3 Thompson, K.A., Valone, S.M., Ellis, W.P., Taylor, T.N., Maggiore, C.J.: J. Vac. Sci. Technol. A **2** (1984) 1005.
84T4 Tolk, N.H., Willerding, B., Steininger, H., Heiland, W., Snowdon, K.J.: Nucl. Instrum. Methods Phys. Res. B **2** (1984) 488.
84T5 Törnqvist, E., Adams, E.D., Copel, M., Gustafsson, T., Graham, W.R.: J. Vac. Sci. Technol. A **2** (1984) 939.
84T6 Tromp, R.M., Smit, L., van der Veen, J.F.: Phys. Rev. B **30** (1984) 6235.
84T7 Tromp, R.M., van Loenen, E.J.: Phys. Rev. B **30** (1984) 7352.
84V1 van Zoest, J.M., van der Meij, C.E., Fluit, J.M.: Nucl. Instrum. Methods Phys. Res. B **2** (1984) 406.
84V2 Vyatkin, E.G., Taratin, A.M., Vorobiev, S.A.: Radiat. Eff. **82** (1984) 97.
84W1 Willerding, B., Heiland, W., Snowdon, K.J.: Phys. Rev. Lett. **53** (1984) 2031.
84W2 Willerding, B., Steininger, H., Snowdon, K.J., Heiland, W.: Nucl. Instrum. Methods Phys. Res. B **2** (1984) 453.
84Y Yamamura, Y., Takeuchi, W.: Radiat. Eff. **82** (1984) 73.
85A1 Aono, M., Souda, R.: Jpn. J. Appl. Phys. **24** (1985) 1249.
85A2 Aono, M., Souda, R., Oshima, C., Ishizawa, Y.: J. Vac. Sci. Technol. A **3** (1985) 1514.
85B Balashova, L.L., Garin, S.N., Molchanov, V.A.: Appl. Phys. A **37** (1985) 171.
85C1 Chang, C.-C., DeLouise, L.A., Winograd, N., Garrison, B.J.: Surf. Sci. **154** (1985) 22.
85C2 Culbertson, R.J., Barrett, J.H., Withrow, S.P.: Surf. Sci. **157** (1985) 451.
85D1 de Zwart, S.T., Drentje, A.G., Boers, A.L.: Nucl. Instrum. Methods Phys. Res. B **9** (1985) 608.
85D2 Dodonov, A.I., Molchanov, V.A.: Radiat. Eff. **90** (1985) 85.
85D3 Dodonov, A.I., Mashkova, E.S.: Radiat. Eff. **90** (1985) 93.
85E Engelmann, G., Taglauer, E., Jackson, D.P.: Surf. Sci. **162** (1985) 921.
85F Frenken, J.W.M., van der Veen, J.F.: Phys. Rev. Lett. **54** (1985) 134.
85G Graser, W., Varelas, C.: Surf. Sci. **157** (1985) 74.
85H1 Heiland, W., Snowdon, K.J., in: The Physics of Ionized Gases-84. Popovic, M.M., Krstic, P. (eds.), Singapore: World Scientific 1985, p. 415.
85H2 Hellings, G.J.A., Ottevanger, H., Boelens, S.W., Knibbeler, C.L.C.M., Brongersma, H.H.: Surf. Sci. **162** (1985) 913.
85H3 Hemme, H., Heiland, W.: Nucl. Instrum. Methods Phys. Res. B **9** (1985) 41.
85H4 Higginbottom, P., Homer, J., O'Connor, D.J., MacDonald, R.J.: Appl. Surf. Sci. **22/23** (1985) 100.
85I Ito, T., Gibson, W.M.: Surf. Sci. **161** (1985) 115.
85J Jakas, M.M., Harrison, D.E.: Surf. Sci. **149** (1985) 500.
85L Landuyt, J.P., De Gryse, R., Van der Elstraeten, L., Vennik, J.: Surf. Sci. **152/153** (1985) 114.
85M Mashkova, E.S., Molchanov, V.A.: Medium-Energy Ion Reflection from Solids. Agranovich, V.M., Maradudin, A.A. (eds.), Modern Problems in Condensed Matter Sciences, Vol. 11, Amsterdam: North-Holland 1985, p. 127.
85N1 Narusawa, T., Kobayashi, K.L.I., Nakashima, H.: Jpn. J. Appl. Phys. **24** (1985) L98.
85N2 Niehus, H., Comsa, G.: Surf. Sci. **152/153** (1985) 93.
85O Overbury, S.H., Huntley, D.R.: Phys. Rev. B **32** (1985) 6278.
85S1 Sato, K., Kono, S., Teruyama, T., Higashiyama, K., Sagawa, T.: Surf. Sci. **158** (1985) 644.
85S2 Smit, L., Derry, T.E., van der Veen, J.F.: Surf. Sci. **150** (1985) 245.
85S3 Smit, L., Tromp, R.M., van der Veen, J.F.: Surf. Sci. **163** (1985) 315.
85S4 Smit, L., van der Veen, J.F., in: Proc. 17th Intern. Conf. Physics of Semiconductors, San Francisco. New York: Springer 1985, p. 81.
85S5 Souda, R., Aono, M., Oshima, C., Otani, S., Ishizawa, Y.: Surf. Sci. **150** (1985) L59.
85T1 Taglauer, E.: Appl. Phys. A **38** (1985) 161.

85T2 Takeuchi, W., Yamamura, Y.: Surf. Sci. **154** (1985) 1.
85T3 Tromp, R.M., van Loenen, E.J.: Surf. Sci. **155** (1985) 441.
85V1 van der Veen, J.F.: Surf. Sci. Rep. **5** (1985) 199.
85V2 van Zoest, J.M., van der Mey, C.E., Fluit, J.M., Niehaus, A.: Surf. Sci. **152/153** (1985) 106.
85W1 Willerding, B., Snowdon, K., Heiland, W.: Z. Phys. B **59** (1985) 435.
85W2 Withrow, S.P., Barrett, J.H., Culbertson, R.J.: Surf. Sci. **161** (1985) 584.
85Z1 Zaima, S., Shibata, Y., Adachi, H., Oshima, C., Otani, S., Aono, M., Ishizawa, Y.: Surf. Sci. **157** (1985) 380.
85Z2 Ziegler, J.F., Biersack, J.P., Littmark, U.: The Stopping and Range of Ions in Solids, Vol. 1. New York: Pergamon Press 1985.
86A1 Akazawa, H., Kusunoki, I., Murata, Y.: Surf. Sci. **177** (1986) 577.
86A2 Alkemade, P.F.A., Turkenburg, W.C., van der Weg, W.F.: Nucl. Instrum. Methods Phys. Res. B **15** (1986) 126.
86B Balashova, L.L.: Moscow Univ. Phys. Bull. (English Transl.) **41** (1986) 113.
86C1 Chang, C.S., Knipping, U., Tsong, I.S.T.: Nucl. Instrum. Methods Phys. Res. B **18** (1986) 11.
86C2 Copel, M., Gustafsson, T.: Phys. Rev. Lett. **57** (1986) 723.
86C3 Copel, M., Gustafsson, T., Graham, W.R., Yalisove, S.M.: Phys. Rev. B **33** (1986) 8110.
86C4 Culbertson, R.J., Kuk, Y., Feldman, L.C.: Surf. Sci. **167** (1986) 127.
86D Derry, T.E., Smit, L., van der Veen, J.F.: Surf. Sci. **167** (1986) 502.
86E Engelmann, G., Taglauer, E., Jackson, D.P.: Nucl. Instrum. Methods Phys. Res. B **13** (1986) 240.
86F1 Fauster, T., Metzner, M.H.: Surf. Sci. **166** (1986) 29.
86F2 Feldman, L.C., Mayer, J.W.: Fundamentals of Surface and Thin Film Analysis. New York: North-Holland 1986, p. 100.
86F3 Frenken, J.W.M., Marée, P.M.J., van der Veen, J.F.: Phys. Rev. B **34** (1986) 7506.
86F4 Frenken, J.W.M., Tromp, R.M., van der Veen, J.F.: Nucl. Instrum. Methods Phys. Res. B **17** (1986) 334.
86F5 Frenken, J.W.M., van der Veen, J.F., Barnett, R.N., Landman, U., Cleveland, C.L.: Surf. Sci. **172** (1986) 319.
86G Graham, W.R., Yalisove, S.M., Adams, E.D., Gustafsson, T., Copel, M., Törnqvist, E.: Nucl. Instrum. Methods Phys. Res. B **16** (1986) 383.
86J Jin, H.-S., Ito, T., Gibson, W.M.: J. Vac. Sci. Technol. A **4** (1986) 1385.
86M1 Mannami, M., Kimura, K., Nakanishi, K., Nishimura, A.: Nucl. Instrum. Methods Phys. Res. B **13** (1986) 587.
86M2 Möller, J., Niehus, H., Heiland, W.: Surf. Sci. **166** (1986) L111.
86M3 Möller, J., Snowdon, K.J., Heiland, W., Niehus, H.: Surf. Sci. **178** (1986) 475.
86M4 Moore, J.A., Jackman, T.E., Davies, J.A., Stevens, J.L.E.: Nucl. Instrum. Methods Phys. Res. B **17** (1986) 250.
86N1 Nelson, G.C.: J. Vac. Sci. Technol. A **4** (1986) 1567.
86N2 Niehus, H.: Surf. Sci. **166** (1986) L107.
86N3 Niehus, H., Comsa, G.: Nucl. Instrum. Methods Phys. Res. B **13** (1986) 213.
86N4 Niehus, H., Comsa, G.: Nucl. Instrum. Methods Phys. Res. B **15** (1986) 122.
86O1 Ohtsuki, Y.H., Kawai, R., Tange, K.: Nucl. Instrum. Methods Phys. Res. B **13** (1986) 193.
86O2 Oku, T., Kanasaki, J., Matsunami, N., Itoh, N., Matsuda, K., Aoki, M.: Nucl. Instrum. Methods Phys. Res. B **15** (1986) 142.
86O3 Overbury, S.H.: Surf. Sci. **175** (1986) 123.
86R Ross, G.G., Terreault, B.: Nucl. Instrum. Methods Phys. Res. B **15** (1986) 146.
86S1 Shul'ga, V.I.: Radiat. Eff. **100** (1986) 71.
86S2 Shul'ga, V.I.: Phys. Status Solidi (b) **134** (1986) 87.
86S3 Smit, L., van der Veen, J.F.: Surf. Sci. **166** (1986) 183.
86S4 Souda, R., Aono, M.: Nucl. Instrum. Methods Phys. Res. B **15** (1986) 114.

86S5 Souda, R., Aono, M., Oshima, C., Otani, S., Ishizawa, Y.: Nucl. Instrum. Methods Phys. Res. B **15** (1986) 138.
86S6 Souda, R., Aono, M., Oshima, C., Otani, S., Ishizawa, Y.: Surf. Sci. **176** (1986) 657.
86T1 Takeuchi, W., Yamamura, Y.: Surf. Sci. **169** (1986) 365.
86T2 Tenner, A.D., Gillen, K.T., Horn, T.C.M., Los, J., Kleyn, A.W.: Surf. Sci. **172** (1986) 90.
86T3 Tenner, A.D., Saxon, R.P., Gillen, K.T., Harrison, D.E., Horn, T.C.M., Kleyn, A.W.: Surf. Sci. **172** (1986) 121.
86V Visscher, B., Boers, A.L., Verheij, L.K., Poelsema, B.: Appl. Surf. Sci. **26** (1986) 121.
86W Willerding, B., Snowdon, K.J., Imke, U., Heiland, W.: Nucl. Instrum. Methods Phys. Res. B **13** (1986) 614.
86Y1 Yalisove, S.M., Graham, W.R., Adams, E.D., Copel, M., Gustafsson, T.: Surf. Sci. **171** (1986) 400.
86Y2 Yarmoff, J.A., Blumenthal, R., Williams, R.S.: Surf. Sci. **165** (1986) 1.
86Y3 Yarmoff, J.A., Cyr, D.M., Huang, J.H., Kim, S., Williams, R.S.: Phys. Rev. B **33** (1986) 3856.
86Y4 Yarmoff, J.A., Williams, R.S.: J. Vac. Sci. Technol. A **4** (1986) 1274.
86Y5 Yarmoff, J.A., Williams, R.S.: Rev. Sci. Instrum. **57** (1986) 433.
86Y6 Yarmoff, J.A., Williams, R.S.: Surf. Sci. **166** (1986) 101.
86Y7 Yarmoff, J.A., Williams, R.S.: Surf. Sci. **165** (1986) L73.
87A1 Aizawa, T., Tsuno, T., Daimon, H., Ino, S.: Phys. Rev. B **36** (1987) 9107.
87A2 Alkemade, P.F.A., Turkenburg, W.C., van der Weg, W.F.: Surf. Sci. **189/190** (1987) 161.
87A3 Alkemade, P.F.A., Turkenburg, W.C., van der Weg, W.F.: Nucl. Instrum. Methods Phys. Res. B **28** (1987) 161.
87A4 Aono, M., Souda, R.: Nucl. Instrum. Methods Phys. Res. B **27** (1987) 55.
87C1 Chang, C.S., Knipping, U., Tsong, I.S.T.: Nucl. Instrum. Methods Phys. Res. B **28** (1987) 493.
87C2 Copel, M., Fenter, P., Gustafsson, T.: J. Vac. Sci. Technol. A **5** (1987) 742.
87D1 Derks, H., Hemme, H., Heiland, W., Overbury, S.H.: Nucl. Instrum. Methods Phys. Res. B **23** (1987) 374.
87D2 Derks, H., Möller, J., Heiland, W.: Surf. Sci. **188** (1987) L685.
87E Engelmann, G., Englert, W., Kato, S., Taglauer, E., Jackson, D.P.: Nucl. Instrum. Methods Phys. Res. B **26** (1987) 522.
87F1 Fauster, T., Schneider, R., Dürr, H., Engelmann, G., Taglauer, E.: Surf. Sci. **189/190** (1987) 610.
87F2 Frenken, J.W.M., Huussen, F., van der Veen, J.F.: Phys. Rev. Lett. **58** (1987) 401.
87G Ghrayeb, R., Purushotham, M., Hou, M., Bauer, E.: Phys. Rev. B **36** (1987) 7364.
87H1 Heiland, W., Imke, U., Schubert, S., Snowdon, K.J.: Nucl. Instrum. Methods Phys. Res. B **27** (1987) 167.
87H2 Holub-Krappe, E., Horn, K., Frenken, J.W.M., Krans, R.L., van der Veen, J.F.: Surf. Sci. **188** (1987) 335.
87K1 Kimura, K., Mannami, M.: Nucl. Instrum. Methods Phys. Res. B **27** (1987) 442.
87K2 Kimura, K., Hasegawa, M., Mannami, M.: Phys. Rev. B **36** (1987) 7.
87K3 Kimura, K., Nishimura, A., Mannami, M.: Surf. Sci. **183** (1987) L313.
87M Mashkova, E.S., Molchanov, V.A.: Radiat. Eff. **102** (1987) 125.
87N Niehus, H.: J. Vac. Sci. Technol. A **5** (1987) 751.
87O1 O'Connor, D.J.: J. Phys. E **20** (1987) 437.
87O2 Overbury, S.H.: Nucl. Instrum. Methods Phys. Res. B **27** (1987) 65.
87P Pluis, B., Denier van der Gon, A.W., Frenken, J.W.M., van der Veen, J.F.: Phys. Rev. Lett. **59** (1987) 2678.
87S1 Smith, R.J., Whang, C.N., Xu Mingde, Worthington, M., Hennessy, C., Kim, M., Holland, N.: Rev. Sci. Instrum. **58** (1987) 2284.
87S2 Smith, R.J., Hennessy, C., Kim, M.W., Whang, C.N., Worthington, M., Xu Mingde: Phys. Rev. Lett. **58** (1987) 702.

87S3 Souda, R., Aono, M., Oshima, C., Otani, S., Ishizawa, Y.: Surf. Sci. **179** (1987) 199.

87Y Yalisove, S.M., Graham, W.R.: Surf. Sci. **183** (1987) 556.

88A Ackermans, P.A.J., van der Meulen, P.F.H.M., Ottevanger, H., van Straten, F.E., Brongersma, H.H.: Nucl. Instrum. Methods Phys. Res. B **35** (1988) 541.

88C Chang, C.-C., Winograd, N., Garrison, B.J.: Surf. Sci. **202** (1988) 309.

88D1 Daley, R.S., Williams, R.S.: J. Vac. Sci. Technol. A **6** (1988) 808.

88D2 Derks, H., Möller, J., Heiland, W., in: The Structure of Surfaces II. van der Veen, J.F., Van Hove, M.A. (eds.), Springer Series in Surface Science, Vol. 11, Berlin: Springer 1988, p. 496.

88D3 Dodonov, A.I., Mashkova, E.S., Molchanov, V.A.: Radiat. Eff. **107** (1988) 27.

88E Eklund, E.A., Daley, R.S., Huang, J.H., Williams, R.S., in: The Structure of Surfaces II. van der Veen, J.F., Van Hove, M.A. (eds.), Springer Series in Surface Science, Vol. 11, Berlin: Springer 1988, p. 75.

88F1 Fauster, T.: Vacuum **38** (1988) 129.

88F2 Fauster, T., Hartwig, D., Dürr, H.: Appl. Phys. A **45** (1988) 63.

88F3 Fenter, P., Gustafsson, T.: Phys. Rev. B **38** (1988) 10197.

88F4 Frenken, J.W.M.: Nucl. Instrum. Methods Phys. Res. B **33** (1988) 884.

88F5 Fujii, Y., Kimura, K., Hasegawa, M., Suzuki, M., Susuki, Y., Mannami, M.: Nucl. Instrum. Methods Phys. Res. B **33** (1988) 405.

88G1 Gotoh, M., Matsunami, N., Kitoh, K., Tanaka, S., Yamamura, Y.: Nucl. Instrum. Methods Phys. Res. B **33** (1988) 415.

88G2 Gustafsson, T., Copel, M., Fenter, P., in: The Structure of Surfaces II. van der Veen, J.F., Van Hove, M.A. (eds.), Springer Series in Surface Science, Vol. 11, Berlin: Springer 1988, p. 110.

88H1 Hetterich, W., Derks, H., Heiland, W.: Nucl. Instrum. Methods Phys. Res. B **33** (1988) 401.

88H2 Hetterich, W., Derks, H., Heiland, W.: Appl. Phys. Lett. **52** (1988) 371.

88K1 Kanasaki, J., Matsunami, N., Itoh, N., Oku, T., Kitoh, K., Aoki, M., Matsuda, K.: Nucl. Instrum. Methods Phys. Res. B **33** (1988) 619.

88K2 Katayama, M., Nomura, E., Kanekama, N., Soejima, H., Aono, M.: Nucl. Instrum. Methods Phys. Res. B **33** (1988) 857.

88K3 Kato, M., Iitaka, T., Ohtsuki, Y.H.: Nucl. Instrum. Methods Phys. Res. B **33** (1988) 432.

88K4 Kato, M., Williams, R.S., Aono, M.: Nucl. Instrum. Methods Phys. Res. B **33** (1988) 462.

88K5 Kimura, K., Fujii, Y., Hasegawa, M., Susuki, Y., Mannami, M.: Phys. Rev. B **38** (1988) 1052.

88K6 Kimura, K., Hasegawa, M., Fujii, Y., Suzuki, M., Susuki, Y., Mannami, M.: Nucl. Instrum. Methods Phys. Res. B **33** (1988) 358.

88M1 MacDonald, R.J., O'Connor, D.J., Wilson, J., Shen, Y.G.: Nucl. Instrum. Methods Phys. Res. B **33** (1988) 446.

88M2 Mannami, M., Fujii, Y., Kimura, K.: Surf. Sci. **204** (1988) 213.

88M3 Marée, P.M.J., Nakagawa, K., van der Veen, J.F., Tromp, R.M.: Phys. Rev. B **38** (1988) 1585.

88M4 Mullins, D.R., Overbury, S.H.: J. Vac. Sci. Technol. A **6** (1988) 811.

88M5 Mullins, D.R., Overbury, S.H.: Surf. Sci. **199** (1988) 141.

88N1 Nakamatsu, H., Sudo, A., Kawai, S.: Surf. Sci. **194** (1988) 265.

88N2 Niehus, H.: Nucl. Instrum. Methods Phys. Res. B **33** (1988) 876.

88P Pfandzelter, R., Schuster, M.: Nucl. Instrum. Methods Phys. Res. B **33** (1988) 898.

88S1 Shoji, F., Nakayama, Y., Oura, K., Hanawa, T.: Nucl. Instrum. Methods Phys. Res. B **33** (1988) 420.

88S2 Snowdon, K.J., O'Connor, D.J., MacDonald, R.J.: Phys. Rev. Lett. **61** (1988) 1760.

88S3 Souda, R., Aono, M., in: The Structure of Surfaces II. van der Veen, J.F., Van Hove, M.A. (eds.), Springer Series in Surface Science, Vol. 11, Berlin: Springer 1988, p. 581.

88S4 Sumitomo, K., Oura, K., Katayama, I., Shoji, F., Hanawa, T.: Nucl. Instrum. Methods Phys. Res. B **33** (1988) 871.

88S5 Suzuki, M., Kawauchi, H., Kimura, K., Mannami, M.: Surf. Sci. **204** (1988) 223.

88T1 Taglauer, E., Beckschulte, M., Margraf, R., Mehl, D.: Nucl. Instrum. Methods Phys. Res. B **35** (1988) 404.

88T2 Takeuchi, W., Yamamura, Y.: Nucl. Instrum. Methods Phys. Res. B **33** (1988) 425.

88Y Yalisove, S.M., Graham, W.R.: J. Vac. Sci. Technol. A **6** (1988) 588.

89A1 Aono, M., Katayama, M., Nomura, E., Chassé, T., Choi, D., Kato, M.: Nucl. Instrum. Methods Phys. Res. B **37/38** (1989) 264.

89A2 Aristarkhova, A.A., Volkov, S.S., Trukhin, V.V., Shchuppe, G.N.: Sov. Tech. Phys. Lett. (English Transl.) **15** (1989) 780.

89B1 Balashova, L.L.: Radiat. Eff. Express **2** (1989) 141.

89B2 Beuken, J.-M., Pierson, E., Bertrand, P.: Surf. Sci. **223** (1989) 201.

89C Chini, T.K., Ghose, D.: Nucl. Instrum. Methods Phys. Res. B **42** (1989) 293.

89D1 Daley, R.S., Huang, J.H., Williams, R.S.: Surf. Sci. **215** (1989) 281.

89D2 Derks, H., Närmann, A., Heiland.W.: Nucl. Instrum. Methods Phys. Res. B **44** (1989) 125.

89E Evstifeev, V.V., Ivanov, I.V.: Surf. Sci. **217** (1989) L373.

89G Grizzi, O., Shi, M., Bu, H., Rabalais, J.W., Hochmann, P.: Phys. Rev. B **40** (1989) 10127.

89H1 Heiland, W.: Vacuum **39** (1989) 367.

89H2 Hetterich, W., Heiland, W.: Surf. Sci. **210** (1989) 129.

89H3 Hou, M.: Vacuum **39** (1989) 309.

89K Kasi, S.R., Kang, H., Sass, C.S., Rabalais, J.W.: Surf. Sci. Rep. **10** (1989) 1.

89M1 McEachern, R.L., Goodstein, D.M., Cooper, B.H.: Phys. Rev. B **39** (1989) 10503.

89M2 Mendenhall, M.H., Weller, R.A.: Nucl. Instrum. Methods Phys. Res. B **40/41** (1989) 1239.

89M3 Monreal, R., Flores, F., Närmann, A., Heiland, W., Schubert, S., Echenique, P.M.: Radiat. Eff. Defects Solids **109** (1989) 75.

89M4 Monreal, R., Goldberg, E.C., Flores, F., Närmann, A., Derks, H., Heiland, W.: Surf. Sci. **211/212** (1989) 271.

89N1 Närmann, A., Derks, H., Heiland, W., Monreal, R., Goldberg, E., Flores, F.: Surf. Sci. **217** (1989) 255.

89N2 Niehus, H., in: The Physics of Ionized Gases-88. Tanovic, L., Konjevic, N., Tanovic, N. (eds.), New York: Nova Science 1989, p. 257.

89N3 Nienhaus, H., Zimny, R., Winter, H.: Radiat. Eff. Defects Solids **109** (1989) 1.

89P Pluis, B., Taylor, T.N., Frenkel, D., van der Veen, J.F.: Phys. Rev. B **40** (1989) 1353.

89R Rabalais, J.W., Grizzi, O., Shi, M., Bu, H.: Phys. Rev. Lett. **63** (1989) 51.

89S1 Shoji, F., Oura, K., Hanawa, T.: Nucl. Instrum. Methods Phys. Res. B **36** (1989) 23.

89S2 Snowdon, K.J., O'Connor, D.J., MacDonald, R.J.: Radiat. Eff. Defects Solids **109** (1989) 25.

89S3 Snowdon, K.J., O'Connor, D.J., MacDonald, R.J.: Radiat. Eff. Defects Solids **109** (1989) 33.

89S4 Stensgaard, I., Purcell, K.G., King, D.A.: Phys. Rev. B **39** (1989) 897.

89V1 van de Riet, E., Fluit, J.M., Niehaus, A.: Surf. Sci. **211/212** (1989) 106.

89V2 van de Riet, E., Smeets, J.B.J., Fluit, J.M., Niehaus, A.: Surf. Sci. **214** (1989) 111.

89V3 Veje, E., Winter, H., Hagedorn, H.: Radiat. Eff. Defects Solids **109** (1989) 47.

89X Xu, C., O'Connor, D.J.: Nucl. Instrum. Methods Phys. Res. B **42** (1989) 251.

89Z Zimny, R., Nienhaus, H., Winter, H.: Radiat. Eff. Defects Solids **109** (1989) 9.

90B1 Boerma, D.O.: Nucl. Instrum. Methods Phys. Res. B **50** (1990) 77.

90B2 Boerma, D.O., Labohm, F., Reinders, J.A.: Nucl. Instrum. Methods Phys. Res. B **50** (1990) 291.

90B3 Bu, H., Shi, M., Masson, F., Rabalais, J.W.: Surf. Sci. **230** (1990) L140.

90C1 Cheng, H.-S., Chui, Z.-X., Xu, H.-J., Yao, X.-W., Yang, F.-J.: Nucl. Sci. Techn. **1** (1990) 105.

90C2 Cheng, H.-S., Chui, Z.-X., Xu, H.-J., Yao, X.-W., Yang, F.-J.: Nucl. Instrum. Methods Phys. Res. B **45** (1990) 424.

90C3 Cox, D.F., Fryberger, T.B.: Surf. Sci. **227** (1990) L105.

90D1 Daley, R.S., Farrelly, D., Williams, R.S.: Surf. Sci. **234** (1990) 355.

90D2 Daley, R.S., Williams, R.S.: Nucl. Instrum. Methods Phys. Res. B **45** (1990) 380.

90 D3 Deckers, S., Habraken, F.H.P.M., van der Weg, W.F., Denier van der Gon, A.W., Pluis, B., van der Veen, J.F., Baudoing, R.: Phys. Rev. B **42** (1990) 3253.

90D4 Denier van der Gon, A.W., Smith, R.J., Gay, J.M., O'Connor, D.J., van der Veen, J.F.: Surf. Sci. **227** (1990) 143.

90D5 Derks, H., Hetterich, W., van de Riet, E., Niehus, H., Heiland, W.: Nucl. Instrum. Methods Phys. Res. B **48** (1990) 315.

90G Grizzi, O., Shi, M., Bu, H., Rabalais, J.W.: Rev. Sci. Instrum. **61** (1990) 740.

90H Hara, S., Slijkerman, W.F.J., van der Veen, J.F., Ohdomari, I., Misawa, S., Sakuma, E., Yoshida, S.: Surf. Sci. **231** (1990) L196.

90K Kasi, S.R., Rabalais, J.W.: Radiat. Eff. Defects Solids **112** (1990) 119.

90L Los, J., Geerlings, J.J.C.: Phys. Rep. **190** (1990) 133.

90N1 Närmann, A., Monreal, R., Echenique, P.M., Flores, F., Heiland, W., Schubert, S.: Phys. Rev. Lett. **64** (1990) 1601.

90N2 Närmann, A., Schmidt, K., Heiland, W., Monreal, R., Flores, F., Echenique, P.M.: Nucl. Instrum. Methods Phys. Res. B **48** (1990) 378.

90N3 Niehof, A., Heiland, W.: Nucl. Instrum. Methods Phys. Res. B **48** (1990) 306.

90N4 Niehus, H., Raunau, W., Besocke, K., Spitzl, R., Comsa, G.: Surf. Sci. **225** (1990) L8.

90O1 O'Connor, D.J., King, B.V., MacDonald, R.J., Shen, Y.G., Xu, C.: Aust. J. Phys. **43** (1990) 601.

90O2 Overbury, S.H., Mullins, D.R., Wendelken, J.F.: Surf. Sci. **236** (1990) 122.

90P1 Pfandzelter, R., Lee, J.W.: Nucl. Instrum. Methods Phys. Res. B **45** (1990) 641.

90P2 Pluis, B., Denier van der Gon, A.W., van der Veen, J.F., Riemersma, A.J.: Surf. Sci. **239** (1990) 265.

90R1 Rabalais, J.W.: Science **250** (1990) 521.

90R2 Ross, G.G., Leblanc, L.: Nucl. Instrum. Methods Phys. Res. B **48** (1990) 134.

90S1 Shao, Q., Cheng, H.-S., Xu, H.-J.: Nucl. Tech. **13** (1990) 91.

90S2 Shi, M., Bu, H., Rabalais, J.W.: Phys. Rev. B **42** (1990) 2852.

90S3 Shoji, F., Kashihara, K., Hanawa, T., Oura, K.: Nucl. Instrum. Methods Phys. Res. B **47** (1990) 1.

90S4 Shul'ga, V.I.: Phys. Chem. Mech. Surfaces (English Transl.) **5** (1990) 1003.

90S5 Snowdon, K.J., O'Connor, D.J., Kato, M., MacDonald, R.J.: Nucl. Instrum. Methods Phys. Res. B **48** (1990) 327.

90S6 Souda, R., Aizawa, T., Oshima, C., Ishizawa, Y.: Nucl. Instrum. Methods Phys. Res. B **45** (1990) 364.

90S7 Spitzl, R., Niehus, H., Comsa, G.: Rev. Sci. Instrum. **61** (1990) 760.

90S8 Susuki, Y., Mukai, H., Kimura, K., Mannami, M.: Nucl. Instrum. Methods Phys. Res. B **48** (1990) 347.

90S9 Susuki, Y., Mukai, H., Kimura, K., Mannami, M.: J. Phys. Soc. Jpn. **59** (1990) 1211.

90V1 van de Riet, E., Fluit, J.M., Niehaus, A.: Surf. Sci. **231** (1990) 368.

90V2 van de Riet, E., Derks, H., Heiland, W.: Surf. Sci. **234** (1990) 53.

90V3 Verbist, G., Devreese, J.T., Brongersma, H.H.: Surf. Sci. **233** (1990) 323.

90W1 Watson, P.R.: J. Phys. Chem. Ref. Data **19** (1990) 85.

90W2 Williams, R.S., Kato, M., Daley, R.S., Aono, M.: Surf. Sci. **225** (1990) 355.

90W3 Winter, H., Remillieux, J., Poizat, J.C.: Nucl. Instrum. Methods Phys. Res. B **48** (1990) 382.

90W4 Wouters, P.A.A.F., Zeijlmans van Emmichoven, P.A., Fluit, J.M., Niehaus, A.: Meas. Sci. Technol. **1** (1990) 41.

90X Xu, C., O'Connor, D.J.: Nucl. Instrum. Methods Phys. Res. B **51** (1990) 278.

90Y1 Yamamura, Y., Matsunami, N., Kitoh, K.: Nucl. Instrum. Methods Phys. Res. B **45** (1990) 403.

90Y2 Yanagisawa, J., Yoshimori, A.: Surf. Sci. **231** (1990) 297.

90Y3 Young, V.Y., Hoflund, G.B., Miller, A.C.: Surf. Sci. **235** (1990) 60.

90Z Zimny, R., Nienhaus, H., Winter, H.: Nucl. Instrum. Methods Phys. Res. B **48** (1990) 361.

91A Adler, D.L., Cooper, B.H.: Phys. Rev. B **43** (1991) 3876.
91B1 Bu, H., Shi, M., Rabalais, J.W.: Surf. Sci. **244** (1991) 96.
91B2 Bu, H., Shi, M., Rabalais, J.W.: Nucl. Instrum. Methods Phys. Res. B **61** (1991) 337.
91D1 Denier van der Gon, A.W., Gay, J.M., Frenken, J.W.M., van der Veen, J.F.: Surf Sci. **241** (1991) 335.
91D2 Denier van der Gon, A.W., van Pinxteren, H.M., Frenken, J.W.M., van der Veen, J.F.: Surf. Sci. **244** (1991) 259.
91D3 Denier van der Gon, A.W., Frenkel, D., Frenken, J.W.M., Smith, R.J., Stoltze, P.: Surf. Sci. **256** (1991) 385.
91E Eckstein, W.: Computer Simulation of Ion-Solid Interactions. Gonser, U., Mooradian, A., Osgood, R.M., Panish, M.B., Sakaki, H. (eds.), Spinger Series in Materials Science, Vol. 10, Berlin: Springer 1991.
91F Fujii, Y., Fujiwara, S., Kimura, K., Mannami, M.: Nucl. Instrum. Methods Phys. Res. B **58** (1991) 18.
91G1 Gruich, D.D., Dzhurakhalov, A.A., Tukhtaev, M.A., Umarov, F.F.: Phys. Chem. Mech. Surfaces (English Transl.) **6** (1991) 1136.
91G2 Guseva, M.B., Babaev, V.G., Ubaidullaev, R.R., An, S.Z.: Vacuum **42** (1991) 613.
91H1 Haner, A.N., Ross, P.N., Bardi, U.: Surf. Sci. **249** (1991) 15.
91H2 Hetterich, W., Höfner, C., Heiland, W.: Surf. Sci. **251/252** (1991) 731.
91H3 Heiland, W.: Surf. Sci. **251/252** (1991) 942.
91H4 Heiland, W., in: Interaction of Charged Particles with Solids and Surfaces. Gras-Marti, A., Flores, F., Urbassek, H.M., Arista, N.R. (eds.), NATO ASI, Series B, Vol. 97, New York: Plenum Press 1991, p. 253.
91H5 Hildner, M.L., Daley, R.S., Felter, T.E., Estrup, P.J.: J. Vac. Sci. Technol. A **9** (1991) 1604.
91H6 Hetterich, W., Heiland, W.: Surf. Sci. **258** (1991) 307.
91H7 Hsu, C.C., Rabalais, J.W.: Surf. Sci. **256** (1991) 77.
91J Jiang, Q.T., Fenter, P., Gustafsson, T.: Phys. Rev. B **44** (1991) 5773.
91K Kamiya, I., Katayama, M., Nomura, E., Aono, M.: Surf. Sci. **242** (1991) 404.
91M1 Masson, F., Bu, H., Shi, M., Rabalais, J.W.: Surf. Sci. **249** (1991) 313.
91M2 Masson, F., Rabalais, J.W.: Surf. Sci. **253** (1991) 245.
91M3 Masson, F., Rabalais, J.W.: Chem. Phys. Lett. **179** (1991) 63.
91M4 Mix, W., Rechtien, J.-H., Danailov, D., Snowdon, K.J.: Nucl. Instrum. Methods Phys. Res. B **58** (1991) 365.
91N1 Nakagawa, S.T.: Radiat. Eff. Defects Solids **116** (1991) 21.
91N2 Närmann, A.: Mod. Phys. Lett. B **5** (1991) 561.
91R1 Rabalais, J.W.: J. Vac. Sci. Technol. A **9** (1991) 1293.
91R2 Rechtien, J.-H., Mix, W., Snowdon, K.J.: Surf. Sci. **248** (1991) 27.
91S1 Schippers, S., Oelschig, S., Heiland, W., Folkerts, L., Morgenstern, R., Eeken, P., Urazgil'din, I.F., Niehaus, A.: Surf. Sci. **257** (1991) 289.
91S2 Souda, R., Aizawa, T., Hayami, W., Otani, S., Ishizawa, Y.: Surf. Sci. **241** (1991) 190.
91T Tappe, W., Niehof, A., Schmidt, K., Heiland, W.: Europhys. Lett. **15** (1991) 405.
91V1 van de Riet, E., Deckers, S., Habraken, F.H.P.M., Niehaus, A.: Surf. Sci. **243** (1991) 49.
91V2 van de Riet, E., Niehaus, A.: Surf. Sci. **243** (1991) 43.
91V3 van Slooten, U., Horn, T.C.M., Kleyn, A.W.: Surf. Sci. **243** (1991) 12.
91W1 Winter, H., Kirsch, R., Poizat, J.C., Remillieux, J.: Phys. Rev. A **43** (1991) 1660.
91W2 Winter, H.: Comments At. Mol. Phys. **26** (1991) 287.
91W3 Wyputta, F., Zimny, R., Winter, H.: Nucl. Instrum. Methods Phys. Res. B **58** (1991) 379.
91X1 Xu, C., O'Connor, D.J.: Nucl. Instrum. Methods Phys. Res. B **53** (1991) 315.
91X2 Xu, C., O'Connor, D.J.: Nucl. Instrum. Methods Phys. Res. B **53** (1991) 326.

7. Interaction of atoms with surfaces

7.1 Introduction

7.1.1 Preliminary remarks

He diffraction from the (100) surfaces of ionic crystals was first seen by Stern in 1929 [29S1]. The phenomenon of selective adsorption, or resonant scattering into surface-bound states, was observed by Frisch and Stern in 1933 [33F1] and explained by Lennard-Jones and Devonshire in 1937 [37L1]. However, the development of thermal energy atom scattering (TEAS), and of He atom scattering (HAS) in particular, as a surface investigation method, proceeded slowly for the next 35 years. During this period many experiments were carried out in the classical scattering regime, mostly to study the average transfer in the atom-surface collision. The surface rainbow effect, which is significant for surface structure analysis, was recognized by McClure [70M1] in the experiments by Smith et al. [69S1]. These developments are reviewed in the book by Goodman and Wachman [76G1].

The power of HAS as a technique for surface studies was gradually increased by many improvements, mainly:

(a) the use of supersonic nozzle beams of high intensity and excellent energy resolution, instead of effusive sources of low intensity with a Maxwellian velocity distribution;

(b) the availability of ultra-high vacuum, which significantly improved the signal-to-noise ratio and enabled the study of many well-characterized surfaces. The remaining limitation of HAS is in the detection of the scattered atoms. The most common detectors involve ionization of He by electron bombardement, followed by the highly efficient counting of the He^+ ions. Unfortunately, the ionization efficiency does not surpass 10^{-5}. Bolometric detectors have comparable sensitivity at best.

By 1975, high-quality diffraction data for He and Ne could be obtained by Boato et al. [76B1]. They were interpreted on the basis of the hard wall model and the eikonal approximation by Levi and collaborators [75G1, 76C2].

This theoretical framework has been used for the interpretation of many structural studies of ordered surfaces using He diffraction (see [82E1] for a review).

High-resolution measurements of surface phonon dispersion curves using a He atom time-of-flight (HATOF or TOF) apparatus were reported by Brusdeylins, Doak, and Toennies in 1980 [80B1]. Since then inelastic HAS has become a major tool for the study of surface dynamics. The data are interpreted with the help of surface lattice dynamics calculations and of inelastic atom-surface scattering theory, as discussed earlier by Cabrera, Celli, and Manson [69C1, 71M1].

With the advent of well-characterized surfaces, it became possible to study steps, terraces, and surface disorder. These developments are reviewed in the book by Poelsema and Comsa [89P1], who introduced the useful notion of the cross section of a surface defect. (See also chapter 2.3 of subvolume III/24a). The use of HAS in conjunction with other techniques, in particular direct imaging by Scanning Tunneling Microscopy, is increasingly fruitful in these studies.

Much of the usefulness of He scattering to surface science stems from data analysis by simple kinematics: the two-dimensional diffraction pattern from an ordered surface directly gives the reciprocal lattice vectors of the surface; the broadening of the diffraction peaks indicates the type and amount of disorder; dips or peaks in the intensity of a diffracted beam as a function of the incidence angles (θ_i, ϕ_i), or of the incident energy E_i, give the surface bound state energies, ε_n; and sharp peaks in the time-of-flight (TOF) spectra directly give the dispersion curves $\omega(\boldsymbol{q}_{\parallel}, j)$ of the surface phonon branches. These phenomena are not unique to TEAS and are discussed in other parts of this volume under general headings, or in connection with electron, ion, and X-ray scattering. However, for a full analysis of the data a knowledge of the He-surface interaction is useful, and often essential. Conversely, information obtained by beam scattering is useful in the study of physisorption.

Section 7.1.2 deals with atom-surface potentials, as determined mostly by elastic scattering. Section 7.1.3 deals with the information obtained by He scattering on the surface phonon spectra. In section 7.2 results for the atom-surface potentials are then given for each compound where they are available. Surface phonon spectra are also given, and reference is made to section 4.1 by R. Wallis and S.Y. Tong, subvolume III/24b.

7.1.2 Elastic scattering: the atom-surface potential

The atom-surface interaction consists of a short-range repulsion V_{rep} and a long range attraction V_{att}, with V_{rep} coming predominantly from the top atom layer on the surface and V_{att} almost entirely from van der Waals dispersion force. As shown in Fig. 1 the salient features of the total potential $V = V_{att} + V_{rep}$ are the surface well of average depth D, with surface bound states ε_n, and the surface corrugation h_{pv}. D and ε_n are determined from resonant scattering, as discussed in section 7.1.2.1, and the corrugation h_{pv} is determined from the strength of the diffraction peaks.

Semi-empirical forms of $V(\boldsymbol{r})$ that fit the observed bound states and corrugation have been widely used and are discussed in section 7.1.2.2. In the simplest approximation V_{rep} is replaced by a corrugated hard wall and V_{att} is neglected, except of course when dealing with the surface-bound states. We review the predictions of the hard wall models in section 7.1.2.3.

The fitting of empirical parameters to experimental data is expensive and model-dependent. A priori calculations of the atom-surface interaction have then come increasingly into play. If nothing else, they have suggested more realistic empirical models, with fewer adjustable parameters. We discuss these theories in section 7.1.2.4 for the static potential and in section 7.1.3.3 for the dynamic interaction.

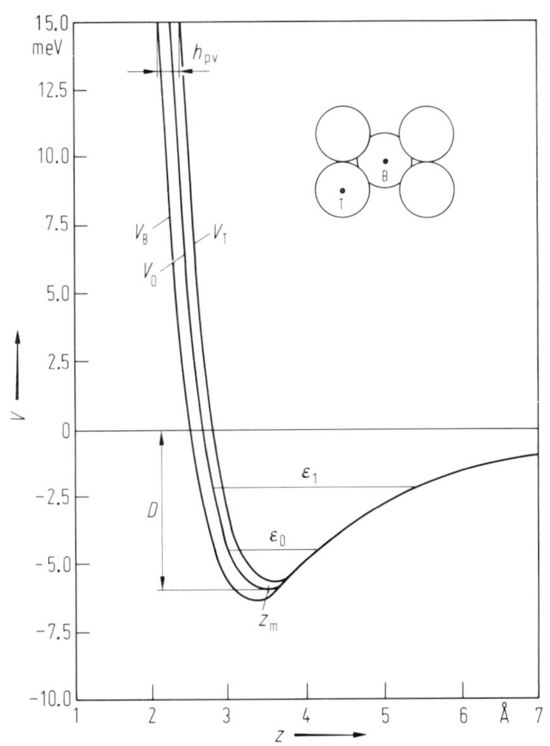

Fig. 1. Typical atom-surface potential V (He-Ag(110), adapted from [87E1]). The surface unit cell is shown in the inset. The potential at the top V_T and bottom V_B registries, and the laterally averaged potential V_0 are plotted as a function of the distance z from the top layer of surface nuclei. D and z_m are the depth and position of the minimum of $V_0(z)$. The first two surface-bound energy levels, ε_0 and ε_1, are also shown. For this potential, the peak-to-valley height of the surface corrugation, h_{pv}, is only weakly energy-dependent for positive energies in the range shown.

7.1.2.1 Kinematics and resonant scattering

The kinematic parameters of elastic scattering will be denoted as follows: $\boldsymbol{k}_i$ is the incident atom's wave vector, with components $(\boldsymbol{k}_{i\|}, k_{iz})$, $k_{iz} < 0$; $\boldsymbol{k}_f$ is the final wave vector, with components $(\boldsymbol{k}_{f\|}, k_{fz})$, $k_{fz} > 0$. Following standard notation, the z axis is perpendicular to the mean surface, pointing outwards, and the "lateral" vector $\boldsymbol{k}_\|$ is in the xy plane.

For diffraction through the surface reciprocal lattice vector $\boldsymbol{g}_\|$,

$$\boldsymbol{k}_{f\|} = \boldsymbol{k}_{i\|} + \boldsymbol{g}_\| \tag{1}$$

and energy conservation requires that $k_{fz} = k_z(\boldsymbol{g}_\|)$, with

$$k_z(\boldsymbol{g}_\|)^2 = k_i^2 - (\boldsymbol{k}_{i\|} + \boldsymbol{g}_\|)^2 . \tag{2}$$

In particular, the perpendicular wave vector of the specular beam is $k_z(0)$, and $k_{iz} = -k_z(0)$. For an atom of mass m, the incident and final energies are $\varepsilon_i = \hbar^2 k_i^2/2m$ and $\varepsilon_f = \hbar^2 k_f^2/2m$.

The intensity of the diffracted beams, including the specular, as a function of ε_i, or of the incidence angles θ_i and ϕ_i, often displays sharp features, including isolated minima (selective adsorption) and maxima. All beams are affected for the same conditions of incidence, which suggests that a resonance process is involved. At resonance, the atom is trapped in a surface-bound state for times of the order of 10^{-12} seconds. In the surface-bound state the lateral momentum $\boldsymbol{k}_{i\|} + \boldsymbol{g}_\|$ is so large that $k_z(\boldsymbol{g}_\|)^2$, according to Eq. (2), is negative, and the corresponding perpendicular energy is approximately equal to a bound state energy ε_n of the laterally averaged surface potential. The total energy is still positive, however, and equal to ε_i; the lost perpendicular energy is transferred to lateral motion. For a molecule, it is also possible to have resonances mediated by a rotational excitation. The general resonance condition is that, for some $\boldsymbol{g}_\|$:

$$\varepsilon_n = \varepsilon_i - \frac{\hbar^2}{2m}(\boldsymbol{k}_{i\|} + \boldsymbol{g}_\|)^2 - \Delta\varepsilon_J , \tag{3}$$

where $\Delta\varepsilon_J = \varepsilon_{rot}(J_f) - \varepsilon_{rot}(J_i)$ is the energy of a rotational transition. Being given by kinematics alone, this condition allows a precise determination of the bound state energies ε_n, directly from experiment. The only approximation made in Eq. (3) is that the lateral motion in the surface-bound state is free and that each resonance can be treated separately. In reality, there are energy bands for lateral motion corresponding to each ε_n and band gaps may appear whenever two free-particle bands cross.

7.1.2.2 Semi-empirical atom-surface potentials

A general expansion of the static potential, which comes naturally for coupled channels calculations, is

$$V(\boldsymbol{r}) = \sum_{\boldsymbol{g}_\|} V_{\boldsymbol{g}_\|}(z)\, e^{i\boldsymbol{g}_\| \cdot \boldsymbol{r}_\|} . \tag{4}$$

(We follow the standard notation where the z axis is perpendicular to the mean surface, pointing outwards, and the "lateral" vector $\boldsymbol{r}_\|$ is in the xy plane.) The laterally averaged potential $V_0(z)$ has often been approximated by a Morse potential,

$$V_0(z) = D\,[e^{-\beta z} - 2\,e^{-\beta z/2}] \tag{5}$$

although this does not reproduce the correct asymptotic behavior of $V_0(z)$ for large z. D is the surface well depth and β is the "softness parameter". If the attractive well can be neglected, $V_0(z)$ is simply approximated by an exponential.

The simplest form of $V_0(z)$ that approaches the correct z^{-3} van der Waals behavior for large z is

$$V_0(z) = A\,e^{-\beta z} - C_3/z^3 . \tag{6}$$

A disadvantage of Eq. (6) is that it contains 3 parameters, compared to 2 in the Morse potential Eq. (5); however, C_3 may be independently known (see section 7.1.2.4.2). It is also possible to rewrite Eq. (6) as

$$V_0(z) = \frac{D}{(u/3) - 1}\left(e^{u-x} - \frac{u^4}{3x^3}\right), \tag{7}$$

where $x = \beta z$, and $u = \beta z_m$ is the value of x that corresponds to the well bottom. Comparison of Eqs. (6) and (7) gives

$$A = D\,e^u/[(u/3) - 1] \quad \text{and} \quad C_3\,\beta^3 = Du^4/(u-3).$$

It has been found empirically that for most atom-surface potentials u is around $u = 5.5$. Then V_0/D is a universal function of βz, as for the Morse potential, and a law of corresponding states is valid [91 V1].

A purely empirical three-parameter form that fits the bound states as well as Eq. (6) is:

$$V_0(z) = D\left[\left(1 + \frac{\lambda z}{2p}\right)^{-2p} - 2\left(1 + \frac{\lambda z}{p}\right)^{-p}\right]. \tag{8}$$

For $\boldsymbol{g}_{\parallel} \neq 0$ there is good justification, as we discuss below, for taking

$$V_{\boldsymbol{g}_{\parallel}}(z) = V_{\boldsymbol{g}_{\parallel}} \exp[-\beta(\boldsymbol{g}_{\parallel})\,z], \tag{9}$$

where $\beta(\boldsymbol{g}_{\parallel}) \cong [\beta^2 + g_{\parallel}^2]^{1/2}$. The parameters D (or A and C_3), β, and $V_{\boldsymbol{g}_{\parallel}}$ are adjusted to reproduce the experimental diffracted intensities, with good success for simple surfaces [82L2]. The experimental intensities depend on the surface temperature T through an effective Debye-Waller factor, which is discussed in section 7.1.3.1. The data should be extrapolated to $T = 0$ for comparison with the intensities calculated using $V(\boldsymbol{r})$.

7.1.2.3 Hard wall models

Hard wall models are still widely used, especially for the analysis of angular distributions at fixed incidence parameters (ε_i, θ_i, ϕ_i). Deviations from the hard wall predictions are often called "softness corrections". They can be substantial enough to make it worth abandoning the hard wall picture, except for comparison [79A1]. In particular, the effective hard wall itself varies with incidence parameters (see Eq. (10) below).

For elastic scattering the total potential $V = V_{\text{att}} + V_{\text{rep}}$ is just a function of the coordinates $\boldsymbol{r} = (\boldsymbol{r}_{\parallel}, z)$ of the incident atom. The effective position of the hard wall, $\zeta(\boldsymbol{r}_{\parallel})$, is given by the locus of classical turning points

$$V(\boldsymbol{r}_{\parallel}, \zeta(\boldsymbol{r}_{\parallel})) = \varepsilon_{zi}, \tag{10}$$

where $\varepsilon_{zi} = \varepsilon_i \cos^2\theta_i$ is the incident perpendicular energy. The shape of the hard-wall corrugation profile $\zeta(\boldsymbol{r}_{\parallel})$ gives a physical picture of the surface.

For inelastic scattering, the total potential is also a function of the displacements of the atoms in the solid from equilibrium, denoted collectively by $\boldsymbol{u}(t): V = V(\boldsymbol{r}; \boldsymbol{u}(t))$. The effective position of the vibrating hard wall, $\zeta(\boldsymbol{r}_{\parallel}, t)$, is defined by the time-dependent generalization of Eq. (10),

$$V(\boldsymbol{r}_{\parallel}, \zeta(\boldsymbol{r}_{\parallel}, t); \boldsymbol{u}_l(t)) = \varepsilon_{zi}. \tag{11}$$

The usual assumption is that, to a first approximation, only the displacements $\boldsymbol{u}(t)$ of the top surface layer determine the time dependence of ζ.

The hard wall model [75G1] has been quite successful in explaining the dependence of the diffracted intensities on θ_i, ϕ_i and ε_i with a very simple expression for $\zeta(\boldsymbol{r}_{\|})$: often just the first term in a Fourier series expansion is enough. For instance, in the classic case of the alkali halide (001) surfaces one can use

$$\zeta(\boldsymbol{r}_{\|}) = h[\cos(2\pi x/a) + \cos(2\pi y/a)], \tag{12}$$

where a is the nearest neighbor distance and $4h$ is the peak-to-valley height of the corrugation. Similarly for the basal plane of graphite [78C2]

$$\zeta(\boldsymbol{r}_{\|}) = h[\cos(2\pi x/a) + \cos(2\pi y/a) + \cos(2\pi(x-y)/a)]. \tag{13}$$

In such cases it is clearly more practical to treat h as an adjustable parameter than to try to calculate it from (nearly) first principles.

The fitting procedure to determine the parameters in $\zeta(\boldsymbol{r}_{\|})$ is very efficient [82E1] if one can use the eikonal approximation for the scattering amplitude $S_{\boldsymbol{g}_{\|}}$. The complete eikonal formula [76C2, 78H2] is $S_{\boldsymbol{g}_{\|}} = P_{\boldsymbol{g}_{\|}} I_{\boldsymbol{g}_{\|}}$, where

$$P_{\boldsymbol{g}_{\|}} = \frac{1 + \cos\theta_i \cos\theta_{\boldsymbol{g}_{\|}} - \sin\theta_i \sin\theta_{\boldsymbol{g}_{\|}} \cos(\phi_i - \phi_{\boldsymbol{g}_{\|}})}{\cos\theta_i(\cos\theta_i + \cos\theta_{\boldsymbol{g}_{\|}})} \tag{14}$$

is a kinematic prefactor of order unity that depends on the initial angles (θ_i, ϕ_i) and on the final angles $(\theta_{\boldsymbol{g}_{\|}}, \phi_{\boldsymbol{g}_{\|}})$, and

$$I_{\boldsymbol{g}_{\|}} = -\int \frac{d^2\boldsymbol{r}_{\|}}{A_c} \, e^{-i[\Delta p \zeta(\boldsymbol{r}_{\|}) + \boldsymbol{g}_{\|}\cdot\boldsymbol{r}_{\|}]}. \tag{15}$$

Here A_c is the area of the surface unit cell and Δp is the perpendicular momentum transfer in the hard wall collision. The effect of V_{att} is fairly well simulated by changing ε_{iz} to $\varepsilon_{iz} + D$, where D is an effective surface well depth (generally larger than the true well depth). This is known as the Beeby correction [71B1]; physically, it means that the He atom is accelerated by V_{att} on its way to the hard-wall collision with the surface, and is decelerated on its way out. Then $\Delta p = |p_i| + |p_{\boldsymbol{g}_{\|}}|$, where $\hbar^2 p_i^2 = 2m(\varepsilon_{iz} + D)$, and similarly for $p_{\boldsymbol{g}_{\|}}$. The explicit eikonal result for the simple corrugation Eq. (12) is expressed in terms of Bessel functions as

$$I_{\boldsymbol{g}_{\|}} = -i^{-|m|-|n|} J_{|m|}(h\Delta^p) J_{|n|}(h\Delta p) \tag{16}$$

for $\boldsymbol{g}_{\|} = (2\pi/a)(m, n)$.

A feature of surface scattering that is naturally explained by the corrugated hard wall model is the surface rainbow. Classically, the rainbow angles θ_{max} and θ_{min} correspond to the maximum possible deflections and the scattered intensity vanishes for polar angles greater than θ_{max} or smaller than θ_{min}. The rainbow deflections $\theta_{max} - \theta_i$ and $\theta_i - \theta_{min}$ are simply twice the maximum inclination angles of the effective corrugation profile, which is the intersection of $\zeta(\boldsymbol{r}_{\|})$ with the scattering plane. A saddle-point evaluation of the eikonal integral Eq. (15) reproduces this classical result; a remnant of the rainbow is often clearly discernible in the full eikonal result. The positions of the rainbow peaks offer the simplest rough estimates of the parameters appearing in $\zeta(\boldsymbol{r}_{\|})$, such as h in Eqs. (12) and (13).

At first sight it appears that a detailed knowledge of $V(\boldsymbol{r})$ is required to predict the resonance shapes, but in fact the simple model [78W1, 79G1] of a structureless well in front of a corrugated hard wall produces good predictions. A resonance with a surface-bound state appears as a dip or a peak, or more generally as an asymmetric Wigner-Fano shape, depending [79C2, 80H1] on the relative phase of the direct and resonant contributions to the total scattering amplitude. If the two are in phase a maximum results, if they are in antiphase there will be a minimum, unless the resonant amplitude is greater than twice the direct amplitude, in which case a maximum results. For the corrugation Eq. (12) the direct amplitude is given explicitly by Eq. (16) and the resonant amplitude involves the product of initial-to-bound and bound-to-final amplitudes of the same form; the resulting predictions of maxima and minima have been well verified [81F1].

7.1.2.4 Theory of the elastic atom-surface potential

It is common practice to divide the atom-surface interaction into an attractive part V_{att} due to van der Waals forces and a repulsive part V_{rep} due to the overlap of the surface wave functions with the closed-shell He orbitals. The theory has been developed for the static potential, but the results can be taken over directly to the dynamic interaction, within the usual Born-Oppenheimer approximation.

7.1.2.4.1 The static repulsive potential

It has been shown by several methods that to a good approximation

$$V_{rep}(\boldsymbol{r}) = A\rho(\boldsymbol{r}), \tag{17}$$

where $\rho(\boldsymbol{r})$ is the unperturbed electronic density of the surface and the constant A depends only weakly on the nature of the surface.

In the case of metal surfaces different approaches yield values of A ranging from $250 \cdots 500$ eV (a.u.)^3. Esbjerg and Norskov [80E1] and Stott and Zaremba [80S2] were the first to obtain Eq. (17) for a He atom immersed in a homogeneous electron gas with a positive uniform charge background (the jellium model). The "best" value of A calculated from this effective medium approach is 329 eV (a.u.)^3 [84M1]. However, the He atom lies outside the jellium background in a region of varying $\rho(\boldsymbol{r})$, and as a result the effective A changes from $329 \cdots 255$ eV (a.u.)^3. Hartree-Fock calculations for He interacting with a metal atom cluster can alo be fitted to Eq. (17) with $A = 373$ eV (a.u.)^3 [84B2, 85B2].

A different approach starts with the He atom well outside the surface and treats it as a perturbation on the surface electrons. This approach was successfully used by Zaremba and Kohn [77Z1] to predict the laterally averaged interaction with metal surfaces and was perfected by Harris and Liebsch [82H1, 82H2] and by Nordlander and Harris [83N1, 84N1], who obtain $A \cong 500$ eV (a.u.)^3. It is also possible to infer V_{rep} from e-He scattering data. In the simplest approximation, that goes back to Fermi [34F1], the interaction between the He atom at $\boldsymbol{r}$ and the electron at $\boldsymbol{r}_e$ is represented by the pseudopotential $(2\pi\hbar^2 a_s/m)\,\delta(\boldsymbol{r}_e - \boldsymbol{r})$, where a_s is the s-wave scattering length. This gives $A = 2\pi\hbar^2 a_s/m \cong 200$ eV (a.u.)^3. However, one should exclude the contribution to a_s coming from the long-range polarization force, because this force gives rise to the van der Waals attraction [85C4]. Further, the e-He scattering data must be continued to negative energies of a few eV, because the surface electrons are bound. The resulting best estimate of A [85T1, 87T1] is about 500 eV (a.u.)^3.

For insulators, one can start directly from pairwise repulsive interactions between the He atom and the surface constituents, which in first approximation can be taken to be undistorted atoms in the case of graphite [78C1, 79C1, 79V1, 80C1, 80V1, 81C3, 81V1] and of noble gas overlayers [86J1, 87J2], and undistorted ions in the case of LiF (100) [85C2, 86F1]. In these cases $\rho(\boldsymbol{r})$ is by construction the sum of the densities of the constituents and Eq. (17) turns out to be valid within the uncertainties of the calculation. The fact that one can take undistorted ions for LiF (100), however, results from accidental cancellation of several corrections; in the other ionic crystal surface that has been studied in detail, NaCl (100), ion distortion is important [86H3].

The main trouble with Eq. (17) ist that, while it works well for the laterally averaged atom-surface potential, it generally overestimates the effective corrugation for metal surfaces. Actually, the correct corrugation is obtained if V_{att} is neglected altogether [81H1], but this is of course inconsistent. The effect of V_{att} is to move the classical turning point of the collision closer to the surface, where the variation of the density is greater [83G1]. Takada and Kohn [85T1] have pointed out that in the pseudo-potential appraoch the discrepancy could in principle be removed by the contribution of p-wave scattering; however, this correction turns out to be small [85T2, 85T3]. Corrections due to the presence of empty states above the Fermi level have been proposed by Annett and Haydock [84A1, 84A2] and are still a subject of controversy; Annett [87A1] finds that they are sufficient to remove the discrepancy with experiment, while Harris and Zaremba [85H1] find them negligible. It is possible that the error comes from uncertainties in the separation of V_{rep} from the attractive part of the potential, to be discussed below. It could also come from a failure of the methods used to compute $\rho(\boldsymbol{r})$ in the region of interest for the scattering of 20 ⋯ 60 meV He atoms ($z = 3 \cdots 4$ Å). Surprisingly, the full calculation of $\rho(\boldsymbol{r})$ gives a result that, in the region of interest, is practically indistinguishable [84B1] from a simple superposition of atomic densities, as long as charge transfer in the surface region is negligible, or suitably included.

As a practical matter the laterally averaged He-surface repulsion is generally well approximated by

$$V_{\text{rep}}(z) = U_{\text{rep}}\, \mathrm{e}^{-\beta z} \tag{18}$$

which is consistent with a pairwise sum of Yukawa potentials $V_\kappa(r) = U_\kappa\, \mathrm{e}^{-\beta r}/\beta r$ over a monatomic top layer (see section 7.1.3.3.1), with $U_{\text{rep}} = 2\pi U_\kappa/\beta^2 A_{\text{c}}$. However, it is more accurate to approximate the pairwise repulsion by the Born-Mayer form $U_\kappa\, \mathrm{e}^{-\beta r}$, which then gives

$$V_{\text{rep}}(z) = (2\pi U_\kappa/\beta^2 A_{\text{c}})(1 + \beta z)\, \mathrm{e}^{-\beta z}. \tag{19}$$

A_{c} is the surface area per atom. If V_{att} is neglected, the potential in Eq. (18) defines the simple soft wall model, and for any exponential-type potential β is called the softness parameter.

It seems prudent at present to regard the corrugations of metal surfaces as adjustable parameters to be determined experimentally. If one wants to preserve the relation $V_{\text{rep}}(\boldsymbol{r}) = A\rho(\boldsymbol{r})$ with $\rho(\boldsymbol{r})$ a superposition of atomic charge densities, one must assume non-spherical atoms. Dondi et al. [88D1] are able to fit the corrugation of Ag(110) by superimposing atomic potentials of the form

$$U_\kappa \exp[-(\beta_x^2 x^2 + \beta_y^2 y^2 + \beta_z^2 z^2)^{1/2}]$$

with $\beta_x = 2.03$ Å^{-1}, $\beta_y = 2.17$ Å^{-1}, $\beta_z = 2.78$ Å^{-1}. Here the x direction is along the close-packed (110) atom rows. For comparison, the Herman-Skillman tables [63H1] give $\beta = 2.42$ Å^{-1}.

7.1.2.4.2 The static attractive potential

V_{att} is usually dominated by the van der Waals dispersion force. At large distances z from the surface (but not large enough that retardation effects come into play), $V_{\text{att}} = -C_3/z^3$. The exact value of C_3 is given by the Lifshitz formula

$$C_3 = \frac{\hbar}{4\pi} \int_0^\infty \frac{\varepsilon(\mathrm{i}u) - 1}{\varepsilon(\mathrm{i}u) + 1}\, \alpha_{\text{He}}(\mathrm{i}u)\, \mathrm{d}u, \tag{20}$$

where ε is the dielectric function of the surface material and α_{He} is the He atom polarizability. The extension of ε and α_{He} to imaginary frequencies $\mathrm{i}u$ is easily performed if these functions are known for real frequencies; further, the integral is numerically convenient because $\varepsilon(\mathrm{i}u)$ and $\alpha(\mathrm{i}u)$ are real, non-negative, and monotonically decreasing functions of u. Thus C_3 can be accurately evaluated.

For a crystal consisting of localized electronic units (atoms, molecules, or ions) having polarizability α, the crudest way to compute V_{att} is to sum pairwise interactions $-c_6/|\boldsymbol{r} - \boldsymbol{r}_l|^6$, where

$$c_6 = \frac{3\hbar}{\pi} \int_0^\infty \alpha(\mathrm{i}u)\, \alpha_{\text{He}}(\mathrm{i}u)\, \mathrm{d}u. \tag{21}$$

The sum of $-c_6/|\boldsymbol{r} - \boldsymbol{r}_l|^6$ over a monatomic plane gives a laterally averaged interaction $-C_4/(z - z_l)^4$, with $C_4 = (\pi/2A_{\text{c}})c_6$. A_{c} is the surface area per atom. The sum over equally spaced planes of $-C_4/(z - z_l)^4$ gives asymptotically $-C_3/(z - d/2)^3$, where the origin of z is on the top atom plane and d is the interplanar spacing. The value of C_3 is $C_4/3d$, or $(\pi/6A_{\text{c}}d)c_6$, in agreement with Eq. (20) in the low-density limit where $\varepsilon = 1 + 4\pi\alpha/A_{\text{c}}d$. However, $-C_3/(z - d/2)^3$ gives a poor approximation to the sum over planes in the region of interest ($z \cong 3$ Å); a better approximation is

$$V_{\text{att}}(z) = -C_3/(z + d/2)^3 - C_4/z^4 \tag{22}$$

(with $C_4 = 3dC_3$), which treats the top atom layer separately.

A systematic improvement of the pairwise sum formula can be obtained by including n-body forces, beginning with the Axelrod-Teller three body interaction [87K1, 69N1, 88C1]. The result is approximately of the form of Eq. (22), with C_3 now given by Eq. (20) and C_4 in general different from $3dC_3$. One can then write, up to terms of order z^{-5}

$$V_{att}(z) = -C_3/(z - z_{vW})^3 . \tag{23}$$

The value of z_{vW} for a variety of surfaces (jellium model and noble metals) was first given by Zaremba and Kohn [76Z1]. A later paper by Liebsch [86L1] suggested that the exact values of z_{vW} are uniformly less – typically about 20% closer to the surface – than calculated by Zaremba and Kohn. Based on these values, the currently accepted z_{vW} for the noble metals is close to $d/2$, which, in the jellium model, is the jellium edge. The form of Eq. (23) has become customary for metal surfaces, although the divergence at $z = z_{vW}$ is unphysical.

Corrections to Eq. (22) or (23) are of three kinds:

(i) multipolar interactions, (ii) charge overlap, and (iii) lateral variation.

(i) Multipolar corrections begin with $-C_5/z^5$, which comes from two terms [86H4], He quadrupole-surface dipole and He dipole-surface quadrupole. The latter is generally more important and hard to estimate. Here again, the pairwise summation model gives explicit results. Pairwise multipolar terms $-c_n/r^n$, with n = 8, 10, 12, are routinely included in interatomic potentials. They lead to $-C_5/z^5$ and higher inverse power terms in the atom-surface potential.

ii) The effect of charge overlap is to reduce the attraction at close range. In pairwise interactions it is found that the reduction of $-c_n/|\boldsymbol{r} - \boldsymbol{r}_l|^n$ is well described by the Tang-Toennies "damping factor" [84T1]

$$f_n(x) = 1 - e^{-x} \sum_{k=0}^{n} \frac{x^k}{k!} , \tag{24}$$

where $x = \beta|\boldsymbol{r} - \boldsymbol{r}_l|$ and β is approximately the same parameter that describes the fall-off of V_{rep}. Other "damping factors" have been proposed, but Eq. (24) works as well as well as any and does not introduce new parameters. A similar form of "damping" for the jellium model leads to [84N1]

$$V_{att}(z) = -C_3 f_2(\beta|z - z_{vW}|)/|z - z_{vW}|^3 . \tag{25}$$

iii) The lateral variation of V_{att} for a sum of pairwise interactions is caused almost entirely by the atoms in the top surface layer and falls off like $e^{-g_{\parallel} z}$. The dipole-dipole term takes the form [74S1]

$$-(C_4/2) \sum_{\boldsymbol{g}_{\parallel} \neq 0} (g_{\parallel}/2z)^2 e^{i\boldsymbol{g}_{\parallel}\cdot\boldsymbol{r}_{\parallel}} K_2(g_{\parallel} z), \tag{26}$$

where K_2 is a modified Bessel function and $C_4 = \pi c_6/A_c$. A formula of the same type, with $C_4/2$ replaced by an appropriate $C_{\boldsymbol{g}_{\parallel}}$, is obtained from the generalization of Eq. (23) to a corrugated surface [82H3, 84L1].

7.1.2.4.3 The total static potential

In order to obtain relatively small effects such as the corrugation of metal surfaces, it may not be a good idea to separate V_{static} into V_{att} and V_{rep}. Around the classical turning point of a typical He-metal collision there is a large cancellation of V_{att} and V_{rep}, which tends to amplify the uncertainties.

A direct calculation of the total V_{static} is feasible, using density functional theory in the local density approximation, for a closed shell atom interacting with a jellium surface. Lang and Norskov [83L1] find by this method very reasonable shapes of V_{static} for $z \leq z_0$, where z_0 is the bottom of the attractive well. The values of the well depth D are also in fair agreement with the values obtained by Zaremba and Kohn [77Z1] and by others [82H2], using the separate evaluation of V_{att} and V_{rep}, as discussed above. Of course for $z > z_0$ the Lang-Norskov potential decays exponentially, rather than as z^{-3}.

7.1.3 Inelastic scattering: surface phonons

He scattering has become a major experimental tool for the study of surface dynamics. It is especially suited to the determination of the phonon spectra throughout the surface Brillouin zone up to energies of about 30 meV. This energy region includes acoustic and low-lying optical phonon branches on clean surfaces, as well as the entire vibrational spectrum of physisorbed noble gas overlayers, for instance. Another major technique to study surface dynamics throughout the surface Brillouin zone is electron energy loss spectroscopy (EELS) (see chapter 6.2). The advantages of He scattering are that it works as well for insulators as for metals, while EELS is limited to good conductors, and that it has better energy resolution (typically 1 meV, compared to 5 meV for EELS, but both are still improving). On the other hand, atom-surface scattering produces an increasing amount of multiphonon processes for increased energy exchange, mostly because the incident atom energy must be correspondingly increased; as a consequence the scattering data approach a classical limit that blurs the details of surface dynamics above 20 ··· 30 meV. By contrast, in EELS the incident electron energy is fixed at several eV, making the entire vibrational spectrum of clean surfaces and of chemisorbed species accessible at essentially the same resolution.

This section focuses on time-of-flight (TOF) experiments using He beams and on the discussion of the inelastic He-surface interaction for clean, unreconstructed surfaces. Reviews covering some of the same subjects include those by Goodman and Wachman [76G1]. Engel and Rieder [82E1], Toennies [84T2], Celli [84C2], Barker and Auerbach [85B1], and Bortolani and Levi [86B2], as well as several articles in the book edited by Benedek and Valbusa [82B1].

In addition to He TOF scattering, other atom beam techniques used in the study of surface phonon spectra include He beam scattering with a LiF crystal analyzer [71W1, 74M1, 81M1, 81M2, 83M1] and Ne beam scattering [82S1, 83S1]. Most of what follows also applies to the analysis of these experiments, although we do not consider in detail the Ne-surface potential.

7.1.3.1 Kinematics

Most of the data analysis requires nothing more than simple kinematics (Fig. 2), namely the conservation laws for energy and for the lateral components of momentum in the atom-surface collision. The wave vector of the incident atom is denoted by $\boldsymbol{k}_i = (\boldsymbol{k}_{i\parallel}, k_{iz})$, with $k_{iz} < 0$, and the corresponding initial energy is $\varepsilon_i = \hbar^2 k_i^2/2m$; the final wave vector is $\boldsymbol{k}_f = (\boldsymbol{k}_{f\parallel}, k_{fz})$, with $k_{fz} > 0$, and the corresponding final energy is $\varepsilon_f = \hbar^2 k_f^2/2m$. The energy gain $\Delta\omega$ and the lateral momentum transfer $\hbar\Delta\boldsymbol{k}_\parallel$ are given by

$$\hbar\Delta\omega = \varepsilon_f - \varepsilon_i \tag{27}$$

$$\Delta\boldsymbol{k}_\parallel = \boldsymbol{k}_{f\parallel} - \boldsymbol{k}_{i\parallel}\,. \tag{28}$$

We use the notation of section 7.1.2.1 and $\varepsilon_f = \hbar^2 k_f/2m$ to denote the final energy. Monoenergetic, well collimated He beams have a well-defined $\boldsymbol{k}_i$; the final angles θ_f and ϕ_f are set by the detector's position, and time-of-flight (TOF) measurements give ε_f. From this information $\delta\omega$ *and* $\Delta\boldsymbol{k}_\parallel$ are uniquely inferred, using Eqs. (27) and (28).

On a given TOF run, the values of $\Delta\omega$ and $\Delta\boldsymbol{k}_\parallel$ for different arrival times (different k_f) lie on a parabola called the scan curve. In the usual case of in-plane scattering ($\phi_f = \phi_i$), Eq. (28) reduces to $\Delta\boldsymbol{k}_\parallel = k_f \sin\theta_f - k_i \sin\theta_i$ (see Fig. 2) and Eq. (27) gives the scan curve

$$\hbar\Delta\omega = \varepsilon_i \left[\frac{(\sin\theta_i + \Delta k_\parallel/k_i)^2}{\sin^2\theta_f} - 1\right]. \tag{29}$$

Examples of scan curves can be seen in Fig. 12.

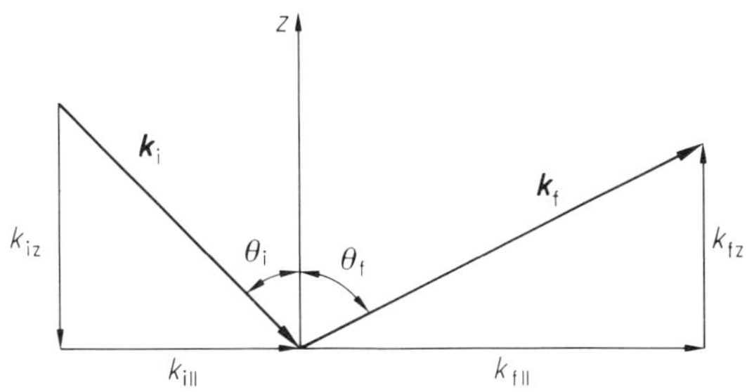

Fig. 2. Kinematics of in plane inelastic scattering.

The phonon frequencies of the semi-infinite crystal are denoted by $\omega(\boldsymbol{q}_\parallel, j)$, where $\boldsymbol{q}_\parallel$ is the lateral phonon momentum and j stands for all other phonon quantum numbers. Quite generally, $\omega(\boldsymbol{q}_\parallel + \boldsymbol{g}_\parallel, j) = \omega(\boldsymbol{q}_\parallel, j)$ and $\omega(-\boldsymbol{q}_\parallel, j) = \omega(\boldsymbol{q}_\parallel, j)$ [71M2]. It is convenient to take $\omega(\boldsymbol{q}_\parallel, j)$ to be positive for phonon absorption (annihilation) and negative for phonon emission (creation). It is also understood that the wave vector of an absorbed phonon is $\boldsymbol{q}_\parallel + \boldsymbol{g}_\parallel$, that of an emitted phonon is $-\boldsymbol{q}_\parallel - \boldsymbol{g}_\parallel$. With these conventions both creation and annihilation processes can be displayed on the same graph, as in Fig. 12, and in every case for one-phonon processes $\Delta\boldsymbol{k}_\parallel = \boldsymbol{q}_\parallel + \boldsymbol{g}_\parallel$ and $\Delta\omega = \omega(\boldsymbol{q}_\parallel, j)$. Thus the dispersion relations of surface phonons are directly determined by the location of peaks in the one-phonon cross-sections.

From the point of view of surface phonon spectroscopy, multiphonon processes, which are difficult to disentangle, are in general a nuisance. An exception is the case of a non-dispersive surface mode (an Einstein oscillator), where of course multiphonon processes appear simply as harmonics. Empirically, a TOF spectrum where one-phonon processes are dominant is often easy to recognize from the fact that the scattered intensity is consistently larger in regions of the $(\omega, \Delta k_\parallel)$ plane where one expects to find the surface-projected frequencies of the bulk phonons, assumed to be known. These are the shaded regions in Figs. 14, 20 and 31 of section 7.2, for instance (see also Fig. 8 of chapter 4.1). Outside these regions one may see sharp peaks, corresponding to surface phonons, that can usually be associated with an edge of the bulk phonon continuum.

Multi-phonon processes generally cause a reduction of the one-phonon cross-sections. This reduction is analogous to the Debye-Waller reduction of (elastic) diffraction due to the competition of inelastic processes (involving one or more phonons). Thus when the Debye-Waller factor e^{-2W} is small, or $W > 1$, for a typical momentum transfer, most one-phonon processes are swamped out. A rough criterion, suggested by Weare [74W1], is obtained by using $2W = \Delta k^2 \langle u^2 \rangle / 3$, where Δk can be approximated by its value for specular scattering, $\Delta k = 2|k_{iz}|$, and $\langle u^2 \rangle$ is an effective mean square thermal displacement of the surface atoms. In the Debye model with characteristic temperature T_D, $\langle u^2 \rangle = 9\hbar^2 T / M k_B T_D^2$, where k_B is Boltzmann's constant, M is an average mass of the surface atoms, and T is the surface temperature; it is supposed that $T > T_D/5$, so that quantum corrections are negligible. As discussed again in section 7.1.3.2, T_D is close to the bulk Debye temperature. The criterion for one-phonon processes to be prominent, $W < 1$, can then be written as

$$24\,(m/M)\,T\,\varepsilon_i \cos^2\theta_i < k_B T_D^2\,. \tag{30}$$

7.1.3.1.1 Kinematic focussing

For in-plane one-phonon scattering, the kinematic relations can be summarized as follows: the contribution of the j-th phonon branch is proportional to $\delta(\Delta\omega - \omega(\Delta k_\parallel, j))$, where $\Delta\omega$ is related to $\Delta k_\parallel$ by the scan curve, Eq. (29).

For the surface-projected bulk phonon branches, the index j stands for the quasi-continuous perpendicular wave number q_z and a discrete band index, such as longitudinal acoustic (LA). In this case the δ-functions for neighboring values of q_z overlap, giving a continuous distribution of the scattered one-phonon intensity.

On the other hand, a surface phonon branch, such as the Rayleigh branch, corresponds to a discrete value of j and makes a δ-like contribution to the scattered intensity, with integrated strength proportional to

$$I_j(\theta_f) = \frac{\partial \Delta\omega/\partial \Delta k_{\parallel}}{|\partial \Delta\omega/\partial \Delta k_{\parallel} - \partial\omega(\Delta k_{\parallel}, j)/\partial \Delta k_{\parallel}|} \tag{31}$$

which diverges when the scan curve osculates the phonon dispersion curve. This condition is known as kinematic focusing [75B1].

Many observations of kinematic focusing have been reported [83B1, 83B6]. The effect is not restricted to the surface phonon branches and can lead to enhancements in the continuum spectrum, as pointed out by Nichols and Weare [85N1]. The extension to out-of-plane scattering has been discussed by Avila and Lagos [81A1], who have proposed the use of kinematic focusing to measure phonon dispersion relations throughout the Brillouin zone without a TOF apparatus.

7.1.3.2 Dynamical theory

Explicit formulae for inelastic processes in atom-surface scattering were first obtained by Jackson and Mott [32J1] and by Lennard-Jones, Strachan and Devonshire [35L1, 37D1, 37L1, 37S1], and have been extended by many authors. The quantity of interest is a reflection coefficient that has been averaged over the initial phonon distribution and integrated over all phonon processes that lead to a given $\boldsymbol{k}_f$ (or at least over the processes that are included in the theory). This quantity, denoted by $R(\boldsymbol{k}_f, \boldsymbol{k}_i)$, is state-to-state between the experimentally measurable states of the He atom. By summing $R(\boldsymbol{k}_f, \boldsymbol{k}_i)$ over a small range of $\boldsymbol{k}_f$, and dividing by that range, one obtains the differential reflection coefficient $\mathrm{d}R(\boldsymbol{k}_f, \boldsymbol{k}_i)/\mathrm{d}^3 k_f$. In an experiment, $\mathrm{d}^3 k_f$ is determined by the final energy range $\mathrm{d}\varepsilon_f$ and by the solid angle $\mathrm{d}\Omega_f$ subtended by the detector. By using $\mathrm{d}^3 k_f = k_f^2\,\mathrm{d}k_f\,\mathrm{d}\Omega_f$ and $\mathrm{d}\varepsilon_f = \hbar^2 k_f\,\mathrm{d}k_f/m$, one obtains:

$$\frac{\mathrm{d}R}{\mathrm{d}\varepsilon_f\,\mathrm{d}\Omega_f} = \frac{m k_f}{\hbar^2}\frac{\mathrm{d}R}{\mathrm{d}^3 k_f}. \tag{32}$$

7.1.3.2.1 The distorted wave Born approximation (DWBA)

The extraction of phonon spectra from TOF data relies mostly on one-phonon processes, for which the first order DWBA is appropriate. The DWBA results should be multiplied by a Debye-Waller attenuation factor e^{-2W} with $2W \cong \langle \Delta\boldsymbol{k}\cdot\boldsymbol{u}\rangle^2$, as discussed in connection with Eq. (30). The DWBA gives the reflection coefficient to first order in the displacements $\boldsymbol{u}_{l\kappa}$ of the crystal atoms from their equilibrium positions $\boldsymbol{r}_{l\parallel} + \boldsymbol{r}_\kappa$. Here $\boldsymbol{r}_{l\parallel}$ denotes a surface net vector and $\boldsymbol{r}_\kappa$ gives the position of the crystal atoms within a surface unit cell. It must be understood that this cell is large in the z direction, extending from $z = 0$ to $z = -L_z$. In the harmonic approximation, $\boldsymbol{u}_{l\kappa}$ can be expanded in normal modes as

$$\boldsymbol{u}_{l\kappa} = \sum_{\boldsymbol{q}_{\parallel} j}\left(\frac{\hbar}{2M_c N_s|\omega(\boldsymbol{q}_{\parallel}, j)|}\right)^{\frac{1}{2}} e^{i\boldsymbol{q}_{\parallel}\cdot\boldsymbol{r}_{l\parallel}}\,\boldsymbol{e}_\kappa(\boldsymbol{q}_{\parallel}, j)\,[a(\boldsymbol{q}_{\parallel}, j) + a^{\dagger}(-\boldsymbol{q}_{\parallel}, j)], \tag{33}$$

where $a(\boldsymbol{q}_{\parallel}, j)$ and $a^{\dagger}(-\boldsymbol{q}_{\parallel}, j)$ are the usual annihilation and creation operators, N_s is the number of cells in the surface area L_s^2, and $\boldsymbol{e}_\kappa(\boldsymbol{q}_{\parallel}, j)$ is a polarization vector, normalized to

$$\sum_\kappa M_\kappa \boldsymbol{e}_\kappa(\boldsymbol{q}_{\parallel}, j)^* \cdot \boldsymbol{e}_\kappa(\boldsymbol{q}_{\parallel}, j') = M_c\,\delta_{jj'}, \tag{34}$$

where M_κ is the mass of the κ-th atom and M_c is the mass in the surface unit cell.

For any atom-surface interaction $V(\boldsymbol{r}, \{\boldsymbol{u}_{l\kappa}\})$, expansion to first order in $\boldsymbol{u}_{l\kappa}$ gives

$$V(\boldsymbol{r}, \{\boldsymbol{u}_{l\kappa}\}) = V_{\text{static}}(\boldsymbol{r}) + \sum_{l\kappa} \boldsymbol{F}_{l\kappa}(\boldsymbol{r}) \cdot \boldsymbol{u}_{l\kappa}\,, \tag{35}$$

where $\boldsymbol{F}_{l\kappa}(\boldsymbol{r})$ is the force exerted on the He atom by the atom, or ion, at $\boldsymbol{r}_{l\|} + \boldsymbol{r}_\kappa$.

To be definite, assume that the atom-surface potential V is a sum of two-body potentials V_κ that depend only on the separation between the position of the He atom, $\boldsymbol{r}$, and the positions of the atoms (or ions) in the crystal:

$$V(\boldsymbol{r}, \{\boldsymbol{u}_{l\kappa}\}) = \sum_{l\kappa} V_\kappa(\boldsymbol{r} - \boldsymbol{r}_{l\|} - \boldsymbol{r}_\kappa - \boldsymbol{u}_{l\kappa})\,. \tag{36}$$

The atoms (or ions) do not deform when displaced. Then we have explicitly:

$$V_{\text{static}}(\boldsymbol{r}) = \sum_{l\kappa} V_\kappa(\boldsymbol{r} - \boldsymbol{r}_{l\|} - \boldsymbol{r}_\kappa) \tag{37}$$

$$\boldsymbol{F}_{l\kappa}(\boldsymbol{r}) = -\nabla V_\kappa(\boldsymbol{r} - \boldsymbol{r}_{l\|} - \boldsymbol{r}_\kappa)\,. \tag{38}$$

The quantity that appears in the DWBA one-phonon matrix element, Eq. (41) below, is the lateral Fourier transform of these forces

$$\boldsymbol{F}_\kappa(\boldsymbol{k}_\| \,|\, z) = -\int \mathrm{d}^2 r_\| \,\nabla V_\kappa(\boldsymbol{r} - \boldsymbol{r}_\kappa)\, \mathrm{e}^{-\mathrm{i}\boldsymbol{k}_\| \cdot \boldsymbol{r}_\|} \tag{39}$$

which can be written in terms of the lateral Fourier transform of $V_\kappa(\boldsymbol{r} - \boldsymbol{r}_\kappa)$:

$$\boldsymbol{F}_\kappa(\boldsymbol{k}_\| \,|\, z) = (-\mathrm{i}\boldsymbol{k}_\|, \partial/\partial z)\, V_\kappa(\boldsymbol{k}_\| \,|\, z)\,. \tag{40}$$

The eigenfunctions of V_{static} are the distorted waves to be used in the DWBA. They are the product of a phonon wave function and an eigenfunction of the He atom; we need the eigenfunction $\chi_\mathrm{i}(\boldsymbol{r})^{(+)}$ that corresponds to the incoming (incident) wave $\exp(\mathrm{i}\boldsymbol{k}_\mathrm{i} \cdot \boldsymbol{r})$ and the eigenfunction $\chi_\mathrm{f}(\boldsymbol{r})^{(-)}$ that corresponds to the single outgoing wave $\exp(\mathrm{i}\boldsymbol{k}_\mathrm{f} \cdot \boldsymbol{r})$ with all the incoming waves necessary to make up this outgoing wave. The superscripts $(+)$ and $(-)$ explicitly denote the boundary conditions (given incoming wave and given outgoing wave, respectively).

The matrix element of the force between the He atom and the κ-th atom is (to first order in the DWBA)

$$\boldsymbol{F}_\kappa(\boldsymbol{k}_\mathrm{f}, \boldsymbol{k}_\mathrm{i}) = \sum_{\boldsymbol{g}'_\| \boldsymbol{g}''_\|} \int \mathrm{d}z \chi(\boldsymbol{g}''_\|, k_{\mathrm{f}z} \,|\, z)^{(-)*} \boldsymbol{F}_\kappa(\boldsymbol{k}_{\mathrm{f}\|} - \boldsymbol{k}_{\mathrm{i}\|} - \boldsymbol{g}'_\| + \boldsymbol{g}''_\| \,|\, z)\, \chi^{(+)}(\boldsymbol{g}'_\|, k_{\mathrm{i}z} \,|\, z) \tag{41}$$

and the differential reflection coefficient, Eq. (30), can be written as

$$\frac{\mathrm{d}R(\boldsymbol{k}_\mathrm{f}, \boldsymbol{k}_\mathrm{i})}{\mathrm{d}\varepsilon_\mathrm{f}\,\mathrm{d}\Omega_\mathrm{f}} = \frac{m^2 k_\mathrm{f}}{8\pi^2 \hbar^3 M_\mathrm{c} A_\mathrm{c} k_\mathrm{i} \cos\theta_\mathrm{i}} \cdot$$

$$\sum_j \frac{n(\omega(\boldsymbol{q}_\|, j))}{\omega(\boldsymbol{q}_\|, j)}\, \delta(\varepsilon_\mathrm{f} - \varepsilon_\mathrm{i} - \hbar\omega(\boldsymbol{q}_\|, j)) \left| \sum_\kappa \boldsymbol{e}_\kappa(\boldsymbol{q}_\|, j) \cdot \boldsymbol{F}_\kappa(\boldsymbol{k}_\mathrm{f}, \boldsymbol{k}_\mathrm{i}) \right|^2, \tag{42}$$

where $\boldsymbol{q}_\|$ differs from $\boldsymbol{k}_{\mathrm{f}\|} - \boldsymbol{k}_{\mathrm{i}\|}$ by a reciprocal lattice vector. Both emission and absorption processes are included in this formula if it is understood that $\omega(\boldsymbol{q}_\|, j)$ can have either sign: $\delta(\varepsilon_\mathrm{f} - \varepsilon_\mathrm{i} - \hbar\omega(\boldsymbol{q}_\|, j))$ automatically picks the $+$ sign for absorption and the $-$ sign for emission. Further, we have used the fact that $n(-|\omega|) = -[n(|\omega|) + 1]$, where $n(\omega)$ is the Bose factor

$$n(\omega) = \frac{1}{\exp(\hbar\omega/k_B T) + 1}\,. \tag{43}$$

7.1.3.2.2 Relation to phonon density of states and correlation functions

It is often useful, at least conceptually, to write the reflection coefficient in terms of the local phonon density of states for lateral momentum $\hbar \boldsymbol{q}_{\|}$, which is actually a matrix in the layer indices $\kappa\kappa'$ and in the cartesian components $\alpha\beta$:

$$D^{\alpha\beta}_{\kappa\kappa'}(\boldsymbol{q}_{\|}, \omega) = \sum_j \frac{\hbar}{2 M_c A_c \omega} e^{\alpha}_{\kappa}(\boldsymbol{q}_{\|}, j)\, e^{\beta}_{\kappa'}(\boldsymbol{q}_{\|}, j)^* \,\delta(\omega - \omega(\boldsymbol{q}_{\|}, j)). \tag{44}$$

We have:

$$\frac{dR(\boldsymbol{k}_f, \boldsymbol{k}_i)}{d\varepsilon_f\, d\Omega_f} = \frac{m^2 k_f}{4\pi^2 \hbar^4 k_i \cos\theta_i} \sum_{\kappa\kappa'} \sum_{\alpha\beta} D^{\alpha\beta}_{\kappa\kappa'}(\Delta \boldsymbol{k}_{\|}, \Delta\omega)\, n(\Delta\omega)\, F^{\alpha}_{\kappa}(\boldsymbol{k}_f, \boldsymbol{k}_i)\, F^{\beta}_{\kappa'}(\boldsymbol{k}_f, \boldsymbol{k}_i)^*, \tag{45}$$

where $\Delta \boldsymbol{k}_{\|} = \boldsymbol{k}_{f\|} - \boldsymbol{k}_{i\|}$ and $\hbar\Delta\omega = \varepsilon_f - \varepsilon_i$. If only the atoms in the top layer ($\kappa = 0$) contribute appreciably to the scattering, the reflection coefficient is proportional to the surface-projected phonon density of states $D^{\alpha\beta}_{00}(\Delta \boldsymbol{k}_{\|}, \Delta\omega)$.

Another physically meaningful quantity is the displacement correlation function

$$\langle \boldsymbol{u}^{\alpha}_{l\kappa}(0)\, \boldsymbol{u}^{\beta}_{l'\kappa'}(t)\rangle = \frac{A_c}{N_s} \sum_{\boldsymbol{q}_{\|}} \int d\omega\, n(\omega)\, e^{i\boldsymbol{q}_{\|}\cdot(\boldsymbol{r}_{l\|} - \boldsymbol{r}_{l'\|})}\, e^{-i\omega t} D^{\alpha\beta}_{\kappa\kappa'}(\boldsymbol{q}_{\|}, \omega). \tag{46}$$

The He reflection coefficient is related to this correlation function in the same way as a neutron cross section, for instance, is related to the correlation function in the bulk:

$$\frac{dR(\boldsymbol{k}_f, \boldsymbol{k}_i)}{d\varepsilon_f\, d\Omega_f} = \frac{m^2 k_f}{4\pi^2 \hbar^4 M_c A_c k_i \cos\theta_i} \sum_{\boldsymbol{r}_{l\|}} \int dt\, e^{-i\boldsymbol{q}_{\|}\cdot \boldsymbol{r}_{l\|}}\, e^{i(\varepsilon_f - \varepsilon_i)t/\hbar}\,.$$

$$\sum_{\kappa\kappa'} \sum_{\alpha\beta} \langle \boldsymbol{u}^{\alpha}_{l\kappa}(0)\, \boldsymbol{u}^{\beta}_{0\kappa'}(t)\rangle\, F^{\alpha}_{\kappa}(\boldsymbol{k}_f, \boldsymbol{k}_i)\, F^{\beta}_{\kappa'}(\boldsymbol{k}_f, \boldsymbol{k}_i)^*. \tag{47}$$

This expression, like the earlier ones, simplifies if only the surface atoms contribute ($\kappa = \kappa' = 0$), if there is enough symmetry that correlations between different components vanish (only $\beta = \alpha$ remains), and even more for Einstein oscillators (only the $\boldsymbol{r}_{l\|} = 0$ term contributes and the scattered distribution is that due to independent oscillators in the presence of the average surface potential).

7.1.3.3 The inelastic atom-surface interaction

In the previous section, we have expressed all the one-phonon DWBA results in terms of the eigenfunctions $\chi^{\pm}$ of the static potential V_{static} and of the force $\boldsymbol{F}_{l\kappa}$ between the He atom and the surface atom at $\boldsymbol{r}_{l\kappa} = \boldsymbol{r}_{l\|} + \boldsymbol{r}_{\kappa}$. V_{static} and $\boldsymbol{F}_{l\kappa}$ should be derived from the same model of the atom-surface interaction. A sum of rigid pairwise potentials provides such a model and leads to Eqs. (37) and (38). Each of these expressions is the first term in an expansion that may include significant three-body and four-body terms. In the case of simple insulators one knows, at least in principle, how to compute the various terms; in the case of metals there are still many open questions. We focus here on the semiempirical formulae that have been used extensively for the calculation of scattered inelastic intensities by Bortolani, Franchini, Santoro, and collaborators [85B3].

7.1.3.3.1 The dynamic repulsion and the cutoff factor

In most of the inelastic scattering calculations performed to date, V_{rep} has been approximated by a pairwise sum of Born-Mayer or Yukawa potentials, or even more drastically by a vibrating wall, soft or hard. The dynamical effects of V_{att} can be fully included in the calculation if V_{att} is represented by a pairwise sum, but it is usually assumed with some justification (see below) that V_{att} does not contribute appreciably to the inelastic matrix elements, except at small lateral momentum transfer $\Delta k_{\parallel}$. Since the prime interest is in the study of surface phonon spectra throughout the surface Brillouin zone, the dynamical V_{att} can then be neglected altogether. We therefore proceed by considering at first V_{rep} alone.

According to the discussion of section 7.1.3.2, what is needed are the matrix elements of the lateral Fourier transform of the pair potential, Eq. (41). For a Yukawa potential $V_\kappa(r) = U_\kappa\, e^{-\beta r}/\beta r$

$$V_\kappa(\boldsymbol{p}_{\parallel}\,|\,z) = \frac{2\pi}{\beta\beta(p_{\parallel})}\, U_\kappa\, e^{-\beta(p_{\parallel})z}, \tag{48}$$

where

$$\beta(p_{\parallel}) = \sqrt{\beta^2 + p_{\parallel}^2}\,. \tag{49}$$

For a Born-Mayer potential $V_\kappa(r) = U_\kappa\, e^{-\beta r}$

$$V_\kappa(\boldsymbol{p}_{\parallel}\,|\,z) = \frac{2\pi\beta}{\beta(p_{\parallel})^3}\,[1 + \beta(p_{\parallel})z]\, U_\kappa\, e^{-\beta(p_{\parallel})z}. \tag{50}$$

For $p_{\parallel} \ll \beta$ one has $\beta(p_{\parallel}) \cong \beta + p_{\parallel}^2/2\beta$ and Eq. (48) and (50), neglecting the $p_{\parallel}$ dependence of the prefactor, give

$$V_\kappa(\boldsymbol{p}_{\parallel}\,|\,z) = V_\kappa(0)\,|\,z)\, e^{-p_{\parallel}^2 z/2\beta}, \tag{51}$$

where $V_\kappa(0\,|\,z)/A_c$ is in fact the same as the laterally averaged $V_{rep}(z)$. The total interaction, to first order in the phonon displacement u_{lz} and neglecting $\boldsymbol{u}_{l\parallel}$, is then approximately the same as

$$V_{rep}(z - (2\pi z/\beta A_c) \sum_l \exp[-\beta(\boldsymbol{r}_{\parallel} - \boldsymbol{r}_l)^2/2z]\, u_{lz} \tag{52}$$

for a monatomic surface with negligible static corrugation. This formula is valid as long as $z \gg |\boldsymbol{r} - \boldsymbol{r}_l|$ for the atoms that contribute appreciably to the pairwise sum over l. Physically this means that the classical turning point z_t of the He-surface collision is not smaller than the interatomic distance on the surface. If z is replaced by z_t in the "influence factor" $\exp[-\beta(\boldsymbol{r}_{\parallel} - \boldsymbol{r}_l)^2/2z]$, Eq. (52) describes a "rippling soft wall", which in the limit $\beta \to \infty$ becomes a "rippling hard wall". The influence factor describes how the displacement of the turning point is affected by the displacement of the underlying surface atoms. The corresponding factor in momentum space is the "cutoff factor" $\exp(-p_{\parallel}^2 z_t/2\beta)$. The influence factor in coordinate space averages over the displacements of neighboring surface atoms; the cuttoff factor in momentum space reduces the contribution to the scattering of short-wavelength phonons. The underlying physical effect was recognized and discussed qualitatively by Hoinkes et al. [72H1], Armand et al. [77A1], and Levi [79L1, 79L2]. The fall off of the scattered one-phonon intensity as $\exp(-q_{\parallel}^2/q_c^2)$, with $q_c^2 \cong \beta/z_t$, describes rather well the decay of the intensity with momentum transfer for scattering from a surface with negligible static corrugation.

7.1.3.3.2 Dynamical effects of the attractive potential

The dynamical effects of the static part of V_{att} are fully included in calculations using Eqs. (42) and (47) with the correct wave functions $\chi^{\pm}$ of V_{static}. It is, however, common practice to keep only V_{rep} throughout the calculation and to account for V_{att} approximately by applying the "Beeby correction" [72B1, 72B2], i.e. by increasing the perpendicular He momentum from k_z to

$$\sqrt{k_z^2 + 2mD/\hbar^2}\,. \tag{53}$$

Here D is supposed to be the surface well depth, but comparison with full DWBA calculations shows that it can be considerably larger, especially at low indicent He energies [83A1]. This is a reasonable approximation for weakly corrugated surfaces, but does not account for the occurrence of inelastic resonant scattering.

Three types of resonant one-phonon scattering processes can occur on a corrugated surface. In a process of the first type the He atom is elastically scattered (diffracted) from the initial state into a surface-bound state, and then inelastically scattered out of it. This process occurs when the initial energy and angles are close to the resonance condition

$$\varepsilon_i = \varepsilon_n + (\hbar^2/2m)(\boldsymbol{k}_{i\parallel} + \boldsymbol{g}_{\parallel})^2 . \tag{54}$$

Here ε_n (< 0) is the binding energy of a surface-bound state and $\boldsymbol{g}_{\parallel}$ is a surface reciprocal lattice vector. In a process of the second type the He atom is inelastically scattered from the initial state into a surface-bound state, and then elastically scattered out of it. This process occurs when the final energy and angles are close to the resonance condition

$$\varepsilon_f = \varepsilon_{n'} + (\hbar^2/2m)(\boldsymbol{k}_{f\parallel} + \boldsymbol{g}'_{\parallel})^2 . \tag{55}$$

In a process of the third type both Eq. (54) and Eq. (55) are approximately satisfied and an inelastic one-phonon transition occurs between the surface-bound states of binding energies ε_n and $\varepsilon_{n'}$. This process usually causes a large increase of the inelastic cross-section.

The resonance manifests itself as a sharp variation of the total inelastic intensity. With TOF analysis, the effect of inelastic resonances in LiF, for instance, is very sharp [81B2, 81B3, 82L3, 83L3]. The theory of the inelastic resonance line shape and of its magnitude has been worked out by Cantini and Tatarek [81C1] in the model of a vibrating hard wall with a stationary attractive potential in front of it. Detailed applications of this model have been made to the analysis of inelastic resonances in LiF [83E1]. Resonances are useful to enhance the scattering from surface phonon branches that are weakly coupled to the He atom; however, one must be careful not to mistake a bound-state resonance for a peak in the surface phonon density of states.

7.2 Phonon spectra and surface potential data

We give now a listing of the phonon spectra of clean surfaces that have been obtained by inelastic He scattering and of the He-surface potential parameters for these surfaces. Potentials for H_2, D_2, HD and Ne, mostly derived from bound state resonance measurements, are also listed. Representative surface potentials, TOF data, and scan curves are shown. The symbols are:

ε_n: bound state energies, Eq. (13);
C_3: Van der Waals coefficient, Eq. (20);
D: surface well depth;
z_m: distance from the top atom layer to the minimum of the surface potential;
z_{vW}: van der Waals distance, Eq. (23);
h_{pv}: peak-to-valley height of the effective surface corrugation. For the hard wall corrugations described by Eqs. (12) and (13), $h_{pv} = 4h$ and $6h$, respectively. The effective h_{pv} varies with incident perpendicular energy E_{zi}; the values quoted are for $E_{zi} \cong 30$ meV, if available.

The surface Brillouin zone of fcc lattices is shown in Fig. 3.

Some of the figures for the phonon spectra have been reported also in chapter 4.1. In such cases mention is given in the figure caption. The reader is reminded to check in chapter 4.1, subvolume III/24b, for possible additional information.

Elements and compounds are listed alphabetically.

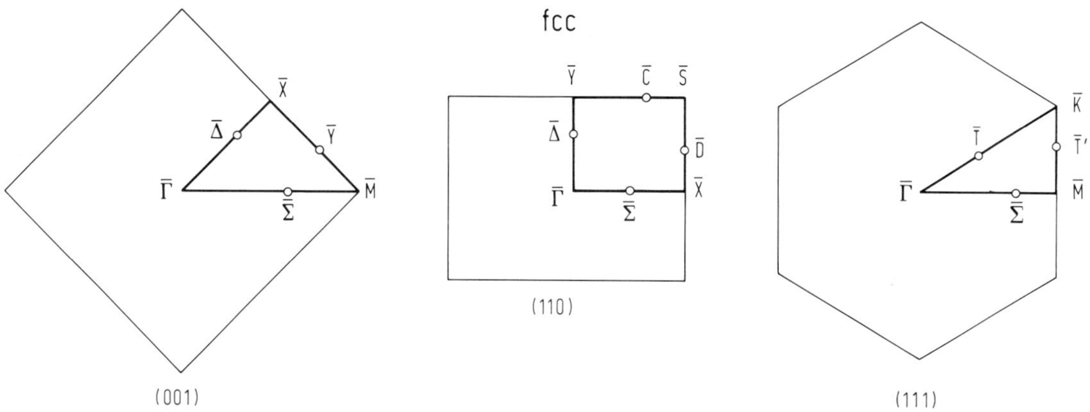

Fig. 3. Surface Brillouin zones for (001), (110) and (111) faces of fcc lattice.

Ag (110)

Surface lattice: rectangular; $a_x = 2.88$ Å, $a_y = 4.08$ Å.
Surface Brillouin zone: see Fig. 3.
D_2 potential: $\varepsilon_n = 27.5, 20.0, 14.3, 9.8, 6.3, 3.8, 2.0, 0.7$ meV [85C3]; same rotationally averaged $V_0(z)$ as for H_2.
H_2 potential: [85C3, 91V1]: $\varepsilon_n = 25.3, 16.4, 9.7, 5.3, 2.5, 1.1$ meV, $C_3 = 714$ meV · Å^3, $D = 31.7$ meV, $z_m = 3.27$ Å, $z_{vW} = 0.755$ Å, $h_{pv} \cong 0.1$ Å.
He potential: $\varepsilon_n = 4.44, 2.18, 0.92, 0.31, 0.1$ meV [86D1], $C_3 = 249$ meV · Å [91V1], $D = 6.0$ meV [91V1], $z_{vW} = 0.755$ Å [92C1], $h_{pv} = 0.22$ Å [92C1]. $V_{g_{\parallel}}(z)$ has been determined using a sum of distorted pairwise potentials, Eq. (19) [92C1]. See also Fig. 4 and Table 1.
Phonon spectra: see Figs. 5 and 6. Franchini et al. [89F1] infer a 40% reduction of the radial force constant between neighboring surface atoms along the close-packed $\langle 1\bar{1}0 \rangle$ direction $\bar{\Gamma}\bar{X}$.

Table 1. Parameters of the atom-rigid metal surface potentials

	He-Cu (110)	He-Cu (111)	He-Ag (110)	He-Ag (111)
Surface averaged part $V_{00}(z)$:				
position of the minimum z_{min} [Å]	3.28	3.41	3.44	3.57
well depth D [meV]	5.79	6.63	5.94	6.97
Peak-to-peak corrugation height at $V = 63$ meV [Å]				
along ⟨001⟩	0.151	...	0.301	...
along ⟨110⟩	0.025	0.012	0.069	0.043
total	0.168	0.014	0.349	0.049

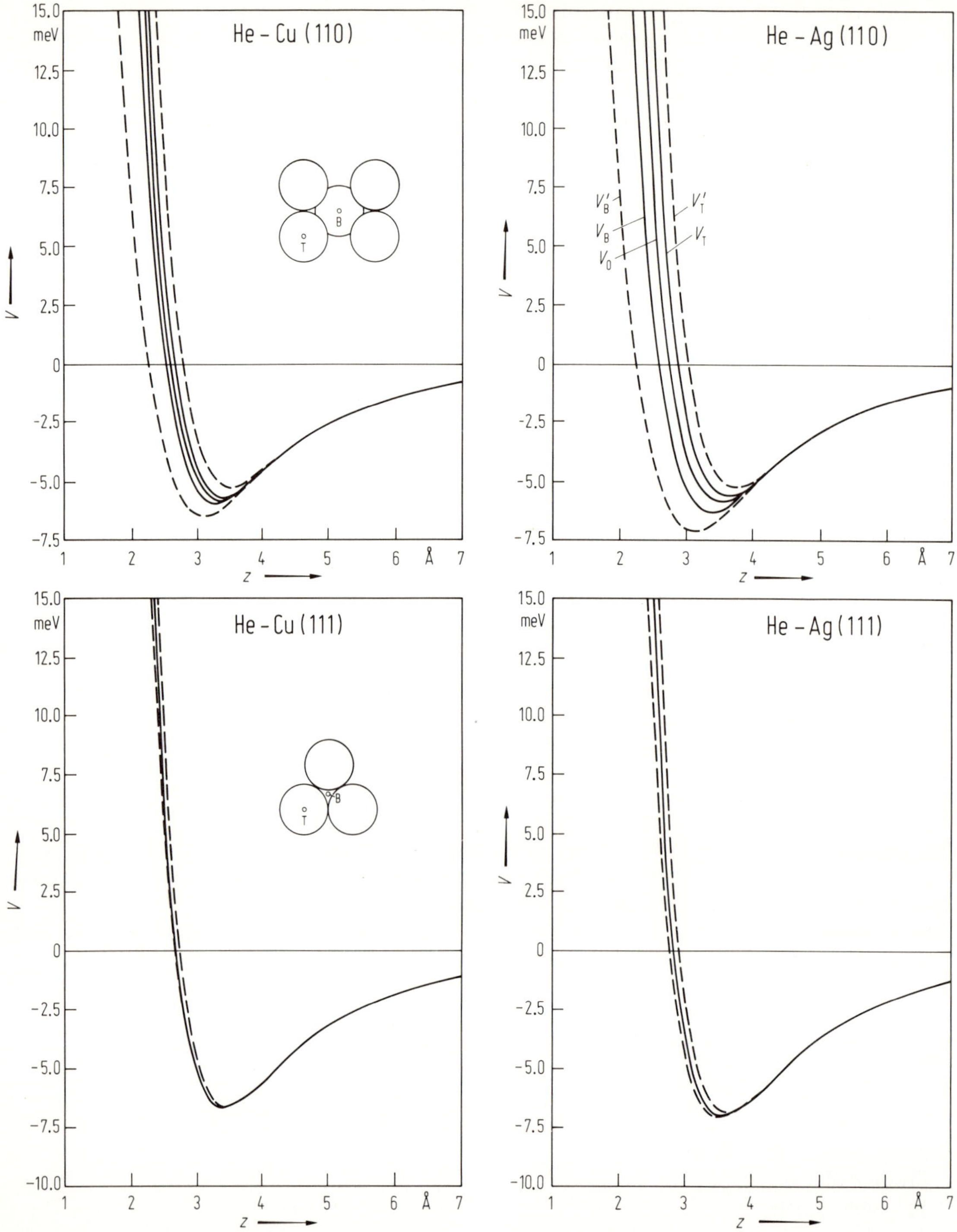

Fig. 4. Ag, Cu. Calculated potential curves for the atomic-static surface interactions He-Cu(110), He-Ag (110), He-Cu(111), and He-Ag(111). The full curves include the Takada-Kohn correction V_T, V_B, whereas for the dashed curves this correction has not been taken into account V_T', V_B'. The central curves denote the surface-averaged parts $V_0(z)$, to the right the potential at the top position (T), to the left that at the bottom position (B) is plotted as function of the distance z from the first layer of nuclei. In the case of the (111) surfaces the corrugation of the potential including the Takada-Kohn correction is very small so that the three full curves fall together in the scale of representaion [86E1].

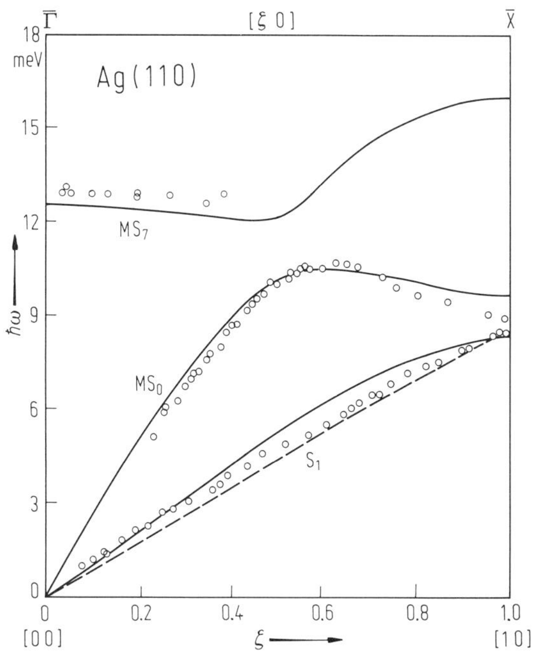

Fig. 5. Ag. Surface phonon dispersion curves in the $\overline{\Gamma}\overline{X}$ direction measured by HATOF (circles: experimental points, solid curves: LDM fit, dashed line: ECT fit) [87B7]. See also Fig. 3, chapter 4.1.

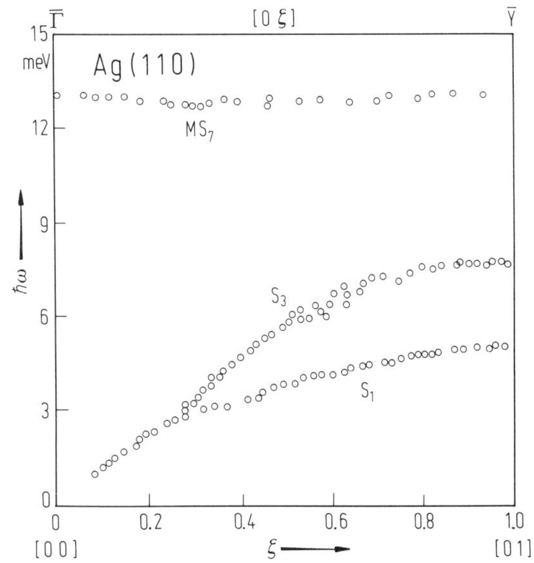

Fig. 6. Ag. Surface phonon dispersion curves in the $\overline{\Gamma}\overline{Y}$ direction measured by HATOF (circles: experimental points) [89T1]. See also Fig. 4, chapter 4.1.

Ag(111)

Surface lattice: hexagonal; $a = 2.88$ Å.
Surface Brillouin zone: see Fig. 3.
D_2 potential [83Y1, 91V1]: $\varepsilon_n = 27.5$, 21.0, 15.2, 14.4, 10.1, 6.5, 3.9, 2.4, 1.2 meV, same rotationally averaged $V_0(z)$ as for H_2.
H_2 potential: [83Y1, 85Y1, 85Y2]: $\varepsilon_n = 25.5$, 16.6, 9.9, 5.4, 2.5, 1.1 meV, $C_3 = 714$ meV · Å [91V1], $D = 32.5$ meV, $z_m = 2.5$ Å, $z_{vW} = 0.2$ Å, $h_{pv} = 0.18$ Å [76B2, 83L2].
He potential: in the absence of data on ε_n, C_3, D, z_m, and z_{vW} should be taken to be the same as for Ag(110), $h_{pv} = 0.05$ Å at 60 meV [86E1]. See also Fig. 4.
Phonon spectra: see Fig. 7. From the presence of a longitudinal resonance (LR), Bortolani et al. [84B3] infer a 40% reduction of the radial force constant between neighboring surface atoms.
Debye-Waller exponent: $2W$ (at $E_i = 63$ meV) = $5.5\,T/1000$, $T_D = 247$ K [80H1, 89B1].
Additional references: [76B2], [83L2]

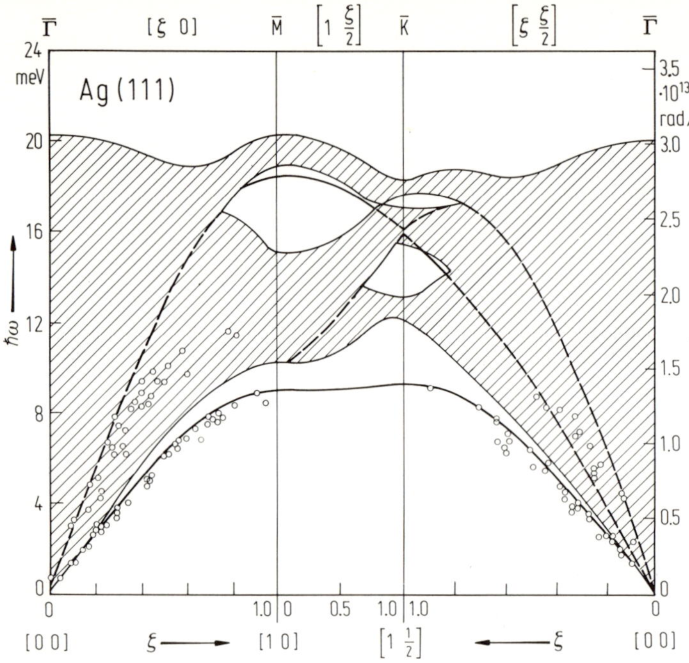

Fig. 7. Ag. Surface phonon dispersion curves measured by HATOF (circles: experimental points, solid and dashed curves: LDM fit, shaded region: bulk phonon modes) [85H3]. See also Fig. 5, chapter 4.1.

Al(110)

Surface lattice: rectangular; $a_x = 2.85$ Å, $a_y = 4.04$ Å.
Surface Brillouin zone: see Fig. 3.
He potential: $C_3 = 202$ meV · Å^3 [91V1], $\varepsilon_0 = 3.65$ meV, $D = 5.2$ meV [91V1], $z_m = 4.47$ Å, $z_{vW} = 1.71$ Å [77Z1, 76B1], $h_{pv} \leq 0.01$ Å [76B1].
Surface phonons: see Fig. 8. No change of the surface force constants is indicated by the data. *A priori* calculations of the surface phonon spectrum have been performed [86H2].

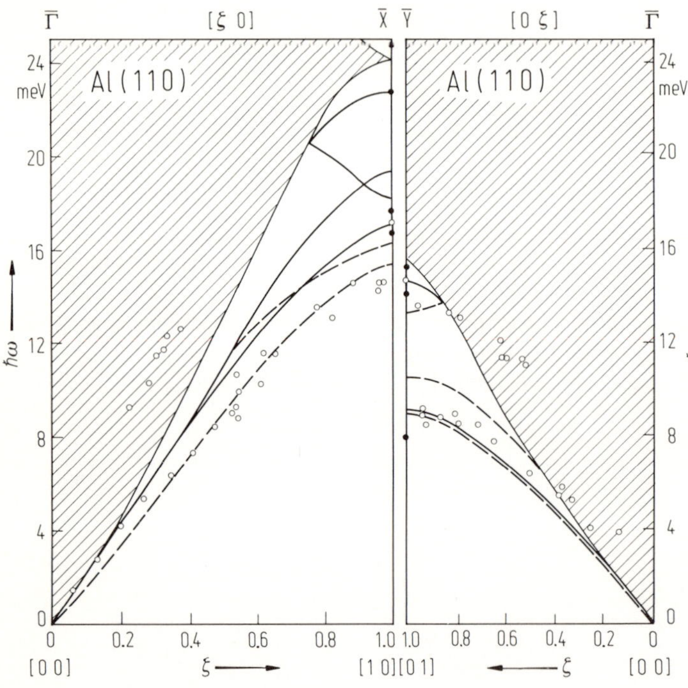

Fig. 8. Al. Surface phonon dispersion curves measured by HATOF (open circles: experimental points, dashed curves: LDM fit with bulk force constants, solid curves: LDM fit with modified surface force constants, full circles: SCP calculation, shaded region: bulk phonon modes) [87T2]. See also Fig. 7, chapter 4.1.

Al(111)

Surface lattice: hexagonal; $a^{*} = 2.8$ Å.
Surface Brillouin zone: see Fig. 3.
Potential: same (from theory) as for (110). No evidence of corrugation.
Surface phonons: see Fig. 8 of chapter 4.1, subvolume III/24b. No change of the surface force constants is indicated by the data.

Au(111)

Surface lattice: hexagonal; $a = 2.88$ Å. This surface undergoes a $23 \times \sqrt{3}$ reconstruction, consisting of alternating stripes with fcc and hcp termination.
Surface Brillouin zone (unreconstructed): see Fig. 3.
He potential: $C_3 = 274$ meV · Å^3, $\varepsilon_0 = 5.91$ meV, $D = 8.0$ meV, $z_m = 3.98$ Å, $z_{vW} = 1.34$ Å [77Z1, 76B1], $h_{pv} \leq 0.01$ Å [76B1].
Surface phonons: see Fig. 9. Harten et al. [85H3, 83A2] have carried out measurements along the $\langle 11\bar{2} \rangle$ direction $\overline{\Gamma \text{M}}$ and along the $\langle 110 \rangle$ direction $\overline{\Gamma \text{K}}$. In the interpretation of these data the surface reconstruction has been ignored, or it has been assumed that the phonon spectra are a superposition of the spectra obtained with fcc and hcp terminations. A truncated bulk model with force constants fitted to bulk data [83A2] fails to reproduce the strong longitudinal resonance observed in both the $\overline{\Gamma \text{M}}$ and the $\overline{\Gamma \text{X}}$ directions. Santoro et al. [84B4, 87S1] are able to reproduce the TOF spectra by using a model with modified surface force constants, the main modification being a 70% reduction of the radial force constant between neighboring surface atoms.

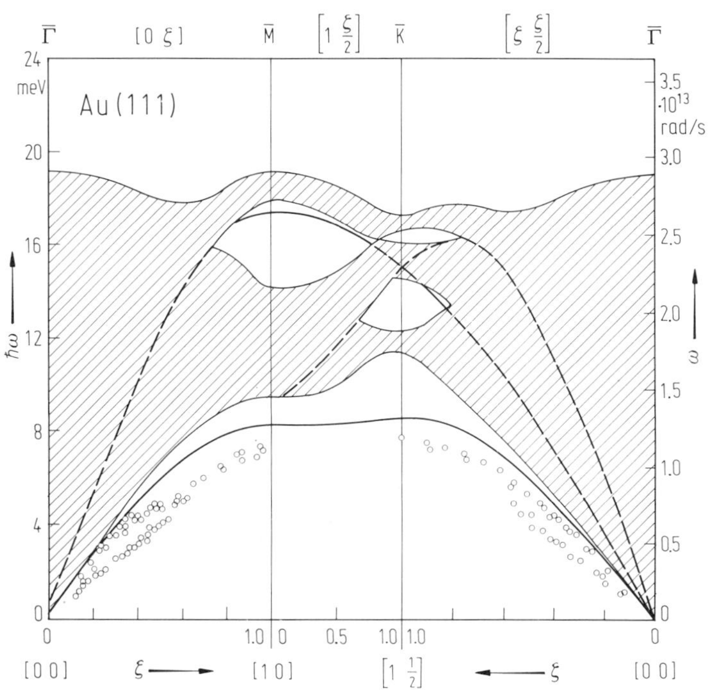

Fig. 9. Au. Surface phonon dispersion curves measured by HATOF (circles: experimental points, solid and dashed lines: LDM fit, shaded region: bulk phonon modes) [85H3]. See also Fig. 10, chapter 4.1.

C(0001)

Surface lattice: Hexagonal; $a = 2.46$ Å.
Surface Brillouin zone: it is the same as that of the (111) surface of the fcc, which is shown in Fig. 3.
D_2 potential: $\varepsilon_n = 44.6, 33.8, 23.1, 15.4, 10.0, 6.3, 3.7, 1.9$ meV [80M1, 91V1]; same rotationally averaged $V_0(z)$ as for H_2.
H_2 potential: $\varepsilon_n = 41.6, 26.4, 15.3, 7.9, 3.6, 1.4$ meV [80M1, 91V1], $C_3 = 520$ meV · Å^3 [91V1], $D = 51.7$ meV [91V1], $z_m = 2.87$ Å [91V1], $z_{vW} = 1.04$ Å [91V1], h_{pv} not determined, but strong diffraction is seen.
He potential: $C_3 = 180$ meV · Å^3, best estimate from [91V1], $\varepsilon_n = 12.42, 6.59, 3.05, 1.24, 0.47, 0.16$ meV [86C1], see also [79D1], $D = 16.6$ meV, best estimate from [91V1], $z_m = 2.65$ Å, best estimate from [91V1], $z_{vW} = 1.22$ Å [84L3], $h_{pv} = 0.22$ Å [78B2, 84L3]. See also Fig. 10.
Surface phonon spectra for both the ⟨100⟩ and the ⟨110⟩ directions were measured for the Rayleigh mode over most of the Brillouin zone and for the ⟨110⟩ direction some longitudinal results were obtained [92S1]. There is good agreement with theoretical predictions [81D1] and with the EELS data by Oshima et al. [88O1], who also observe higher modes, Fig. 11 (see also Fig. 11 in chapter 4.1, subvolume III/24b).
Additional References: [92S1, 88O1].

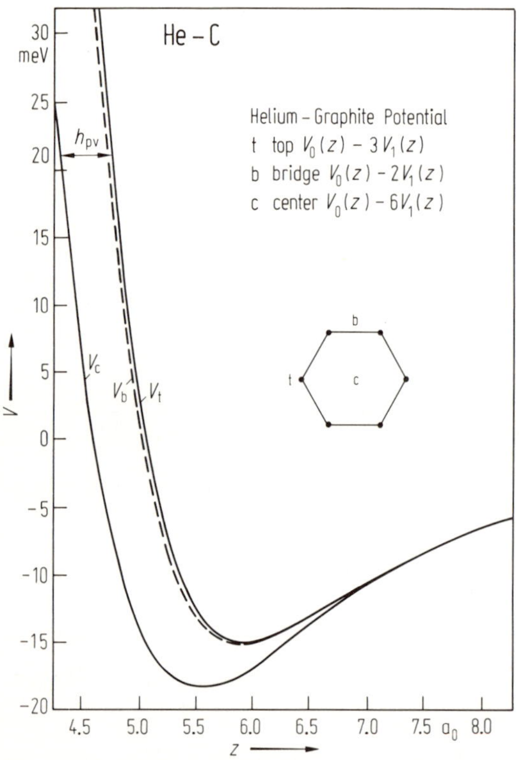

Fig. 10. C. Helium-graphite potential as function of distance from the surface for top, center and bridge positions. The corrugation h_{pv} at 21 meV is $0.45\,a_0$ [84L3].

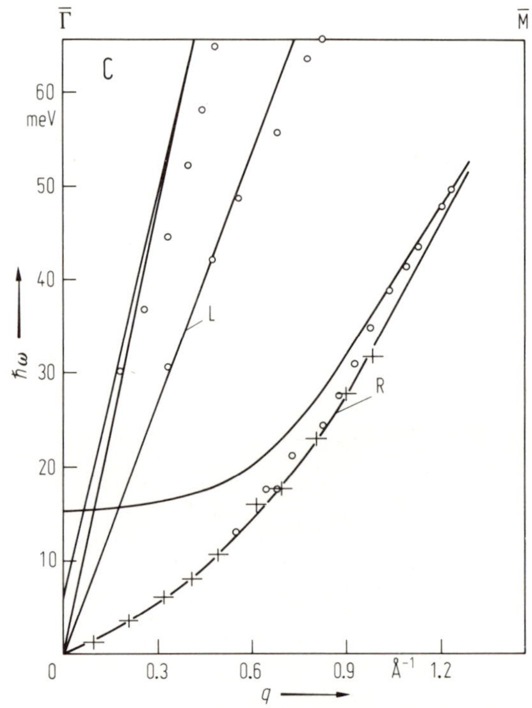

Fig. 11. C. A comparison of surface dispersion curves for graphite at low energies for the entire $\bar{\Gamma}\bar{M}$ direction. The open circles show the results of Oshima et al. [88O1] taken by EELS; the crosses are from typical HAS experiments [86B1] and the solid lines from theory [81D1]. The L and R label the longitudinal and Rayleigh modes, respectively.

Cu(110)

Surface lattice: rectangular; $a_x = 2.55$ Å, $a_y = 3.61$ Å.
Surface Brillouin zone: see Fig. 3.
He potential: $\varepsilon_n = 4.52, 2.2, 1.0, 0.3$ meV [82P1], $C_3 = 235$ meV · Å^3 [91V1], $D = 6.27$ meV, $z_m = 3.65$ Å, $z_{vW} = 1.21$ Å [86L1], $h_{pv} = 0.15$ Å along ⟨110⟩ [87E1]. (see also Table 1, p. 293).

Surface phonons: see Figs. 12 and 13; see also Figs. 13 and 14 of chapter 4.1, subvolume III/24b.

Surface phonon spectra have been obtained by He scattering using a LiF crystal analyzer by Mason et al. [83M1] and have been interpreted by Black et al. [87B4]. More complete spectra, covering both the ⟨1$\bar{1}$0⟩ direction $\overline{\Gamma X}$ and the ⟨001⟩ direction $\overline{\Gamma Y}$, have been obtained more recently by Zeppenfeld et al. [88Z1] using TOF measurements (Fig. 12). A truncated bulk model with force constants fitted to bulk data reproduces the spectra fairly well; an excellent fit is obtained by small modifications of the surface force constants corresponding mainly to a surface stress normal to the surface and pointing towards the crystal (Fig. 13).

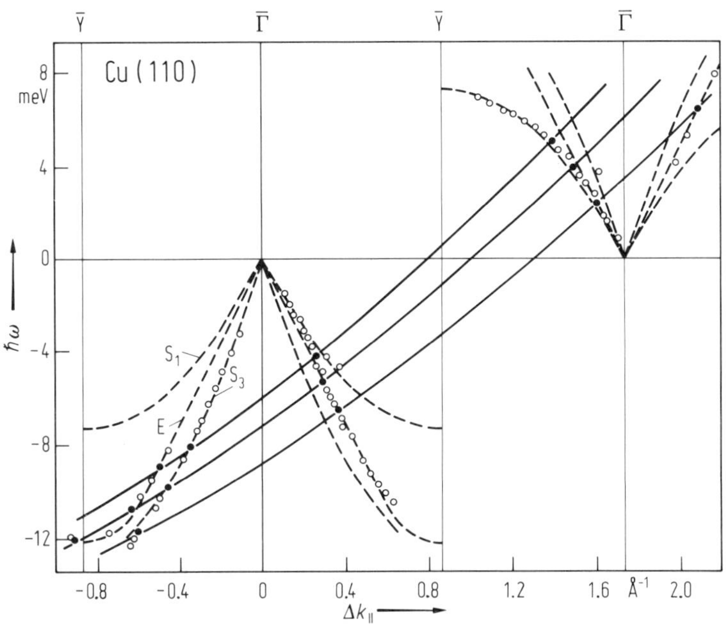

Fig. 12. Cu. Scan curves (solid lines), and the resulting phonon dispersion curves (circles and dashed lines) in the (001) direction [88Z1].

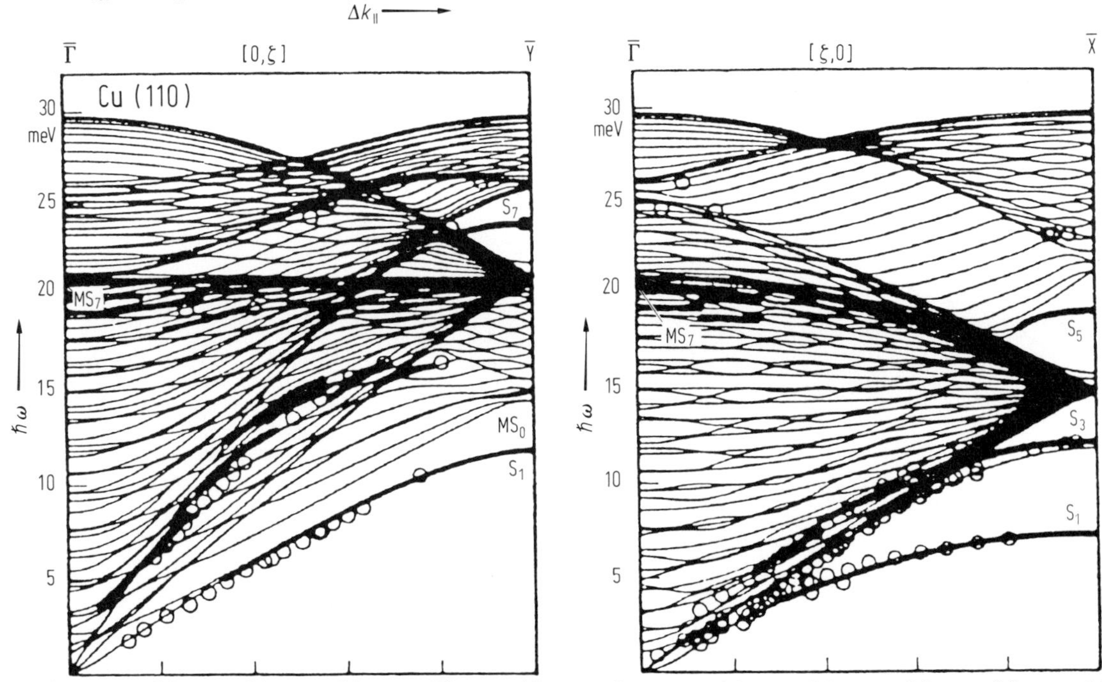

Fig. 13. Cu. Phonon dispersion curves of a 30-layer Cu(110) slab along the ⟨1$\bar{1}$0⟩ and ⟨001⟩ directions, using modified surface force constants for the best fit to experimental data (open circles). Thick lines indicate modes that contribute significantly to the z-polarized or sagittal vibrational amplitude of the surface layer [88Z1].

Cu(111)

Surface lattice: hexagonal; $a = 2.55$ Å.
Surface Brillouin zone: see Fig. 3.
He potential: surface well parameters presumed to be nearly the same as for Cu(110), $h_{pv} = 0.012$ Å [87E1]. (See also Table 1, p. 293).
Surface phonons: see Fig. 14.

Surface phonon spectra have been obtained by Harten et al. [85H3] from He TOF measurements along the ⟨11$\bar{2}$⟩ direction $\overline{\Gamma}\overline{M}$ and along the ⟨110⟩ direction $\overline{\Gamma}\overline{K}$ (Fig. 3). A truncated bulk model with force constants fitted to bulk data [83A2] fails to reproduce the strong longitudinal resonance observed in both directions. Bortolani et al. [89B2] are able to reproduce the TOF spectra using a model with modified surface force constants, the main modification being a 50–70% reduction of the radial force constant between neighboring surface atoms. However, their model is inconsistent with electron scattering data [88M3]: these show a much weaker longitudinal resonance than the He data and are best fitted by a stiffening of the force constant between the two top atom layers, along with a smaller reduction of the force constant between surface neighbors (15–30% instead of 50–70%). Calculations by Jayanthi et al. [87J1], using parameters determined independently from the He TOF data, also give a surface force constant softening of about 30%. The discrepancy is at present unresolved, but it raises some doubt about the correctness of the DWBA matrix elements, since Bortolani et al. [89B2] rely on the fit of the scattered intensities to assess the strength and polarization of the longitudinal resonance. On the other hand, Eichenauer et al. [86E1] have obtained good fits to the intensities of the Rayleigh phonon peaks.

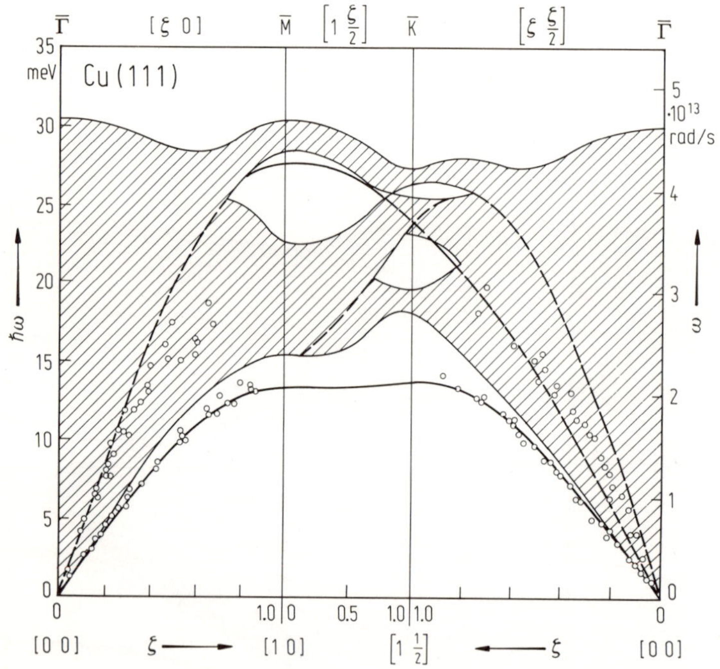

Fig. 14. Cu. Surface phonon dispersion curves measured by HATOF (circles: experimental points, solid and dashed lines: LDM fit, shaded region: bulk phonon modes) [85H3]. See also Fig. 15, chapter 4.1.

Cu(11n), n = 3, 5, 7

These are regularly stepped surfaces. He bound state resonances have been observed [82P1] at $\varepsilon_n(113) = 4.6, 2.2, 1.0$ meV, $\varepsilon_n(115) = 4.45, 2.1, 0.9, 0.35$ meV, $\varepsilon_n(117) = 4.1, 2.1$ meV. Within error, C_3, D, z_m, z_{vW} are about the same as for Cu(110).

GaAs(110)

Surface lattice: rectangular; $a_x = 4.00$ Å, $a_y = 5.65$ Å.
Surface Brillouin zone: see Fig. 3. The surface undergoes a well-studied 1×1 reconstruction (relaxation) as shown in Fig. 15.
He potential: the surface is strongly corrugated and displays many overlapping resonances in He scattering. Despite considerable effort [81C2, 82L1], the assignment of bound state energies is still uncertain. A theoretically computed potential and a potential that fits some of the diffraction data are shown in Fig. 16 [82L1]. These potentials are roughly consistent with $\varepsilon_0 = 4.2$ meV, $D = 7$ meV, $C_3 = 473$ meV · Å^3, $z_m = 3.5$ Å, $z_{vW} = -0.2$ Å, $h_{pv}\langle 001 \rangle = 1.1$ Å, $h_{pv}\langle 110 \rangle = 0.3$ Å. However, the theoretical value of C_3 is 173 meV · Å^3.
Surface phonons: see Fig. 17 [90S4]. There are three surface modes in the $\langle 100 \rangle$, one in the $\langle 1\bar{1}1 \rangle$ and three modes in the $\langle \bar{1}10 \rangle$ directions, respectively [87H1, 87D2]. An analysis based on the bond charge model provides good agreement with the experiment [90S4].
Additional reference: [81C3].

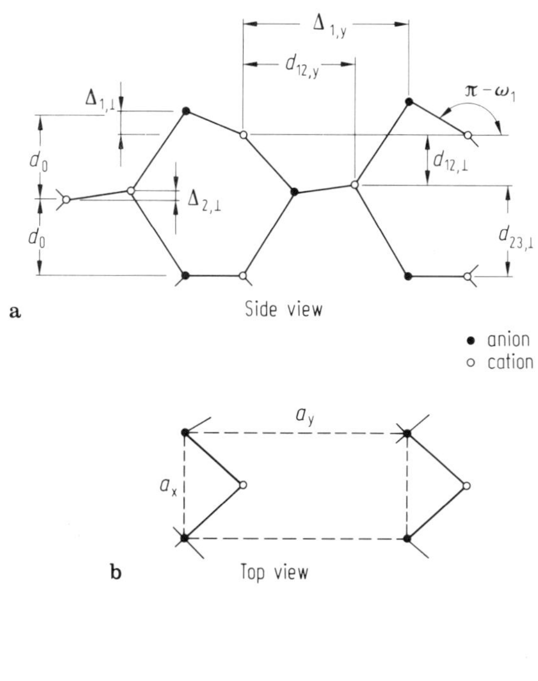

Fig. 15a, b. GaAs. Side **a** and top **b** views of GaAs(110), showing the reconstruction with a tilt angle ω_1 of about 30° [85D].

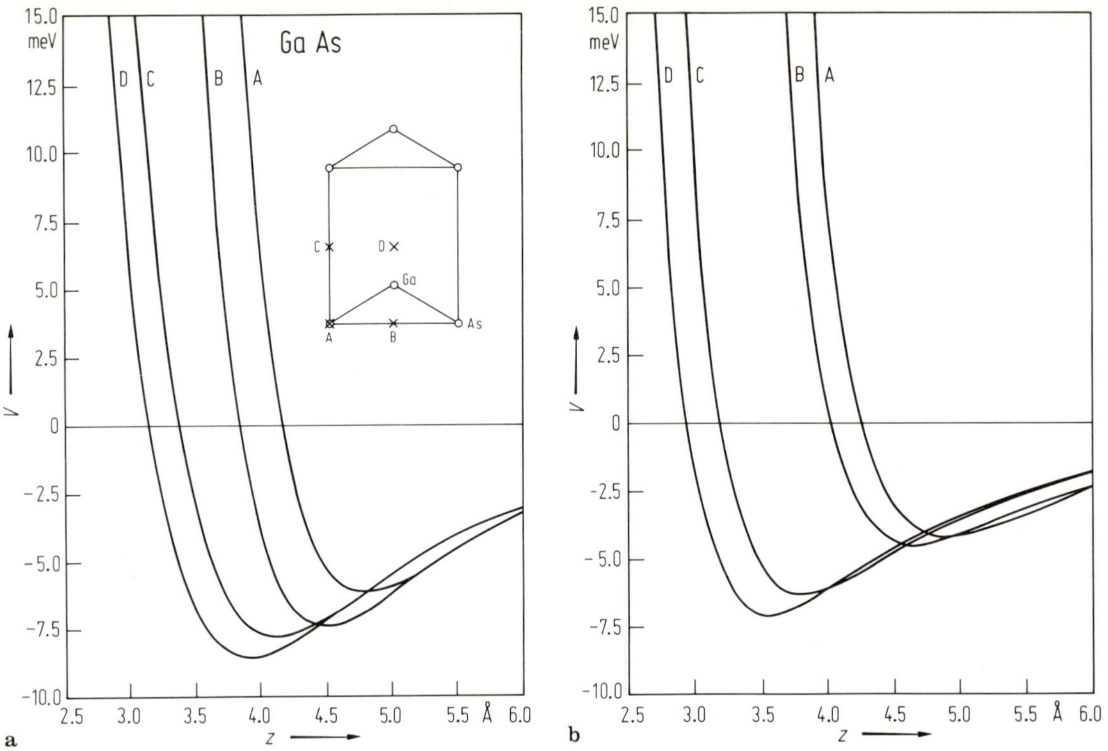

Fig. 16a, b. GaAs. Comparison of empirical potential **a** with the model potential **b** constructed from atomic As charge densities with no Ga core. These are plots of potential vs. height above the surface at symmetry points in the surface unit cell. The origin of z for the model potential is the center of the As atom before reconstruction (it moves upward 0.4 Å upon reconstruction). The origin of z for the empirical potential has been adjusted to obtain agreement with the model potential. [82L1].

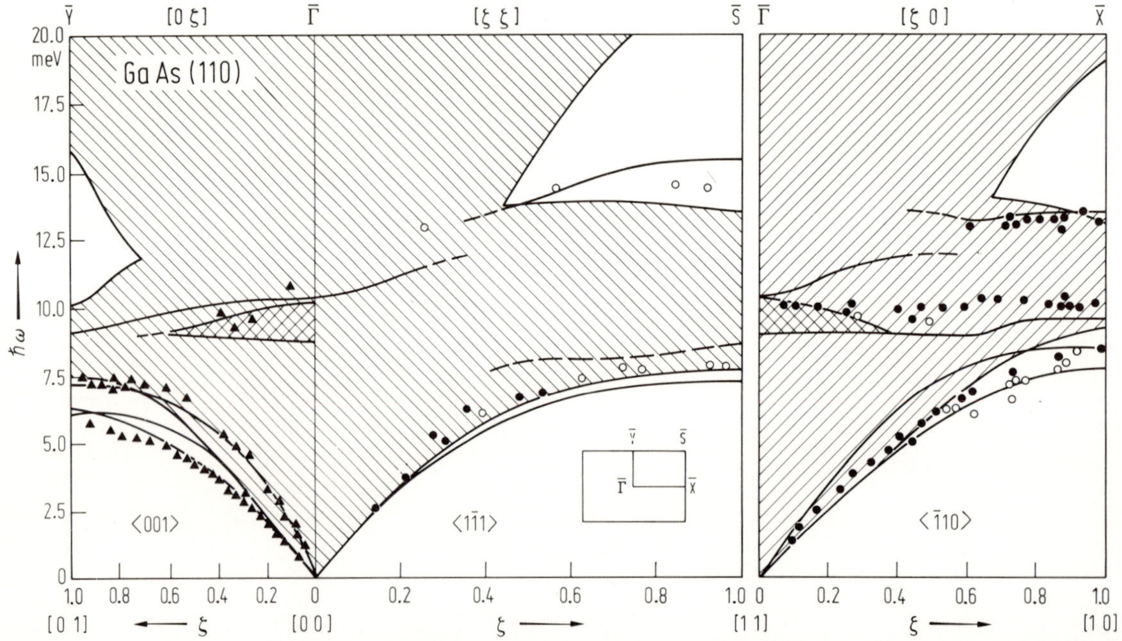

Fig. 17. GaAs. Phonon spectrum of GaAs(110). Shaded areas represent the surface projection of the bulk bands, solid lines and dashed lines correspond to surface modes and surface resonances, respectively, computed using a bond charge model [90S4]. Solid circles are experimental data from [87H1], open circles from a subsequent reanalysis of the same data, triangles from [87D2].

GaSe(0001)

Surface lattice: hexagonal; $a = 3.75$ Å.
Surface Brillouin zone: it is the same as that of the (111) surface of the fcc, which appears in Fig. 3.
Surface phonons: see Fig. 18.

Surface dispersion curves were measured in the ⟨100⟩ and ⟨110⟩ directions with five different modes being observed [86B3]. An identification of the modes suggested they were the Rayleigh mode, the longitudinal mode, a possible crossing mode, and two optical modes. Several non-identified points are also shown and there is considerable softening. A model fit has been given by Miglio and Colombo [88M3].

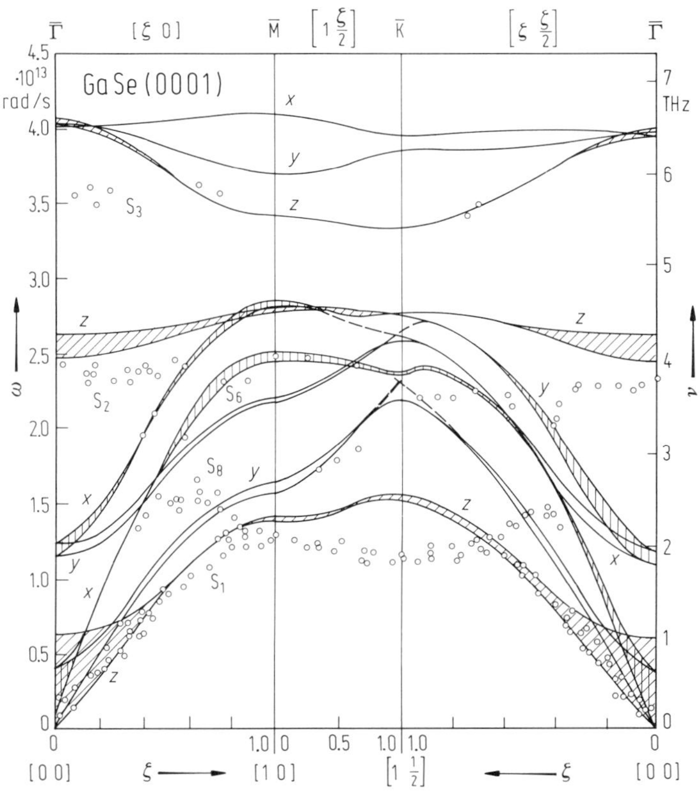

Fig. 18. GaSe. Surface phonon dispersion curves measured by HATOF, (circles: experimental points, solid lines: LDM fit, oblique shaded regions: vertically polarized bulk modes, vertical shaded regions: longitudinally polarized bulk modes) [86B3]. See also Fig. 17, chapter 4.1.

KBr(001)

Surface lattice: square; $a = 4.66$ Å.
Surface Brillouin zone: see Fig. 3.
He potential: not known in detail. Similar to other alkali halide (001) surfaces, with, roughly, $h_{pv} = 0.3$ Å [89C2].
Surface phonons: see Fig. 19.

The measurements by Chern et al. [89C2] provide clear evidence of four different modes: the Rayleigh mode, the longitudinal acoustic mode, a crossing mode, and a gap-localized S_2 optical mode, which is affected by a LA-LO hybridization in the ⟨100⟩ $\overline{\Gamma M}$ direction. A theoretical explanation of detailed features of the spectrum has been given [90S1].

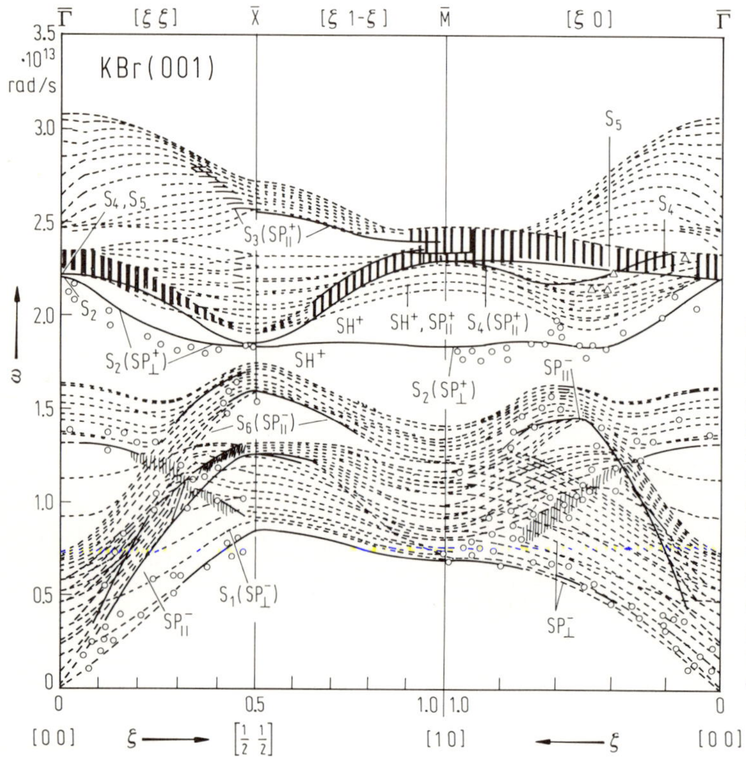

Fig. 19. KBr. Surface phonon dispersion curves measured by HATOF (circles: experimental points for strong intensities, triangles: experimental points for weak intensities, solid lines: LDM calculations for surface modes, dashed lines: LDM calculations for surface-projected bulk modes, $SP_\perp$ and $SP_\parallel$: sagittal-plane vibrations perpendicular and parallel to the surface, respectively, SH: shear-horizontal vibration, + or −: positive or negative ion carries out the vibration [89C2]. See also Fig. 19, chapter 4.1.

KCl(001)

Surface lattice: square; $a = 4.45$ Å.
Surface Brillouin zone: see Fig. 3.
He potential: not known in detail. Similar to other alkali halide (001) surfaces. From semiempirical interpolations [91V1] $C_3 = 93$ meV · Å^3, $D = 9.1$ meV.
Surface phonons: see Fig. 20.

Surface phonon spectra have been obtained from He TOF measurements along the ⟨100⟩ direction by Brusdeylins et al. [83B6]. The Rayleigh mode is folded back at the zone boundary as a result of the near-equality of the K and Cl masses. The results compare well with the lattice dynamical calculations of Benedek and Galimberti [78B1].

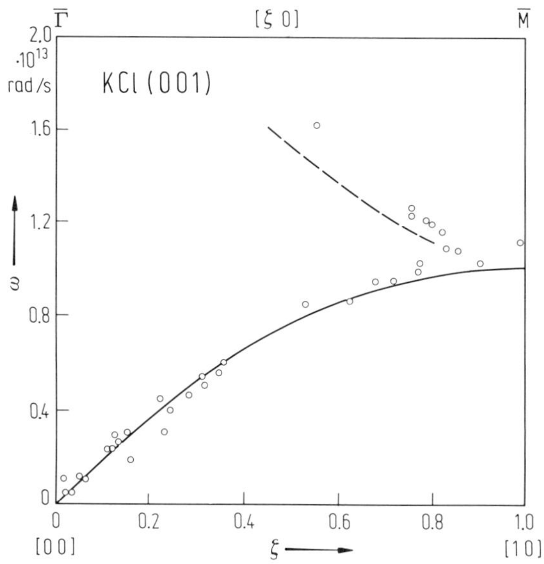

Fig. 20. KCl. Surface phonon dispersion curves measured by HATOF (circles: experimental points, solid and dashed lines: LDM calculations for modes) [83B6]. See also Fig. 20, chapter 4.1.

LiF(001)

Surface lattice: square; $a = 2.84$ Å.
Surface Brillouin zone: see Fig. 3.
H_2 potential [91V1]: $C_3 = 261$ meV · Å^3, $\varepsilon_0 = 17.3$ meV.
He potential: (fit by [85C2] to the data [78D1]) $C_3 = 137$ meV · Å^3, $\varepsilon_n(\mathrm{He}^4) = 5.90, 2.46, 0.78, 0.21$ meV [78D1], $\varepsilon_n(\mathrm{He}^3) = 5.59, 2.00$ meV [78D1], $D = 8.72$ meV, $z_m = 3.04$ Å, $z_{vW} = 1.0$ Å, $h_{pv} = 0.52$ Å [88E1]. The "best estimate" values of [91V1] are $C_3 = 93$ meV · Å^3, $\varepsilon_0 = 5.9$ meV, $D = 8.5$ meV, $z_m = 2.98$ Å, $z_{vW} = 0.79$ Å. See also Fig. 21.
Ne potential [91V1]: $C_3 = 192$ meV · Å^3 [91V1], $\varepsilon_0 = 11.7$ meV [91V1], $D = 13.5$ meV [91V1], $h_{pv} = 0.54$ Å [86B2].
Surface phonons: see Figs. 21 and 22 of chapter 4.1, subvolume III/24b and the data by Bracco et al. [87B6] shown in Fig. 22.

Surface phonon spectra were obtained from He TOF measurements along the ⟨100⟩ $\overline{\Gamma\mathrm{M}}$ direction for this surface by Brusdeylins, Doak, and Toennies [81B3, 83B6] in the pioneering experiments in this field. Detailed calculations of He TOF intensities in the DWBA by Eichenauer and Toennies [86E1] compare rather well with experiment. The effect of surface resonances has been repeatedly discussed and the resulting enhancement of phonon transitions has been computed [85N1, 87B6]. Complete spectra along the ⟨110⟩ $\overline{\Gamma\mathrm{X}}$ direction and along the ⟨100⟩ $\overline{\Gamma\mathrm{M}}$ direction have been obtained by Bracco et al. [87B6]. They compare rather well with recent lattice dynamical calculations by Kress et al. [87K2] (Fig. 22).
Additional reference [76B1].

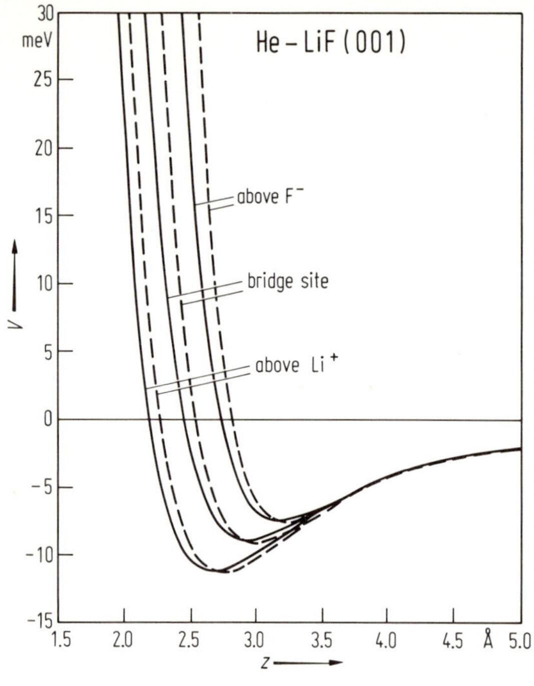

Fig. 21. LiF. The He-LiF(001) potential as function of the distance z from the center of the first surface ion. The potentials are plotted for three sites within the same unit cell: above a positive ion (lower curve), above a negative ion (upper curve), and above a bridge position halfway between two like ions (central curve). The solid and dashed curves are for different models. [88E1].

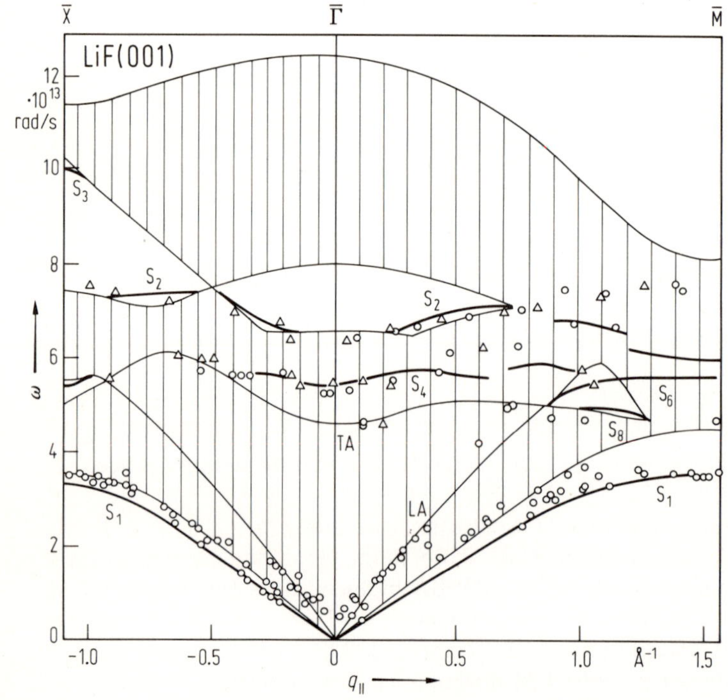

Fig. 22. LiF. Phonon spectra along the $\overline{\Gamma X}\langle 110\rangle$ direction and the $\overline{\Gamma M}\langle 100\rangle$ direction for LiF(001). Circles and triangles refer to phonon peaks detected in non-resonant and resonant conditions, respectively. Surface phonon dispersion curves for sagittal polarization (bold lines) and surface-projected bulk bands (hatched regions) are reported as calculated by Kress et al. [87K2]. From Bracco et al. [87B6].

MgO(001)

Surface lattice: square; $a = 2.98$ Å.
Surface Brillouin zone: see Fig. 3.
He potential: $C_3 = 151$ meV · Å^3 [91V1], $\varepsilon_0 = 5.5$ meV, $D = 7.5$ meV, $z_m = 3.6$ Å, $z_{vW} = 0.76$ Å, $h_{pv} = 0.47$ Å [83B2, 83B7].
Surface phonons: see Fig. 23 of chapter 4.1, subvolume III/24b and Fig. 23. A Rayleigh mode has been measured by He TOF scattering, first in the ⟨100⟩ $\overline{\Gamma}\overline{M}$ direction by Brusdeylins et al. [83B7], then also in the ⟨110⟩ $\overline{\Gamma}\overline{X}$ direction [89J1, 89M1, 91B1]. Results from [91B1].
Additional reference: [84C1].

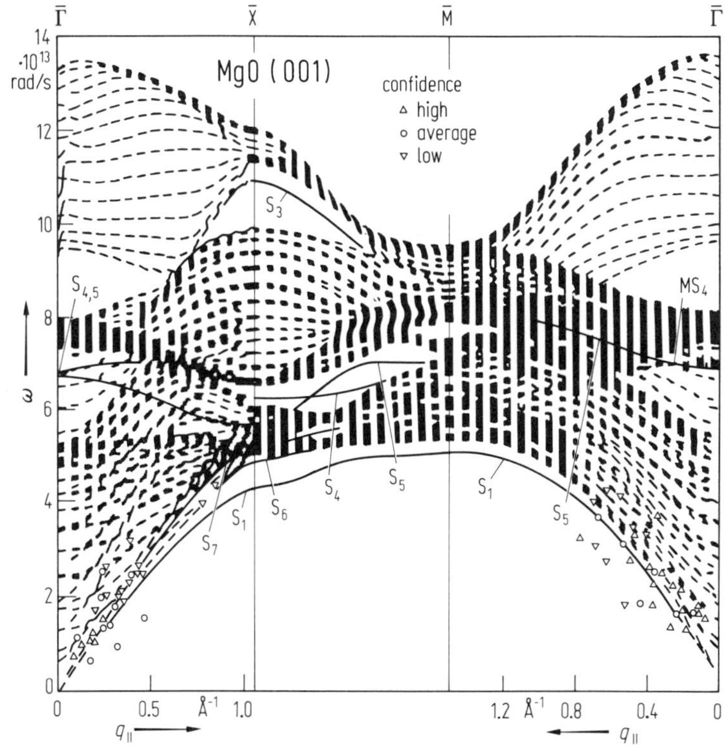

Fig. 23. MgO. Theoretical surface phonon dispersion curves in the $\overline{\Gamma}\overline{X}$, $\overline{\Gamma}\overline{M}$ and $\overline{X}\overline{M}$ directions. [91B1] with inelastic HAS data (symbols).

Mo(001)

Surface lattice: square, $a = 3.15$ Å.
Surface Brillouin zone: same as for fcc (001), see Fig. 3.
He potential: $D = 3.7$ meV [91V1], $z_m = 3.86$ Å [91V1].
Surface phonons: see Figs. 24 and 25 [90H1].

A structural phase transition is driven by a phonon mode softening at the zone boundary. This soft phonon mode continues toward the center of the Brillouin zone and mixes with the Rayleigh phonon in an avoided crossing.
Additional reference: [89H1].

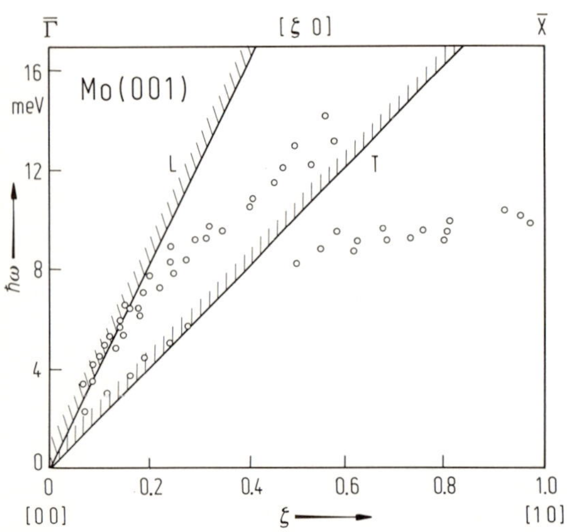

Fig. 24. Mo. Surface phonon dispersion curves in the $\bar{\Gamma}\bar{X}$ direction measured by HATOF (circles: experimental points, solid line labelled L: bulk longitudinal mode threshold, solid line labelled T: bulk transverse mode threshold) [90H1]. See also Fig. 24, chapter 4.1.

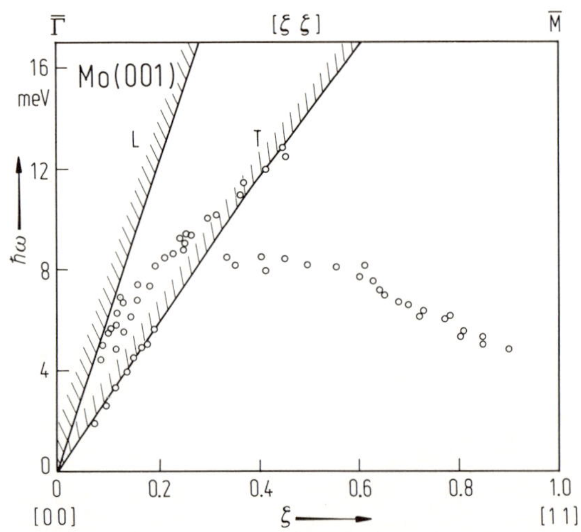

Fig. 25. Mo. Surface phonon dispersion curves in the $\bar{\Gamma}\bar{M}$ direction measured by HATOF (circles: experimental points at surface temperature 800 K, solid line labelled L: bulk transverse mode threshold) [90H1]. See also Fig. 25, chapter 4.1.

NaCl(001)

Surface lattice: square, $a = 3.98$ Å.
Surface Brillouin zone: see Fig. 3.
He potential: $C_3 = 106$ meV · Å^3 [91V1], $\varepsilon_n(He^4) = 4.1$, 1.5, 0.31 meV [84L2], $D = 6.1$ meV [91V1], $z_m = 3.49$ Å [91V1], $z_{vW} = 1.03$ Å [91V1], $h_{pv} = 0.85$ Å [88E1]. See Fig. 26.
Surface phonons: see Figs. 27 and 28.

The spectrum shows Rayleigh modes and longitudinal modes for both the ⟨100⟩ $\bar{\Gamma}\bar{M}$ direction and the ⟨110⟩ $\bar{\Gamma}\bar{X}$ direction [83B2, 90S2, 91S1]. Many selective absorption features (surface bound state resonances) appear in the TOF data and must be carefully separated from the phonon peaks [91S1].

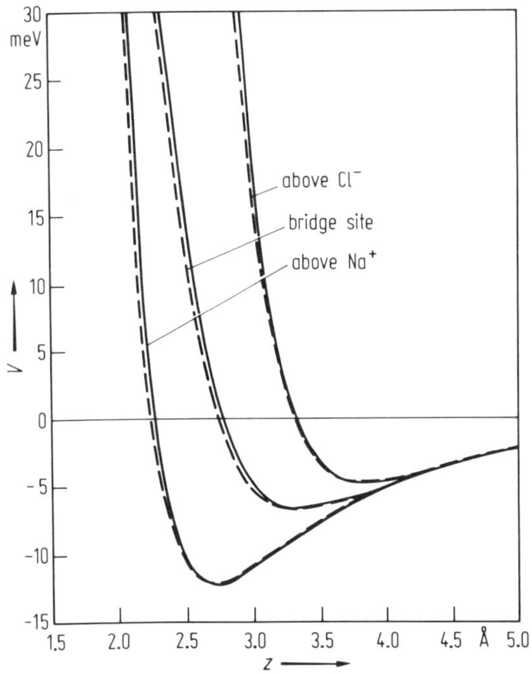

Fig. 26. NaCl. The He-NaCl(100) potential, [88E1]. Further explanation, see Fig. 21.

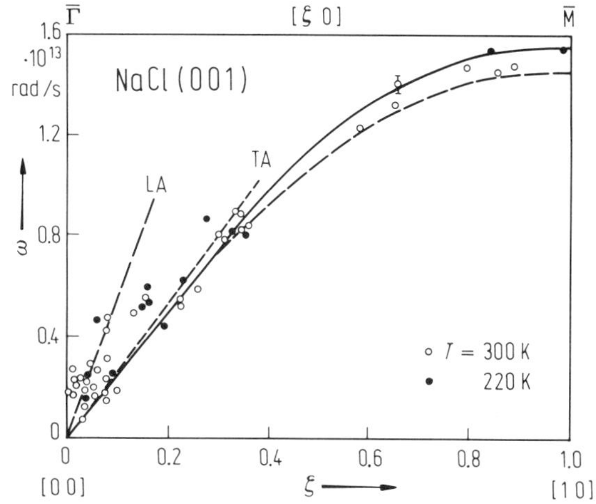

Fig. 27. NaCl. Surface phonon dispersion curves in the $\bar{\Gamma}\bar{M}$ direction measured by HATOF (symbols: experimental points at surface temperature of 300 K and of 220 K, solid and dashed lines: LDM calculations for surface modes, LA and TA: ECT dispersion curves for bulk longitudinal acoustic and transverse acoustic modes) [83B2]. See also Fig. 26, chapter 4.1.

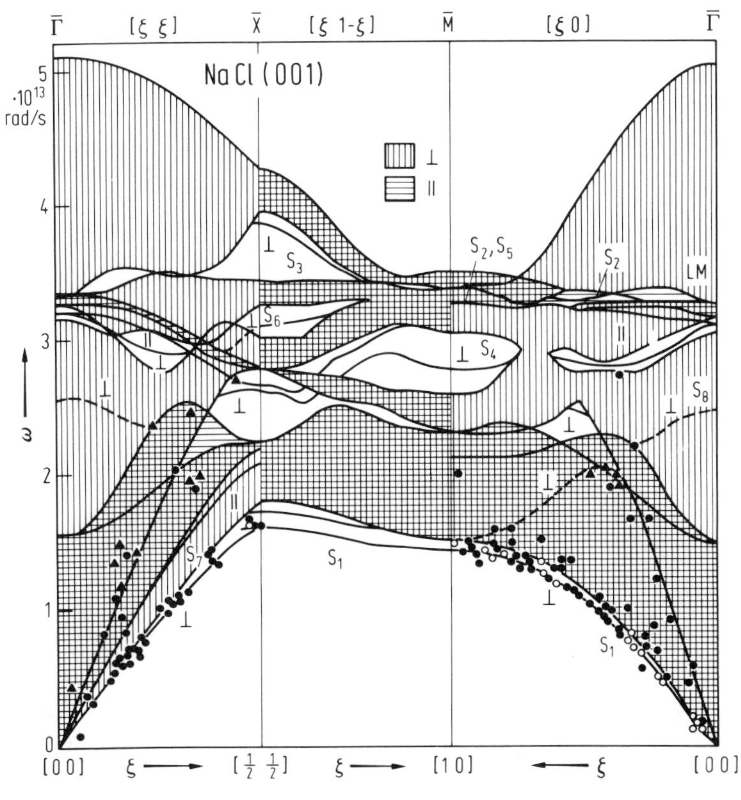

Fig. 28. NaCl. Phonon dispersion curves for NaCl(001) [90S2]. The measured surface phonon peaks are indicated by open circles [83B2] and closed circles [90S2]. Close triangles indicate weaker peaks. The calculated bulk and surface phonon spectra were obtained by a Green's function method.

NaF(001)

Surface lattice: square; $a = 3.27$ Å.

Surface Brillouin zone: see Fig. 3.

He potential: $C_3 = 73$ meV · Å^3 [91V1], $\varepsilon_n(\mathrm{He}^4) = 4.92$, 1.87, 0.54 meV [78D1], $\varepsilon_n(\mathrm{He}^3) = 4.50$, 1.38 meV [78D1], $D = 7.58$ meV [88E1], $z_m = 3.00$ Å, $h_{pv} = 0.45$ Å [88E1]. The minimum of the potential is on the bridge site. See Fig. 29.

Surface phonons: see Fig. 30 [84T2].

Surface phonon spectra were obtained from He TOF measurements along the ⟨100⟩ $\overline{\Gamma}\overline{M}$ direction by Brusdeylins et al. [85B7] and then also along the ⟨110⟩ $\overline{\Gamma}\overline{X}$ direction [87B1]. It was possible in this case to observe optical modes using the enhancement effect of a bound state resonance (see Fig. 29). The data show branch-folding effects due to the nearly identical masses of Na and F, and are generally in good agreement with lattice dynamical calculations [82B2].

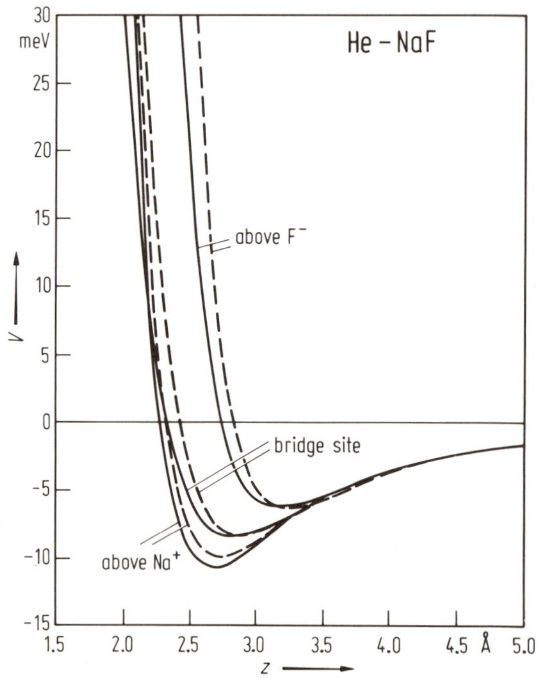

Fig. 29. NaF. The He-NaF potential [88E1]. For further explanations, see Fig. 21.

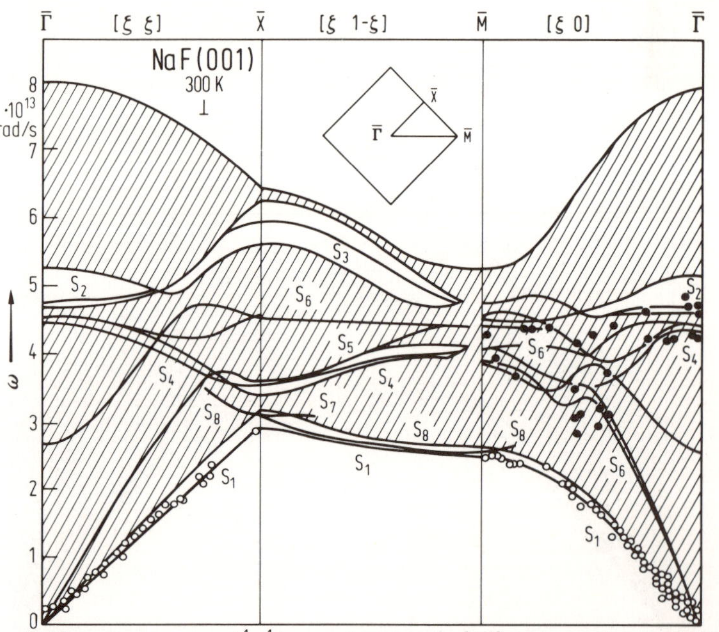

Fig. 30. NaF. Phonon spectra along the $\overline{\Gamma}\overline{X}$⟨110⟩ direction and the $\overline{\Gamma}\overline{M}$⟨100⟩ direction for NaF(001). Open and filled circles are from He TOF data at $\varepsilon_i = 20$ meV and $60 \cdots 100$ meV, respectively. They are compared with calculations for sagittal plane modes [82B2]. From [84T2].

NaI(001)

Surface lattice: square; $a = 4.58$ Å.
Surface Brillouin zone: see Fig. 3.
He potential: not known in detail. Peak-to-valley corrugation estimated to be $h_{pv} = 1.2$ Å [90B1].
Surface phonons: see Fig. 31.

The phonon spectrum has a large gap between the acoustic and optical bands due to the large mass difference between Na and I. The measured surface modes all lie below this gap and no surface relaxation effects are observed; the results agree with theoretical predictions [90B1].

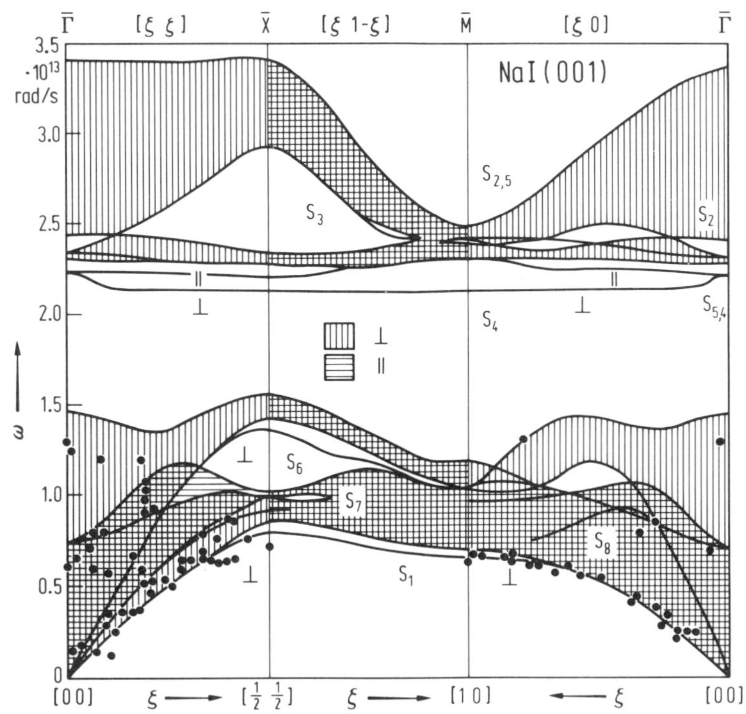

Fig. 31. NaI. Surface phonon spectra along the $\bar{\Gamma}\bar{X}$, $\bar{\Gamma}\bar{M}$ and $\bar{X}\bar{M}$ directions for NaI(001) [90B1].

Nb(001)

Surface lattice: square; $a = 3.30$ Å.
Surface Brillouin zone: see Fig. 3.
Surface phonons: see Fig. 32 [91B1].

Surface phonon spectra were obtained from He TOF measurements along the ⟨100⟩$\overline{\Gamma M}$ direction and the ⟨110⟩ $\overline{\Gamma X}$ direction by Hulpke et al. [90H2]. A Rayleigh mode and a longitudinal resonance are seen in both directions. A good fit to the data is obtained with a slab calculation using strongly reduced surface constants [92H1].

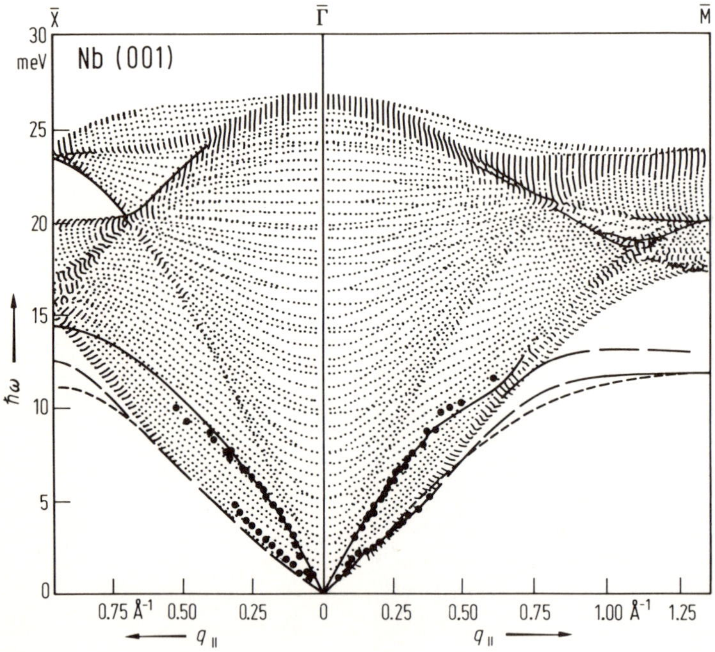

Fig. 32. Nb. Theoretical surface dispersion curves (dots and lines) of Nb(001) with inelastic HAS data (circles) [91B1].

Ni(001)

Surface lattice: square; $a = 2.49$ Å.
Surface Brillouin zone: see Fig. 3.
He potential: $C_3 = 218$ meV · Å^3 [85S1]. Theory gives $D = 4.2$ meV, $z_m = 3.76$ Å [76K1].
Surface phonons: see Fig. 33.

The surface phonon spectrum has been thoroughly studied by EELS [86R1] and is well reproduced by a truncated bulk force constants model with a 20% stiffening of the force between the two top layers [82B3]. He TOF data along the (001) direction $\overline{\Gamma}\overline{M}$ have been obtained by Berndt et al. [87B3]. As shown in the figure, there is good agreement between the phonon spectra determined by the two techniques, EELS and He TOF.

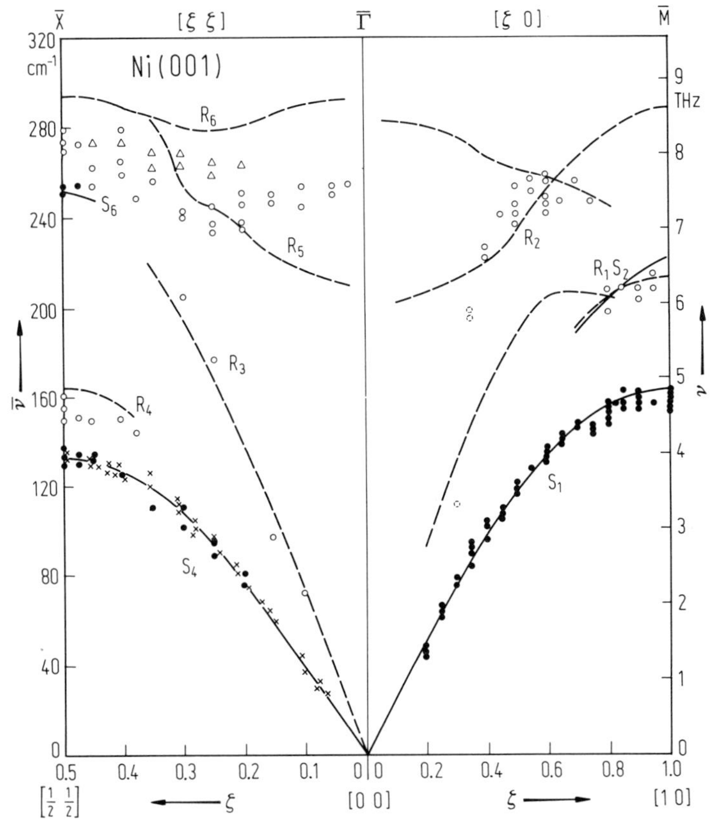

Fig. 33. Ni. Surface phonon dispersion curves of Ni(001) measured by EELS (full circles S_i: experimental points for surface phonons, open circles R_i: experimental points for surface resonances, solid and dashed lines: LDM fits) [86R1]. See also Fig. 29, chapter 4.1.

NiO(001)

Surface lattice: square; $a = 2.95$ Å, but the chemical cell 1×1 is different from the magnetic cell 2×1. See Fig. 34 for magnetic/chemical cell.
Surface Brillouin zone of the chemical cell: see Fig. 3.
He potential: $C_3 = 220$ meV · Å^3 [91V1], $\varepsilon_n = 7.9$, 4.0, 1.6 meV [79C1], $D = 10.1$ meV [79C1], detailed shape of potential not determined, $h_{pv} = 0.3$ Å from hard-wall fit of data [79C1].
Surface phonons: see Fig. 35.

Surface phonon spectra were obtained from He TOF measurements along the ⟨100⟩ $\overline{\Gamma M}$ direction and the ⟨110⟩ $\overline{\Gamma X}$ by Brug et al. [92B1]. The measurements were taken for the paramagnetic phase above the Néel temperature.

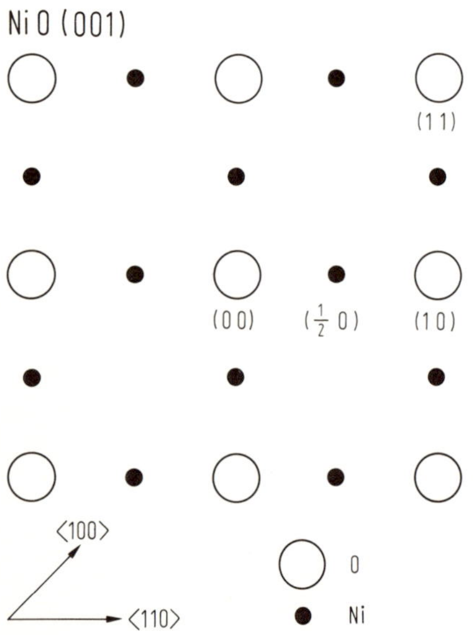

Fig. 34. NiO. Magnetic and chemical reciprocal lattice of NiO (001) [91B1].

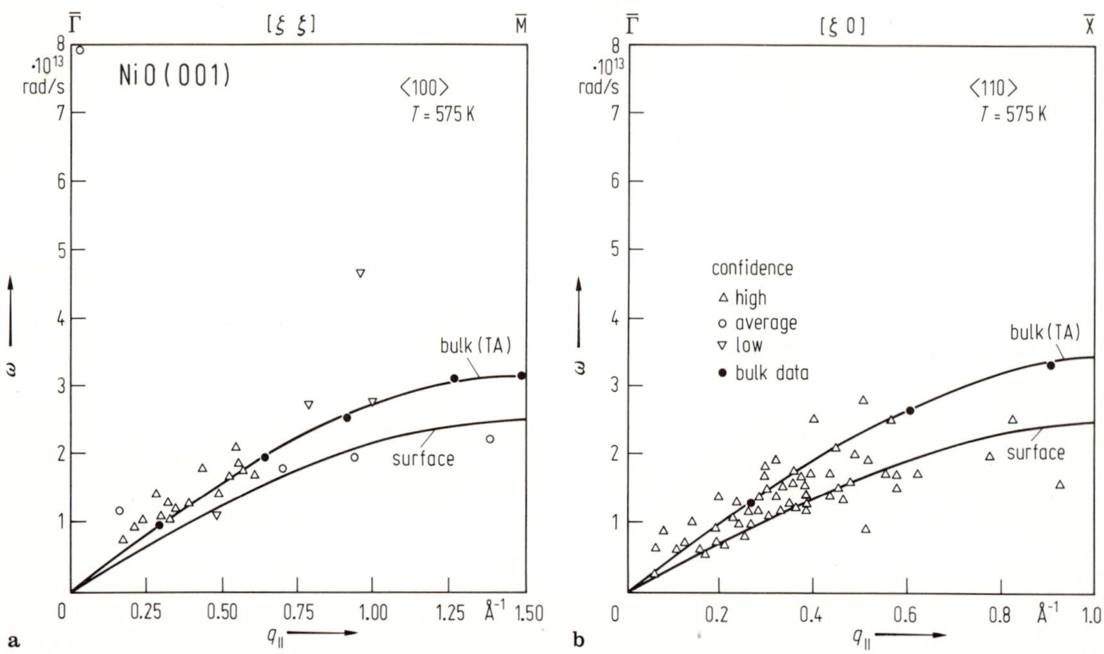

Fig. 35a, b. NiO. Bulk and Surface Dispersion Curves of NiO(001) for **a** ⟨100⟩ direction and **b** ⟨110⟩ direction [91B1].

Pd(110)

Surface lattice: rectangular; $a_x = 2.75$ Å, $a_y = 3.88$ Å.
Surface Brillouin zone: see Fig. 3.
He potential [91V1]: $C_3 = 2.11$ meV · Å^3, $\varepsilon_0 = 6.42$ meV, $D = 8.05$ meV, $z_m = 3.60$ Å.
Surface phonons: see Fig. 36.

Surface phonon spectra have been obtained by Lahee, Toennies and Wöll [87L1] from He TOF measurements along the ⟨1$\bar{1}$0⟩ direction $\overline{\Gamma X}$ and along the ⟨001⟩ direction $\overline{\Gamma Y}$. A truncated bulk model with force constants fitted to bulk data does not reproduce well several features of the observed spectra (dashed curves). A modified set of force constants that gives the best to the TOF data [87L1] includes a radial force constant between neighboring surface atoms reduced by 30% and much weakened surface second-neighbor interactions. These modifications give rise respectively to the A_1 mode along the close-packed ⟨1$\bar{1}$0⟩ rows and to the rather small gap between the S_1 and S_2 modes in the ⟨001⟩ direction (solid curves).

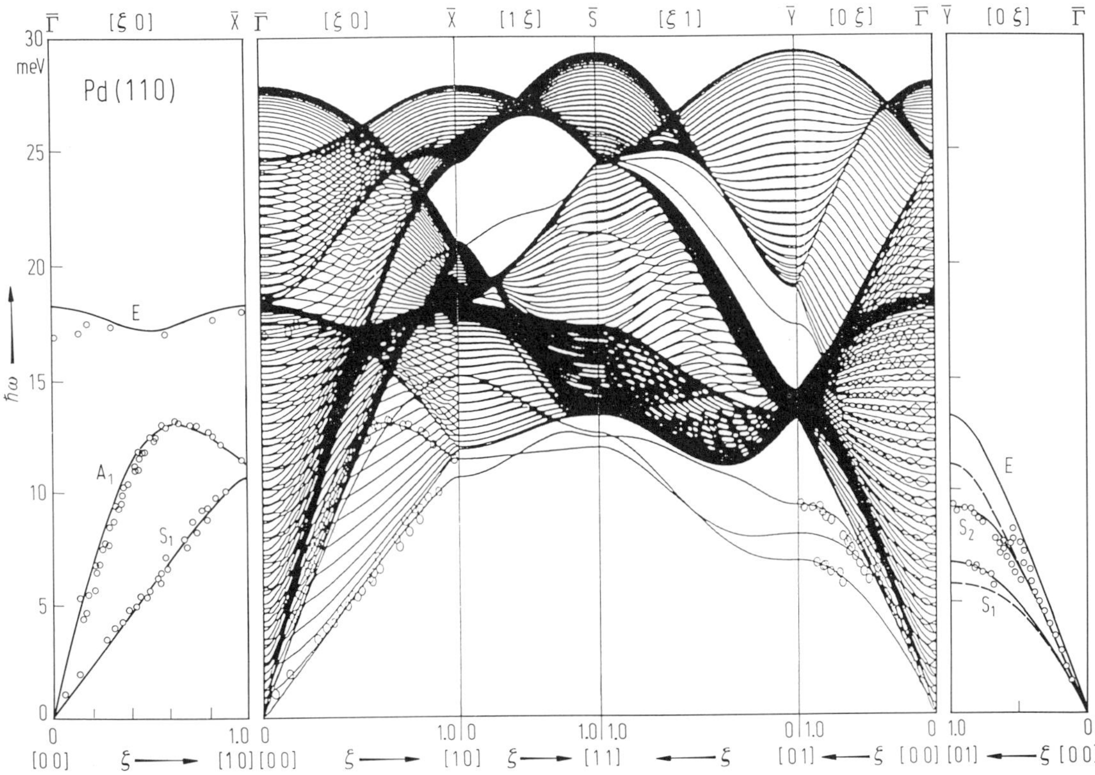

Fig. 36. Pd. Phonon spectra of Pd(110) [87L1]. See text. See also Fig. 34, chapter 4.1.

Pt(110) 1 × 2

Surface lattice: rectangular; $a_x = 2.77$ Å, $a_y = 3.92$ Å. See Fig. 37 [89K1].
Surface Brillouin zone: see Fig. 3.
He potential: $C_3 = 251$ meV · Å^3 [91V1], $\varepsilon_n = 7.4$, 5.1, 3.1, 1.5 meV [90K1], $D = 8.8$ meV [90K1], $h_{pv} = 1.5$ Å [89K1].

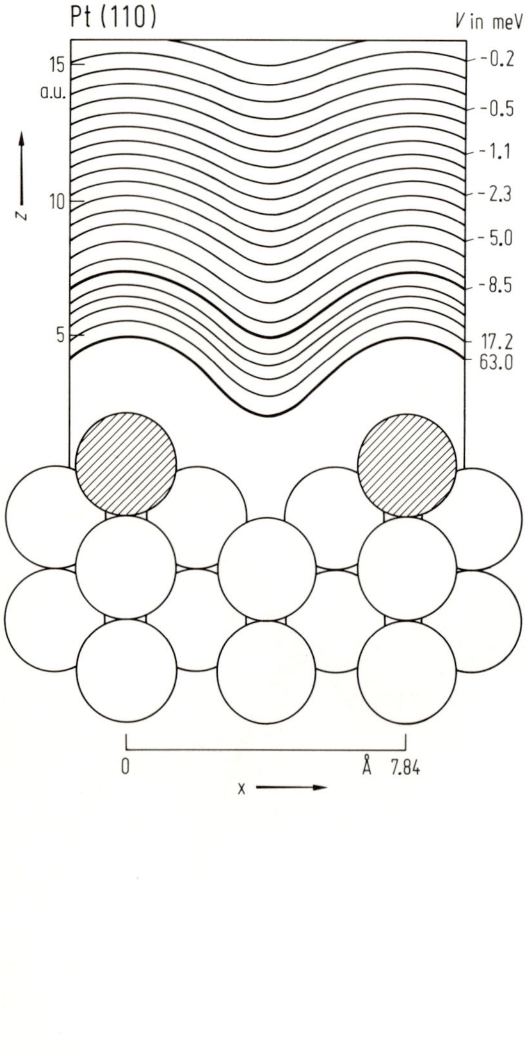

Fig. 37. Pt. Lower part: side view of a hard sphere model of the Pt(110) 1 × 2 missing now reconstructed surface. Upper part: the He-Pt(110) 1 × 2 interaction potential contours as observed from the diffraction patterns. The contours corresponding to the incoming energy of 63 meV and the potential minimum are emphasized [89K1].

Pt (111)

Surface lattice: hexagonal; $a = 2.77$ Å.
Surface Brillouin zone: see Fig. 3.
He potential: $C_3 = 251$ meV · Å^3 [91V1], ε_0, D presumed to be nearly the same as for Pt(110).
Surface phonons: see Figs. 38 and 39.

Surface phonon spectra have been obtained from He TOF measurements along the $\langle 11\bar{2}\rangle$ direction $\overline{\Gamma}\overline{\mathrm{M}}$ and along the $\langle 110\rangle$ direction $\overline{\Gamma}\overline{\mathrm{K}}$ by the groups in Göttingen [85H2], Jülich [86K1, 86K2], and Köln [86N1]. The data from the three groups are mutually consistent, but the interpretation differs somewhat. Harten et al. [85H2] and Neuhaus et al. [86N1] emphasize the existence of a surface Kohn anomaly in the Rayleigh branch (see Fig. 39) while Kern et al. [86K1] present a comprehensive study of the surface phonon spectrum. There is general agreement that a truncated bulk model with force constants fitted to bulk data fails to reproduce the strong longitudinal resonance observed in both the $\overline{\Gamma}\overline{\mathrm{M}}$ and the $\overline{\Gamma}\overline{\mathrm{K}}$ directions, and that there must be a 40% reduction of the radial force constant between neighboring surface atoms. However, Kern et al. [86K1] explain their data by using a model with nearest neighbor force constants only, while Bortolani et al. [89B2] use a much more elaborate model to reproduce the Göttingen TOF spectra.

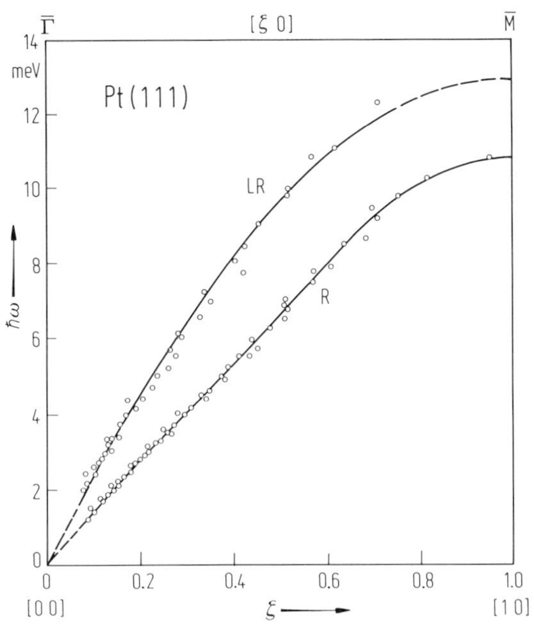

Fig. 38. Pt. Surface phonon dispersion curves measured by HATOF for the $\overline{\Gamma}\overline{\mathrm{M}}$ direction for Pt(111) (circles: experimental points for surface temperature 400 K, solid and dashed lines: LDM fit, R: Rayleigh mode, LR: longitudinal resonance) [85H2]. See also Fig. 35, chapter 4.1.

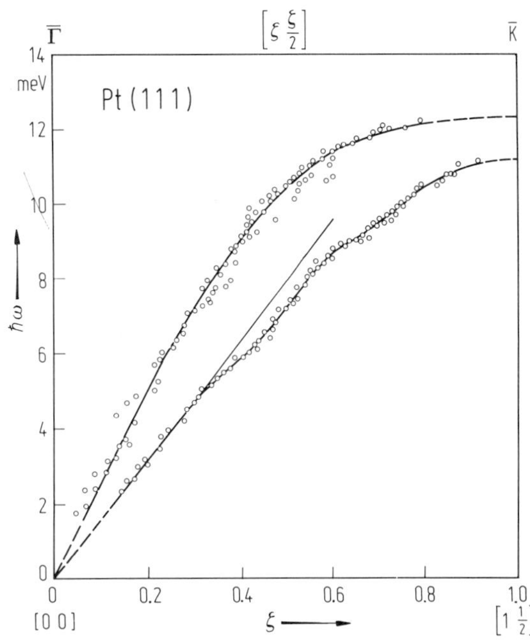

Fig. 39. Pt. Rayleigh mode dispersion curves measured by HATOF for the $\overline{\Gamma}\overline{\mathrm{K}}$ direction for Pt(111) (circles: experimental points for surface temperature 160 K, solid and dashed lines: LDM fit) [85H2]. See also Fig. 36, chapter 4.1.

RbBr(001)

Surface lattice: square; $a = 4.85$ Å.
Surface Brillouin zone: see Fig. 3.
He potential: not known in detail. Peak-to-valley corrugation height estimated to be $h_{pv} = 0.4$ Å [89C3].
Surface phonons: see Fig. 40. Relaxation of the surface is predicted [87D1], but the observed phonon spectrum is consistent with an unrelaxed surface. A crossing mode and the Rayleigh mode are seen in both the ⟨100⟩ $\overline{\Gamma M}$ direction and the ⟨110⟩ $\overline{\Gamma X}$ direction [89C3].
Additional reference: [90S1].

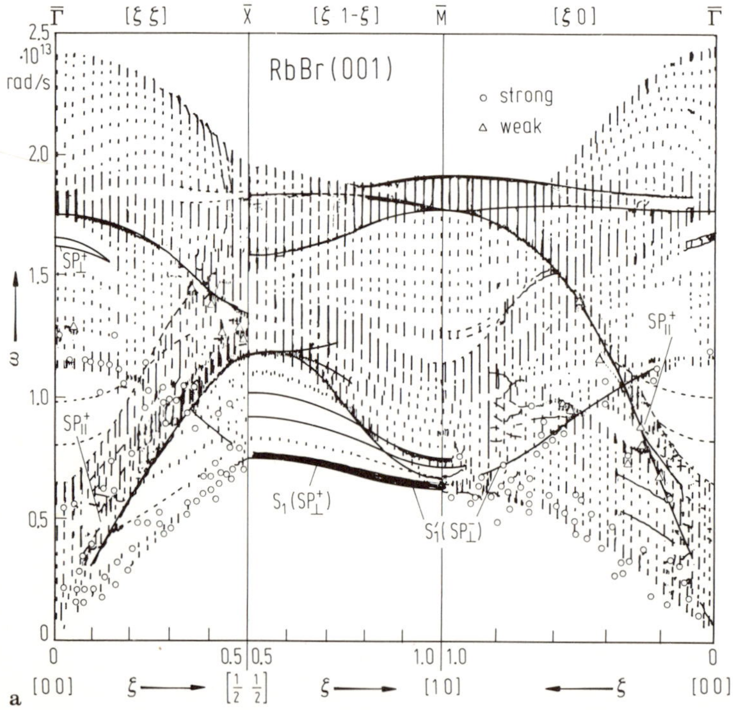

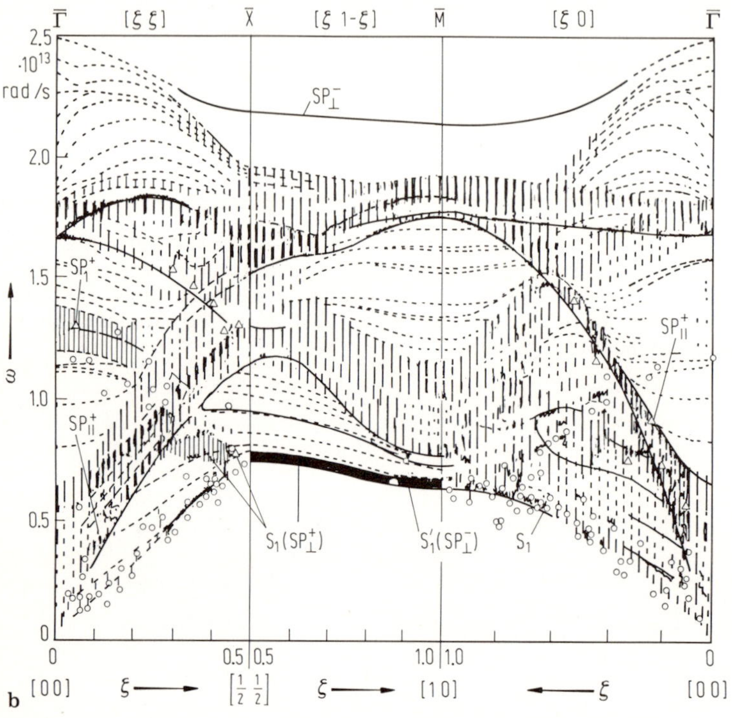

Fig. 40a, b. RbBr. Surface phonon dispersion curves measured by HATOF for RbBr(001) (circles: experimental points for strong intensities, triangles: experimental points for weak intensities, solid lines: LDM calculations for surface phonons for **a** unrelaxed surface and **b** relaxed surface, dashed lines: LDM calculations for surface-projected bulk phonons, $SP_\perp$ and $SP_\parallel$: sagittal-plane vibrations perpendicular and parallel to the surface, respectively, SH: shear-horizontal vibration, + or −: positive or negative ion carries out the vibration in the surface layer) [89C3]. See also Fig. 37a, b, chapter 4.1.

RbCl(001)

Surface lattice: square; $a = 4.65$ Å.
Surface Brillouin zone: see Fig. 3.
He potential: not known in detail. Peak-to-valley corrugation estimated to be $h_{pv} = 0.15$ Å [89C2].
Surface phonons: See Fig. 41.

The measurements by Chern et al. [89C1] clearly show the Rayleigh mode for both the ⟨100⟩ $\overline{\Gamma M}$ direction and the ⟨110⟩ $\overline{\Gamma X}$ direction. The results agree with calculations for an unrelaxed surface. Several isolted points are also seen at higher frequencies.

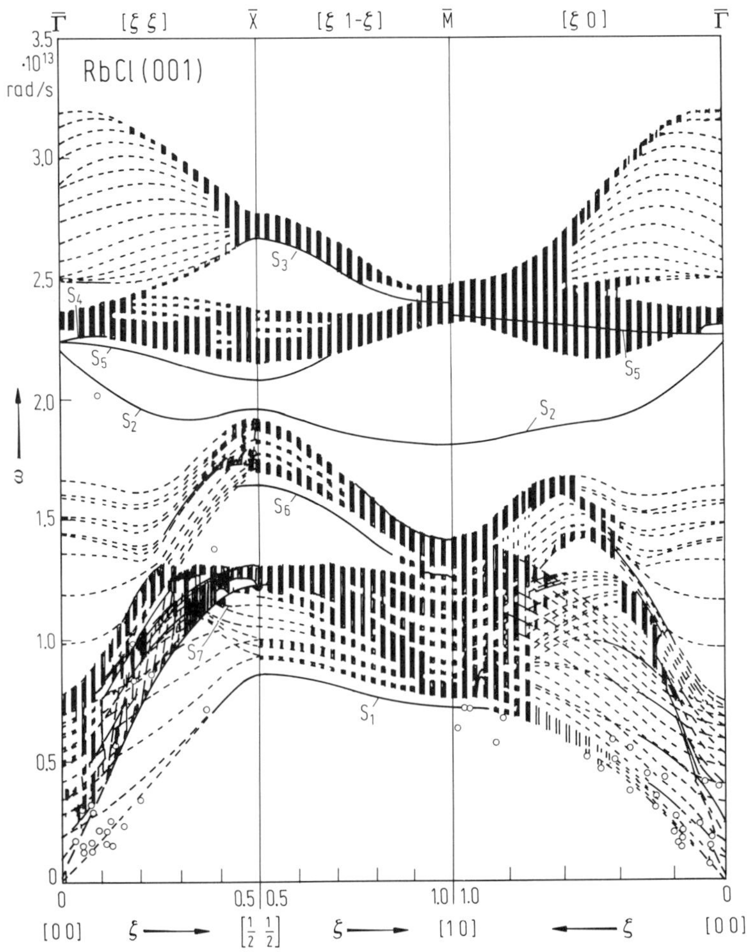

Fig. 41. RbCl. Surface phonon dispersion curves measured by HATOF for RbCl(001) (circles: experimental points, solid lines: LDM calculations for surface phonons, dashed lines: LDM calculations for surface-projected bulk phonons) [89C1]. S_i: surface states. See also Fig. 38, chapter 4.1.

RbI(001)

Surface lattice: square; $a = 5.19$ Å.
Surface Brillouin zone: see Fig. 3.
Surface phonons: five modes have been observed on this surface by Safron et al. [90S3]: (1) the Rayleigh mode, (2) a crossing mode, (3) a longitudinal acoustic mode, (4) an unidentified mode at the top of the accoustic band in the ⟨100⟩ direction near the $\overline{M}$ point, and (5) a mode lying above the optical band edge in the ⟨100⟩ $\overline{\Gamma}\overline{M}$ direction. Mode (4) is suggested to be a large amplitude, second layer vibration; the detection of a second layer mode by He scattering would represent a novel feature. Mode (5) is believed to be due to surface relaxation [87D1].

Si(111)2×1

Surface lattice: rectangular; $a_{\langle 110\rangle} = 3.84$ Å, $a_{\langle 112\rangle} = 6.65$ Å.
Surface Brillouin zone: see inset in Fig. 42.
He potential: $C_3 = 174$ meV · Å^3 for Si(100) [91V1].
Surface phonons: see Fig. 42.

The phonon spectrum shows a Rayleigh mode and a flat band at about 10 meV [86H1]. Various interpretations of the nature of this band have been proposed. A bond charge model gives a good account of the data [89M2].

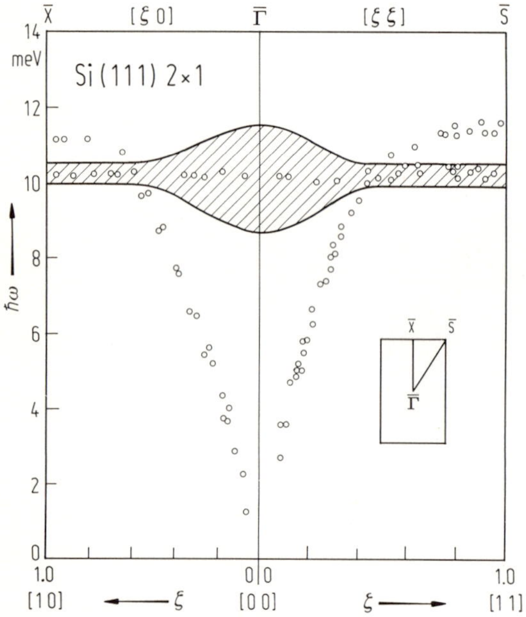

Fig. 42. Si. Surface phonon dispersion curves measured by HATOF for Si(111) 2×1 for surface temperature 300 K (circles: experimental points, shaded region: full width at half maximum for the measured peaks due to the 10.5 meV mode). The inset shows the SBZ [86H1]. See also Fig. 39, chapter 4.1.

TaS_2(0001)(1T phase)

Surface lattice: hexagonal; $a = 3.381$ Å. A CDW reconstruction with $a_{cdw} = \sqrt{13}a$ is present below 183 K.
Surface Brillouin zone: hexagonal, with superlattice spots of the CDW.

He potential: the corrugation amplitude reflects the presence of a superlattice distortion connected with the formation of a charge density wave. The form of the surface corrugation is well described by $z = 2\zeta_0[\cos(g(4x-y)) + \cos(g(x+3y) + \cos(g(3x-4y))] + 2\zeta_1[\cos(gx) + \cos(gy) + \cos(g(x-y))]$ with $\zeta_0 = 0.055$ Å and $\zeta_1 = 0.04$ Å [89B3]. The well depth is $D = 8.7$ meV [89B3].

Surface phonon spectra have been obtained by Brusdeylins et al. [90B2] along the ⟨100⟩ symmetry directions of the lattice and of the superlattice for the low temperature commensurate $\sqrt{13} \times \sqrt{13}$ R 13.9° commensurate phase. The observed CDW amplitude in this material is large and affects the surface phonon dispersion curves more markedly than in the case of 2H-$TaSe_2$. A simple model gives the main features of the measurement.

2H-$TaSe_2$(0001)

Surface lattice: hexagonal; $a = 3.44$ Å.
Surface Brillouin zone: hexagonal.
Surface phonons: see Figs. 43 and 44.

This crystal surface was studied extensively by the Göttingen group. Surface phonon spectra were obtained by He atom scattering along both high symmetry directions both above and below the temperature of the charge density wave at 122 K. The Rayleigh mode and optical modes were observed and softening was seen to take place primarily in the Rayleigh mode in the ⟨100⟩ direction [87B2, 88B1, 88B2, 90B3].

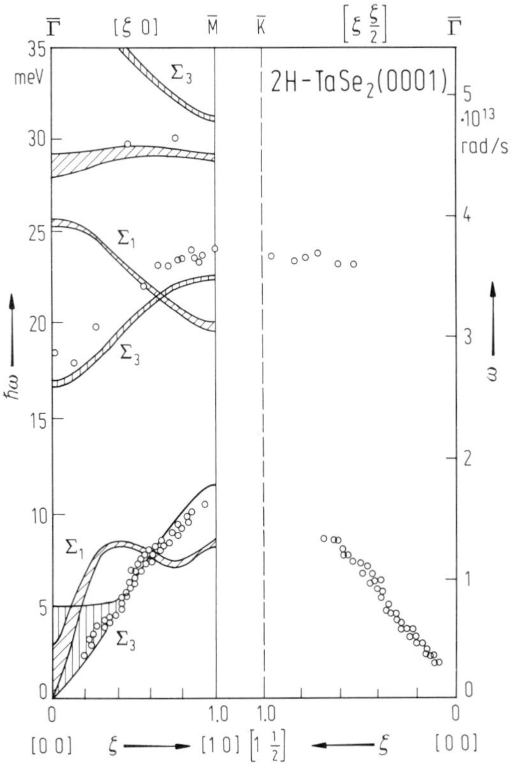

Fig. 43. $TaSe_2$. Surface phonon dispersion curves measured by HATOF for 2H-$TaSe_2$(0001) (circles: experimental points at surface temperature 140 K, vertical shaded regions: LDM calculation of bulk phonon modes with Σ_3 symmetry, oblique shaded regions: LDM calculation of bulk phonon modes with Σ_1 symmetry) [90B3]. See also Fig. 41, chapter 4.1.

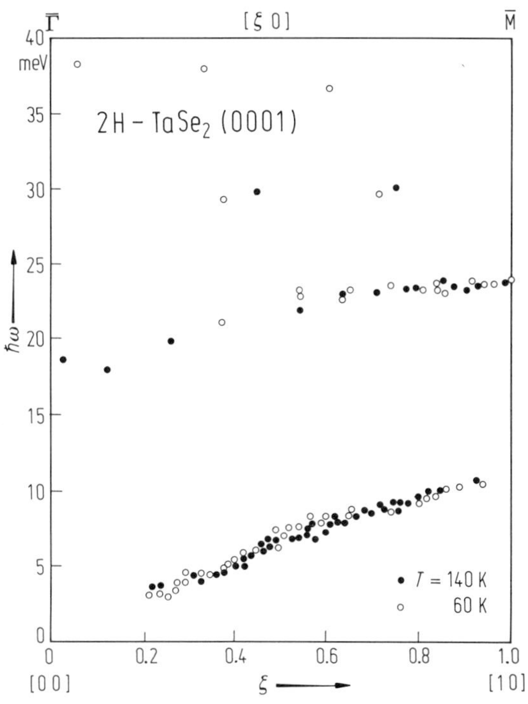

Fig. 44. $TaSe_2$. Surface phonon dispersion curves in the Γ̄M̄ direction measured by HATOF for 2H-$TaSe_2$(0001) (experimental points at surface temperature 140 K and 60 K) [90B3]. See also Fig. 42, chapter 4.1.

W(001)

Surface lattice: square; $a = 3.16$ Å.
Surface Brillouin zone: see Fig. 3.
He potential: for W(110), which should be similar, $\varepsilon_0 = 5.6$ meV, $C_3 = 265$ meV · Å^3 [91V1].
Surface phonons: see Fig. 45.

Surface phonon spectra have been obtained from He TOF measurements along the ⟨110⟩ direction $\overline{\Gamma}\overline{M}$ by Ernst et al. [87E2]. A temperature dependent softening of the Rayleigh phonon branch near the zone boundary signals the onset of a $(\sqrt{2} \times \sqrt{2})$ R 34° reconstruction below 300 K.
Additional reference: [90H2].

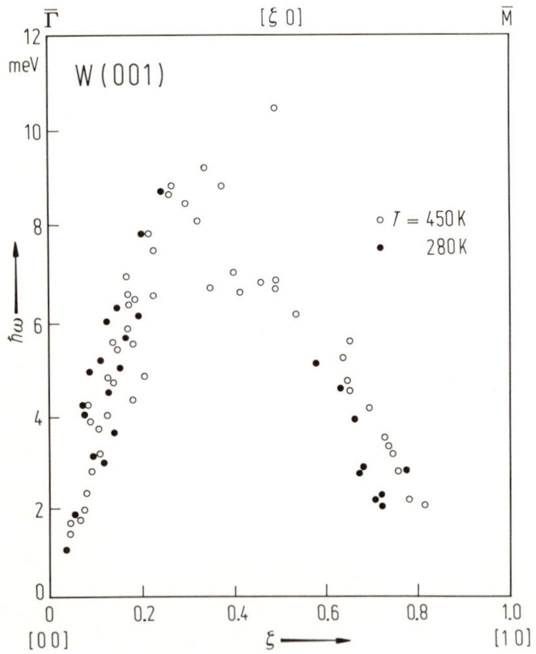

Fig. 45. W. Surface phonon dispersion curves measured by HATOF for W(001) (experimental points for surface temperature 450 K and 280 K) [87E2]. See also Fig. 45, chapter 4.1.

7.3 References for 7

29S1 Stern, O.: Naturwissenschaften **17** (1929) 391.
32J1 Jackson, J. M., Mott, N. F.: Proc. R. Soc. London A **137** (1932) 703.
33F1 Frisch, R., Stern, O.: Z. Phys. **84** (1933) 430.
34F1 Fermi, E.: Nuovo Cimento **11** (1934) 157.
35L1 Lennard-Jones, J. E., Strachan, C.: Proc. R. Soc. London A **150** (1935) 442.
37D1 Devonshire, A. F.: Proc. R. Soc. London A **158** (1937) 269.
37L1 Lennard-Jones, J. E., Devonshire, A. F.: Proc. R. Soc. London A **158** (1937) 253.
37S1 Strachan, C.: Proc. R. Soc. London A **158** (1937) 591.
63H1 Herman, F., Skillman, S.: Atomic Structure Calculations. Englewood Cliffs: Prentice-Hall 1963.
69C1 Cabrera, N., Celli, V., Manson, R.: Phys. Rev. Lett. **22** (1969) 346.
69N1 Nijboer, B. R. A., Renne, M. J.: Chem. Phys. Lett. **2** (1969) 35.
69S1 Smith, J. N., O'Keefe, D. R., Saltsburg, H., Palmer, R. L.: J. Chem. Phys. **50** (1969) 4667.
70M1 McClure, J. D.: J. Chem. Phys. **52** (1970) 2712.
71B1 Beeby, J. L.: J. Phys. C **4** (1971) L359.
71M1 Manson, R., Celli, V.: Surf. Sci. **24** (1971) 495.
71M2 Maradudin, A. A., Montroll, E. W., Weiss, C. H., Ipatova, I. P.: Theory of Lattice Dynamics in the Harmonic Approximation, 2nd ed., New York, London: Academic Press 1971.
71W1 Williams, B. R.: J. Chem. Phys. **55** (1971) 3220.
72B1 Beeby, J. L.: J. Phys. C **5** (1972) 3438.
72B2 Beeby, J. L.: J. Phys. C **5** (1972) 3457.
72G1 Goodman, F. O.: Surf. Sci. **30** (1972) 1.
72H1 Hoinkes, H., Nahr, H., Wilsch, H.: Surf. Sci. **33** (1972) 516.
73G1 Garibaldi, U., Levi, A. C., Spadacini, R., Tommei, G. E.: Surf. Sci. **38** (1973) 269.
74C1 Clementi, E., Roetti, C.: At. Data Nucl. Data Tables **14** (1974) 177.
74M1 Mason, B. F., Williams, B. R.: J. Chem. Phys. **61** (1974) 2765.
74S1 Steele, W. A.: The Interaction of Gases with Solid Surfaces. New York: Pergamon 1974, p. 22.
74W1 Weare, J. H.: J. Chem. Phys. **61** (1974) 2900.
75B1 Benedek, G.: Phys. Rev. Lett. **35** (1975) 234.
75G1 Garibaldi, U., Levi, A. C., Spadacini, R., Tommei, G. E.: Surf. Sci. **48** (1975) 649.
75L1 Lagos, M., Birstein, L.: Surf. Sci. **52** (1975) 391.
76B1 Boato, G., Cantini, P., Mattera, L.: Surf. Sci. **55** (1976) 141.
76B2 Boato, G., Cantini, P., Tatarek, R.: J. Phys. F **6** (1976) L237.
76C1 Cantini, P., Felcher, G. P., Tatarek, R.: Phys. Rev. Lett. **37** (1976) 606.
76C2 Chiroli, C., Levi, A. C.: Surf. Sci. **59** (1976) 325.
76G1 Goodman, F. O., Wachman, H. Y.: Dynamics of Gas-Surface Scattering. New York: Academic Press 1976.
76K1 Kleiman, G. G., Landman, U.: Solid State Commun. **18** (1976) 819.
76Z1 Zaremba, E., Kohn, W.: Phys. Rev. B **13** (1976) 2270.
77A1 Armand, G., Lapujoulade, J., Lejav, Y.: Surf. Sci. **63** (1977) 143.
77C1 Cantini, P., Felcher, G. P., Tatarek, R., in: Proceedings of the Seventh International Vacuum Congress and the Third International Conference on Solid Surfaces. Dobrozemsky, R., et al. (eds.), Vienna 1977, p. 1357.
77L1 Lagos, M.: Surf. Sci. **65** (1977) 124.
77Z1 Zaremba, E., Kohn, W.: Phys. Rev. B **15** (1977) 1769.
78B1 Benedek, G., Galimberti, F.: Surf. Sci. **71** (1978) 87.
78B2 Boato, G., Cantini, P., Tatarek, L.: Phys. Rev. Lett. **40** (1978) 887.
78C1 Carlos, W. E., Cole, M. W.: Surf. Sci. **77** (1978) L173.
78C2 Cole, M. W., Frankl, D. R.: Surf. Sci. **70** (1978) 585.
78D1 Derry, G., Wesner, D., Krishnaswamy, S. V., Frankl, D. R.: Surf. Sci. **74** (1978) 245.
78H1 Harvie, E., Weare, J. H.: Solid State Commun. **27** (1978) 1297.
78H2 Hill, N. R., Celli, V.: Surf. Sci. **75** (1978) 577.

78W1 Wolfe, K., Weare, J. H.: Phys. Rev. Lett. **41** (1978) 1663.
79A1 Armand, G., Manson, J. R.: Surf. Sci. **80** (1979) 532.
79B1 Benedek, G., Garcia, N.: Surf. Sci. **80** (1979) 543.
79B2 Brenig, W.: Z. Phys. B **36** (1979) 81.
79C1 Carlos, W. E., Cole, M. W.: Phys. Rev. Lett. **43** (1979) 697.
79C2 Celli, V., Garcia, N., Hutchison, J.: Surf. Sci. **87** (1979) 112.
79D1 Derry, G. N., Wesner, D., Carlos, W. E., Frankl, D. R.: Surf. Sci. **89** (1979) 629.
79G1 Garcia, N., Goodman, F. O., Celli, V., Hill, M. R.: Phys. Rev. B **19** (1979) 1908.
79L1 Levi, A. C.: Nuovo Cimento B **54** (1979) 357.
79L2 Levi, A. C., Suhl, H.: Surf. Sci. **88** (1979) 221.
79M1 Marvin, A. M., Toigo, F.: Nuovo Cimento B **53** (1979) 25.
79V1 Vidali, G., Cole, M. W., Schwartz, C.: Surf. Sci. **87** (1979) L273.
80B1 Brusdeylins, G., Doak, R. B., Toennies, J. P.: Phys. Rev. Lett. **44** (1980) 1417.
80C1 Carlos, W. E., Cole, M. W.: Surf. Sci. **91** (1980) 339.
80E1 Esbjerg, N., Norskov, J. K.: Phys. Rev. Lett. **45** (1980) 807.
80H1 Hutchison, J., Celli, V.: Surf. Sci. **93** (1980) 263.
80M1 Mattera, L., Rosatelli, F., Salvo, C., Tommasini, F., Valbusa, U., Vidali, G.: Surf. Sci. **93** (1980) 515.
80S1 Sedlmeir, R., Brenig, W.: Z. Phys. B **36** (1980) 245.
80S2 Stott, M. J., Zaremba, E.: Phys. Rev. B **22** (1980) 1564.
80V1 Vidali, G., Cole, M. W.: Phys. Rev. B **22** (1980) 4661.
81A1 Avila, R., Lagos, M.: Surf. Sci. **103** (1981) L104.
81B1 Benedek, G., Garcia, N.: Surf. Sci. **103** (1981) L143.
81B2 Brusdeylins, G., Doak, R. B., Toennies, J. P.: J. Chem. Phys. **75** (1981) 1784.
81B3 Brusdeylins, G., Doak, R. B., Toennies, J. P.: Phys. Rev. Lett. **46** (1981) 437.
81C1 Cantini, P., Tatarek, R.: Phys. Rev. B **23** (1981) 3030.
81C2 Cardillo, M. J., Becker, B. E., Sibener, S. J., Miller, D. R.: Surf. Sci. **107** (1981) 469.
81C3 Cole, M. W., Frankl, D. R., Goodstein, D. L.: Rev. Mod. Phys. **53** (1981) 199.
81D1 de Rouffignac, E., Alldredge, G. P., de Wette, F. W.: Phys. Rev. B **23** (1981) 4208.
81F1 Frankl, D. R., Vidali, G.: Phys. Rev. B **24** (1981) 1798.
81H1 Hamann, D. R.: Phys. Rev. Lett. **46** (1981) 1227.
81M1 Mason, B. F., Williams, B. R.: Phys. Rev. Lett. **46** (1981) 1138.
81M2 Mason, B. F., Williams, B. R.: J. Chem. Phys. **75** (1981) 2199.
81M3 Meyer, H.-D.: Surf. Sci. **104** (1981) 117.
81S1 Stutzki, J., Brenig, W.: Z. Phys. B **45** (1981) 49.
81V1 Vidali, G., Cole, M. W.: Surf. Sci. **110** (1981) 10.
82B1 Benedek, G., Valbusa, U.: Dynamics of Gas-Surface Interaction. Berlin: Springer-Verlag 1982.
82B2 Benedek, G., Brivio, G. P., Miglio, L., Velasco, V. R.: Phys. Rev. B **26** (1982) 497.
82B3 Black, J. E., Campbell, D. A., Wallis, R. F.: Surf. Sci. **115** (1982) 161.
82B4 Brako, R., Newns, D. M.: Phys. Rev. Lett. **48** (1982) 1859.
82B5 Brako, R., Newns, D. M.: Surf. Sci. **117** (1982) 42.
82B6 Brako, R.: Surf. Sci. **123** (1982) 439.
82E1 Engel, T., Rieder, K. H., in: Springer Tracts in Modern Physics, Vol. 91, Structural Studies of Surfaces, Höhler, G. (ed.), Berlin: Springer-Verlag 1982, p. 56.
82H1 Harris, J., Liebsch, A.: Phys. Rev. Lett. **49** (1982) 341.
82H2 Harris, J., Liebsch, A.: J. Phys. C **15** (1982) 2275.
82H3 Hill, N. R., Haller, M., Celli, V.: Chem. Phys. **73** (1982) 363.
82L1 Laughlin, R. B.: Phys. Rev. B **25** (1982) 2222.
82L2 Liebsch, A., Harris, J., Salanon, B., Lapujoulade, J.: Surf. Sci. **123** (1982) 338.
82L3 Lilienkamp, G., Toennies, J. P.: Phys. Rev. B **26** (1982) 4752.
82P1 Perreau, J., Lapujoulade, J.: Surf. Sci. **122** (1982) 341.
82S1 Semerad, E., Horl, E. M.: Surf. Sci. **115** (1982) 346.
82T1 Toennies, J. P., Lilienkamp, G.: Phys. Rev. (Rapid Communications) B **26** (1982) 4572.
83A1 Armand, G., Manson, J. R.: J. Phys. Paris **44** (1983) 473.

83A2 Armand, G.: Solid State Commun. **48** (1983) 261.
83B1 Benedek, G., Brusdeylins, J. P., Toennies, J. P.: Phys. Rev. B **27** (1983) 2488.
83B2 Benedek, G., Brusdeylins, G., Doak, R. B., Skofronick, J. G., Toennies, J. P.: Phys. Rev. B **28** (1983) 2104.
83B3 Bortolani, V., Franchini, A., Garcia, N., Nizzoli, F., Santoro, G.: Phys. Rev. B **28** (1983) 7358.
83B4 Bortolani, V., Franchini, A., Nizzoli, F., Santoro, G., Benedek, G., Celli, V.: Surf. Sci. **128** (1983) 249.
83B5 Bortolani, V., Franchini, A., Nizzoli, F., Santoro, G., Benedek, G., Celli, V., Garcia, N.: Solid State Commun. **48** (1983) 1045.
83B6 Brusdeylins, G., Doak, R. B., Toennies, J. P.: Phys. Rev. B **27** (1983) 3662.
83B7 Brusdeylins, G., Doak, R. B., Skofronick, J. G., Toennies, J. P.: Surf. Sci. **128** (1983) 191.
83D1 Doak, R. B., Harten, U., Toennies, J. P.: Phys. Rev. Lett. **51** (1983) 578.
83E1 Evans, D., Celli, V., Benedek, G., Toennies, J. P., Doak, R. B.: Phys. Rev. Lett. **50** (1983) 1854.
83G1 Garcia, N., Barker, J. A., Rieder, K. H.: Solid State Commun. **45** (1983) 567.
83L1 Lang, N. D., Norskov, J. K.: Phys. Rev. B **27** (1983) 4612.
83L2 Liebsch, A., Harris, J.: Surf. Sci. **130** (1983) L349.
83L3 Lilienkamp, G., Toennies, J. P.: J. Chem. Phys. **78** (1983) 5210.
83M1 Mason, B. F., McGreer, K., Williams, B. R.: Surf. Sci. **130** (1983) 282.
83N1 Nordlander, P.: Surf. Sci. **126** (1983) 675.
83S1 Semerad, E., Horl, E. M.: Surf. Sci. **126** (1983) 661.
83Y1 Yu, C.-F., Whaley, K. B., Hogg, C. S., Sibener, S. J.: Phys. Rev. Lett. **51** (1983) 2210.
84A1 Annett, J. F., Haydock, R.: Phys. Rev. Lett. **53** (1984) 838.
84A2 Annett, J. F., Haydock, R.: Phys. Rev. B **29** (1984) 3773.
84B1 Barker, J. A., Garcia, N., Batra, I. P., Baumberger, M.: Surf. Sci. **141** (1984) L317.
84B2 Batra, I. P.: Surf. Sci. **148** (1984) 1.
84B3 Bortolani, V., Franchini, A., Nizzoli, F., Santoro, G.: Phys. Rev. Lett. **52** (1984) 429.
84B4 Bortolani, V., Santoro, G., Harten, U., Toennies, J. P.: Surf. Sci. **148** (1984) 82.
84C1 Cantini, P., Cevasco, E.: Surf. Sci. **148** (1984) 37.
84C2 Celli, V., in: Many-Body Phenomena at Surfaces. Langreth, D., Suhl, H. (eds.), Orlando: Academic Press 1984, p. 315.
84C3 Celli, V., Benedek, G., Harten, U., Toennies, J. P., Doak, R. B., Bortolani, V.: Surf. Sci. **143** (1984) L376.
84E1 Eichenauer, D., Toennies, J. P., in: Dynamics at Surfaces. Pullman, B., et al. (eds.), Dordrecht: D. Reidel Publishing Co. 1984, p. 1.
84L1 Lahee, A. M., Allison, W., Willis, R. F.: Surf. Sci. **147** (1984) L630.
84L2 Leung, W. Y., Larese, J. Z., Frankl, D. R.: Surf. Sci. **143** (1984) L398.
84L3 Liebsch, A., Harris, J., Weinert, M.: Surf. Sci. **145** (1984) 207.
84M1 Manninen, M., Norskov, J. K., Puska, M. J., Umrigar, C.: Phys. Rev. B **29** (1984) 2314.
84M2 Meyer, H.-D., Levine, R. D.: Chem. Phys. **85** (1984) 189.
84N1 Nordlander, P. Harris, J.: J. Phys. C **17** (1984) 1141.
84T1 Tang, K. T., Toennies, J. P.: J. Chem. Phys. **80** (1984) 3726.
84T2 Toennies, J. P.: J. Vac. Sci. Technol. A **2** (1984) 1055.
85B1 Barker, J. A., Auerbach, D. J.: Surf. Sci. Rep. **4** (1985) 1.
85B2 Batra, I. P., Bagus, P. S., Barker, J. A.: Phys. Rev. B **31** (1985) 1737.
85B3 Bortolani, V., Franchini, A., Santoro, G., in: Dynamical Phenomena at Surfaces, Interfaces, and Superlattices. Nizzoli, F., Rieder, K. H., Willis, R. F. (eds.), Berlin: Springer-Verlag 1985, p. 92.
85B4 Bortolani, V., Franchini, A., Nizzoli, F., Santoro, G., in: Electronic Structure, Dynamics, and Quantum Structural Properties of Condensed Matter. Devreese, J. T., Van Camp, P. (eds.), New York: Plenum Press 1985, p. 401.
85B5 Bortolani, V., Santoro, G., in: Phonon Physics. Kollar, J., Kroo, N., Menyhard, M., Siklos, T. (eds.), Singapore: World Scientific 1985, p. 566.
85B6 Bortolani, V., Franchini, A., Nizzoli, F., Santoro, G.: Surf. Sci. **152** (1985) 811.
85B7 Brusdeylins, G., Rechsteiner, R., Skofronick, J. G., Toennies, J. P., Benedek, G., Miglio, L.: Phys. Rev. Lett. **54** (1985) 466.

85C1 Celli, V., Maradudin, A. A.: Phys. Rev. B **31** (1985) 825.
85C2 Celli, V., Eichenauer, D., Kaufhold, A., Toennies, J. P.: J. Chem. Phys. **83** (1985) 2504.
85C3 Chiesa, M., Mattera, L., Musenich, R., Salvo, C.: Surf. Sci. **151** (1985) 145.
85C4 Cole, M. W., Toigo, F.: Phys. Rev. B **31** (1985) 727.
85D Duke, C. B., Paton, A.: Surf. Sci. **164** (1985) L797.
85H1 Harris, J., Zaremba, E.: Phys. Rev. Lett. **55** (1985) 1940C.
85H2 Harten, U., Toennies, J. P., Wöll, Ch., Zhang, G.: Phys. Rev. Lett. **55** (1985) 2308.
85H3 Harten, U., Toennies, J. P., Wöll, Ch.: Faraday Discuss. Chem. Soc. **80** (1985) 137.
85M1 Marvin, A. M., Celli, V., Toigo, F.: Surf. Sci. **154** (1985) 121.
85N1 Nichols, W. L., Weare, J. H.: Phys. Rev. Lett. **56** (1985) 753.
85S1 Schwartz, C., Le Roy, R. J.: Surf. Sci. **166** (1985) L141.
85T1 Takada, Y., Kohn, W.: Phys. Rev. Lett. **54** (1985) 470.
85T2 Takada, Y., Kohn, W.: Phys. Rev. Lett. **55** (1985) 141C.
85T3 Tersoff, J.: Phys. Rev. Lett. **55** (1985) 140C.
85Y1 Yu, C.-F., Whaley, K. B., Hogg, C. S., Sibener, S. J.: J. Chem. Phys. **83** (1985) 4217.
85Y2 Yu, C.-F., Whaley, K. B., Hogg, C. S., Sibener, S. J.: J. Chem. Phys. **83** (1985) 4235.
86A1 Annett, J. F., Haydock, R.: Phys. Rev. B **34** (1986) 6860.
86A2 Annett, J. F., Haydock, R.: Phys. Rev. Lett. **57** (1986) 1382.
85A3 Armand, G., Manson, J. R., Jayanthi, C. S.: Phys. Rev. B **34** (1986) 6627.
86B1 Benedek, G., Brusdeylins, G., Heimlich, C., Toennies, J. P., Valbusa, U.: Surf. Sci. **178** (1986) 545.
86B2 Bortolani, V., Levi, A. C.: Riv. Nuovo Cimento Soc. Ital. Fis. **9** (1986) 1.
86B3 Brusdeylins, G., Rechsteiner, R., Skofronick, J. G., Toennies, J. P., Benedek, G., Miglio, L.: Phys. Rev. B **34** (1986) 902.
86C1 Chung, S., Kara, A., Frankl, D. R.: Surf. Sci. **171** (1986) 45.
86D1 Dondi, M. G., Mattera, L., Terreni, S., Tommasini, F., Linke, U.: Phys. Rev. B **34** (1986) 5897.
86E1 Eichenauer, D., Toennies, J. P.: J. Chem. Phys. **85** (1986) 532.
86F1 Fowler, P. W., Hutson, J. M.: Phys. Rev. B **33** (1986) 3724.
86H1 Harten, U., Toennies, J. P., Wöll, Ch.: Phys. Rev. Lett. **57** (1986) 2947.
86H2 Ho, K. M., Bohnen, K. P.: Phys. Rev. Lett. **56** (1986) 934.
86H3 Hutson, J. M., Fowler, P. W.: Surf. Sci. **173** (1986) 337.
86H4 Hutson, J. M., Fowler, P. W., Zaremba, E.: Surf. Sci. **175** (1986) L775.
86J1 Jonsson, H., Weare, J.: Phys. Rev. Lett. **57** (1986) 412.
86K1 Kern, K., David, R., Palmer, R. L., Comsa, G., Rahman, T. S.: Phys. Rev. B **33** (1986) 4334.
86K2 Kern, K., David, R., Palmer, R. L., Comsa, G., Rahman, T. S.: Surf. Sci. **178** (1986) 537.
86L1 Liebsch, A.: Phys. Rev. B **33** (1986) 7249.
86N1 Neuhaus, D., Joo, F., Feuerbacher, B.: Surf. Sci. **165** (1986) L90.
86R1 Rocca, M., Lehwald, S., Ibach, H., Rahman, T. S.: Surf. Sci. **171** (1986) 632.
86S1 Stiles, M. D., Wilkins, J. W., Persson, M.: Phys. Rev. B **34** (1986) 4490.
87A1 Annett, J. F.: Phys. Rev. B **35** (1987) 7826.
87B1 Benedek, G., Miglio, L., Brusdeylins, G., Skofronick, J. G., Toennies, J. P.: Phys. Rev. B **35** (1987) 6593.
87B2 Benedek, G., Miglio, L., Skofronick, J. G., Brusdeylins, G., Heimlich, C., Toennies, J. P.: J. Vac. Sci. Technol. A **5** (1987) 1093.
87B3 Berndt, R., Toennies, J. P., Wöll, Ch.: J. Electron Spectrosc. Relat. Phenom. **44** (1987) 183.
87B4 Black, J. E., Franchini, A., Bortolani, V., Santoro, G., Wallis, R. F.: Phys. Rev. B **36** (1987) 2996.
87B5 Bortolani, V., Franchini, A., Santoro, G.: Surf. Sci. **189/190** (1987) 675.
87B6 Bracco, G., D'Avanzo, M., Salvo, C., Tatarek, R., Terreni, S.: Surf. Sci. **189/190** (1987) 684.
87B7 Bracco, G., Tatarek, R., Tommasini, F., Linke, U., Persson, M.: Phys. Rev. B **36** (1987) 2928.
87D1 de Wette, F. W., Kulkarni, A. D., Schröder, U., Kress, W.: Phys. Rev. B **35** (1987) 2476.
87D2 Doak, R. B, Nguyen, D. B.: J. Electron Spectrosc. Relat. Phenom. **44** (1987) 205.
87E1 Eichenauer, D., Harten, U., Toennies, J. P., Celli, V.: J. Chem. Phys. **86** (1987) 3693.
87E2 Ernst, H.-J., Hulpke, E., Toennies, J. P.: Phys. Rev. Lett. **58** (1987) 1941.
87H1 Harten, U., Toennies, J. P.: Europhys. Lett. **4** (1987) 833.

87I1 Idiodi, J., Bortolani, V., Franchini, A., Santoro, G., Celli, V.: Phys. Rev. B **35** (1987) 6029.
87J1 Jayanthi, C. S., Bilz, H., Kress, W., Benedek, G.: Phys. Rev. Lett. **59** (1987) 795.
87J2 Jonsson, H., Weare, J.: J. Chem. Phys. **86** (1987) 3711.
87K1 Kim, H.-Y., Cole, M. W.: Phys. Rev. B **35** (1987) 3990.
87K2 Kress, W., de Wette, F. W., Kulkarni, A. D., Schröder, U.: Phys. Rev. B. **35** (1987) 2467.
87L1 Lahee, A. M., Toennies, J. P., Wöll, Ch.: Surf. Sci. **191** (1987) 529.
87M1 Manson, J. R., Armand, G.: J. Vac. Sci. Technol. A **5** (1987) 448.
87S1 Santoro, G., Franchini, A., Bortolani, V., Harten U., Toennies, J. P., Wöll, Ch.: Surf. Sci. **183** (1987) 180.
87T1 Takada, Y., Kohn, W.: Phys. Rev. B **37** (1987) 826.
87T2 Toennies, J. P., Wöll, Ch.: Phys. Rev. B **36** (1987) 4475.
88A1 Armand, G., Manson, J. R.: Phys. Rev. B **37** (1988) 4363.
88B1 Benedek, G., Miglio, L., Brusdeylins, G., Heimlich, C., Skofronick, J. G., Toennies, J. P.: Europhys. Lett. **5** (1988) 253.
88B2 Benedek, G., Brusdeylins, G., Heimlich, C., Miglio, L., Skofronick, J. G., Toennies, J. P., Vollmer, R.: Phys. Rev. Lett. **60** (1988) 1037.
88C1 Cheng, E., Cole, M. W.: Phys. Rev. B **38** (1988) 987.
88D1 Dondi, M. G., Terreni, S., Tommasini, F., Linke, U.: Phys. Rev. B **37** (1988) 8034.
88E1 Eichenauer, D., Toennies, J. P.: Surf. Sci. **197** (1988) 267.
88L1 Lock, A., Toennies, J. P., Wöll, Ch., Bortolani, V., Franchini, A., Santoro, G.: Phys. Rev. B **37** (1988) 7087.
88M1 Manson, J. R., Armand, G.: Surf. Sci. **195** (1988) 513.
88M2 Miglio, L., Colombo, L.: Phys. Rev. B **37** (1988) 3025.
88M3 Mohamad, M. H., Kesmodel, L. L., Hall, B. M., Mills, D. L.: Phys. Rev. B **37** (1988) 2763.
88O1 Oshima, C., Aizawa, T., Souda, R., Ishizawa, Y., Sumiyoshi, Y.: Solid State Commun. **65** (1988) 1601.
88S1 Stiles, M. D., Wilkins, J. W.: Phys. Rev. B **37** (1988) 7306.
88Z1 Zeppenfeld, P., Kern, K., David, R., Kuhnke, K., Comsa, G.: Phys. Rev. B **38** (1988) 12329.
89A1 Armand, G.: J. Phys. Paris **50** (1989) 1493.
89B1 Bortolani, V., Celli, V., Franchini, A., Idiodi, J., Santoro, G., Kern, K., Poelsema, B., Comsa, G.: Surf. Sci. **208** (1989) 1.
89B2 Bortolani, V., Franchini, A., Santoro, G., Toennies, J. P., Wöll, Ch., Zhang, G.: Phys. Rev. B **40** (1989) 3524.
89B3 Brusdeylins. G., Heimlich, C., Toennies, J. P.: Surf. Sci. **211/212** (1989) 98.
89C1 Chern, G., Brug, W. P., Safron, S. A., Skofronick, J. G.: J. Vac. Sci. Technol. A **7** (1989) 2094.
89C2 Chern, G., Skofronick, J. G., Brug, W. P., Safron, S. A.: Phys. Rev. B **39** (1989) 12828.
89C3 Chern, G., Skofronick, J. G., Brug, W. P., Safron, S. A.: Phys. Rev. B **39** (1989) 12838.
89F1 Franchini, A., Santoro, G., Bortolani, V., Wallis, R. F.: Phys. Rev. B **39** (1989) 12139.
89H1 Hulpke, E., Smilgies, D. M.: Phys. Rev. B **40** (1989) 1338.
89J1 Jung, D. R., Mahgerefteh, M., Frankl, D. R.: Phys. Rev. B **39** (1989) 11164.
89K1 Kirsten, E., Rieder, K. H.: Surf. Sci. **222** (1989) L837.
89M1 Mahgerefteh, M., Jung, D. R., Frankl, D. R.: Phys. Rev. B **39** (1989) 3900.
89M2 Miglio, L., Santini, P., Ruggerone, P., Benedek, G.: Phys. Rev. Lett. **62** (1989) 3070.
89P1 Poelsema, B., Comsa, G., in: Springer Tracts in Modern Physics, Vol. **115**, Scattering of Thermal Energy Atoms, Höhler, G. (ed.), Berlin: Springer-Verlag 1989.
89T1 Tatarek, R., Bracco, G., Tommasini, F., Franchini, A., Bortolani, V., Santoro, G., Wallis, R. F.: Surf. Sci. **211/212** (1989) 314.
90B1 Brug, W. P., Chern, G., Duan, J., Safron, S. A. Skofronick, J. G., Benedek, G.: J. Vac. Sci. Technol. A **8** (1990) 2632.
90B2 Brusdeylins, G., Hofmann, F., Ruggerone, P., Toennies, J. P., Vollmer, R., Benedek, G., Skofronick, J. G., in: Phonons 89. Hunklinger, S., Ludwig, W., Weiss, G. (eds.), Singapore: World Scientific 1990, p. 892.
90B3 Brusdeylins, G., Heimlich, C., Skofronick, J. G., Toennies, J. P., Vollmer, R., Benedek, G., Miglio, L.: Phys. Rev. B **41** (1990) 5707.

90H1 Hulpke, E., Smilgies, D. M.: Phys. Rev. B **42** (1990) 9203.
90H2 Hulpke, E.: J. Electron Spectrosc. Relat. Phenom. **54/55** (1990) 299.
90K1 Kirsten, E., Parschau, G., Rieder, K. H.: Phys. Rev. B **41** (1990) 539.
90S1 Safron, S. A., Chern, G., Brug, W. P., Skofronick, J. G., Benedek, G.: Phys. Rev. B **41** (1990) 10146.
90S2 Safron, S. A., Brug, W. P., Chern, G., Duan, J., Skofronick, J. G.: J. Vac. Sci. Technol. A **8** (1990) 2627.
90S3 Safron, S. A., Brug, W. P., Bishop, G. G., Duan, J., Chern, G., Skofronick, J. G.: J. Electron Spectrosc. Relat. Phenom. **54/55** (1990) 343.
90S4 Santini, P., Miglio, L., Benedek, G., Harten, U., Ruggerone, P., Toennies, J. P.: Phys. Rev. B **42** (1990) 11942.
91B1 Brug, W. P.: The Surface Lattice Dynamics of Crystalline Ionic Insulators via He Atom Scattering. Ph.D. Thesis Dept. of Chem., Florida State Univ., Tallahassee, FL 32306 USA, 1991. Available from University Microfilms, Ann Arbor, MI.
91S1 Safron, S. A., Brug, W. P., Bishop, G. G., Chern, G., Derrick, M. E., Duan, J., Deweese, M. E., Skofronick, J. G.: J. Vac. Sci. Technol. A **9** (1991) 1657.
91V1 Vidali, G., Ihm, G., Kim, H.-Y., Cole, M. W.: Surf. Sci. Rep. **12** (1991) 135.
92B1 Brug, W. P., Chern, G., Duan, J., Bishop, G. G., Safron, S. A., Skofronick, J. G.: J. Vac. Sci. Technol. A **10** (1992) 2222.
92C1 Cortona, P., Dondi, M. G., Lausi, A., Tommasini, F.: Surf. Sci. **261** (1992) L35.
92H1 Hulpke, E., Hüppauff, M., Smilgies, D.-M., Kulkarni, A. D., de Wette, F. W.: Phys. Rev. B **45** (1992) 1820.
92S1 Skofronick, J. G., Toennies, J. P., in: Surface Properties of Layered Structures. Benedek, G. (ed.), Dordrecht: Kluwer Academic publishers 1992.